T0311646

Computational Models for Turbulent Reacting Flows

This book presents the current state of the art in computational models for turbulent reacting flows, and analyzes carefully the strengths and weaknesses of the various techniques described. The focus is on formulation of practical models as opposed to numerical issues arising from their solution.

A theoretical framework based on the one-point, one-time joint probability density function (PDF) is developed. It is shown that all commonly employed models for turbulent reacting flows can be formulated in terms of the joint PDF of the chemical species and enthalpy. Models based on direct closures for the chemical source term as well as transported PDF methods, are covered in detail. An introduction to the theory of turbulence and turbulent scalar transport is provided for completeness.

The book is aimed at chemical, mechanical, and aerospace engineers in academia and industry, as well as developers of computational fluid dynamics codes for reacting flows.

RODNEY O. FOX received his Ph.D. from Kansas State University, and is currently the Herbert L. Stiles Professor in the Chemical Engineering Department at Iowa State University. He has held visiting positions at Stanford University and at the CNRS Laboratory in Rouen, France, and has been an invited professor at ENSIC in Nancy, France; Politecnico di Torino, Italy; and Aalborg University, Denmark. He is the recipient of a National Science Foundation Presidential Young Investigator Award, and has published over 70 scientific papers.

CAMBRIDGE SERIES IN CHEMICAL ENGINEERING

Titles in the Series:

Diffusion: Mass Transfer in Fluid Systems, Second Edtion, E. L. Cussler

Principles of Gas-Solid Flows, Liang-Shih Fan and Chao Zhu

Modeling Vapor-Liquid Equilibria: Cubic Equations of State and their Mixing Rules, Hasan Orbey and Stanley I. Sandler

Advanced Transport Phenomena, John C. Slattery

Parametric Sensitivity in Chemical Systems, Arvind Varma, Massimo Morbidelli and Hua Wu

Chemical Engineering Design and Analysis, T. Michael Duncan and Jeffrey A. Reimer

Chemical Product Design, E. L. Cussler and G. D. Moggridge

Catalyst Design: Optimal Distribution of Catalyst in Pellets, Reactors, and Membranes, Massimo Morbidelli, Asterios Gavriilidis and Arvind Varma

Process Control: A First Course with MATLAB, Pao C. Chau

Computational Models for Turbulent Reacting Flows, Rodney O. Fox

Computational Models for Turbulent Reacting Flows

Rodney O. Fox
Herbert L. Stiles Professor of Chemical Engineering
Iowa State University

CAMBRIDGE
UNIVERSITY PRESS

University Printing House, Cambridge CB2 8BS, United Kingdom

Cambridge University Press is part of the University of Cambridge.

It furthers the University's mission by disseminating knowledge in the pursuit of education, learning and research at the highest international levels of excellence.

www.cambridge.org
Information on this title: www.cambridge.org/9780521659079

First published 2003

A catalogue record for this publication is available from the British Library

Library of Congress Cataloguing in Publication data
Fox, Rodney O., 1959–
Computational models for turbulent reacting flows / Rodney O. Fox.
p. cm. – (Cambridge series in chemical engineering)
Includes bibliographical references and index.
ISBN 0 521 65049 6 – ISBN 0 521 65907 8 (paperback)
1. Turbulence – Mathematical models. 2. Combustion – Mathematical models. 3. Fluid dynamics – Mathematical models. I. Title. II. Series.
QA913.F677 2003
660′.284–dc21 2003048570

ISBN 978-0-521-65049-6 Hardback
ISBN 978-0-521-65907-9 Paperback

à Roberte

Contents

Preface

In setting out to write this book, my main objective was to provide a reasonably complete introduction to computational models for turbulent reacting flows for students, researchers, and industrial end-users new to the field. The focus of the book is thus on the *formulation* of models as opposed to the numerical issues arising from their solution. Models for turbulent reacting flows are now widely used in the context of computational fluid dynamics (CFD) for simulating chemical transport processes in many industries. However, although CFD codes for non-reacting flows and for flows where the chemistry is relatively insensitive to the fluid dynamics are now widely available, their extension to reacting flows is less well developed (at least in commercial CFD codes), and certainly less well understood by potential end-users. There is thus a need for an introductory text that covers all of the most widely used reacting flow models, and which attempts to compare their relative advantages and disadvantages for particular applications.

The primary intended audience of this book comprises graduate-level engineering students and CFD practitioners in industry. It is assumed that the reader is familiar with basic concepts from chemical-reaction-engineering (CRE) and transport phenomena. Some previous exposure to theory of turbulent flows would also be very helpful, but is not absolutely required to understand the concepts presented. Nevertheless, readers who are unfamiliar with turbulent flows are encouraged to review Part I of the recent text *Turbulent Flows* by Pope (2000) before attempting to tackle the material in this book. In order to facilitate this effort, I have used the same notation as Pope (2000) whenever possible. The principal differences in notation occur in the treatment of multiple reacting scalars. In general, vector/matrix notation is used to denote the collection of thermodynamic variables (e.g., concentrations, temperature) needed to describe a reacting flow. Some familiarity with basic linear algebra and elementary matrix operations is assumed.

The choice of models to include in this book was dictated mainly by their ability to treat the wide range of turbulent reacting flows that occur in technological applications of interest to chemical engineers. In particular, models that cannot treat 'general' chemical kinetics have been excluded. For example, I do not discuss models developed for premixed turbulent combustion based on the 'turbulent burning velocity' or on the 'level-set' approach. This choice stems from my desire to extend the CRE approach for modeling reacting flows to be compatible with CFD codes. In this approach, the exact treatment of the chemical kinetics is the *sine qua non* of a good model. Thus, although most of the models discussed in this work can be used to treat non-premixed turbulent combustion, this will not be our primary focus. Indeed, in order to keep the formulation as simple as possible, all models are presented in the context of constant-density flows. In most cases, the extension to variable-density flows is straightforward, and can be easily undertaken after the reader has mastered the application of a particular model to constant-density cases.

In order to compare various reacting-flow models, it is necessary to present them all in the same conceptual framework. In this book, a statistical approach based on the one-point, one-time joint probability density function (PDF) has been chosen as the common theoretical framework. A similar approach can be taken to describe turbulent flows (Pope 2000). This choice was made due to the fact that nearly all CFD models currently in use for turbulent reacting flows can be expressed in terms of quantities derived from a joint PDF (e.g., low-order moments, conditional moments, conditional PDF, etc.). Ample introductory material on PDF methods is provided for readers unfamiliar with the subject area. Additional discussion on the application of PDF methods in turbulence can be found in Pope (2000). Some previous exposure to engineering statistics or elementary probability theory should suffice for understanding most of the material presented in this book.

The material presented in this book is divided into seven chapters and two appendices. Chapter 1 provides background information on turbulent reacting flows and on the two classical modeling approaches (chemical-reaction-engineering and fluid-mechanical) used to describe them. The chapter ends by pointing out the similarity between the two approaches when dealing with the effect of molecular mixing on chemical reactions, especially when formulated in a Lagrangian framework.

Chapter 2 reviews the statistical theory of turbulent flows. The emphasis, however, is on collecting in one place all of the necessary concepts and formulae needed in subsequent chapters. The discussion of these concepts is necessarily brief, and the reader is referred to Pope (2000) for further details. It is, nonetheless, essential that the reader become familiar with the basic scaling arguments and length/time scales needed to describe high-Reynolds-number turbulent flows. Likewise, the transport equations for important one-point statistics in inhomogeneous turbulent flows are derived in Chapter 2 for future reference.

Chapter 3 reviews the statistical description of scalar mixing in turbulent flows. The emphasis is again on collecting together the relevant length and time scales needed to describe turbulent transport at high Reynolds/Schmidt numbers. Following Pope (2000), a model scalar energy spectrum is constructed for stationary, isotropic scalar fields. Finally, the transport equations for important one-point scalar statistics in inhomogeneous turbulent mixing are derived in Chapter 3.

In order to model turbulent reacting flows accurately, an accurate model for turbulent transport is required. In Chapter 4 I provide a short introduction to selected computational models for *non-reacting* turbulent flows. Here again, the goal is to familiarize the reader with the various options, and to collect the most important models in one place for future reference. For an in-depth discussion of the physical basis of the models, the reader is referred to Pope (2000). Likewise, practical advice on choosing a particular turbulence model can be found in Wilcox (1993).

With regards to reacting flows, the essential material is presented in Chapters 5 and 6. Chapter 5 focuses on reacting flow models that can be expressed in terms of Eulerian (as opposed to Lagrangian) transport equations. Such equations can be solved numerically using standard finite-volume techniques, and thus can be easily added to existing CFD codes for turbulent flows. Chapter 6, on the other hand, focuses on *transported PDF* or *full PDF* methods. These methods typically employ a Lagrangian modeling perspective and 'non-traditional' CFD methods (i.e., Monte-Carlo simulations). Because most readers will not be familiar with the numerical methods needed to solve transported PDF models, an introduction to the subject is provided in Chapter 7.

Chapter 5 begins with an overview of chemical kinetics and the chemical-source-term closure problem in turbulent reacting flows. Based on my experience, closure methods based on the moments of the scalars are of very limited applicability. Thus, the emphasis in Chapter 5 is on presumed PDF methods and related closures based on conditioning on the mixture fraction. The latter is a non-reacting scalar that describes mixing between non-premixed inlet streams. A general definition of the mixture-fraction vector is derived in Chapter 5. Likewise, it is shown that by using a so-called 'mixture-fraction' transformation it is possible to describe a turbulent reacting flow by a reduced set of scalars involving the mixture-fraction vector and a 'reaction-progress' vector. Assuming that the mixture-fraction PDF is known, we introduce closures for the reaction-progress vector based on chemical equilibrium, 'simple' chemistry, laminar diffusion flamelets, and conditional moment closures. Closures based on presuming a form for the PDF of the reacting scalars are also considered in Chapter 5.

Chapter 6 presents a relatively complete introduction to transported PDF methods for turbulent reacting flow. For these flows, the principal attraction of transported PDF methods is the fact that the highly non-linear chemical source term is treated without closure. Instead, the modeling challenges are shifted to the molecular mixing model, which describes the combined effects of turbulent mixing (i.e., the scalar length-scale distribution) and molecular diffusion on the joint scalar PDF. Because the transported PDF treatment of turbulence is extensively discussed in Pope (2000), I focus in Chapter 6 on modeling issues associated with molecular mixing. The remaining sections in Chapter 6 deal with Lagrangian PDF methods, issues related to estimation of statistics based on 'particle' samples, and with tabulation methods for efficiently evaluating the chemical source term.

Chapter 7 deviates from the rest of the book in that it describes computational *methods* for 'solving' the transported PDF transport equation. Although Lagrangian PDF codes are generally preferable to Eulerian PDF codes, I introduce both methods and describe their relative advantages and disadvantages. Because transported PDF codes are less developed than standard CFD methods, readers wishing to utilize these methods should consult the literature for recent advances.

The material covered in the appendices is provided as a supplement for readers interested in more detail than could be provided in the main text. Appendix A discusses the derivation of the spectral relaxation (SR) model starting from the scalar spectral transport equation. The SR model is introduced in Chapter 4 as a non-equilibrium model for the scalar dissipation rate. The material in Appendix A is an attempt to connect the model to a more fundamental description based on two-point spectral transport. This connection can be exploited to extract model parameters from direct-numerical simulation data of homogeneous turbulent scalar mixing (Fox and Yeung 1999).

Appendix B discusses a new method (DQMOM) for solving the Eulerian transported PDF transport equation without resorting to Monte-Carlo simulations. This offers the advantage of solving for the joint composition PDF introduced in Chapter 6 using standard finite-volume CFD codes, without resorting to the chemical-source-term closures presented in Chapter 5. Preliminary results found using DQMOM are quite encouraging, but further research will be needed to understand fully the range of applicability of the method.

I am extremely grateful to the many teachers, colleagues and graduate students who have helped me understand and develop the material presented in this work. In particular, I would like to thank Prof. John C. Matthews of Kansas State University who, through his rigorous teaching style, attention to detail, and passion for the subject of transport phenomena, first planted the seed in the author that has subsequently grown into the book that you have before you. I would also like to thank my own students in the graduate courses that I have offered on this subject who have provided valuable feedback about the text. I want especially to thank Kuochen Tsai and P. K. Yeung, with whom I have enjoyed close collaborations over the past several years, and Jim Hill at Iowa State for his encouragement to undertake the writing of this book. I would also like to acknowledge the important contributions of Daniele Marchisio in the development of the DQMOM method described in Appendix B.

For his early support and encouragement to develop CFD models for chemical-reaction-engineering applications, I am deeply indebted to my post-doctoral advisor, Jacques Villermaux. His untimely death in 1997 was a great loss to his friends and family, as well as to the profession.

I am also deeply indebted to Stephen Pope in many different ways, starting from his early encouragement in 1991 to consider PDF methods as a natural modeling framework for describing micromixing in chemical reactors. However, I am particularly grateful that his text on turbulent flows appeared before this work (relieving me of the arduous task of covering this subject in detail!), and for his generosity in sharing early versions of his text, as well as his LaTeX macro files and precious advice on preparing the manuscript.

Beginning with a Graduate Fellowship, my research in turbulent reacting flows has been almost continuously funded by research grants from the US National Science Foundation. This long-term support has made it possible for me to pursue novel research ideas outside the traditional modeling approach used by chemical reaction engineers. In hindsight, the application of CFD to chemical reactor design and analysis appears to be a rather natural idea. Indeed, all major chemical producers now use CFD tools routinely to solve day-to-day engineering problems. However, as recently as the 1990s the gap between chemical reaction engineering and fluid mechanics was large, and only through a sustained effort to understand both fields in great detail was it possible to bridge this gap. While much research remains to be done to develop a complete set of CFD tools for chemical reaction engineering (most notably in the area of *multiphase* turbulent reacting flows), one is certainly justified in pointing to computational models for turbulent reacting flows as a highly successful example of fundamental academic research that has led to technological advances in real-world applications. Financial assistance provided by my industrial collaborators: Air Products, BASF, BASELL, Dow Chemical, DuPont, and Fluent, is deeply appreciated.

I also want to apologize to my colleagues in advance for not mentioning many of their excellent contributions to the field of turbulent reacting flows that have appeared over the last 50 years. It was my original intention to include a section in Chapter 1 on the history of turbulent-reacting-flow research. However, after collecting the enormous number of articles that have appeared in the literature to date, I soon realized that the task would require more time and space than I had at my disposal in order to do it justice. Nonetheless, thanks to the efforts of Jim Herriott at Iowa State, I have managed to include an extensive Reference section that will hopefully serve as a useful starting point for readers wishing to delve into the history of particular subjects in greater detail.

Finally, I dedicate this book to my wife, Roberte. Her encouragement and constant support during the long period of this project and over the years have been invaluable.

1

Turbulent reacting flows

1.1 Introduction

At first glance, to the uninitiated the subject of turbulent reacting flows would appear to be relatively simple. Indeed, the basic governing principles can be reduced to a statement of conservation of chemical species and energy ((1.28), p. 16) and a statement of conservation of fluid momentum ((1.27), p. 16). However, anyone who has attempted to master this subject will tell you that it is in fact quite complicated. On the one hand, in order to understand how the fluid flow affects the chemistry, one must have an excellent understanding of turbulent flows and of turbulent mixing. On the other hand, given its paramount importance in the determination of the types and quantities of chemical species formed, an equally good understanding of chemistry is required. Even a cursory review of the literature in any of these areas will quickly reveal the complexity of the task. Indeed, given the enormous research production in these areas during the twentieth century, it would be safe to conclude that no one could simultaneously master all aspects of turbulence, mixing, and chemistry.

Notwithstanding the intellectual challenges posed by the subject, the main impetus behind the development of computational models for turbulent reacting flows has been the increasing awareness of the impact of such flows on the environment. For example, incomplete combustion of hydrocarbons in internal combustion engines is a major source of air pollution. Likewise, in the chemical process and pharmaceutical industries, inadequate control of product yields and selectivities can produce a host of undesirable byproducts. Even if such byproducts could all be successfully separated out and treated so that they are not released into the environment, the economic cost of doing so is often prohibitive. Hence, there is an ever-increasing incentive to improve industrial processes and devices in order for them to remain competitive in the marketplace.

Given their complexity and practical importance, it should be no surprise that different approaches for dealing with turbulent reacting flows have developed over the last 50 years. On the one hand, the chemical-reaction-engineering (CRE) approach came from the application of chemical kinetics to the study of chemical reactor design. In this approach, the details of the fluid flow are of interest only in as much as they affect the product yield and selectivity of the reactor. In many cases, this effect is of secondary importance, and thus in the CRE approach greater attention has been paid to other factors that directly affect the chemistry. On the other hand, the fluid-mechanical (FM) approach developed as a natural extension of the statistical description of turbulent flows. In this approach, the emphasis has been primarily on how the fluid flow affects the rate of chemical reactions. In particular, this approach has been widely employed in the study of combustion (Rosner 1986; Peters 2000; Poinsot and Veynante 2001; Veynante and Vervisch 2002).

In hindsight, the primary factor in determining which approach is most applicable to a particular reacting flow is the characteristic time scales of the chemical reactions relative to the turbulence time scales. In the early applications of the CRE approach, the chemical time scales were larger than the turbulence time scales. In this case, one can safely ignore the details of the flow. Likewise, in early applications of the FM approach to combustion, all chemical time scales were assumed to be much smaller than the turbulence time scales. In this case, the details of the chemical kinetics are of no importance, and one is free to concentrate on how the heat released by the reactions interacts with the turbulent flow. More recently, the shortcomings of each of these approaches have become apparent when applied to systems wherein some of the chemical time scales overlap with the turbulence time scales. In this case, an accurate description of both the turbulent flow and the chemistry is required to predict product yields and selectivities accurately.

With these observations in mind, the reader may rightly ask 'What is the approach used in this book?' The accurate answer to this question may be 'both' or 'neither,' depending on your perspective. From a CRE perspective, the methods discussed in this book may appear to favor the FM approach. Nevertheless, many of the models find their roots in CRE, and one can argue that they have simply been rewritten in terms of detailed transport models that can be solved using computational fluid dynamics (CFD) techniques (Fox 1996a; Harris *et al.* 1996; Ranada 2002). Likewise, from an FM perspective, very little is said about the details of turbulent flows or the computational methods needed to study them. Instead, we focus on the models needed to describe the source term for chemical reactions involving *non-premixed* reactants. Moreover, for the most part, density variations in the fluid due to mixing and/or heat release are not discussed in any detail. Otherwise, the only criterion for including a particular model in this book is the requirement that it must be able to handle detailed chemistry. This criterion is motivated by the need to predict product yield and selectivity accurately for finite-rate reactions.

At first glance, the exclusion of premixed reactants and density variations might seem to be too drastic. (Especially if one equates 'turbulent reacting flows' with 'combustion.'[1])

[1] Excellent treatments of modern approaches to combustion modeling are available elsewhere (Kuznetsov and Sabel'nikov 1990; Warnatz *et al.* 1996; Peters 2000; Poinsot and Veynante 2001).

However, if one looks at the complete range of systems wherein turbulence and chemistry interact, one will find that many of the so-called 'mixing-sensitive' systems involve liquids or gas-phase reactions with modest density changes. For these systems, a key feature that distinguishes them from classical combusting systems is that the reaction rates are fast regardless of the temperature (e.g., acid–base chemistry). In contrast, much of the dynamical behavior of typical combusting systems is controlled by the fact that the reactants do not react at ambient temperatures. Combustion can thus be carried out in either premixed or non-premixed modes, while mixing-sensitive reactions can only be carried out in non-premixed mode. This distinction is of considerable consequence in the case of premixed combustion. Indeed, models for premixed combustion occupy a large place unto themselves in the combustion literature. On the other hand, the methods described in this book will find utility in the description of non-premixed combustion. In fact, many of them originated in this field and have already proven to be quite powerful for the modeling of diffusion flames with detailed chemistry.

In the remainder of this chapter, an overview of the CRE and FM approaches to turbulent reacting flows is provided. Because the description of turbulent flows and turbulent mixing makes liberal use of ideas from probability and statistical theory, the reader may wish to review the appropriate appendices in Pope (2000) before starting on Chapter 2. Further guidance on how to navigate the material in Chapters 2–7 is provided in Section 1.5.

1.2 Chemical-reaction-engineering approach

The CRE approach for modeling chemical reactors is based on mole and energy balances, chemical rate laws, and idealized flow models.[2] The latter are usually constructed (Wen and Fan 1975) using some combination of plug-flow reactors (PFRs) and continuous-stirred-tank reactors (CSTRs). (We review both types of reactors below.) The CRE approach thus avoids solving a detailed flow model based on the momentum balance equation. However, this simplification comes at the cost of introducing unknown model parameters to describe the flow rates between various sub-regions inside the reactor. The choice of a particular model is far from unique,[3] but can result in very different predictions for product yields with complex chemistry.

For isothermal, first-order chemical reactions, the mole balances form a system of *linear* equations. A non-ideal reactor can then be modeled as a collection of *Lagrangian* fluid elements moving *independently* through the system. When parameterized by the amount of time it has spent in the system (i.e., its residence time), each fluid element behaves as a batch reactor. The species concentrations for such a system can be completely characterized by the inlet concentrations, the chemical rate constants, and the residence time distribution (RTD) of the reactor. The latter can be found from simple tracer experiments carried out under identical flow conditions. A brief overview of RTD theory is given below.

[2] In CRE textbooks (Hill 1977; Levenspiel 1998; Fogler 1999), the types of reactors considered in this book are referred to as *non-ideal*. The flow models must take into account fluid-mixing effects on product yields.

[3] It has been described as requiring 'a certain amount of art' (Fogler 1999).

For non-isothermal or non-linear chemical reactions, the RTD no longer suffices to predict the reactor outlet concentrations. From a Lagrangian perspective, local *interactions* between fluid elements become important, and thus fluid elements cannot be treated as individual batch reactors. However, an accurate description of fluid-element interactions is strongly dependent on the underlying fluid flow field. For certain types of reactors, one approach for overcoming the lack of a detailed model for the flow field is to input empirical flow correlations into so-called *zone* models. In these models, the reactor volume is decomposed into a finite collection of well mixed (i.e., CSTR) zones connected at their boundaries by molar fluxes.[4] (An example of a zone model for a stirred-tank reactor is shown in Fig. 1.5.) Within each zone, all fluid elements are assumed to be identical (i.e., have the same species concentrations). Physically, this assumption corresponds to assuming that the chemical reactions are slower than the local *micromixing time*.[5]

For non-linear chemical reactions that are fast compared with the local micromixing time, the species concentrations in fluid elements located in the same zone cannot be assumed to be identical (Toor 1962; Toor 1969; Toor and Singh 1973; Amerja *et al.* 1976). The canonical example is a non-premixed acid–base reaction for which the reaction rate constant is essentially infinite. As a result of the infinitely fast reaction, a fluid element can contain either acid or base, but not both. Due to the chemical reaction, the local fluid-element concentrations will therefore be different depending on their stoichiometric excess of acid or base. Micromixing will then determine the rate at which acid and base are transferred between fluid elements, and thus will determine the mean rate of the chemical reaction.

If all chemical reactions are fast compared with the local micromixing time, a non-premixed system can often be successfully described in terms of the *mixture fraction*.[6] The more general case of *finite-rate* reactions requires a detailed description of micromixing or, equivalently, the interactions between local fluid elements. In the CRE approach, micromixing is modeled using a Lagrangian description that follows individual fluid elements as they flow through the reactor. (Examples of micromixing models are discussed below.) A key parameter in such models is the micromixing time, which must be related to the underlying flow field.

For canonical turbulent flows (Pope 2000), the flow parameters required to complete the CRE models are readily available. However, for the complex flow fields present in most chemical reactors, the flow parameters must be found either empirically or by solving a CFD turbulence model. If the latter course is taken, the next logical step would be to attempt to reformulate the CRE model in terms of a set of transport equations that can be added to the CFD model. The principal complication encountered when following this path is the fact that the CRE models are expressed in a Lagrangian framework, whilst the CFD models are expressed in an *Eulerian* framework. One of the main goals of this book

4 The zones are thus essentially identical to the *finite volumes* employed in many CFD codes.
5 The micromixing time has an exact definition in terms of the rate of decay of *concentration fluctuations*.
6 The mixture fraction is defined in Chapter 5.

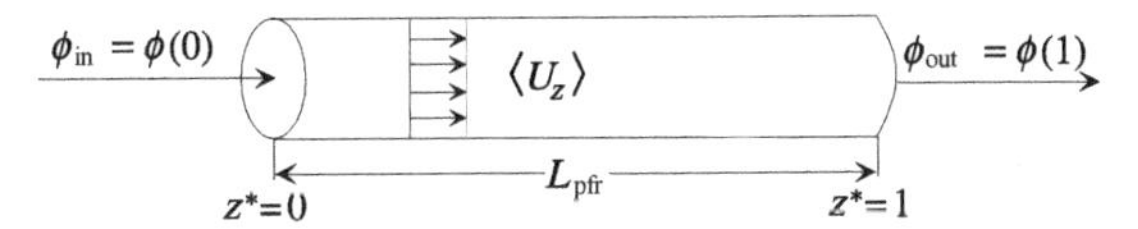

Figure 1.1. Sketch of a plug-flow reactor.

is thus to demonstrate how the two approaches can be successfully combined when both are formulated in terms of an appropriate statistical theory.

In the remainder of this section, we will review those components of the CRE approach that will be needed to understand the modeling approach described in detail in subsequent chapters. Further details on the CRE approach can be found in introductory textbooks on chemical reaction engineering (e.g., Hill 1977; Levenspiel 1998; Fogler 1999).

1.2.1 PFR and CSTR models

The PFR model is based on turbulent pipe flow in the limit where axial dispersion can be assumed to be negligible (see Fig. 1.1). The mean residence time τ_{pfr} in a PFR depends only on the mean axial fluid velocity $\langle U_z \rangle$ and the length of the reactor L_{pfr}:

$$\tau_{pfr} \equiv \frac{L_{pfr}}{\langle U_z \rangle}. \tag{1.1}$$

Defining the dimensionless axial position by $z^* \equiv z/L_{pfr}$, the PFR model for the species concentrations ϕ becomes[7]

$$\frac{d\phi}{dz^*} = \tau_{pfr}\mathbf{S}(\phi) \quad \text{with} \quad \phi(0) = \phi_{in} = \text{inlet concentrations}, \tag{1.2}$$

where $\mathbf{S}$ is the *chemical source term*. Given the inlet concentrations and the chemical source term, the PFR model is readily solved using numerical methods for initial-value problems to find the outlet concentrations $\phi(1)$.

The PFR model ignores mixing between fluid elements at *different* axial locations. It can thus be rewritten in a Lagrangian framework by substituting $\alpha = \tau_{pfr} z^*$, where α denotes the elapsed time (or *age*) that the fluid element has spent in the reactor. At the end of the PFR, all fluid elements have the same age, i.e., $\alpha = \tau_{pfr}$. Moreover, at every point in the PFR, the species concentrations are uniquely determined by the age of the fluid particles at that point through the solution to (1.2).

In addition, the PFR model assumes that mixing between fluid elements at the *same* axial location is infinitely fast. In CRE parlance, all fluid elements are said to be *well micromixed*. In a tubular reactor, this assumption implies that the inlet concentrations are uniform over the cross-section of the reactor. However, in real reactors, the inlet streams are often segregated (non-premixed) at the inlet, and a finite time is required as they move down the reactor before they become well micromixed. The PFR model can be easily

[7] The notation is chosen to be consistent with that used in the remainder of the book. Alternative notation is employed in most CRE textbooks.

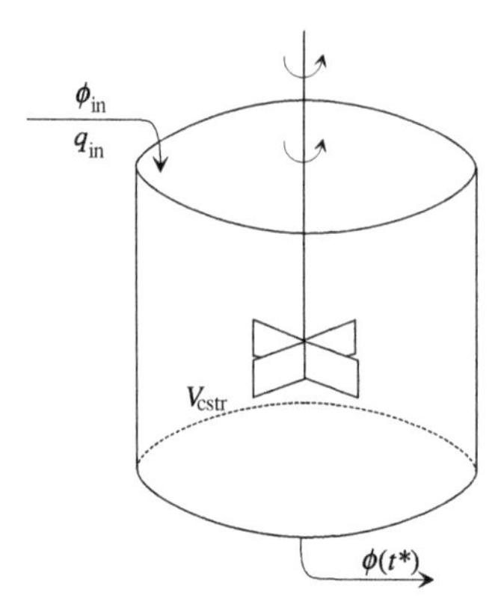

Figure 1.2. Sketch of a continuous-stirred-tank reactor (CSTR).

extended to describe radial mixing by introducing a micromixing model. We will look at a *poorly micromixed* PFR model below.

The CSTR model, on the other hand, is based on a stirred vessel with continuous inflow and outflow (see Fig. 1.2). The principal assumption made when deriving the model is that the vessel is stirred vigorously enough to eliminate all concentration gradients inside the reactor (i.e., the assumption of *well stirred*). The outlet concentrations will then be identical to the reactor concentrations, and a simple mole balance yields the CSTR model equation:

$$\frac{d\boldsymbol{\phi}}{dt^*} = \tau_{cstr}\mathbf{S}(\boldsymbol{\phi}) + \boldsymbol{\phi}_{in} - \boldsymbol{\phi}. \tag{1.3}$$

The CSTR mean residence time is defined in terms of the inlet flow rate q_{in} and the reactor volume V_{cstr} by

$$\tau_{cstr} \equiv \frac{V_{cstr}}{q_{in}}, \tag{1.4}$$

and the dimensionless time t^* is defined by $t^* \equiv t/\tau_{cstr}$. At steady state, the left-hand side of (1.3) is zero, and the CSTR model reduces to a system of (non-linear) equations that can be solved for $\boldsymbol{\phi}$.

The CSTR model can be derived from the fundamental scalar transport equation (1.28) by integrating the spatial variable over the entire reactor volume. This process results in an integral for the volume-average chemical source term of the form:

$$\int_{V_{cstr}} \mathbf{S}(\boldsymbol{\phi}(\mathbf{x}, t))\, d\mathbf{x} = V_{cstr}\mathbf{S}(\boldsymbol{\phi}(t)), \tag{1.5}$$

where the right-hand side is found by invoking the assumption that $\boldsymbol{\phi}$ is independent of $\mathbf{x}$. In the CRE parlance, the CSTR model applies to a reactor that is both *well macromixed* and *well micromixed* (Fig. 1.3). The well macromixed part refers to the fact that a fluid element's location in a CSTR is *independent* of its age.[8] This fact follows from the well

[8] The PFR is thus not well macromixed since a fluid element's location in a PFR is a linear function of its age.

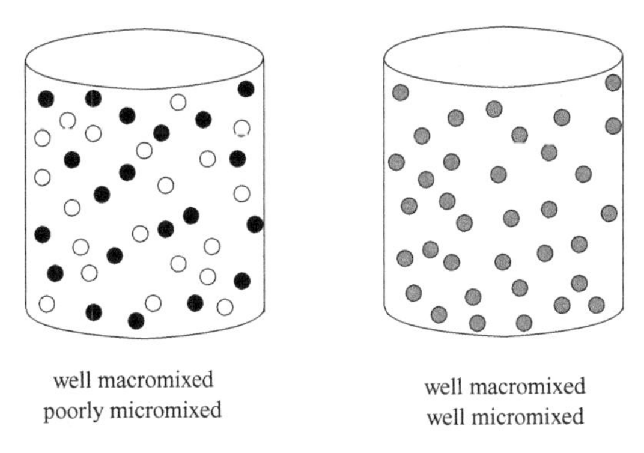

Figure 1.3. Sketch of a poorly micromixed versus a well micromixed CSTR.

stirred assumption, but is not equivalent to it. Indeed, if fluid elements inside the reactor did not interact due to micromixing, then the fluid concentrations ϕ would depend only on the age of the fluid element. Thus, the CSTR model also implies that the reactor is well micromixed.[9] We will look at the extension of the CSTR model to well macromixed but poorly micromixed systems below.

The applicability of the PFR and CSTR models for a particular set of chemical reactions depends on the characteristic time scales of reaction rates relative to the mixing times. In the PFR model, the only relevant mixing times are the ones that characterize radial dispersion and micromixing. The former will be proportional to the integral time scale of the turbulent flow,[10] and the latter will depend on the inlet flow conditions but, at worst, will also be proportional to the turbulence integral time scale. Thus, the PFR model will be applicable to chemical reaction schemes[11] wherein the shortest chemical time scale is greater than or equal to the turbulence integral time scale.

On the other hand, for the CSTR model, the largest time scale for the flow will usually be the *recirculation time*.[12] Typically, the recirculation time will be larger than the largest turbulence integral time scale in the reactor, but smaller than the mean residence time. Chemical reactions with characteristic time scales larger than the recirculation time can be successfully treated using the CSTR model. Chemical reactions that have time scales intermediate between the turbulence integral time scale and the recirculation time should be treated by a CSTR zone model. Finally, chemical reactions that have time scales smaller than the turbulence integral time scale should be described by a micromixing model.

9 In the statistical theory of fluid mixing presented in Chapter 3, well macromixed corresponds to the condition that the scalar means $\langle \phi \rangle$ are independent of position, and well micromixed corresponds to the condition that the scalar variances are null. An equivalent definition can be developed from the residence time distribution discussed below.

10 In Chapter 2, we show that the turbulence integral time scale can be defined in terms of the turbulent kinetic energy k and the turbulent dissipation rate ε by $\tau_u = k/\varepsilon$. In a PFR, τ_u is proportional to $D/\langle U_z \rangle$, where D is the tube diameter.

11 The chemical time scales are defined in Chapter 5. In general, they will be functions of the temperature, pressure, and local concentrations.

12 Heuristically, the recirculation time is the average time required for a fluid element to return to the impeller region after leaving it.

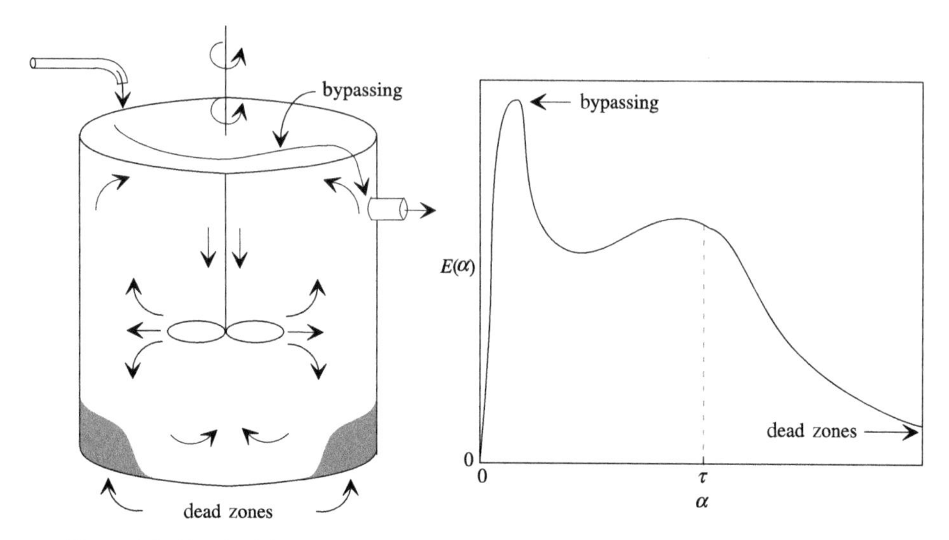

Figure 1.4. Sketch of the residence time distribution (RTD) in a non-ideal reactor.

1.2.2 RTD theory

In the CRE literature, the residence time distribution (RTD) has been shown to be a powerful tool for handling isothermal first-order reactions in arbitrary reactor geometries. (See Nauman and Buffham (1983) for a detailed introduction to RTD theory.) The basic ideas behind RTD theory can be most easily understood in a Lagrangian framework. The residence time of a fluid element is defined to be its age α as it leaves the reactor. Thus, in a PFR, the *RTD function* $E(\alpha)$ has the simple form of a delta function:

$$E_{\text{pfr}}(\alpha) = \delta(\alpha - \tau_{\text{pfr}}), \tag{1.6}$$

i.e., all fluid elements have identical residence times. On the other hand, in a CSTR, the RTD function has an exponential form:[13]

$$E_{\text{cstr}}(\alpha) = \frac{1}{\tau_{\text{cstr}}} \exp\left(-\frac{\alpha}{\tau_{\text{cstr}}}\right). \tag{1.7}$$

RTD functions for combinations of ideal reactors can be constructed (Wen and Fan 1975) based on (1.6) and (1.7). For non-ideal reactors, the RTD function (see example in Fig. 1.4) can be measured experimentally using passive tracers (Levenspiel 1998; Fogler 1999), or extracted numerically from CFD simulations of time-dependent passive scalar mixing.

In this book, an alternative description based on the *joint probability density function* (PDF) of the species concentrations will be developed. (Exact definitions of the joint PDF and related quantities are given in Chapter 3.) The RTD function is in fact the PDF of the fluid-element ages as they leave the reactor. The relationship between the PDF description and the RTD function can be made transparent by defining a fictitious chemical species

[13] The outflow of a CSTR is a *Poisson process*, i.e., fluid elements are randomly selected regardless of their position in the reactor. The *waiting time* before selection for a Poisson process has an exponential probability distribution. See Feller (1971) for details.

ϕ_τ whose inlet concentration is null, and whose chemical source term is $S_\tau = 1$. Owing to turbulent mixing in a chemical reactor, the PDF of ϕ_τ will be a function of the *composition-space variable* ψ, the spatial location in the reactor $\mathbf{x}$, and time t. Thus, we will denote the PDF by $f_\tau(\alpha; \mathbf{x}, t)$. The PDF of ϕ_τ *at the reactor outlet*, $\mathbf{x}_{\text{outlet}}$, is then equal to the *time-dependent* RTD function:

$$E(\alpha, t) = f_\tau(\alpha; \mathbf{x}_{\text{outlet}}, t). \tag{1.8}$$

At steady state, the PDF (and thus the RTD function) will be independent of time. Moreover, the *internal-age distribution* at a point $\mathbf{x}$ inside the reactor is just $I(\alpha; \mathbf{x}, t) = f_\tau(\alpha; \mathbf{x}, t)$. For a *statistically homogeneous* reactor (i.e., a CSTR), the PDF is independent of position, and hence the steady-state internal-age distribution $I(\alpha)$ will be independent of time and position.

One of the early successes of the CRE approach was to show that RTD theory suffices to treat the special case of *non-interacting* fluid elements (Danckwerts 1958). For this case, each fluid element behaves as a *batch reactor*:

$$\frac{\mathrm{d}\boldsymbol{\phi}_{\text{batch}}}{\mathrm{d}\alpha} = \mathbf{S}(\boldsymbol{\phi}_{\text{batch}}) \quad \text{with} \quad \boldsymbol{\phi}_{\text{batch}}(0) = \boldsymbol{\phi}_{\text{in}}. \tag{1.9}$$

For fixed initial conditions, the solution to this expression is uniquely defined in terms of the age, i.e., $\boldsymbol{\phi}_{\text{batch}}(\alpha)$. The *joint composition PDF* $f_\phi(\boldsymbol{\psi}; \mathbf{x}, t)$ *at the reactor outlet* is then uniquely defined in terms of the time-dependent RTD distribution:[14]

$$f_\phi(\boldsymbol{\psi}; \mathbf{x}_{\text{outlet}}, t) = \int_0^\infty \delta(\boldsymbol{\psi} - \boldsymbol{\phi}_{\text{batch}}(\alpha)) E(\alpha, t)\, \mathrm{d}\alpha, \tag{1.10}$$

where the multi-variable delta function is defined in terms of the product of single-variable delta functions for each chemical species by

$$\delta(\boldsymbol{\psi} - \boldsymbol{\phi}) \equiv \prod_\beta \delta(\psi_\beta - \phi_\beta). \tag{1.11}$$

For the general case of *interacting* fluid elements, (1.9) and (1.10) no longer hold. Indeed, the correspondence between the RTD function and the composition PDF breaks down because the species concentrations inside each fluid element can no longer be uniquely parameterized in terms of the fluid element's age. Thus, for the general case of complex chemistry in non-ideal reactors, a mixing theory based on the composition PDF will be more powerful than one based on RTD theory.

The utility of RTD theory is best illustrated by its treatment of *first-order* chemical reactions. For this case, each fluid element can be treated as a batch reactor.[15] The concentration

[14] At steady state, the left-hand side of this expression has independent variables $\boldsymbol{\psi}$. For fixed $\boldsymbol{\psi} = \boldsymbol{\psi}^*$, the integral on the right-hand side sweeps over all fluid elements in search of those whose concentrations $\boldsymbol{\phi}_{\text{batch}}$ are equal to $\boldsymbol{\psi}^*$. If these fluid elements have the same age (say, $\alpha = \alpha^*$), then the joint PDF reduces to $f_\phi(\boldsymbol{\psi}^*; \mathbf{x}_{\text{outlet}}) = E(\alpha^*)$, where $E(\alpha^*)\, \mathrm{d}\alpha^*$ is the fraction of fluid elements with age α^*.

[15] Because the outlet concentrations will not depend on it, micromixing between fluid particles can be neglected. The reader can verify this statement by showing that the micromixing term in the poorly micromixed CSTR and the poorly micromixed PFR falls out when the mean outlet concentration is computed for a first-order chemical reaction. More generally, one can show that the chemical source term appears in closed form in the transport equation for the scalar means.

of a chemical species in a fluid element then depends only on its age through the solution to the batch-reactor model:

$$\frac{\mathrm{d}\phi}{\mathrm{d}\alpha} = -k\phi \quad \text{with} \quad \phi(0) = \phi_{\text{in}}, \tag{1.12}$$

i.e.,

$$\phi(\alpha) = \phi_{\text{in}} \mathrm{e}^{-k\alpha}. \tag{1.13}$$

In RTD theory, the concentrations at the reactor outlet are found by averaging over the ages of all fluid elements leaving the reactor:[16]

$$\phi_{\text{out}} = \int_0^\infty \phi(\alpha) E(\alpha)\,\mathrm{d}\alpha. \tag{1.14}$$

Thus, for first-order reactions, exact solutions can be found for the outlet concentration, e.g., from (1.13):

$$\left(\frac{\phi_{\text{out}}}{\phi_{\text{in}}}\right)_{\text{pfr}} = \mathrm{e}^{-k\tau_{\text{pfr}}} \quad \text{and} \quad \left(\frac{\phi_{\text{out}}}{\phi_{\text{in}}}\right)_{\text{cstr}} = \frac{1}{1 + k\tau_{\text{cstr}}}.$$

For higher-order reactions, the fluid-element concentrations no longer obey (1.9). Additional terms must be added to (1.9) in order to account for micromixing (i.e., local fluid-element interactions due to molecular diffusion). For the poorly micromixed PFR and the poorly micromixed CSTR, extensions of (1.9) can be employed with (1.14) to predict the outlet concentrations in the framework of RTD theory. For non-ideal reactors, extensions of RTD theory to model micromixing have been proposed in the CRE literature. (We will review some of these micromixing models below.) However, due to the non-uniqueness between a fluid element's concentrations and its age, micromixing models based on RTD theory are generally *ad hoc* and difficult to validate experimentally.

1.2.3 Zone models

An alternative method to RTD theory for treating non-ideal reactors is the use of zone models. In this approach, the reactor volume is broken down into well mixed zones (see the example in Fig. 1.5). Unlike RTD theory, zone models employ an Eulerian framework that ignores the age distribution of fluid elements inside each zone. Thus, zone models ignore micromixing, but provide a model for macromixing or large-scale inhomogeneity inside the reactor.

Denoting the *transport rate* of fluid from zone i to zone j by f_{ij}, a zone model can be expressed mathematically in terms of mole balances for each of the N zones:

$$\frac{\mathrm{d}\phi^{(i)}}{\mathrm{d}t} = \sum_{j=0}^{N+1} \left(f_{ji}\phi^{(j)} - f_{ij}\phi^{(i)} \right) + \mathbf{S}(\phi^{(i)}) \quad i = 1, \ldots, N. \tag{1.15}$$

[16] For non-interacting fluid elements, the RTD function is thus equivalent to the joint PDF of the concentrations. In composition space, the joint PDF would lie on a one-dimensional sub-manifold (i.e., have a one-dimensional support) parameterized by the age α. The addition of micromixing (i.e., interactions between fluid elements) will cause the joint PDF to spread in composition space, thereby losing its one-dimensional support.

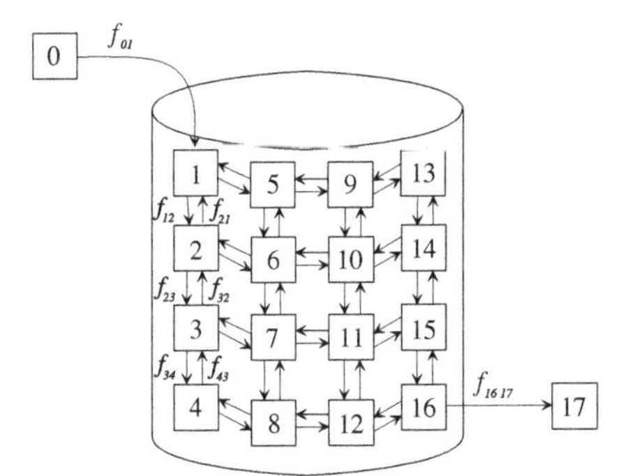

Figure 1.5. Sketch of a 16-zone model for a stirred-tank reactor.

In this expression, the *inlet-zone* ($j = 0$) concentrations are defined by $\phi^{(0)} = \phi_{\text{in}}$, and the *inlet transport rates* are denoted by f_{0i}. Likewise, the *outlet transport rates* are denoted by $f_{i\,N+1}$. Thus, by definition, $f_{i0} = f_{N+1\,i} = 0$.

The transport rates f_{ij} will be determined by the turbulent flow field inside the reactor. When setting up a zone model, various methods have been proposed to extract the transport rates from experimental data (Mann *et al.* 1981; Mann *et al.* 1997), or from CFD simulations. Once the transport rates are known, (1.15) represents a (large) system of coupled ordinary differential equations (ODEs) that can be solved numerically to find the species concentrations in each zone and at the reactor outlet.

The form of (1.15) is identical to the balance equation that is used in finite-volume CFD codes for passive scalar mixing.[17] The principal difference between a zone model and a finite-volume CFD model is that in a zone model the grid can be chosen to optimize the capture of inhomogeneities in the scalar fields *independent* of the mean velocity and turbulence fields.[18] Theoretically, this fact could be exploited to reduce the number of zones to the minimum required to resolve spatial gradients in the scalar fields, thereby greatly reducing the computational requirements.

In general, zone models are applicable to chemical reactions for which local micromixing effects can be ignored. In turbulent flows, the transport rates appearing in (1.15) will scale with the local *integral-scale turbulence frequency*[19] (Pope 2000). Thus, strictly speaking, zone models[20] will be applicable to turbulent reacting flows for which the local chemical time scales are all greater than the integral time scale of the turbulence. For chemical reactions with shorter time scales, micromixing can have a significant impact on the species concentrations in each zone, and at the reactor outlet (Weinstein and Adler 1967; Paul and Treybal 1971; Ott and Rys 1975; Bourne and Toor 1977; Bourne *et al.* 1977; Bourne 1983).

17 In a CFD code, the transport rate will depend on the mean velocity and *turbulent diffusivity* for each zone.
18 The CFD code must use a grid that also resolves spatial gradients in the mean velocity and turbulence fields. At some locations in the reactor, the scalar fields may be constant, and thus a coarser grid (e.g., a zone) can be employed.
19 The integral-scale turbulence frequency is the inverse of the turbulence integral time scale. The turbulence time and length scales are defined in Chapter 2.
20 Similar remarks apply for CFD models that ignore sub-grid-scale mixing. The problem of closing the chemical source term is discussed in detail in Chapter 5.

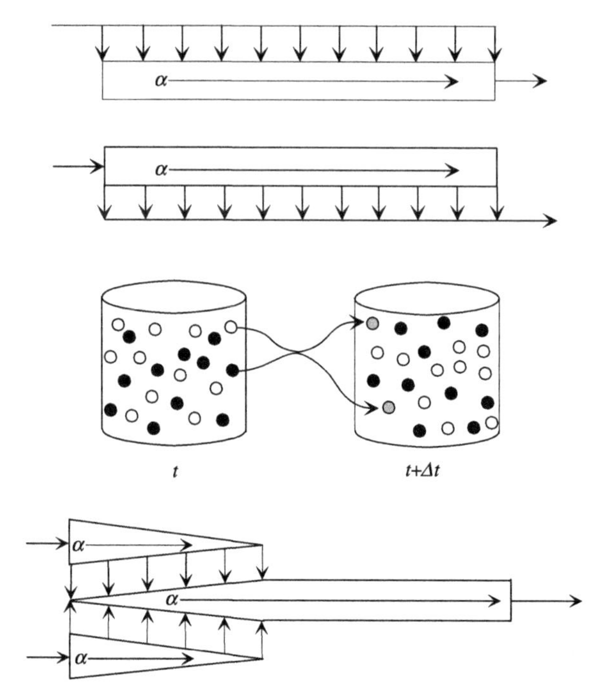

Figure 1.6. Four micromixing models that have appeared in the literature. From top to bottom: maximum-mixedness model; minimum-mixedness model; coalescence-redispersion model; three-environment model.

1.2.4 Micromixing models

Danckwerts (1953) pointed out that RTD theory is insufficient to predict product yields for complex kinetics and noted that a general treatment of this case is extremely difficult (Danckwerts 1957; Danckwerts 1958). Nonetheless, the desire to quantify the degree of segregation in the RTD context has led to a large collection of micromixing models based on RTD theory (e.g., Zwietering 1959; Zwietering 1984). Some of these models are discussed in CRE textbooks (e.g., Fogler 1999). Four examples are shown in Fig. 1.6. Note that these micromixing models do not contain or use any information about the detailed flow field inside the reactor. The principal weakness of RTD-based micromixing models is the lack of a firm physical basis for determining the exchange parameters. We will discuss this point in greater detail in Chapter 3. Moreover, since RTD-based micromixing models do not predict the spatial distribution of reactants inside the reactor, it is impossible to validate fully the model predictions.

Another class of micromixing models is based on *fluid environments* (Nishimura and Matsubara 1970; Ritchie and Tobgy 1979; Mehta and Tarbell 1983a; Mehta and Tarbell 1983b). The basic idea behind these models is to divide *composition space* into a small number of environments that interact due to micromixing. Thus, unlike zone models, which divide up physical space, each environment can be thought of as existing at a particular spatial location with a certain probability. In some cases, the probabilities are fixed (e.g., equal to the inverse of the number of environments). In other cases, the probabilities evolve due to the interactions between environments. In Section 5.10 we will discuss in detail the general formulation of multi-environment micromixing models in the context of CFD models. Here, we will limit our consideration to two simple models: the *interaction by exchange with the mean* (IEM) model for the poorly micromixed PFR and the IEM model for the poorly micromixed CSTR.

The IEM model for a non-premixed PFR employs two environments with probabilities p_1 and $p_2 = 1 - p_1$, where p_1 is the volume fraction of stream 1 at the reactor inlet. In the IEM model, p_1 is assumed to be constant.[21] The concentration in environment n is denoted by $\phi^{(n)}$ and obeys

$$\frac{\mathrm{d}\phi^{(n)}}{\mathrm{d}\alpha} = \frac{1}{t_{\mathrm{iem}}}\left(\langle\phi(\alpha)\rangle - \phi^{(n)}\right) + \mathbf{S}\left(\phi^{(n)}\right) \quad \text{with} \quad \phi^{(n)}(0) = \phi_{\mathrm{in}}^{(n)}, \tag{1.16}$$

where $\phi_{\mathrm{in}}^{(n)}$ is the inlet concentration to environment n, and t_{iem} is the IEM micromixing time. The first term on the right-hand side of (1.16) is a simple linear model for fluid–particle interactions. In this case, all fluid elements with age α are assumed to interact by exchanging matter with a fictitious fluid element whose concentration is $\langle\phi(\alpha)\rangle$.

By definition, averaging (1.16) with respect to the operator $\langle\cdot\rangle$ (defined below in (1.18)) causes the micromixing term to drop out:[22]

$$\frac{\mathrm{d}\langle\phi\rangle}{\mathrm{d}\alpha} = \langle\mathbf{S}\rangle \quad \text{with} \quad \langle\phi(0)\rangle = \langle\phi_{\mathrm{in}}\rangle. \tag{1.17}$$

Note that in order to close (1.16), the micromixing time must be related to the underlying flow field. Nevertheless, because the IEM model is formulated in a Lagrangian framework, the chemical source term in (1.16) appears in closed form. This is not the case for the chemical source term in (1.17).

The mean concentrations appearing in (1.16) are found by averaging with respect to the *internal-age transfer function*[23] $H(\alpha, \beta)$ and the environments:[24]

$$\langle\phi(\alpha)\rangle = \sum_{n=1}^{2} p_n \int_0^\infty \phi^{(n)}(\beta) H(\alpha, \beta)\,\mathrm{d}\beta. \tag{1.18}$$

For the PFR and the CSTR, $H(\alpha, \beta)$ has particularly simple forms:

$$H_{\mathrm{pfr}}(\alpha, \beta) = \delta(\beta - \alpha) \quad \text{and} \quad H_{\mathrm{cstr}}(\alpha, \beta) = E_{\mathrm{cstr}}(\beta). \tag{1.19}$$

[21] If p_1 is far from 0.5 (i.e., non-equal-volume mixing), the IEM model yields poor predictions. Alternative models (e.g., the E-model of Baldyga and Bourne (1989)) that account for the evolution of p_1 should be employed to model non-equal-volume mixing.

[22] In Chapter 6, this is shown to be a general physical requirement for all micromixing models, resulting from the fact that molecular diffusion in a closed system conserves mass. $\langle\phi(\alpha)\rangle$ is the mean concentration with respect to all fluid elements with age α. Thus, it is a *conditional expected value*.

[23] $H(\alpha, \beta)$ is a *weighting kernel* to generate the contribution of fluid elements with age β to the mean concentration at age α. Similarly, in the transported PDF codes discussed in Chapter 6, a *spatial* weighting kernel of the form $h_W(s)$ appears in the definition of the local mean concentrations.

[24] For a CSTR, (1.18) is numerically unstable for small t_{iem} (Fox 1989). For numerical work, it should thus be replaced by an equivalent integro-differential equation (Fox 1991).

Note that the mean concentrations in the PFR are just the volume-averaged concentrations of the two environments with the same age. On the other hand, in the CSTR, the mean concentrations are independent of age (i.e., they are the same at every point in the reactor).

The IEM model can be extended to model unsteady-state stirred reactors (Fox and Villermaux 1990b), and to study micromixing effects for complex reactions using bifurcation theory (Fox and Villermaux 1990a; Fox *et al.* 1990; Fox 1991; Fox *et al.* 1994). Nevertheless, its principal weaknesses when applied to stirred reactors are the need to specify an appropriate micromixing time and the assumption that the mean concentrations are independent of the spatial location in the reactor. However, as discussed in Section 5.10, these shortcomings can be overcome by combining multi-environment micromixing models with CFD models for stirred-tank reactors. A more detailed, but similar, approach based on *transported PDF methods* is discussed in Chapter 6. Both multi-environment CFD models and transported PDF methods essentially combine the advantages of both zone models and micromixing models to provide a more complete description of turbulent reacting flows. An essential ingredient in all approaches for modeling micromixing is the choice of the micromixing time, which we discuss next.

1.2.5 Micromixing time

The micromixing time is a key parameter when modeling fast chemical reactions in non-premixed reactors (Fox 1996a). Indeed, in many cases, the choice of the micromixing time has a much greater impact on the predicted product distribution than the choice of the micromixing model. When combining a CRE micromixing model with a CFD turbulence model, it is thus paramount to understand the relationship between the micromixing time and the scalar dissipation rate.[25] The latter is employed in CFD models for scalar mixing based on the transport equation for the *scalar variance*. The relationship between the micromixing time and the scalar dissipation rate is most transparent for the poorly micromixed PFR. We will thus consider this case in detail using the IEM model.

Consider an inert (non-reacting) scalar ϕ in a poorly micromixed PFR. The IEM model for this case reduces to

$$\frac{\mathrm{d}\phi^{(n)}}{\mathrm{d}\alpha} = \frac{1}{t_{\mathrm{iem}}} \left(\langle \phi(\alpha) \rangle - \phi^{(n)} \right) \quad \text{with} \quad n = 1, 2, \tag{1.20}$$

where

$$\langle \phi(\alpha) \rangle = p_1 \phi^{(1)}(\alpha) + p_2 \phi^{(2)}(\alpha). \tag{1.21}$$

Since the inlet concentrations will have no effect on the final result, for simplicity we let $\phi^{(1)}(0) = 0$ and $\phi^{(2)}(0) = 1$. Applying (1.18) to (1.20), it is easily shown that the scalar mean is constant and given by $\langle \phi(\alpha) \rangle = p_2$.

[25] The scalar dissipation rate is defined in Chapter 3.

The next step is to derive an expression for the scalar variance defined by

$$\langle \phi'^2(\alpha) \rangle \equiv \langle \phi^2(\alpha) \rangle - \langle \phi(\alpha) \rangle^2 \tag{1.22}$$

wherein

$$\langle \phi^2(\alpha) \rangle = \sum_{n=1}^{2} p_n \left(\phi^{(n)}(\alpha) \right)^2 . \tag{1.23}$$

Differentiating (1.22) with respect to α and substituting (1.20) leads to

$$\frac{\mathrm{d}\langle \phi'^2 \rangle}{\mathrm{d}\alpha} = -\frac{2}{t_{\mathrm{iem}}} \langle \phi'^2 \rangle . \tag{1.24}$$

In Section 3.2, we show that under the same conditions the right-hand side of (1.24) is equal to the negative scalar dissipation rate ((3.45), p. 70). Thus, the micromixing time is related to the scalar dissipation rate ε_ϕ and the scalar variance by

$$t_{\mathrm{iem}} = \tau_\phi \equiv \frac{2 \langle \phi'^2 \rangle}{\varepsilon_\phi} . \tag{1.25}$$

Choosing the micromixing time in a CRE micromixing model is therefore equivalent to choosing the scalar dissipation rate in a CFD model for scalar mixing.

In the CRE literature, turbulence-based micromixing models have been proposed that set the micromixing time proportional to the Kolmogorov time scale:

$$t_{\mathrm{iem}} \propto \left(\frac{\nu}{\varepsilon} \right)^{1/2} , \tag{1.26}$$

where ν is the kinematic viscosity of the fluid, and ε is the turbulent dissipation rate. As discussed in detail in Chapter 3, this choice is only valid under *very limited* inlet conditions. In fact, the micromixing time will be strongly dependent on the inlet conditions of the scalar field and the underlying turbulence fields. In CFD models for scalar mixing, the micromixing time is usually found either by assuming that the scalar dissipation rate is controlled by the *rate of scalar energy transfer* from large to small scales (the so-called *equilibrium model*), or by solving a transport equation for ε. We will look at both approaches in Chapters 3 and 4.

1.3 Fluid-mechanical approach

The FM approach to modeling turbulent reacting flows had as its initial focus the description of turbulent combustion processes (e.g., Chung 1969; Chung 1970; Flagan and Appleton 1974; Bilger 1989). In many of the early applications, the details of the chemical reactions were effectively ignored because the reactions could be assumed to be in local chemical equilibrium.[26] Thus, unlike the early emphasis on slow and finite-rate reactions

[26] In other words, all chemical reactions are assumed to occur much faster than micromixing.

in the CRE literature, much of the early FM literature on reacting flows emphasized the modeling of the turbulent flow field and the effects of density changes due to chemical reactions. However, more recently, the importance of finite-rate reactions in combustion processes has become clear. This, in turn, has led to the development of FM approaches that can handle complex chemistry but are numerically tractable (Warnatz *et al.* 1996; Peters 2000).

Like CRE micromixing models, the goal of current FM approaches is the accurate treatment of the chemical source term and molecular mixing. As a starting point, most FM approaches for turbulent reacting flows can be formulated in terms of the joint PDF of the velocity and the composition variables. Thus, many experimental and theoretical studies have reported on velocity and concentration fluctuation statistics in simple canonical flows (Corrsin 1958; Corrsin 1961; Toor 1962; Lee and Brodkey 1964; Keeler *et al.* 1965; Vassilatos and Toor 1965; Brodkey 1966; Gegner and Brodkey 1966; Lee 1966; Corrsin 1968; Toor 1969; Torrest and Ranz 1970; Mao and Toor 1971; Gibson and Libby 1972; Lin and O'Brien 1972; Dopazo and O'Brien 1973; Lin and O'Brien 1974; Dopazo and O'Brien 1976; Hill 1976; Breidenthal 1981; Bennani *et al.* 1985; Lundgren 1985; Koochesfahani and Dimotakis 1986; Hamba 1987; Komori *et al.* 1989; Bilger *et al.* 1991; Guiraud *et al.* 1991; Komori *et al.* 1991a; Komori *et al.* 1991b; Brown and Bilger 1996). In Chapters 2 and 3, we review the statistical description of turbulent flows and turbulent scalar mixing. In the remainder of this section, we give a brief overview of the FM approach to modeling turbulent reacting flows. In the following section, we will compare the similarities and differences between the CRE and FM approaches.

1.3.1 Fundamental transport equations

For the constant-density flows considered in this work,[27] the fundamental governing equations are the Navier–Stokes equation for the fluid velocity $\mathbf{U}$ (Bird *et al.* 2002):

$$\frac{\partial U_i}{\partial t} + U_j \frac{\partial U_i}{\partial x_j} = \nu \frac{\partial^2 U_i}{\partial x_j \partial x_j} - \frac{1}{\rho} \frac{\partial p}{\partial x_i}, \tag{1.27}$$

and the reacting scalar transport equation (ϕ_α represents a chemical species concentration or enthalpy):

$$\frac{\partial \phi_\alpha}{\partial t} + U_j \frac{\partial \phi_\alpha}{\partial x_j} = \Gamma_\alpha \frac{\partial^2 \phi_\alpha}{\partial x_j \partial x_j} + S_\alpha(\boldsymbol{\phi}). \tag{1.28}$$

In interpreting these expressions, the usual summation rules for roman indices apply, e.g., $a_i b_i = a_1 b_1 + a_2 b_2 + a_3 b_3$. Note that the scalar fields are assumed to be *passive*, i.e., $\boldsymbol{\phi}$ does not appear in (1.27).

The fluid density appearing in (1.27) is denoted by ρ and is assumed to be constant. The molecular-transport coefficients appearing in the governing equations are the kinematic

[27] Although this choice excludes combustion, most of the modeling approaches can be directly extended to non-constant-density flows with minor modifications.

viscosity ν, and the molecular and thermal diffusivities Γ_α for the chemical species and enthalpy fields. The pressure field p appearing on the right-hand side of (1.27) is governed by a Poisson equation:

$$\nabla^2 p = -\rho \frac{\partial U_i}{\partial x_j} \frac{\partial U_j}{\partial x_i}. \tag{1.29}$$

This expression is found from (1.27) using the continuity equation for a constant-density flow:

$$\boldsymbol{\nabla} \cdot \mathbf{U} = 0. \tag{1.30}$$

The last term on the right-hand side of (1.28) is the chemical source term. As will be seen in Chapter 5, the chemical source term is often a complex, non-linear function of the scalar fields ϕ, and thus solutions to (1.28) are very different than those for the *inert*-scalar transport equation wherein $\mathbf{S}$ is null.

1.3.2 Turbulence models

Under the operating conditions of most industrial-scale chemical reactors, the solution to (1.27) will be *turbulent* with a large range of length and time scales (Bischoff 1966; McKelvey *et al.* 1975; Brodkey 1984; Villermaux 1991). As a consequence of the complexity of the velocity field, chemical-reactor models based on solving (1.27) directly are computationally intractable. Because of this, in its early stages of development, the FM approach for turbulent mixing was restricted to describing canonical turbulent flows (Corrsin 1951a; Corrsin 1951b; Corrsin 1957; Gibson and Schwarz 1963a; Gibson and Schwarz 1963b; Lee and Brodkey 1964; Nye and Brodkey 1967a; Nye and Brodkey 1967b; Gibson 1968a; Gibson 1968b; Grant *et al.* 1968; Christiansen 1969; Gibson *et al.* 1970), and thus had little impact on CRE models for industrial-scale chemical reactors. However, with the advances in computing technology, CFD has become a viable tool for simulating industrial-scale chemical reactors using *turbulence models* based on the statistical theory of turbulent flows.

The potential economic impact of CFD in many engineering disciplines has led to considerable research in developing Reynolds-averaged Navier–Stokes (RANS) turbulence models (Daly and Harlow 1970; Launder and Spalding 1972; Launder 1991; Hanjalić 1994; Launder 1996) that can predict the mean velocity $\langle \mathbf{U} \rangle$, turbulent kinetic energy k, and the turbulent dissipation rate ε in high-Reynolds-number turbulent flows.[28] These and more sophisticated models are now widely available in commercial CFD codes, and are routinely employed for reactor design in the chemical process industry. For completeness, we review some of the most widely used turbulence models in Chapter 4. A more thorough discussion of the foundations of turbulence modeling can be found in Pope (2000).

Similarly, turbulent scalar transport models based on (1.28) for the case where the chemical source term is null have been widely studied. Because (1.28) in the absence

[28] The experienced reader will recognize these CFD models as the so-called RANS turbulence models.

of chemical reactions is linear in the scalar variable, CFD models for the mean scalar field closely resemble the corresponding turbulence models for k and ε. In Chapter 3, the transport equation for the scalar mean is derived starting from (1.28) using Reynolds averaging. For inert-scalar turbulent mixing, the closure problem reduces to finding an appropriate model for the *scalar flux*. In most CFD codes, the scalar flux is found either by a gradient-diffusion model or by solving an appropriate transport equation. Likewise, scalar fluctuations can be characterized by solving the transport equation of the scalar variance (see Chapter 3). For *reacting-scalar* turbulent mixing, the chemical source term poses novel, and technically more difficult, closure problems.

1.3.3 Chemical source term

Despite the progress in CFD for inert-scalar transport, it was recognized early on that the treatment of turbulent *reacting* flows offers unique challenges (Corrsin 1958; Danckwerts 1958). Indeed, while turbulent transport of an inert scalar can often be successfully described by a small set of *statistical moments* (e.g., $\langle \mathbf{U} \rangle$, k, ε, $\langle \phi \rangle$, and $\langle \phi'^2 \rangle$), the same is not true for scalar fields, which are strongly coupled through the chemical source term in (1.28). Nevertheless, it has also been recognized that because the chemical source term depends only on the *local* molar concentrations $\mathbf{c}$ and temperature T:

$$\mathbf{S}(\phi), \quad \text{where} \quad \phi^{\mathrm{T}} = (c_{\mathrm{A}}, c_{\mathrm{B}}, \ldots, T),$$

knowledge of the one-point, one-time composition PDF $f_{\phi}(\psi; \mathbf{x}, t)$ at all points in the flow will suffice to predict the mean chemical source term $\langle \mathbf{S} \rangle$, which appears in the reacting-scalar transport equation for the scalar means $\langle \phi \rangle$ (Chung 1976; O'Brien 1980; Pope 1985; Kollmann 1990).

As discussed in Chapter 2, a fully developed turbulent flow field contains flow structures with length scales much smaller than the grid cells used in most CFD codes (Daly and Harlow 1970).[29] Thus, CFD models based on *moment methods* do not contain the information needed to predict $f_{\phi}(\psi; \mathbf{x}, t)$. Indeed, only the direct numerical simulation (DNS) of (1.27)–(1.29) uses a fine enough grid to resolve completely all flow structures, and thereby avoids the need to predict $f_{\phi}(\psi; \mathbf{x}, t)$. In the CFD literature, the small-scale structures that control the chemical source term are called sub-grid-scale (SGS) fields, as illustrated in Fig. 1.7.

Heuristically, the SGS distribution of a scalar field $\phi(\mathbf{x}, t)$ can be used to *estimate* the composition PDF by constructing a histogram from all SGS points within a particular CFD grid cell.[30] Moreover, because the important statistics needed to describe a scalar field (e.g., its expected value $\langle \phi \rangle$ or its variance $\langle \phi'^2 \rangle$) are nearly constant on sub-grid

[29] Only direct numerical simulation (DNS) resolves all scales (Moin and Mahesh 1998). However, DNS is computationally intractable for chemical reactor modeling.

[30] The reader familiar with the various forms of averaging (Pope 2000) will recognize this as a spatial average over a locally statistically homogeneous field.

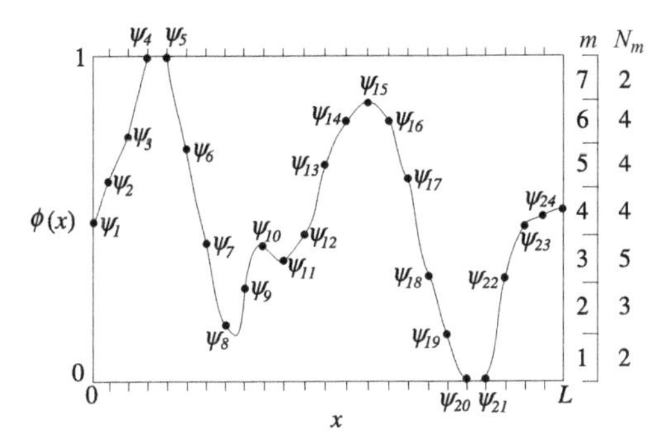

Figure 1.7. Sketch of sub-grid-scale (SGS) distribution of ϕ.

scales, the SGS field can be considered *statistically homogeneous*. This implies that all points sampled from the same CFD grid cell are statistically equivalent, i.e., sampled from the same composition PDF.

As an example of estimating a scalar PDF, consider a *bounded*, one-dimensional scalar field $\phi \in [0, 1]$ defined on $x \in [0, L]$, where L is the CFD grid size as shown in Fig. 1.7. In a CFD calculation, only $\phi(0)$ and $\phi(L)$ would be computed (or, more precisely, the mean values $\langle\phi(0)\rangle$ and $\langle\phi(L)\rangle$ at the grid points). However, if $\phi(x)$ were somehow available for all values of x, a histogram could be constructed as follows:

(i) Choose a fine grid with spacing $l \ll L$, and let $N = 1 + \text{integer}(L/l)$. Sample $\phi(x)$ on the fine grid:

$$\psi_1 = \phi(0),\ \psi_2 = \phi(l),\ \psi_3 = \phi(2l),\ \ldots,\ \psi_N = \phi((N-1)l).$$

(ii) Use the samples $(\psi_1, \ldots, \psi_N)$ to construct a histogram for ϕ:
   (a) Construct M bins in composition space $\psi \in [0, 1]$ with spacing $\Delta = 1/M$.
   (b) Count the number of samples N_m that fall in bin $m \in 1, \ldots, M$.
   (c) Define the value of the histogram at bin m by

$$h(m\Delta) \equiv \frac{N_m}{N\Delta}. \tag{1.31}$$

   (Note that in anticipation of considering the histogram as an approximation of the PDF of ϕ, $h(\psi)$ has been normalized so that its integral over composition space is unity.)

(iii) The histogram can then be plotted versus the mid-point value for each bin as shown in Fig. 1.8.

In the limit where $l \to 0$, the number of samples N will become very large, and the bin spacing Δ can be decreased while keeping N_m large enough to control statistical fluctuations. The histogram then becomes nearly continuous in ψ and can be used to

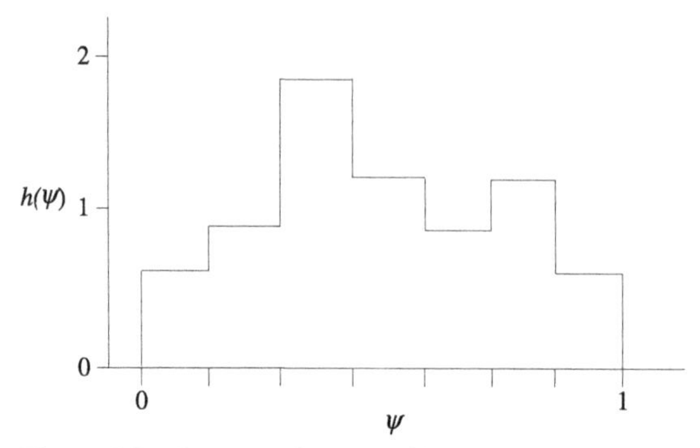

Figure 1.8. Histogram for sub-grid-scale distribution of ϕ based on 24 samples and seven bins.

estimate the PDF of ϕ:

$$\lim_{N,M\to\infty} h(m\Delta) \to \hat{f}_\phi(\psi). \tag{1.32}$$

The true PDF $f_\phi(\psi)$ is defined axiomatically (see Chapter 3), but can be thought of as representing all possible realizations of $\phi(x)$ generated with the same flow conditions (i.e., an *ensemble*). Because $\hat{f}_\phi(\psi)$ has been found based on a single realization, it may or may not be a good approximation for $f_\phi(\psi)$, depending on how well the single realization represents the entire ensemble. Generally speaking, in a turbulent flow the latter will depend on the value of the *integral scale* of the quantity of interest relative to the grid spacing L. For a turbulent scalar field, the integral scale L_ϕ is often approximately equal to L, in which case $\hat{f}_\phi(\psi)$ offers a poor representation of $f_\phi(\psi)$. However, for statistically stationary flows, the estimate can be improved by collecting samples at different times.[31]

For turbulent reacting flows, we are usually interested in chemical reactions involving multiple scalars. As for a single scalar, a histogram can be constructed from multiple scalar fields (Fig. 1.9). For example, if there are two reactants A and B, the samples will be bi-variate:

$$\begin{aligned}
\psi_1 &= (\phi_A(0), \phi_B(0)),\\
\psi_2 &= (\phi_A(l), \phi_B(l)),\\
\psi_3 &= (\phi_A(2l), \phi_B(2l)),\\
&\vdots\\
\psi_N &= (\phi_A((N-1)l), \phi_B((N-1)l)).
\end{aligned}$$

The resultant histogram is also bi-variate, $h_{A,B}(m_A\Delta_A, m_B\Delta_B)$, and can be represented by a contour plot, as shown in Fig. 1.10.

[31] This procedure is widely used when extracting statistical estimates from DNS data.

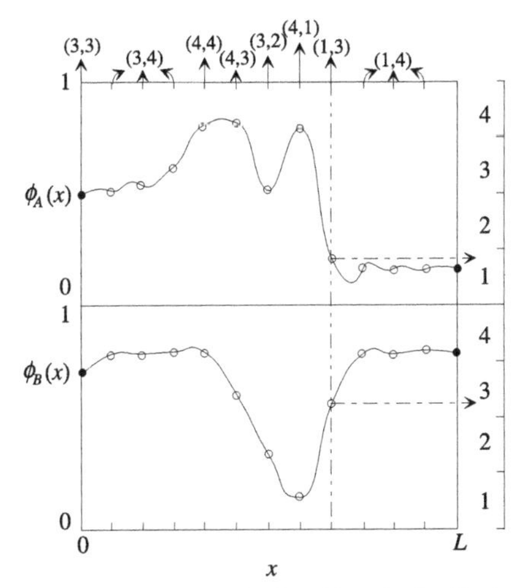

Figure 1.9. Sketch of sub-grid-scale distribution of ϕ_A and ϕ_B. The bin numbers for each sample point are given at the top of the figure.

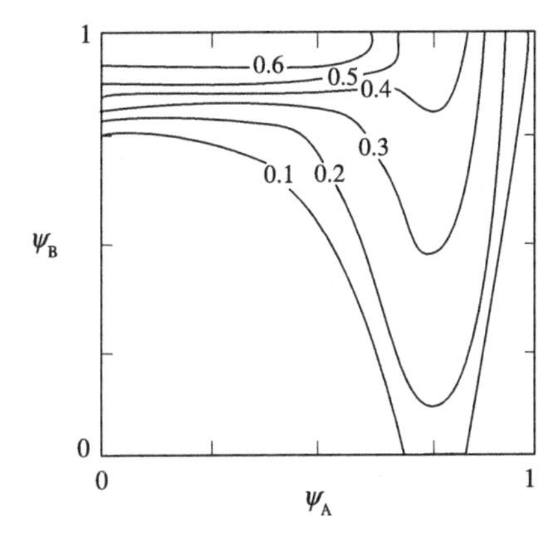

Figure 1.10. Contour plot of the joint histogram for ϕ_A and ϕ_B.

Likewise, in the limit of large numbers of samples and bins, the bi-variate histogram can be used to compute an estimate for the joint PDF of ϕ_A and ϕ_B:

$$\lim_{N, M_A, M_B \to \infty} h_{A,B}(\psi_A, \psi_B) \to \hat{f}_{A,B}(\psi_A, \psi_B). \tag{1.33}$$

We shall see in Chapter 5 that knowledge of $f_{A,B}(\psi_A, \psi_B)$ suffices to close the chemical source term for the isothermal, second-order reaction

$$A + B \xrightarrow{k_1} P.$$

The procedure presented above can be easily extended to estimate the joint PDF of a vector of K composition variables $f_{\phi}(\psi)$. For example, the mean chemical source term $\langle \mathbf{S} \rangle$ inside a CFD grid cell can be estimated by sampling the chemical source term at every point on a finer grid:

$$\begin{aligned}
\mathbf{S}_1 &= \mathbf{S}(\psi_1), \\
\mathbf{S}_2 &= \mathbf{S}(\psi_2), \\
\mathbf{S}_3 &= \mathbf{S}(\psi_3), \\
&\vdots \\
\mathbf{S}_N &= \mathbf{S}(\psi_N);
\end{aligned}$$

and summing over all samples:

$$\begin{aligned}
\langle \mathbf{S} \rangle &\equiv \int_{-\infty}^{+\infty} \cdots \int_{-\infty}^{+\infty} \mathbf{S}(\psi) f_\phi(\psi) \, \mathrm{d}\psi \\
&\approx \sum_{m_1=1}^{M_1} \cdots \sum_{m_K=1}^{M_K} \mathbf{S}(m_1 \Delta_1, \ldots, m_K \Delta_K) h(m_1 \Delta_1, \ldots, m_K \Delta_K) \Delta_1 \cdots \Delta_K \\
&\approx \frac{1}{N} \sum_{n=1}^{N} \mathbf{S}(\psi_n).
\end{aligned} \tag{1.34}$$

The last term in (1.34) follows by approximating $\mathbf{S}$ evaluated at the bin center by its value at a sample point contained in the bin. In the limit where the fine grid becomes infinitely fine, the last term is just the spatial-average chemical source term:

$$\langle \mathbf{S} \rangle_L = \int_0^L \mathbf{S}(\phi(x)) \, \mathrm{d}x = \lim_{N \to \infty} \frac{1}{N} \sum_{n=1}^{N} \mathbf{S}(\psi_n). \tag{1.35}$$

Note that (1.34) defines the mean chemical source term in terms of $f_\phi(\psi)$, and that the latter contains considerably less information than the original scalar fields $\phi(x)$.[32]

In summary, the FM approach to turbulent reacting flows is closely connected (either directly or indirectly) with the determination of the joint composition PDF. As is true for turbulent flows (Pope 2000), statistical models for turbulent reacting flows are best derived axiomatically (e.g., in terms of $\langle \mathbf{S} \rangle$ instead of $\langle \mathbf{S} \rangle_L$). In the examples given above, we assumed that the SGS scalar fields were known, and thus were able to estimate the composition joint PDF using a histogram. The challenge posed in the FM approach to turbulent reacting flows is thus to derive adequate representations for the mean chemical source term consistent with known theoretical constraints and experimental observations *without direct knowledge of the SGS scalar fields*. For this purpose, many *one-point* models (with widely differing degrees of generality) have been proposed and successfully implemented. In Chapter 5, the most general and widely employed models are discussed in some detail. A common feature of all one-point models for turbulent reacting flows is the need for a description of molecular mixing.

[32] For example, all information is lost concerning the relative spatial locations of two random samples. As discussed in Chapter 2, this fact implies that all information concerning the spatial derivatives of the scalar fields is lost when the scalar field is described by its *one-point* joint PDF.

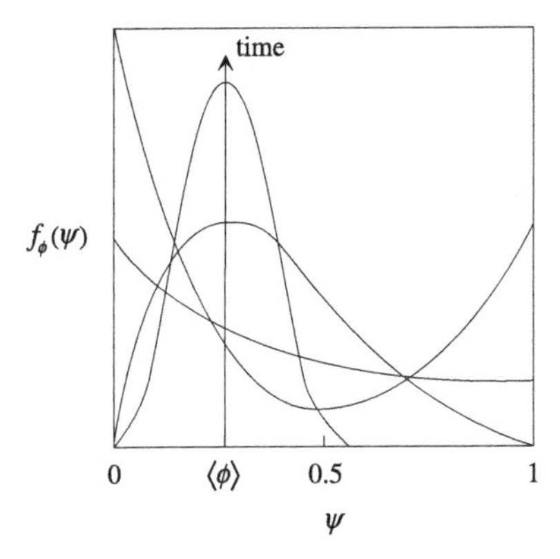

Figure 1.11. A non-premixed scalar PDF as a function of time for inert-scalar mixing. Note that at very short times the PDF is bi-modal since all molecular mixing occurs in thin diffusion layers between regions of pure fluid where $\phi = 0$ or 1. On the other hand, for large times, the scalar PDF is nearly Gaussian.

1.3.4 Molecular mixing

As seen above, the mean chemical source term is intimately related to the PDF of the *concentration fluctuations*. In non-premixed flows, the rate of decay of the concentration fluctuations is controlled by the scalar dissipation rate. Thus, a critical part of any model for chemical reacting flows is a description of how molecular diffusion works to damp out concentration fluctuations at the SGS level.

As an example, consider the Lagrangian formulation of (1.28):

$$\frac{\mathrm{d}\phi^*_\alpha}{\mathrm{d}t} = \langle \Gamma_\alpha \nabla^2 \phi_\alpha | \phi = \phi^* \rangle + S_\alpha(\phi^*). \tag{1.36}$$

The first term on the right-hand side is the expected value of the scalar Laplacian *conditioned* on the scalars having values ϕ^*.[33] An example of the time evolution of the *conditional scalar Laplacian*, corresponding to the scalar PDF in Fig. 1.11, is plotted in Fig. 1.12 for an initially non-premixed inert-scalar field. The closure of the conditional scalar Laplacian is discussed in Chapter 6. For the time being, it suffices to note the similarity between (1.36) and the IEM model, (1.16). Indeed, the IEM model is a closure for the conditional scalar Laplacian, i.e.,

$$\langle \Gamma_\alpha \nabla^2 \phi_\alpha | \phi = \phi^* \rangle = \frac{1}{\tau_\phi}(\langle \phi \rangle - \phi^*), \tag{1.37}$$

which is widely employed in transported PDF simulations of turbulent reacting flows.[34]

In other closures for the chemical source term, a model for the *conditional scalar dissipation rate* $\langle \epsilon_\phi | \phi = \psi \rangle$ is required. (An example is plotted in Fig. 1.13 for the scalar PDF shown in Fig. 1.11.) Like the conditional scalar Laplacian, the conditional scalar

[33] Conditional expectations are defined in Chapter 2.
[34] Note that all terms in (1.37) can be directly extracted from DNS data for turbulent-scalar mixing. Thus, unlike the CRE approach, the FM approach allows for the direct validation of micromixing models.

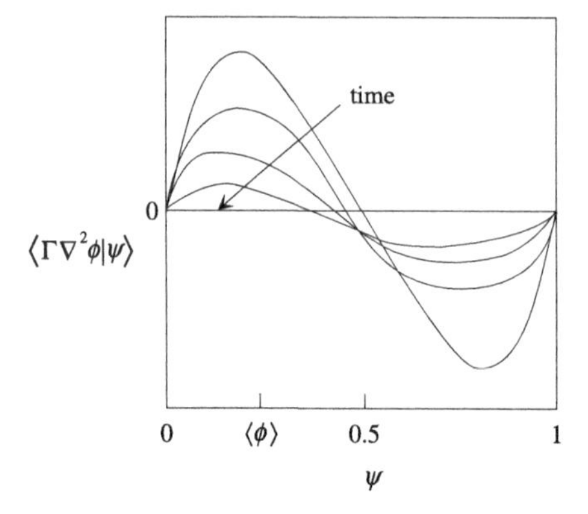

Figure 1.12. The conditional scalar Laplacian $\langle \Gamma \nabla^2 \phi | \psi \rangle$ for the scalar PDF in Fig. 1.11. Note that at very short times $\langle \Gamma \nabla^2 \phi | \psi = 0.5 \rangle = 0$, since all molecular mixing occurs in thin diffusion layers between regions of pure fluid where $\phi = 0$ or 1. On the other hand, for large times, $\langle \Gamma \nabla^2 \phi | \psi = \langle \phi \rangle \rangle = 0$. In this limit, the scalar field is nearly Gaussian, and the conditional scalar Laplacian can be accurately described by the IEM model.

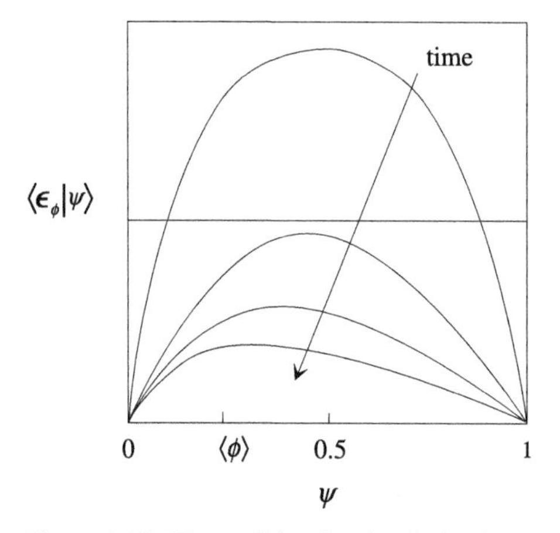

Figure 1.13. The conditional scalar dissipation rate $\langle \epsilon_\phi | \psi \rangle$ for the scalar PDF in Fig. 1.11.

dissipation rate can be extracted from DNS data for model validation. For non-premixed reacting flows, the effects of chemical reactions on the molecular mixing terms are generally ignored (e.g., τ_ϕ in (1.37) is assumed to be the same for all scalars). Nevertheless, one of the great advantages of the FM approach is that assumptions of this type can be verified using DNS data for turbulent reacting flows. Indeed, since the advent of DNS, significant improvements in molecular mixing models have resulted due to model validation studies.

1.4 Relationship between approaches

The relationship between the CRE approach and the FM approach to modeling turbulent reacting flows is summarized in Table 1.1. Despite the obvious and significant differences

Table 1.1. *Relationship between the CRE and FM approaches for modeling the important physical processes present in turbulent reacting flows.*

Physical process	CRE approach	FM approach
Large-scale mixing	RTD theory, zone models	turbulence models for scalar transport
Small-scale segregation	multi-environment micromixing models	composition PDF methods
Molecular-scale mixing	micromixing models, micromixing time	conditional scalar Laplacian, conditional dissipation rate

in the treatment of large-scale mixing, the two approaches share many of the same technical difficulties when treating molecular-scale mixing. Indeed, the micromixing models employed in the CRE approach can be viewed as defining a particular form for the composition PDF employed in the FM approach. For example, as discussed in detail in Section 5.10, a two-environment micromixing model is equivalent to a presumed composition PDF composed of two delta functions. Moreover, both approaches must deal with the fundamental problem of how the scalar fields are affected by the turbulent flow at the level of molecular-scale mixing.

Despite the apparent similarities in treating molecular-scale mixing, the FM approach offers a more complete description of turbulent reacting flows at the level of large-scale mixing.[35] As a direct result (and because it draws heavily on turbulence theory and advances in turbulence modeling), I believe that the FM approach offers the best tools for developing CFD models for describing plant-scale chemical reactors. Thus, a prime goal of this book is to present CFD models for turbulent reacting flows based on the FM approach in sufficient detail such that the reader can gain an appreciation of their individual advantages and limitations. After studying this book, it is my hope that the reader will be able to choose confidently the CFD model that is best suited for a particular turbulent-reacting-flow problem and, when necessary, be able to modify existing models to handle more complex flows.

1.5 A road map to Chapters 2–7

Depending on the background and interests of the reader, the material covered in Chapters 2–7 can be approached in different ways. For example, Chapters 2 and 3 present general background information on turbulent flows and turbulent mixing, respectively. Thus, readers with extensive knowledge of turbulent flows may skip Chapter 2, while readers with expertise in turbulent mixing may skip Chapter 3. However, because the derivations of the Reynolds-averaged transport equations are covered in detail in these

[35] For example, the RTD can be computed from the results of a turbulent-scalar transport model, but not vice versa.

chapters, all readers should review the material presented in Tables 2.4 and 3.3, on pp. 55 and 96, respectively.

A general overview of models for turbulent transport is presented in Chapter 4. The goal of this chapter is to familiarize the reader with the various closure models available in the literature. Because detailed treatments of this material are readily available in other texts (e.g., Pope 2000), the emphasis of Chapter 4 is on presenting the various models using notation that is consistent with the remainder of the book. However, despite its relative brevity, the importance of the material in Chapter 4 should not be underestimated. Indeed, all of the reacting-flow models presented in subsequent chapters depend on accurate predictions of the turbulent flow field. With this caveat in mind, readers conversant with turbulent transport models of non-reacting scalars may wish to proceed directly to Chapter 5.

As discussed in the present chapter, the closure of the chemical source term lies at the heart of models for turbulent reacting flows. Thus, the material on chemical source term closures presented in Chapter 5 will be of interest to all readers. In Chapter 5, attention is given to closures that can be used in conjunction with 'standard' CFD-based turbulence models (e.g., presumed PDF methods). For many readers, these types of closures will be sufficient to model many of the turbulent-reacting-flow problems that they confront in real applications. Moreover, these closures have the advantage of being particularly simple to incorporate into existing CFD codes.

Readers confronted with more difficult reacting-flow problems will be interested in transported PDF methods presented in Chapters 6 and 7. Of all the advantages of transported PDF methods, the most striking is perhaps the fact that the chemical source term is treated exactly. However, as discussed in Chapter 6, this does not imply that the turbulent-reacting-flow model is closed. Instead, the closure problem is moved to the so-called conditional acceleration and diffusion terms that arise in the one-point PDF description. At present, truly general predictive models for these terms have yet to be formulated and thoroughly validated. Nevertheless, existing models can be used successfully for complex reacting-flow calculations, and are thus discussed in some detail. The numerical methods needed to implement transported PDF simulations are introduced in Chapter 7. These differ from 'standard' CFD-based models by their use of Monte-Carlo simulations to represent the flow, and thus require special attention to potential numerical and statistical errors.

2

Statistical description of turbulent flow

In this chapter, we review selected results from the statistical description of turbulence needed to develop CFD models for turbulent reacting flows. The principal goal is to gain insight into the dominant physical processes that control scalar mixing in turbulent flows. More details on the theory of turbulence and turbulent flows can be found in any of the following texts: Batchelor (1953), Tennekes and Lumley (1972), Hinze (1975), McComb (1990), Lesieur (1997), and Pope (2000). The notation employed in this chapter follows as closely as possible the notation used in Pope (2000). In particular, the random velocity field is denoted by $\mathbf{U}$, while the fluctuating velocity field (i.e., with the mean velocity field subtracted out) is denoted by $\mathbf{u}$. The corresponding *sample space variables* are denoted by $\mathbf{V}$ and $\mathbf{v}$, respectively.

2.1 Homogeneous turbulence

At high Reynolds number, the velocity $\mathbf{U}(\mathbf{x}, t)$ is a *random field*, i.e., for fixed time $t = t^*$ the function $\mathbf{U}(\mathbf{x}, t^*)$ varies randomly with respect to $\mathbf{x}$. This behavior is illustrated in Fig. 2.1 for a homogeneous turbulent flow. Likewise, for fixed $\mathbf{x} = \mathbf{x}^*$, $\mathbf{U}(\mathbf{x}^*, t)$ is a *random process* with respect to t. This behavior is illustrated in Fig. 2.2. The meaning of 'random' in the context of turbulent flows is simply that a variable may have a different value each time an experiment is repeated under the same set of flow conditions (Pope 2000). It does not imply, for example, that the velocity field evolves erratically in time and space in an unpredictable fashion. Indeed, due to the fact that it must satisfy the Navier–Stokes equation, (1.27), $\mathbf{U}(\mathbf{x}, t)$ is differentiable in both time and space and thus is relatively 'smooth.'

A third way of looking at the velocity field is from a *Lagrangian* perspective (i.e., following a fluid particle; see Section 6.7 for more details). In this case, the position of the fluid particle $\mathbf{x}^+(t)$ varies with time. The resulting Lagrangian velocity

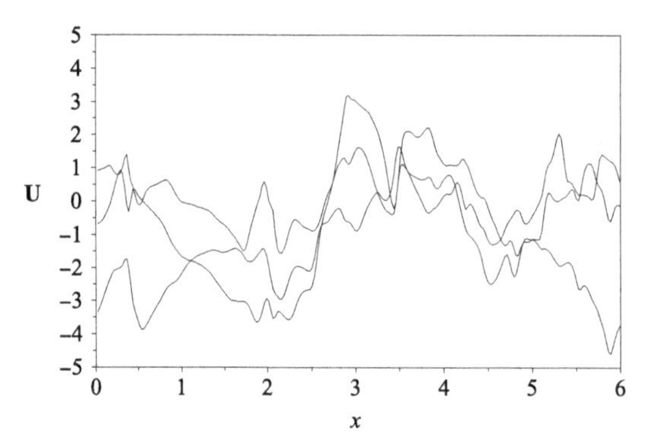

Figure 2.1. Three components of the random field $\mathbf{U}(\mathbf{x}, t^*)$ as a function of $x = x_1$ with fixed $t = t^*$. The velocity was extracted from DNS of isotropic turbulence ($R_\lambda = 140$) with $\langle \mathbf{U} \rangle = \mathbf{0}$. (Courtesy of P. K. Yeung.)

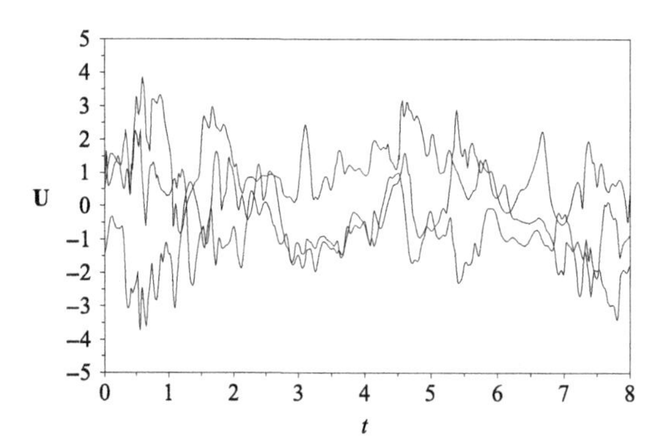

Figure 2.2. Three components of the random process $\mathbf{U}(\mathbf{x}^*, t)$ as a function of t with fixed $\mathbf{x} = \mathbf{x}^*$. The velocity was extracted from DNS of isotropic turbulence ($R_\lambda = 140$) with $\langle \mathbf{U} \rangle = \mathbf{0}$. (Courtesy of P. K. Yeung.)

$\mathbf{U}^+(t) \equiv \mathbf{U}(\mathbf{x}^+(t), t)$ is illustrated in Fig. 2.3.[1] Note that the Lagrangian velocity varies more slowly with time than the Eulerian counterpart shown in Fig. 2.2 (Yeung 2002). This fact has important ramifications on stochastic models for the Lagrangian velocity discussed in Chapter 6.

Much of the theoretical work in turbulent flows has been concentrated on the description of statistically homogeneous turbulence. In a statistically homogeneous turbulent flow, measurable statistical quantities such as the mean velocity[2] or the turbulent kinetic energy are the same at every point in the flow. Among other things, this implies that the turbulence statistics can be estimated using *spatial averages* based on a single realization of the flow.

[1] Additional samples can be found in Yeung (2001) and Yeung (2002).

[2] More precisely, the spatial gradient of the mean velocity is independent of position in a homogeneous turbulent *shear* flow.

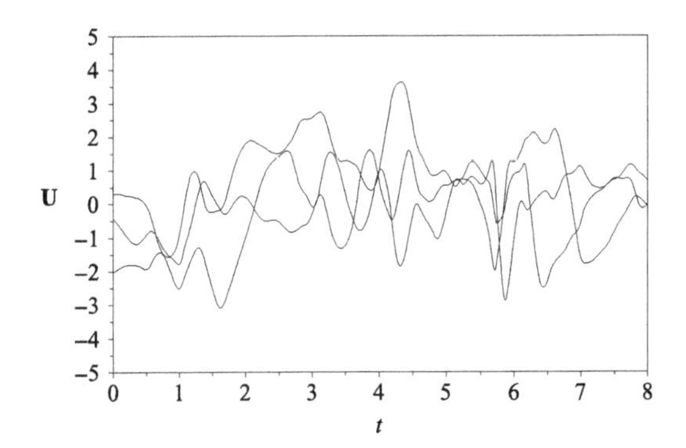

Figure 2.3. Three components of the Lagrangian velocity $\mathbf{U}^{+}(t)$ as a function of t. The velocity was extracted from DNS of isotropic turbulence ($R_\lambda = 140$) with $\langle \mathbf{U} \rangle = \mathbf{0}$. (Courtesy of P. K. Yeung.)

Unlike simple random variables that have no space or time dependence, the statistics of the random velocity field in homogeneous turbulence can be described at many different levels of complexity. For example, a probabilistic theory could be formulated in terms of the *set of functions* $\{\mathbf{U}(\mathbf{x}, t) : (\mathbf{x}, t) \in R^3 \times R\}$.[3] However, from a CFD modeling perspective, such a theory would be of little practical use. Thus, we will consider only *one-point* and *two-point* formulations that describe a homogeneous turbulent flow by the velocity statistics at one or two *fixed* points in space and/or time.

2.1.1 One-point probability density function

For a *fixed point in space* $\mathbf{x}$ and a *given instant* t, the random velocity field $U_1(\mathbf{x}, t)$ can be characterized by a one-point probability density function (PDF) $f_{U_1}(V_1; \mathbf{x}, t)$ defined by[4]

$$f_{U_1}(V_1; \mathbf{x}, t)\,\mathrm{d}V_1 \equiv \mathrm{P}\{V_1 \le U_1(\mathbf{x}, t) < V_1 + \mathrm{d}V_1\}. \tag{2.1}$$

In words, the right-hand side is the probability that the random variable $U_1(\mathbf{x}, t)$ falls between the *sample space* values V_1 and $V_1 + \mathrm{d}V_1$ for different realizations of the turbulent flow.[5] In a homogeneous flow, this probability is independent of $\mathbf{x}$, and thus we can write the one-point PDF as only a function of the sample space variable and time: $f(V_1; t)$.

The above definition can be extended to include an arbitrary number of random variables. For example, the one-point joint velocity PDF $f_{\mathbf{U}}(\mathbf{V}; \mathbf{x}, t)$ describes all three velocity

[3] R denotes the set of real numbers.

[4] A semi-colon is used in the argument list to remind us that V_1 is an independent (sample space) variable, while $\mathbf{x}$ and t are fixed parameters. Some authors refer to $f_{U_1}(V_1; \mathbf{x}, t)$ as the one-point, *one-time* velocity PDF. Here we use 'point' to refer to a space-time point in the four-dimensional space $(\mathbf{x}, t)$.

[5] Because $\mathbf{x}$ and t are fixed, experimentally this definition implies that $U_1(\mathbf{x}, t)$ is the first velocity component measured at a fixed location in the flow at a fixed time instant from the start of the experiment.

components of $\mathbf{U}$ at fixed $(\mathbf{x}, t)$, and is defined by[6]

$$\begin{aligned} f_{\mathbf{U}}(\mathbf{V}; \mathbf{x}, t)\,\mathrm{d}\mathbf{V} &= f(V_1, V_2, V_3; \mathbf{x}, t)\,\mathrm{d}V_1\,\mathrm{d}V_2\,\mathrm{d}V_3 \\ &= \mathrm{P}[\{V_1 \leq U_1(\mathbf{x}, t) < V_1 + \mathrm{d}V_1\} \\ &\quad \cap \{V_2 \leq U_2(\mathbf{x}, t) < V_2 + \mathrm{d}V_2\} \\ &\quad \cap \{V_3 \leq U_3(\mathbf{x}, t) < V_3 + \mathrm{d}V_3\}]. \end{aligned} \tag{2.2}$$

In homogeneous turbulence, the one-point joint velocity PDF can be written as $f_{\mathbf{U}}(\mathbf{V}; t)$, and can be readily measured using hot-wire anemometry or laser Doppler velocimetry (LDV).

In fully developed homogeneous turbulence,[7] the one-point joint velocity PDF is nearly Gaussian (Pope 2000). A Gaussian joint PDF is uniquely defined by a vector of *expected values* $\boldsymbol{\mu}$ and a *covariance matrix* $\mathbf{C}$:

$$f_{\mathbf{U}}(\mathbf{V}; t) = [(2\pi)^3 |\mathbf{C}|]^{-1/2} \exp\left[-\tfrac{1}{2}(\mathbf{V} - \boldsymbol{\mu})^{\mathrm{T}} \mathbf{C}^{-1} (\mathbf{V} - \boldsymbol{\mu})\right]. \tag{2.3}$$

The mean velocity can be computed directly from the joint velocity PDF:[8]

$$\begin{aligned} \mu_i(t) \equiv \langle U_i(\mathbf{x}, t) \rangle &\equiv \iiint_{-\infty}^{+\infty} V_i f_{\mathbf{U}}(\mathbf{V}; t)\,\mathrm{d}\mathbf{V} \\ &= \int_{-\infty}^{+\infty} V_i f_{U_i}(V_i; t)\,\mathrm{d}V_i. \end{aligned} \tag{2.4}$$

The last equality follows from the definition of the one-point *marginal* velocity PDF, e.g.,

$$f_{U_3}(V_3; t) \equiv \iint_{-\infty}^{+\infty} f_{\mathbf{U}}(\mathbf{V}; t)\,\mathrm{d}V_1\,\mathrm{d}V_2. \tag{2.5}$$

Likewise, the velocity *covariance matrix* is defined in terms of the velocity fluctuations

$$u_i = U_i - \langle U_i \rangle, \tag{2.6}$$

and can be computed from the joint velocity PDF:

$$C_{ij}(t) \equiv \langle u_i(\mathbf{x}, t) u_j(\mathbf{x}, t) \rangle \equiv \iiint_{-\infty}^{+\infty} (V_i - \langle U_i \rangle)(V_j - \langle U_j \rangle) f_{\mathbf{U}}(\mathbf{V}; t)\,\mathrm{d}\mathbf{V}. \tag{2.7}$$

The reader familiar with turbulence modeling will recognize the covariance matrix as the *Reynolds stresses*. Thus, for fully developed homogeneous turbulence, knowledge of the mean velocity and the Reynolds stresses completely determines the one-point joint velocity PDF.

[6] The operation $A \cap B$ denotes the intersection of two *events*. $\mathrm{P}[A \cap B]$ is thus the probability that both event A and event B occur simultaneously.

[7] Note that homogeneity rules out wall-bounded turbulent flows (e.g., pipe flow) wherein the velocity PDF can be far from Gaussian.

[8] Note that the $\mathbf{x}$-dependency of $U_i(\mathbf{x}, t)$ drops out due to homogeneity after computing the expected value.

In general, the one-point joint velocity PDF can be used to evaluate the expected value of any arbitrary function $h(\mathbf{U})$ of $\mathbf{U}$ using the definition

$$\langle h(\mathbf{U}) \rangle \equiv \iiint_{-\infty}^{+\infty} h(\mathbf{V}) f_{\mathbf{U}}(\mathbf{V}; t) \, \mathrm{d}\mathbf{V}. \tag{2.8}$$

Moreover it can be employed to define a one-point *conditional* velocity PDF. For example, the conditional PDF of $U_1(\mathbf{x}, t)$ given $U_2(\mathbf{x}, t) = V_2$ and $U_3(\mathbf{x}, t) = V_3$ is defined by

$$f_{U_1|U_2,U_3}(V_1 | V_2, V_3; \mathbf{x}, t) \equiv \frac{f_{\mathbf{U}}(\mathbf{V}; \mathbf{x}, t)}{f_{U_2,U_3}(V_2, V_3; \mathbf{x}, t)} \tag{2.9}$$

where

$$f_{U_2,U_3}(V_2, V_3; \mathbf{x}, t) \equiv \int_{-\infty}^{+\infty} f_{\mathbf{U}}(\mathbf{V}; \mathbf{x}, t) \, \mathrm{d}V_1. \tag{2.10}$$

The conditional PDF can be employed to compute *conditional expected values*. For example, the *conditional mean* of U_1 given $U_2(\mathbf{x}, t) = V_2$ and $U_3(\mathbf{x}, t) = V_3$ is defined by

$$\begin{aligned} \langle U_1(\mathbf{x}, t) | V_2, V_3 \rangle &\equiv \langle U_1(\mathbf{x}, t) | U_2(\mathbf{x}, t) = V_2, U_3(\mathbf{x}, t) = V_3 \rangle \\ &\equiv \int_{-\infty}^{+\infty} V_1 f_{U_1|U_2,U_3}(V_1 | V_2, V_3; \mathbf{x}, t) \, \mathrm{d}V_1. \end{aligned} \tag{2.11}$$

The unconditional mean can be found from the conditional mean by averaging with respect to the joint PDF of U_2 and U_3:

$$\langle U_1(\mathbf{x}, t) \rangle = \iint_{-\infty}^{+\infty} \langle U_1 | V_2, V_3 \rangle f_{U_2,U_3}(V_2, V_3; \mathbf{x}, t) \, \mathrm{d}V_2 \, \mathrm{d}V_3. \tag{2.12}$$

Likewise, the *conditional variance* of $U_1(\mathbf{x}, t)$ given $U_2(\mathbf{x}, t) = V_2$ and $U_3(\mathbf{x}, t) = V_3$ is defined by

$$\begin{aligned} &\langle (U_1 - \langle U_1 | V_2, V_3 \rangle)^2 | V_2, V_3 \rangle \\ &\quad \equiv \int_{-\infty}^{+\infty} (V_1 - \langle U_1 | V_2, V_3 \rangle)^2 f_{U_1|U_2,U_3}(V_1 | V_2, V_3; \mathbf{x}, t) \, \mathrm{d}V_1. \end{aligned} \tag{2.13}$$

Note that, like the unconditional variance, the conditional variance can be expressed as

$$\langle (U_1 - \langle U_1 | V_2, V_3 \rangle)^2 | V_2, V_3 \rangle = \langle U_1^2 | V_2, V_3 \rangle - \langle U_1 | V_2, V_3 \rangle^2. \tag{2.14}$$

For Gaussian random variables, an extensive theory exists relating the joint, marginal, and conditional velocity PDFs (Pope 2000). For example, if the one-point joint velocity PDF is Gaussian, then it can be shown that the following properties hold:

(i) The marginal PDF of U_i is Gaussian. Thus, it is uniquely defined by the mean $\langle U_i \rangle$ and the variance $\langle u_i^2 \rangle$.

(ii) The conditional PDF of U_i given $U_j = V_j$ is Gaussian. Thus, it is uniquely defined by the conditional mean $\langle U_i | V_j \rangle$ and the conditional variance $\langle (U_i - \langle U_i | V_j \rangle)^2 | V_j \rangle$. (No summation is implied on index j.)

(iii) The conditional mean of U_i given $U_j = V_j$ is given by

$$\langle U_i | V_j \rangle = \langle U_i \rangle + \frac{\langle u_i u_j \rangle}{\langle u_j^2 \rangle} (V_j - \langle U_j \rangle). \tag{2.15}$$

(No summation is implied on index j.)

(iv) The conditional variance of U_i given $U_j = V_j$ is given by

$$\langle (U_i - \langle U_i | V_j \rangle)^2 | V_j \rangle = \langle u_i^2 \rangle \left(1 - \rho_{ij}^2\right) \tag{2.16}$$

where the *velocity correlation function* ρ_{ij} is defined by

$$\rho_{ij} = \frac{\langle u_i u_j \rangle}{\langle u_i^2 \rangle^{1/2} \langle u_j^2 \rangle^{1/2}}. \tag{2.17}$$

(No summation is implied on index i or j.)

Despite its widespread use in the statistical description of turbulent reacting flows, the one-point joint velocity PDF does not describe the random velocity field in sufficient detail to understand the physics completely. For example, the one-point description tells us nothing about the statistics of velocity gradients, e.g.,

$$\frac{\partial U_i}{\partial x_1} = \lim_{|x_1 - x_1^*| \to 0} \frac{U_i(x_1, x_2, x_3, t) - U_i(x_1^*, x_2, x_3, t)}{x_1 - x_1^*} \tag{2.18}$$

because *two-point* statistical information at both point (x_1, x_2, x_3, t) and point (x_1^*, x_2, x_3, t) is required. In general, higher-order derivatives or higher-order approximations of the velocity gradient require a *multi-point* joint velocity PDF. For example, the three-point, finite-difference approximation of $\mathbf{\Delta} \equiv \nabla^2 \mathbf{U}$ contains $U_i(\mathbf{x}^*, t) - 2U_i(\mathbf{x}, t) + U_i(\mathbf{x}^\dagger, t)$, where $\mathbf{x}^*$, $\mathbf{x}$, and $\mathbf{x}^\dagger$ are adjacent grid points. A statistical theory for $\mathbf{\Delta}$ would thus require at least a three-point joint velocity PDF. The need for velocity statistics at multiple points leads to a *closure problem* in the PDF description of turbulent flows. Furthermore, even assuming that suitable closures can be found at some multi-point level, the desire to keep CFD models computationally tractable drastically limits the range of choices available. At present, most transported PDF models applicable to *inhomogeneous* flows employ one-point PDF closures. Nonetheless, two-point closures offer a powerful, yet tractable, alternative for homogeneous flows.

2.1.2 Spatial correlation functions

The two-point description of homogeneous turbulence begins with the two-point joint velocity PDF $f_{\mathbf{U},\mathbf{U}^*}(\mathbf{V}, \mathbf{V}^*; \mathbf{x}, \mathbf{x}^*, t)$ defined by

$$\begin{aligned} &f_{\mathbf{U},\mathbf{U}^*}(\mathbf{V}, \mathbf{V}^*; \mathbf{x}, \mathbf{x}^*, t)\, \mathrm{d}\mathbf{V}\, \mathrm{d}\mathbf{V}^* \\ &\quad \equiv \mathrm{P}[\{\mathbf{V} \le \mathbf{U}(\mathbf{x}, t) < \mathbf{V} + \mathrm{d}\mathbf{V}\} \cap \{\mathbf{V}^* \le \mathbf{U}(\mathbf{x}^*, t) < \mathbf{V}^* + \mathrm{d}\mathbf{V}^*\}]. \end{aligned} \tag{2.19}$$

A two-point PDF contains length-scale information through the correlation between the velocities located at two neighboring points $\mathbf{x}$ and $\mathbf{x}^*$. This spatial correlation is of great interest in turbulence theory, and, in homogeneous turbulence, is usually reported in terms of the *two-point velocity correlation function*:

$$R_{ij}(\mathbf{r}, t) \equiv \langle u_i(\mathbf{x}, t) u_j(\mathbf{x}^*, t) \rangle \tag{2.20}$$

where $\mathbf{r} = \mathbf{x}^* - \mathbf{x}$. Note that, from its definition, $R_{ij}(0, t) = \langle u_i u_j \rangle$. Moreover, the two-point velocity correlation function is the starting point for *spectral* theories of turbulence as described in the following section.

In homogeneous *isotropic turbulence*, the two-point velocity correlation function can be expressed (Pope 2000) in terms of the *longitudinal* (f) and *transverse* (g) auto-correlation functions:

$$R_{ij}(\mathbf{r}, t) = \frac{2k}{3} \left(g(r, t)\delta_{ij} + [f(r, t) - g(r, t)] \frac{r_i r_j}{r^2} \right), \tag{2.21}$$

where

$$f(r, t) \equiv \frac{3}{2k} R_{11}(r\mathbf{e}_1, t), \tag{2.22}$$

$$g(r, t) \equiv \frac{3}{2k} R_{22}(r\mathbf{e}_1, t), \tag{2.23}$$

and $\mathbf{e}_1$ is the unit vector in the x_1 direction. However, because the auto-correlation functions in isotropic turbulence are related by

$$g(r, t) = f(r, t) + \frac{r}{2} \frac{\partial}{\partial r} f(r, t), \tag{2.24}$$

$R_{ij}(\mathbf{r}, t)$ is completely determined by the longitudinal auto-correlation function $f(r, t)$.

The auto-correlation functions can be used to define two *characteristic length scales* of an isotropic turbulent flow. The *longitudinal integral scale* is defined by

$$L_{11}(t) \equiv \int_0^\infty f(r, t)\, \mathrm{d}r. \tag{2.25}$$

Likewise, the *transverse integral scale* is defined by

$$L_{22}(t) \equiv \int_0^\infty g(r, t)\, \mathrm{d}r = \frac{1}{2} L_{11}(t). \tag{2.26}$$

These length scales characterize the larger eddies in the flow (hence the name 'integral').

Other widely used length scales are the *Taylor microscales*, which are determined by the behavior of the auto-correlation functions near the origin. These microscales are defined by

$$\lambda_f(t) \equiv \left(-\frac{1}{2} \frac{\partial^2 f}{\partial r^2}(0, t) \right)^{-1/2} \tag{2.27}$$

and

$$\lambda_g(t) \equiv \left(-\frac{1}{2} \frac{\partial^2 g}{\partial r^2}(0, t) \right)^{-1/2} = \frac{1}{\sqrt{2}} \lambda_f(t) \tag{2.28}$$

where λ_f is the *longitudinal Taylor microscale* and λ_g is the *transverse Taylor microscale*.

The relationship between the various length scales can be best understood by looking at their dependence on the *turbulence Reynolds number* defined in terms of the turbulent kinetic energy k, the turbulent dissipation rate ε, and the kinematic viscosity ν by

$$\mathrm{Re}_L \equiv \frac{k^2}{\varepsilon \nu}. \tag{2.29}$$

The turbulence integral scales L_{11} and L_{22} are proportional to a turbulence integral scale L_u defined in terms of k and ε:

$$L_u \equiv \frac{k^{3/2}}{\varepsilon}. \tag{2.30}$$

Likewise, the ratio of the transverse Taylor microscale and the turbulence integral scale can be expressed as

$$\frac{\lambda_g}{L_u} = \sqrt{10}\,\mathrm{Re}_L^{-1/2}. \tag{2.31}$$

On the other hand, the smallest scales in a turbulent flow are proportional to the *Kolmogorov scale* defined by

$$\eta \equiv \left(\frac{\nu^3}{\varepsilon}\right)^{1/4}. \tag{2.32}$$

The ratio of the Kolmogorov scale and the turbulence integral scale can be expressed in terms of the turbulence Reynolds number by

$$\frac{\eta}{L_u} = \mathrm{Re}_L^{-3/4}. \tag{2.33}$$

Note that at high Reynolds number λ_g lies at scales between L_u and η, and thus cannot be given a clear physical interpretation in terms of eddies in the flow. Nevertheless, the Taylor microscale is often used to define the *Taylor-scale Reynolds number*:

$$R_\lambda \equiv \left(\frac{2k}{3}\right)^{1/2} \frac{\lambda_g}{\nu} = \left(\frac{20}{3}\mathrm{Re}_L\right)^{1/2}. \tag{2.34}$$

In simulations of homogeneous turbulence, R_λ is often used to characterize the 'magnitude' of the turbulence.

2.1.3 Temporal correlation functions

The spatial correlation functions are computed from the two-point joint velocity PDF based on two points in space. Obviously, the same idea can be extended to cover two points in time. Indeed, the Eulerian[9] *two-time* joint velocity PDF $f^E_{\mathbf{U},\mathbf{U}^*}(\mathbf{V}, \mathbf{V}^*; \mathbf{x}, t, t^*)$,

[9] The term 'Eulerian' is used to distinguish the case where $\mathbf{x}$ is fixed from the Lagrangian case where $\mathbf{x} = \mathbf{X}^*(t)$ is convected with the flow.

Table 2.1. *The principal length and time scales, and Reynolds numbers characterizing a fully developed turbulent flow defined in terms of the turbulent kinetic energy k, turbulent dissipation rate ε, and the kinematic viscosity ν.*

Note that the characteristic velocity for each scale is found by taking the ratio of the length scale and time scale.

Quantity	Integral scale	Taylor scale	Kolmogorov scale
Length	$L_u = \frac{k^{3/2}}{\varepsilon}$	$\lambda_g = \left(\frac{10k\nu}{\varepsilon}\right)^{1/2}$	$\eta = \left(\frac{\nu^3}{\varepsilon}\right)^{1/4}$
Time	$\tau_u = \frac{k}{\varepsilon}$	$\tau_\lambda = \left(\frac{15\nu}{\varepsilon}\right)^{1/2}$	$\tau_\eta = \left(\frac{\nu}{\varepsilon}\right)^{1/2}$
Reynolds number	$\mathrm{Re}_L = \frac{k^2}{\varepsilon\nu}$	$R_\lambda = \left(\frac{20}{3}\right)^{1/2} \frac{k}{(\varepsilon\nu)^{1/2}}$	$\mathrm{Re}_\eta = 1$

defined by

$$f^E_{\mathbf{U},\mathbf{U}^*}(\mathbf{V}, \mathbf{V}^*; \mathbf{x}, t, t^*)\,\mathrm{d}\mathbf{V}\,\mathrm{d}\mathbf{V}^* \\ \equiv \mathrm{P}[\{\mathbf{V} \le \mathbf{U}(\mathbf{x}, t) < \mathbf{V} + \mathrm{d}\mathbf{V}\} \cap \{\mathbf{V}^* \le \mathbf{U}(\mathbf{x}, t^*) < \mathbf{V}^* + \mathrm{d}\mathbf{V}^*\}], \tag{2.35}$$

is useful for describing *statistically stationary* turbulent flows (Pope 2000). The two-time PDF contains time-scale information and can be used to compute the *velocity time auto-correlation function*:

$$\rho_{ij}(\tau) \equiv \frac{\langle u_i(\mathbf{x}, t) u_j(\mathbf{x}, t^*)\rangle}{\langle u_i u_j\rangle}, \tag{2.36}$$

where $\tau = t^* - t$.

Isotropic turbulence is described by a single-time auto-correlation function $\rho_u(\tau)$. Thus, an *integral time scale* can be defined in terms of the auto-correlation function by

$$\bar{\tau} \equiv \int_0^\infty \rho(\tau)\,\mathrm{d}\tau. \tag{2.37}$$

This time scale is proportional to the *turbulence integral time scale* defined by

$$\tau_u \equiv \frac{k}{\varepsilon}. \tag{2.38}$$

Note that the turbulence integral time scale is related to the *Kolmogorov time scale*

$$\tau_\eta \equiv \left(\frac{\nu}{\varepsilon}\right)^{1/2} \tag{2.39}$$

by

$$\tau_u = \mathrm{Re}_L^{1/2}\tau_\eta. \tag{2.40}$$

The principal length and time scales, and the Reynolds numbers that are used to characterize a fully developed turbulent flow are summarized in Table 2.1. Conversion tables between the time and length scales, written in terms of the turbulence Reynolds number Re_L, are given in Tables 2.2 and 2.3.

Table 2.2. *Conversion table for the turbulent length scales given in Table 2.1.*

Expressed in terms of the turbulence Reynolds number $\mathrm{Re}_L = k^2/\varepsilon\nu$.
For example, $L_u = \mathrm{Re}_L^{3/4}\eta$.

	L_u	λ_g	η
L_u	1	$\mathrm{Re}_L^{1/2}/\sqrt{10}$	$\mathrm{Re}_L^{3/4}$
λ_g	$\sqrt{10}\mathrm{Re}_L^{-1/2}$	1	$\sqrt{10}\mathrm{Re}_L^{1/2}$
η	$\mathrm{Re}_L^{-3/4}$	$\mathrm{Re}_L^{-1/4}/\sqrt{10}$	1

Table 2.3. *Conversion table for the turbulent time scales given in Table 2.1.*

Expressed in terms of the turbulence Reynolds number $\mathrm{Re}_L = k^2/\varepsilon\nu$.
For example, $\tau_u = \mathrm{Re}_L^{1/2}\tau_\eta$.

	τ_u	τ_λ	τ_η
τ_u	1	$\mathrm{Re}_L^{1/2}/\sqrt{15}$	$\mathrm{Re}_L^{1/2}$
τ_λ	$\sqrt{15}\mathrm{Re}_L^{-1/2}$	1	$\sqrt{15}$
τ_η	$\mathrm{Re}_L^{-1/2}$	$1/\sqrt{15}$	1

2.1.4 Turbulent energy spectrum

In homogeneous turbulence, the *velocity spectrum tensor* is related to the spatial correlation function defined in (2.20) through the following Fourier transform pair:

$$\Phi_{ij}(\boldsymbol{\kappa}, t) = \frac{1}{(2\pi)^3} \iiint_{-\infty}^{+\infty} R_{ij}(\mathbf{r}, t)\, \mathrm{e}^{-i\boldsymbol{\kappa}\cdot\mathbf{r}}\, \mathrm{d}\mathbf{r}, \tag{2.41}$$

$$R_{ij}(\mathbf{r}, t) = \iiint_{-\infty}^{+\infty} \Phi_{ij}(\boldsymbol{\kappa}, t)\, \mathrm{e}^{i\boldsymbol{\kappa}\cdot\mathbf{r}}\, \mathrm{d}\boldsymbol{\kappa}. \tag{2.42}$$

This relation shows that for homogeneous turbulence, working in terms of the two-point spatial correlation function or in terms of the velocity spectrum tensor is entirely equivalent. In the turbulence literature, models formulated in terms of the velocity spectrum tensor are referred to as spectral models (for further details, see McComb (1990) or Lesieur (1997)).

The Fourier transform introduces the *wavenumber vector* $\boldsymbol{\kappa}$, which has units of 1/length. Note that, from its definition, the velocity spatial correlation function is related to the Reynolds stresses by

$$R_{ij}(0, t) = \langle u_i u_j \rangle = \iiint_{-\infty}^{+\infty} \Phi_{ij}(\boldsymbol{\kappa}, t)\, \mathrm{d}\boldsymbol{\kappa}. \tag{2.43}$$

Thus, $\Phi_{ij}(\boldsymbol{\kappa}, t)\,\mathrm{d}\boldsymbol{\kappa}$ roughly corresponds to the 'amount' of Reynolds stress located at a point $\boldsymbol{\kappa}$ in wavenumber space at time t.

The *turbulent energy spectrum* is defined in terms of the velocity spectrum tensor by integrating out all directional information:

$$E_u(\kappa, t) = \iiint_{-\infty}^{+\infty} \frac{1}{2} \left[\Phi_{11}(\boldsymbol{\kappa}, t) + \Phi_{22}(\boldsymbol{\kappa}, t) + \Phi_{33}(\boldsymbol{\kappa}, t)\right] \delta(\kappa - |\boldsymbol{\kappa}|)\,\mathrm{d}\boldsymbol{\kappa}, \tag{2.44}$$

where the integral is over shells in wavenumber space located at a distance κ from the origin. The integral is most easily performed in a spherical coordinate system by integrating out all angular dependency at a fixed radius κ from the origin. Note that κ on the left-hand side is now a scalar instead of a vector. The time-dependent turbulent energy spectrum thus depends on two scalars κ and t.

For isotropic turbulence, the velocity spectrum tensor is related to the turbulent energy spectrum by

$$\Phi_{ij}(\boldsymbol{\kappa}, t) = \frac{E_u(\kappa, t)}{4\pi\kappa^2}\left(\delta_{ij} - \frac{\kappa_i \kappa_j}{\kappa^2}\right). \tag{2.45}$$

Although isotropic turbulence is a useful theoretical concept, it rarely exists in nature. We shall see in Chapter 4 that in the presence of a mean velocity gradient the Reynolds stresses are produced in an anisotropic manner, thereby forcing the Reynolds stresses (at least at large scales) to be anisotropic. Nevertheless, at high Reynolds numbers, we can expect the small scales to be nearly isotropic (Pope 2000). In the classical picture of homogeneous turbulence, one speaks of energy produced with a distinct directional orientation (i.e., coherent vortices) at large scales that loses its directional preference as it cascades to small scales where it is dissipated.

By definition, the turbulent kinetic energy k can be found directly from the turbulent energy spectrum by integrating over wavenumber space:

$$k(t) = \int_0^\infty E_u(\kappa, t)\,\mathrm{d}\kappa = \frac{1}{2}\left(\langle u_1^2\rangle + \langle u_2^2\rangle + \langle u_3^2\rangle\right). \tag{2.46}$$

Thus, $E_u(\kappa, t)\,\mathrm{d}\kappa$ represents the amount of turbulent kinetic energy located at wavenumber κ.

For isotropic turbulence, the longitudinal integral length scale L_{11} is related to the turbulent energy spectrum by

$$L_{11}(t) = \frac{3\pi}{4k}\int_0^\infty \frac{E_u(\kappa, t)}{\kappa}\,\mathrm{d}\kappa. \tag{2.47}$$

At high Reynolds numbers, the longitudinal integral length scale is related to the turbulence integral length scale by (Pope 2000)

$$L_u(t) = \lim_{\mathrm{Re}_L \to \infty} \frac{3\pi}{4} L_{11}(t). \tag{2.48}$$

The longitudinal integral length scale can also be used to define a characteristic integral time scale τ_e:

$$\tau_e \equiv \frac{L_{11}}{k^{1/2}}. \tag{2.49}$$

This time scale is usually referred to as the *eddy turnover time*.

One of the cornerstone results from turbulence theory concerns the shape of the *stationary* turbulent energy spectrum $E_u(\kappa)$ (Batchelor 1953; Pope 2000). Based on statistical arguments, Kolmogorov (1941a) and Kolmogorov (1941b)[10] postulated that for large Re_L, $E_u(\kappa)$ for large κ (i.e., for small scales) will be *universal* (i.e., the same for all flows),[11] and depend only on the kinematic viscosity ν and the turbulent energy dissipation rate ε. At sufficiently high Reynolds numbers, the small scales will thus be *locally isotropic*, even for inhomogeneous flows of practical interest. Kolmogorov also postulated that for large Re_L, $E_u(\kappa)$ for intermediate wavenumbers will depend only on ε (the so-called 'inertial range'). These important observations are the starting point for the *equilibrium turbulence models* discussed in Chapter 4. In these models, the small scales are assumed to be in dynamic equilibrium with the large scales, and thus do not need to be modeled explicitly.

The turbulent energy dissipation rate can be expressed in terms of the fluctuating rate-of-strain tensor:

$$s_{ij} \equiv \frac{1}{2}\left(\frac{\partial u_i}{\partial x_j} + \frac{\partial u_j}{\partial x_i}\right) \tag{2.50}$$

by

$$\varepsilon \equiv 2\nu\langle s_{ij}s_{ij}\rangle. \tag{2.51}$$

From the definition of the turbulent energy spectrum, ε is related to $E_u(\kappa, t)$ by

$$\varepsilon(t) = 2\nu \int_0^\infty \kappa^2 E_u(\kappa, t)\, \mathrm{d}\kappa = \int_0^\infty D_u(\kappa, t)\, \mathrm{d}\kappa \tag{2.52}$$

where $D_u(\kappa, t)$ is the *turbulent energy dissipation spectrum*. Both of these definitions clearly illustrate the fact that the turbulent energy dissipation rate is a *small-scale quantity* dominated by the large wavenumber portion of the energy spectrum. However, because the characteristic time scale of a turbulent eddy decreases with increasing wavenumber, the small scales will be in dynamic equilibrium with the large scales in a fully developed turbulent flow. The turbulent energy dissipation rate will thus be equal to the *rate of energy transfer* from large to small scales, and hence a detailed small-scale model for ε is not required in an equilibrium turbulence model.

[10] The turbulence community refers to this work as 'Kolmogorov 41.' A second article (Kolmogorov 1962) referred to as 'Kolmogorov 62' contains the 'refined Kolmogorov hypothesis.'

[11] At high Reynolds number, this would also apply to *inhomogeneous turbulent flows*.

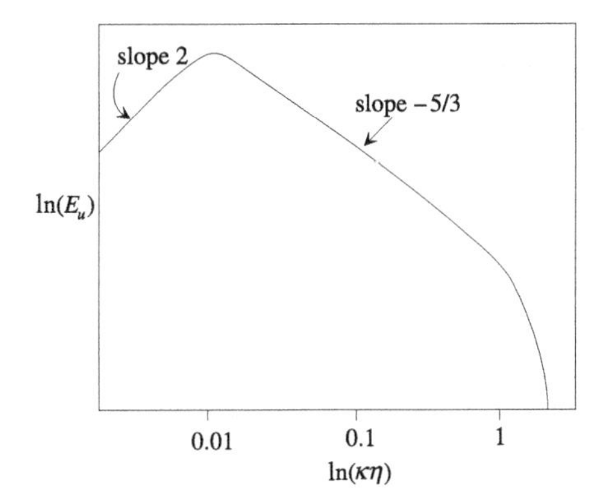

Figure 2.4. Sketch of model turbulent energy spectrum at $R_\lambda = 500$.

2.1.5 Model velocity spectrum

Pope (2000) developed the following model turbulent energy spectrum to describe fully developed homogeneous turbulence:[12]

$$E_u(\kappa) = C\varepsilon^{2/3}\kappa^{-5/3} f_L(\kappa L_u) f_\eta(\kappa\eta), \tag{2.53}$$

where the non-dimensional cut-off functions f_L and f_η are defined by

$$f_L(\kappa L_u) = \left(\frac{\kappa L_u}{[(\kappa L_u)^2 + c_L]^{1/2}} \right)^{5/3+p_0} \tag{2.54}$$

and

$$f_\eta(\kappa\eta) = \exp\left[-\beta\left(\left[(\kappa\eta)^4 + c_\eta^4 \right]^{1/4} - c_\eta \right) \right]. \tag{2.55}$$

The parameter p_0 controls the behavior of the velocity spectrum for small κL_u. The usual choice is $p_0 = 2$ leading to $E_u(\kappa) \sim \kappa^2$ for small κ. The alternative choice $p_0 = 4$ leads to the von Kármán spectrum, which has $E_u(\kappa) \sim \kappa^4$ for small κ. For fixed k, ε, and ν, the model spectrum is determined by setting $C = 1.5$ and $\beta = 5.2$, and the requirement that $E_u(\kappa)$ and $2\nu\kappa^2 E_u(\kappa)$ integrate to k and ε, respectively. The turbulence Reynolds number Re_L, along with k and L_u, also uniquely determines the model spectrum.

The model turbulent energy spectrum for $R_\lambda = 500$ is shown in Fig. 2.4. Note that the turbulent energy spectrum can be divided into roughly three parts:

(i) $(0 \le \kappa \le \kappa_{\mathrm{EI}})$, the *energy-containing range*, near the peak of $E_u(\kappa)$;
(ii) $(\kappa_{\mathrm{EI}} < \kappa < \kappa_{\mathrm{DI}})$, the *inertial range*, where $E_u(\kappa) \sim \kappa^{-5/3}$;
(iii) $(\kappa_{\mathrm{DI}} \le \kappa)$, the *dissipation range*, where $E_u(\kappa)$ falls off exponentially.

[12] Because the flow is assumed to be stationary, the time dependence has been dropped. However, the model spectrum could be used to describe a slowly evolving non-stationary spectrum by inserting $k(t)$ and $\varepsilon(t)$.

Furthermore, the *universal equilibrium range* is composed of the inertial range and the dissipation range. As its name indicates, at high Reynolds numbers the universal equilibrium range should have approximately the same form in all turbulent flows.

Pope (2000) gives the following estimates for the cut-off wavenumbers κ_{EI} and κ_{DI}:[13]

$$\kappa_{\mathrm{EI}} \approx \frac{38}{L_{11}} \tag{2.56}$$

and

$$\kappa_{\mathrm{DI}} \approx \frac{0.1}{\eta}. \tag{2.57}$$

Also note that from the shape of the turbulent energy spectrum, most of the turbulent energy is contained in the energy-containing range. This implies that integral-scale quantities such as k and $\langle u_i u_j \rangle$ are determined primarily by turbulent eddies in the energy-containing range of the energy spectrum. Indeed, it is these scales that are the most flow-dependent, and thus which depend strongly on how the turbulence is generated.

For a homogeneous turbulent *shear* flow, turbulent kinetic energy is added to the turbulent energy spectrum in the energy-containing range by extracting energy from the mean flow field as described by the kinetic energy production term

$$P_k \equiv \langle u_i u_j \rangle S_{ij}, \tag{2.58}$$

where S_{ij} is mean rate-of-strain tensor,

$$S_{ij} \equiv \frac{1}{2}\left(\frac{\partial \langle U_i \rangle}{\partial x_j} + \frac{\partial \langle U_j \rangle}{\partial x_i} \right), \tag{2.59}$$

and the spatial gradients of the mean velocity are constant. In the inertial range, the turbulent kinetic energy from the energy-containing range cascades to smaller scales without significant dissipation or production. In the dissipative range, the turbulent kinetic energy is dissipated by molecular viscosity. The turbulent energy cascade in *non-stationary* homogeneous turbulence can be described by spectral transport equations derived from the Navier–Stokes equation as described in the next section.

As the Reynolds number increases, the separation between the energy-containing and dissipative ranges increases, and turbulent eddies in these two ranges become *statistically independent*. This fact has important ramifications for equilibrium turbulence models because it implies that expected values involving energy-containing range and dissipation range quantities can be treated as statistically independent; e.g., if $i \neq j$, then

$$\langle u_i \nabla^2 u_j \rangle = \langle u_i \rangle \langle \nabla^2 u_j \rangle = 0. \tag{2.60}$$

Moreover, it also implies that the scales in the energy-containing range can be modeled independently of the universal equilibrium range given a model for the flux of energy

[13] The definition of 'high Reynolds number' could thus be a Reynolds number for which $L_u/\eta \geq 380$. Using (2.48), this condition yields $\mathrm{Re}_L \geq 8630$ or $R_\lambda \geq 240$.

through the inertial range. This significant observation is the starting point for the *large-eddy simulation* of turbulent flows discussed in Section 4.2.

2.1.6 Spectral transport

The model turbulent energy spectrum given in (2.53) was introduced to describe fully developed turbulence, i.e., the case where $E_u(\kappa, t)$ does not depend explicitly on t. The case where the turbulent energy spectrum depends explicitly on time can be handled by deriving a transport equation for the velocity spectrum tensor $\Phi_{ij}(\boldsymbol{\kappa}, t)$ starting from the Navier–Stokes equation for homogeneous velocity fields with zero or constant mean velocity (McComb 1990; Lesieur 1997). The resultant expression can be simplified for isotropic turbulence to a transport equation for $E_u(\kappa, t)$ of the form[14]

$$\frac{\partial E_u}{\partial t} = T_u(\kappa, t) - 2\nu\kappa^2 E_u(\kappa, t). \tag{2.61}$$

The second term on the right-hand side of (2.61) can be rewritten in terms of the turbulent energy dissipation spectrum as

$$2\nu\kappa^2 E_u(\kappa, t) = D_u(\kappa, t), \tag{2.62}$$

and thus represents the dissipation of turbulent kinetic energy due to viscous dissipation. Note that, due to its dependence on $\nu\kappa^2$, the dissipation term will be negligible at small wavenumbers, but will rapidly increase for increasing κ.

The first term on the right-hand side of (2.61) is the *spectral transfer function*, and involves two-point correlations between three components of the velocity vector (see McComb (1990) for the exact form). The spectral transfer function is thus *unclosed*, and models must be formulated in order to proceed in finding solutions to (2.61). However, some useful properties of $T_u(\kappa, t)$ can be deduced from the spectral transport equation. For example, integrating (2.61) over all wavenumbers yields the transport equation for the turbulent kinetic energy:

$$\frac{\mathrm{d}k}{\mathrm{d}t} = \int_0^\infty T_u(\kappa, t)\,\mathrm{d}\kappa - \varepsilon = -\varepsilon, \tag{2.63}$$

where the final equality follows from the definition of the turbulent dissipation rate in homogeneous turbulence. Any model for the spectral transfer function must conserve spectral energy:

$$\int_0^\infty T_u(\kappa, t)\,\mathrm{d}\kappa = 0. \tag{2.64}$$

In words, (2.64) implies that $T_u(\kappa, t)$ is responsible for transferring energy between different wavenumbers without changing the total turbulent kinetic energy.

[14] A homogeneous turbulent shear flow will have an additional unclosed production term on the right-hand side (Pope 2000).

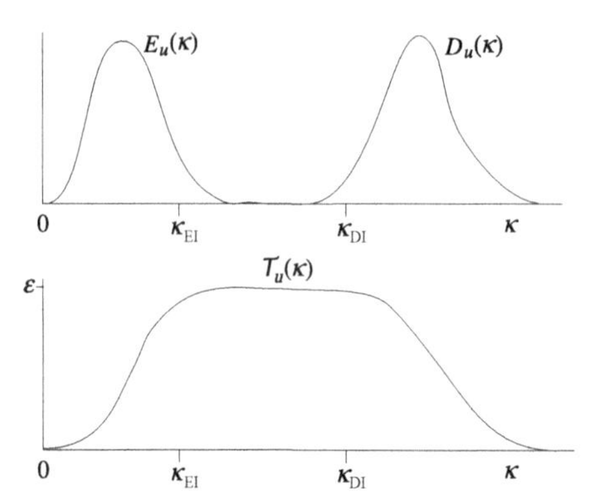

Figure 2.5. Sketch of $E_u(\kappa)$, $D_u(\kappa)$, and $\mathcal{T}_u(\kappa)$ in fully developed turbulence.

In order to understand the role of spectral energy transfer in determining the turbulent energy spectrum at high Reynolds numbers, it is useful to introduce the *spectral energy transfer rate* $\mathcal{T}_u(\kappa, t)$ defined by

$$\mathcal{T}_u(\kappa, t) \equiv -\int_0^{\kappa} T_u(s, t)\, \mathrm{d}s. \tag{2.65}$$

From this definition, it can be observed that $\mathcal{T}_u(\kappa, t)$ is the net rate at which turbulent kinetic energy is transferred from wavenumbers less than κ to wavenumbers greater than κ. In fully developed turbulent flow, the net flux of turbulent kinetic energy is from large to small scales. Thus, the stationary spectral energy transfer rate $\mathcal{T}_u(\kappa)$ will be positive at spectral equilibrium. Moreover, by definition of the inertial range, the net rate of transfer through wavenumbers κ_{EI} and κ_{DI} will be identical in a fully developed turbulent flow, and thus

$$\mathcal{T}_u(\kappa_{\mathrm{EI}}) = \mathcal{T}_u(\kappa_{\mathrm{DI}}) = \varepsilon. \tag{2.66}$$

Representative plots of $E_u(\kappa)$, $D_u(\kappa)$, and $\mathcal{T}_u(\kappa)$ are shown in Fig. 2.5.

Note that $\mathcal{T}_u(\kappa, t)$ and $E_u(\kappa, t)$ can be used to derive a characteristic time scale for spectral transfer at a given wavenumber $\tau_{\mathrm{st}}(\kappa, t)$ defined by

$$\tau_{\mathrm{st}}(\kappa, t) \equiv \frac{\kappa E_u(\kappa, t)}{\mathcal{T}_u(\kappa, t)}. \tag{2.67}$$

In the inertial range of fully developed turbulence, the spectral transfer time scale becomes

$$\tau_{\mathrm{st}}(\kappa) \propto (\varepsilon \kappa^2)^{-1/3} \propto \left(\frac{\kappa_0}{\kappa}\right)^{2/3} \tau_u, \tag{2.68}$$

where $\kappa_0 = 2\pi/L_u$. Thus, as the wavenumber increases, the characteristic time decreases rapidly. This observation is often used to justify the assumption used in one-point turbulence models that the small scales of an *inhomogeneous* turbulent flow are in dynamic equilibrium with the energy-containing scales. Indeed, we will make use of this assumption repeatedly in Section 2.2 to simplify the transport equations for one-point turbulence statistics.

The spectral transport equation can also be used to generate a spectral model for the dissipation rate ε. Multiplying (2.61) by $2\nu\kappa^2$ yields the spectral transport equation for $D_u(\kappa, t)$:

$$\frac{\partial D_u}{\partial t} = 2\nu\kappa^2 T_u(\kappa, t) - 2\nu\kappa^2 D_u(\kappa, t). \tag{2.69}$$

Owing to the presence of $\nu\kappa^2$, the right-hand side of this expression will be negligible for $\kappa < \kappa_{\mathrm{DI}}$. Thus, integration over all wavenumbers results in the following expression for ε:

$$\frac{\mathrm{d}\varepsilon}{\mathrm{d}t} \approx 2 \int_{\kappa_{\mathrm{DI}}}^{\infty} \nu\kappa^2 T_u(\kappa, t)\, \mathrm{d}\kappa - \mathcal{D}_\varepsilon, \tag{2.70}$$

where the dissipation-dissipation rate $\mathcal{D}_\varepsilon$ is defined by

$$\mathcal{D}_\varepsilon \equiv 2 \int_0^{\infty} \nu\kappa^2 D_u(\kappa, t)\, \mathrm{d}\kappa. \tag{2.71}$$

The integral on the right-hand side of (2.70) can be rewritten in terms of $\mathcal{T}_u(\kappa, t)$:

$$\int_{\kappa_{\mathrm{DI}}}^{\infty} \nu\kappa^2 T_u(\kappa, t)\, \mathrm{d}\kappa = - \int_{\kappa_{\mathrm{DI}}}^{\infty} \nu\kappa^2 \frac{\partial \mathcal{T}_u}{\partial \kappa}\, \mathrm{d}\kappa. \tag{2.72}$$

Using integration by parts, the right-hand side can be further simplified to

$$\int_{\kappa_{\mathrm{DI}}}^{\infty} \nu\kappa^2 T_u(\kappa, t)\, \mathrm{d}\kappa = \nu\kappa_{\mathrm{DI}}^2 \mathcal{T}_u(\kappa_{\mathrm{DI}}, t) + 2 \int_{\kappa_{\mathrm{DI}}}^{\infty} \nu\kappa \mathcal{T}_u(\kappa, t)\, \mathrm{d}\kappa, \tag{2.73}$$

or, by using the time scale defined in (2.67), to

$$\int_{\kappa_{\mathrm{DI}}}^{\infty} \nu\kappa^2 T_u(\kappa, t)\, \mathrm{d}\kappa = \nu\kappa_{\mathrm{DI}}^2 \mathcal{T}_u(\kappa_{\mathrm{DI}}, t) + \int_{\kappa_{\mathrm{DI}}}^{\infty} \frac{D_u(\kappa, t)}{\tau_{\mathrm{st}}(\kappa, t)}\, \mathrm{d}\kappa. \tag{2.74}$$

By definition, the dissipation range is dominated by viscous dissipation of Kolmogorov-scale vortices. The characteristic time scale τ_{st} in (2.74) can thus be taken as proportional to the Kolmogorov time scale τ_η, and taken out of the integral. This leads to the final form for (2.70),

$$\frac{\mathrm{d}\varepsilon}{\mathrm{d}t} \approx 2\nu\kappa_{\mathrm{DI}}^2 \mathcal{T}_u(\kappa_{\mathrm{DI}}, t) + 2C_{\mathrm{s}} \left(\frac{\varepsilon}{\nu}\right)^{1/2} \varepsilon - \mathcal{D}_\varepsilon, \tag{2.75}$$

where C_{s} is a constant of order unity. This expression illustrates the importance of the spectral energy transfer rate in determining the dissipation rate in high-Reynolds-number turbulent flows. Indeed, near spectral equilibrium, $\mathcal{T}_u(\kappa_{\mathrm{DI}}, t)$ will vary on time scales on the order of the eddy turnover time τ_{e}, while the characteristic time scale of (2.75) is $\tau_\eta \ll \tau_{\mathrm{e}}$. This implies that $\varepsilon(t)$ will be determined primarily by the 'slow' variations in $\mathcal{T}_u(\kappa_{\mathrm{DI}}, t)$, i.e., by the energy flux from large scales.

2.2 Inhomogeneous turbulence

Most of the results presented thus far in this chapter are strictly valid only for homogeneous turbulence. However, turbulent flows encountered in practice are usually *inhomogeneous*. In order to model inhomogeneous turbulent flows, transport equations are needed to describe how the various turbulence quantities are transported from one location to another in the flow. In this section, transport equations for the most frequently employed turbulence quantities are derived from the fundamental governing equations (1.27) and (1.28). Owing to the non-linearities in the governing equations, this process inevitably introduces unclosed terms that must be modeled. As discussed in Chapter 4, turbulence models are fitted first to data for homogeneous turbulent flows, and then validated using *canonical* inhomogeneous flows of practical importance. Nevertheless, model parameters must often be 'calibrated' to simulate correctly complex flows of industrial interest.

Owing to the complexity of multi-point descriptions, almost all CFD models for complex turbulent flows are based on one-point turbulence statistics. As shown in Section 2.1, one-point turbulence statistics are found by integrating over the velocity sample space, e.g.,

$$\langle \mathbf{U}(\mathbf{x}, t) \rangle \equiv \iiint_{-\infty}^{+\infty} \mathbf{V} f_{\mathrm{U}}(\mathbf{V}; \mathbf{x}, t) \, \mathrm{d}\mathbf{V}. \tag{2.76}$$

Likewise, important joint velocity, composition statistics can be computed from the one-point joint velocity, scalar PDF. For example, the scalar flux is defined by

$$\langle \mathbf{U}(\mathbf{x}, t) \phi(\mathbf{x}, t) \rangle \equiv \int_{-\infty}^{+\infty} \iiint_{-\infty}^{+\infty} \mathbf{V} \psi f_{\mathrm{U},\phi}(\mathbf{V}, \psi; \mathbf{x}, t) \, \mathrm{d}\mathbf{V} \, \mathrm{d}\psi, \tag{2.77}$$

where $f_{\mathrm{U},\phi}(\mathbf{V}, \psi; \mathbf{x}, t)$ is the one-point joint velocity, composition PDF defined in Section 3.2. Furthermore, space and/or time derivatives of mean quantities can be easily related to space and/or time derivatives of $f_{\mathrm{U},\phi}(\mathbf{V}, \psi; \mathbf{x}, t)$. For example, starting from (2.77), the time derivative of the scalar flux is given by

$$\frac{\partial \langle \mathbf{U} \phi \rangle}{\partial t} \equiv \int_{-\infty}^{+\infty} \iiint_{-\infty}^{+\infty} \mathbf{V} \psi \frac{\partial f_{\mathrm{U},\phi}}{\partial t}(\mathbf{V}, \psi; \mathbf{x}, t) \, \mathrm{d}\mathbf{V} \, \mathrm{d}\psi. \tag{2.78}$$

In general, given $f_{\mathrm{U},\phi}(\mathbf{V}, \psi; \mathbf{x}, t)$, transport equations for one-point statistics can be easily derived. This is the approach used in *transported PDF methods* (see Chapter 6). However, the more widely used approach for deriving one-point transport equations is *Reynolds averaging*, and begins by taking the expected value of (1.27) directly. Nevertheless, in order to apply Reynolds averaging correctly, we must first define what is meant by taking the expected value of the *derivative* of a random field, and then show that taking the expected value and taking the derivative commute, i.e.,

$$\left\langle \frac{\partial \mathbf{U}}{\partial t} \right\rangle = \frac{\partial \langle \mathbf{U} \rangle}{\partial t}. \tag{2.79}$$

Note that, unlike derivations that rely on time averages,[15] for which

$$\left\langle \frac{\partial \mathbf{U}}{\partial t} \right\rangle_T \equiv \frac{1}{T} \int_0^T \frac{\partial}{\partial t} \mathbf{U}(\mathbf{x}, t + \tau)\, \mathrm{d}\tau \neq \frac{\partial \langle \mathbf{U} \rangle_T}{\partial t}, \tag{2.80}$$

a properly defined statistical theory avoids the conceptual and mathematical difficulties encountered in choosing T relative to the characteristic time scales of the flow.

2.2.1 Expected values of derivatives

In order to show that the expected-value and derivative operations commute, we begin with the definition of the derivative in terms of a limit:[16]

$$\frac{\partial \mathbf{U}}{\partial t} = \lim_{t^* \to t} \frac{\mathbf{U}(\mathbf{x}, t^*) - \mathbf{U}(\mathbf{x}, t)}{t^* - t}. \tag{2.81}$$

The two random variables on the right-hand side are completely determined by the two-time PDF, $f_{\mathbf{U},\mathbf{U}^*}(\mathbf{V}, \mathbf{V}^*; \mathbf{x}, t, t^*)$. Thus, the expected value of the time derivative can be defined by

$$\begin{aligned} \left\langle \frac{\partial \mathbf{U}}{\partial t} \right\rangle &\equiv \left\langle \lim_{t^* \to t} \frac{\mathbf{U}(\mathbf{x}, t^*) - \mathbf{U}(\mathbf{x}, t)}{t^* - t} \right\rangle \\ &= \lim_{t^* \to t} \frac{\langle \mathbf{U}(\mathbf{x}, t^*) - \mathbf{U}(\mathbf{x}, t) \rangle}{t^* - t}. \end{aligned} \tag{2.82}$$

The expected value appearing in the final expression of (2.82) can be rewritten as

$$\begin{aligned} &\langle \mathbf{U}(\mathbf{x}, t^*) - \mathbf{U}(\mathbf{x}, t) \rangle \\ &\quad \equiv \iiint\limits_{-\infty}^{+\infty} \iiint\limits_{-\infty}^{+\infty} (\mathbf{V}^* - \mathbf{V}) f_{\mathbf{U},\mathbf{U}^*}(\mathbf{V}, \mathbf{V}^*; \mathbf{x}, t, t^*)\, \mathrm{d}\mathbf{V}^*\, \mathrm{d}\mathbf{V} \\ &\quad = \iiint\limits_{-\infty}^{+\infty} \mathbf{V}^* f_{\mathbf{U}}(\mathbf{V}^*; \mathbf{x}, t^*)\, \mathrm{d}\mathbf{V}^* - \iiint\limits_{-\infty}^{+\infty} \mathbf{V} f_{\mathbf{U}}(\mathbf{V}; \mathbf{x}, t)\, \mathrm{d}\mathbf{V} \\ &\quad = \iiint\limits_{-\infty}^{+\infty} \mathbf{V} (f_{\mathbf{U}}(\mathbf{V}; \mathbf{x}, t^*) - f_{\mathbf{U}}(\mathbf{V}; \mathbf{x}, t))\, \mathrm{d}\mathbf{V}, \end{aligned} \tag{2.83}$$

where we have used the following relationships between the one-time and two-time PDFs:[17]

$$f_{\mathbf{U}}(\mathbf{V}^*; \mathbf{x}, t^*) = \iiint\limits_{-\infty}^{+\infty} f_{\mathbf{U},\mathbf{U}^*}(\mathbf{V}, \mathbf{V}^*; \mathbf{x}, t, t^*)\, \mathrm{d}\mathbf{V} \tag{2.84}$$

[15] As discussed by Pope (2000), a time average is a *random variable*, and thus a poor starting point for deriving a mathematically well defined closure theory!

[16] The demonstration is done here for t. However, the same steps can be repeated for each of the components of $\mathbf{x}$.

[17] For the one-point PDF, a $*$ is no longer needed to differentiate between $\mathbf{U}(t)$ and $\mathbf{U}(t^*)$, i.e., $f_{\mathbf{U}}(\mathbf{V}^*; \mathbf{x}, t^*) = f_{\mathbf{U}^*}(\mathbf{V}^*; \mathbf{x}, t^*)$.

and

$$f_{\mathbf{U}}(\mathbf{V};\mathbf{x},t) = \iiint_{-\infty}^{+\infty} f_{\mathbf{U},\mathbf{U}^*}(\mathbf{V},\mathbf{V}^*;\mathbf{x},t,t^*)\,\mathrm{d}\mathbf{V}^*. \tag{2.85}$$

The final result is then obtained by using (2.83) in (2.82):

$$\begin{aligned}
\left\langle \frac{\partial \mathbf{U}}{\partial t} \right\rangle &= \lim_{t^*\to t} \frac{\iiint_{-\infty}^{+\infty} \mathbf{V}\,(f_{\mathbf{U}}(\mathbf{V};\mathbf{x},t^*) - f_{\mathbf{U}}(\mathbf{V};\mathbf{x},t))\,\mathrm{d}\mathbf{V}}{t^* - t} \\
&= \iiint_{-\infty}^{+\infty} \mathbf{V} \lim_{t^*\to t} \frac{f_{\mathbf{U}}(\mathbf{V};\mathbf{x},t^*) - f_{\mathbf{U}}(\mathbf{V};\mathbf{x},t)}{t^* - t}\,\mathrm{d}\mathbf{V} \\
&= \iiint_{-\infty}^{+\infty} \mathbf{V} \frac{\partial f_{\mathbf{U}}}{\partial t}(\mathbf{V};\mathbf{x},t)\,\mathrm{d}\mathbf{V} \\
&= \frac{\partial}{\partial t} \iiint_{-\infty}^{+\infty} \mathbf{V} f_{\mathbf{U}}(\mathbf{V};\mathbf{x},t)\,\mathrm{d}\mathbf{V} \\
&= \frac{\partial \langle \mathbf{U} \rangle}{\partial t},
\end{aligned} \tag{2.86}$$

where the last equality follows from (2.78).

Note that, although linear terms involving only derivatives can be treated exactly in the one-point theory, non-linear terms such as

$$\left\langle U_k \frac{\partial U_j}{\partial x_i} \right\rangle$$

or

$$\left\langle \frac{\partial U_j}{\partial x_i} \frac{\partial U_l}{\partial x_k} \right\rangle$$

cause serious closure problems. Indeed, using the definition of the derivative in terms of a limit, it can easily be shown that the expected value of non-linear derivative terms requires knowledge of at least the two-point joint velocity PDF.

Notwithstanding these difficulties, it is sometimes possible to manipulate a non-linear expression using the usual rules of calculus into a form that can be further simplified using high-Reynolds-number turbulence theory. For example, using

$$\nabla^2(\phi\mathbf{u}) = \mathbf{u}\nabla^2\phi + 2(\boldsymbol{\nabla}\phi)\cdot(\boldsymbol{\nabla}\mathbf{u}) + \phi\nabla^2\mathbf{u}, \tag{2.87}$$

we can rewrite the non-linear derivative term $2\langle\Gamma(\boldsymbol{\nabla}\phi)\cdot(\boldsymbol{\nabla}\mathbf{u})\rangle$ which appears in the scalar-flux transport equation as

$$2\langle\Gamma(\boldsymbol{\nabla}\phi)\cdot(\boldsymbol{\nabla}\mathbf{u})\rangle = \Gamma\nabla^2\langle\phi\mathbf{u}\rangle - \Gamma\langle\mathbf{u}\nabla^2\phi\rangle - \Gamma\langle\phi\nabla^2\mathbf{u}\rangle. \tag{2.88}$$

At high Reynolds numbers, the first term on the right-hand side representing the molecular diffusion of a mean quantity will be negligible compared with the turbulent transport terms. (In a homogeneous flow, this term will be exactly zero.) Likewise, local isotropy and separation of scales at high Reynolds numbers imply that the last two terms will also be negligible. Thus, in a high-Reynolds-number turbulent flow, $\langle \Gamma(\nabla \phi) \cdot (\nabla \mathbf{u}) \rangle$ will be approximately zero, and thus can be neglected in the scalar-flux transport equation ((3.95), p. 83). Similar simplifications are possible for other non-linear terms.

2.2.2 Mean velocity

For a constant-density flow, the continuity equation is linear, and reduces to $\nabla \cdot \mathbf{U} = 0$. Reynolds averaging then yields

$$\langle \nabla \cdot \mathbf{U} \rangle = \nabla \cdot \langle \mathbf{U} \rangle = 0. \tag{2.89}$$

Thus, the mean velocity field – as well as the fluctuation field $\mathbf{u}$ – are solenoidal.

The momentum equation ((1.27), p. 16), on the other hand, contains a non-linear term of the form $\mathbf{U} \cdot \nabla \mathbf{U}$. Using the continuity equation, this term can be rewritten as $\nabla \cdot (\mathbf{UU})$. Thus, Reynolds averaging yields

$$\langle \nabla \cdot (\mathbf{UU}) \rangle = \nabla \cdot \langle \mathbf{UU} \rangle. \tag{2.90}$$

The right-hand side is further simplified by rewriting $\mathbf{U}$ as the sum of $\langle \mathbf{U} \rangle$ and $\mathbf{u}$, and then using the properties of expected values. This process yields

$$\langle U_i U_j \rangle = \langle U_i \rangle \langle U_j \rangle + \langle u_i u_j \rangle. \tag{2.91}$$

With this result, the Reynolds average of the non-linear term in the momentum equation can be written as

$$\left\langle U_j \frac{\partial U_i}{\partial x_j} \right\rangle = \langle U_j \rangle \frac{\partial \langle U_i \rangle}{\partial x_j} + \frac{\partial \langle u_j u_i \rangle}{\partial x_j}. \tag{2.92}$$

The remaining terms in the momentum equation are linear. Thus, Reynolds averaging yields

$$\frac{\partial \langle U_i \rangle}{\partial t} + \langle U_j \rangle \frac{\partial \langle U_i \rangle}{\partial x_j} + \frac{\partial \langle u_j u_i \rangle}{\partial x_j} = -\frac{1}{\rho} \frac{\partial \langle p \rangle}{\partial x_i} + \nu \nabla^2 \langle U_i \rangle. \tag{2.93}$$

The two terms on the right-hand side of this expression appear in closed form. However, the molecular transport term $\nu \nabla^2 \langle U_i \rangle$ is of order Re_L^{-1}, and thus will be negligible at high Reynolds numbers.

Note that (2.93) is unclosed due to the appearance of the Reynolds stresses $\langle u_j u_i \rangle$, which form a symmetric second-order tensor:

$$\langle u_i u_j \rangle = \langle u_j u_i \rangle. \tag{2.94}$$

The diagonal components of the Reynolds stress tensor (e.g., $\langle u_1 u_1 \rangle$) are referred to as the *Reynolds normal stresses*, while the off-diagonal components are referred to as the *Reynolds shear stresses*.

A governing equation for the mean pressure field appearing in (2.93) can be found by Reynolds averaging (1.29). This leads to a Poisson equation of the form

$$-\frac{1}{\rho}\nabla^2 \langle p \rangle = \left\langle \frac{\partial U_i}{\partial x_j} \frac{\partial U_j}{\partial x_i} \right\rangle = \frac{\partial \langle U_i \rangle}{\partial x_j} \frac{\partial \langle U_j \rangle}{\partial x_i} + \left\langle \frac{\partial u_i}{\partial x_j} \frac{\partial u_j}{\partial x_i} \right\rangle. \tag{2.95}$$

As it stands, the last term on the right-hand side of this expression is non-linear in the spatial derivatives and appears to add a new closure problem. However, using the fact that the fluctuation field is solenoidal,

$$\frac{\partial u_i}{\partial x_i} = 0, \tag{2.96}$$

we can write

$$\frac{\partial u_i u_j}{\partial x_i \partial x_j} = \frac{\partial u_j}{\partial x_i} \frac{\partial u_i}{\partial x_j}. \tag{2.97}$$

Thus, (2.95) reduces to

$$-\frac{1}{\rho}\nabla^2 \langle p \rangle = \frac{\partial \langle U_i \rangle}{\partial x_j} \frac{\partial \langle U_j \rangle}{\partial x_i} + \frac{\partial^2 \langle u_i u_j \rangle}{\partial x_j \partial x_j}, \tag{2.98}$$

so that the unclosed term involves only the Reynolds stresses.

In summary, the mean velocity field $\langle \mathbf{U} \rangle$ could be found by solving (2.93) and (2.98) if a closure were available for the Reynolds stresses. Thus, we next derive the transport equation for $\langle u_i u_j \rangle$ starting from the momentum equation.

2.2.3 Reynolds stresses

The transport equation for the Reynolds stresses can be found starting from the governing equation for the velocity fluctuations:

$$\frac{\partial u_i}{\partial t} + U_k \frac{\partial u_i}{\partial x_k} + u_k \frac{\partial \langle U_i \rangle}{\partial x_k} = \nu \nabla^2 u_i - \frac{1}{\rho} \frac{\partial p'}{\partial x_i} + \frac{\partial \langle u_i u_k \rangle}{\partial x_k}. \tag{2.99}$$

In addition, we will need the following relations for the derivatives of $u_i u_j$:

$$\frac{\partial u_i u_j}{\partial t} = u_i \frac{\partial u_j}{\partial t} + u_j \frac{\partial u_i}{\partial t}, \tag{2.100}$$

$$\frac{\partial u_i u_j}{\partial x_k} = u_i \frac{\partial u_j}{\partial x_k} + u_j \frac{\partial u_i}{\partial x_k}, \tag{2.101}$$

and

$$\nabla^2 (u_i u_j) = u_i \nabla^2 u_j + 2 \frac{\partial u_i}{\partial x_k} \frac{\partial u_j}{\partial x_k} + u_j \nabla^2 u_i. \tag{2.102}$$

The first step in the derivation is to multiply (2.99) by u_j. The second step is to replace i with j in (2.99) and to multiply the resultant equation by u_i. The third step is to add the equations found in the first two steps to find:

$$\begin{aligned}\frac{\partial u_i u_j}{\partial t} + U_k \frac{\partial u_i u_j}{\partial x_k} = &- u_j u_k \frac{\partial \langle U_i \rangle}{\partial x_k} - u_i u_k \frac{\partial \langle U_j \rangle}{\partial x_k} \\ &- \frac{1}{\rho}\left(u_j \frac{\partial p'}{\partial x_i} + u_i \frac{\partial p'}{\partial x_j} \right) \\ &+ \nu \nabla^2 (u_i u_j) - 2\nu \frac{\partial u_i}{\partial x_k}\frac{\partial u_j}{\partial x_k} \\ &+ u_j \frac{\partial \langle u_i u_k \rangle}{\partial x_k} + u_i \frac{\partial \langle u_j u_k \rangle}{\partial x_k}.\end{aligned} \tag{2.103}$$

The next step is to Reynolds average (2.103) term by term. The second term on the left-hand side yields

$$\left\langle U_k \frac{\partial u_i u_j}{\partial x_k} \right\rangle = \langle U_k \rangle \frac{\partial \langle u_i u_j \rangle}{\partial x_k} + \frac{\partial \langle u_i u_j u_k \rangle}{\partial x_k}, \tag{2.104}$$

where we have used (2.96) to form the triple-correlation term $\langle u_i u_j u_k \rangle$, which describes how the Reynolds stresses are transported by velocity fluctuations.

The final form for the transport equation for the Reynolds stresses is then given by

$$\frac{\partial \langle u_i u_j \rangle}{\partial t} + \langle U_k \rangle \frac{\partial \langle u_i u_j \rangle}{\partial x_k} + \frac{\partial \langle u_i u_j u_k \rangle}{\partial x_k} = \mathcal{P}_{ij} + \Pi_{ij} + \nu \nabla^2 \langle u_i u_j \rangle - \varepsilon_{ij}. \tag{2.105}$$

The molecular transport term $\nu \nabla^2 \langle u_i u_j \rangle$ is closed, but negligible (order Re_L^{-1}) in high-Reynolds-number turbulent flows. The production term

$$\mathcal{P}_{ij} \equiv -\langle u_i u_k \rangle \frac{\partial \langle U_j \rangle}{\partial x_k} - \langle u_j u_k \rangle \frac{\partial \langle U_i \rangle}{\partial x_k} \tag{2.106}$$

is closed, and represents the source of Reynolds stresses due to the mean velocity gradients.

In a simple shear flow where

$$0 < \frac{\partial \langle U_1 \rangle}{\partial x_2} = \text{constant}$$

is the only non-zero mean velocity gradient, the non-zero components of the production term simplify to

$$\mathcal{P}_{11} = -2\langle u_1 u_2 \rangle \frac{\partial \langle U_1 \rangle}{\partial x_2}, \tag{2.107}$$

and

$$\mathcal{P}_{12} = -\langle u_2 u_2 \rangle \frac{\partial \langle U_1 \rangle}{\partial x_2}. \tag{2.108}$$

Thus, because $\langle u_2 u_2 \rangle$ is always non-negative, $\mathcal{P}_{12}$ will be negative, which forces $\langle u_1 u_2 \rangle$ to also be negative. In turn, the production term $\mathcal{P}_{11}$ will be positive, which forces $\langle u_1 u_1 \rangle$ to be non-zero. As discussed below, some of the energy in $\langle u_1 u_1 \rangle$ will be redistributed to

$\langle u_2 u_2 \rangle$ by Π_{ij}. We can thus conclude that, even in a simple homogeneous shear flow, the Reynolds stresses will be *anisotropic* for all Reynolds numbers.

The unclosed velocity–pressure-gradient term in (2.105) is defined by

$$\Pi_{ij} \equiv -\frac{1}{\rho}\left\langle u_i \frac{\partial p}{\partial x_j} + u_j \frac{\partial p}{\partial x_i} \right\rangle, \tag{2.109}$$

and describes correlations between velocity fluctuations and the fluctuating pressure field. Insight into the physical significance of this term can be gained by further decomposing it into two new terms, i.e.,

$$\Pi_{ij} = \mathcal{R}_{ij} - \frac{\partial \mathcal{T}^{(p)}_{kij}}{\partial x_k}. \tag{2.110}$$

The pressure–rate-of-strain tensor $\mathcal{R}_{ij}$ is defined by

$$\mathcal{R}_{ij} \equiv \left\langle \frac{p}{\rho} \left(\frac{\partial u_i}{\partial x_j} + \frac{\partial u_j}{\partial x_i} \right) \right\rangle. \tag{2.111}$$

Note that $\mathcal{R}_{ii} = 0$ due to the continuity equation. Thus, the pressure–rate-of-strain tensor's role in a turbulent flow is to redistribute turbulent kinetic energy among the various components of the Reynolds stress tensor. The pressure-diffusion term $\mathcal{T}^{(p)}_{kij}$ is defined by

$$\mathcal{T}^{(p)}_{kij} \equiv \frac{1}{\rho}\langle u_i p \rangle \delta_{jk} + \frac{1}{\rho}\langle u_j p \rangle \delta_{ik}, \tag{2.112}$$

and is responsible for spatial transport of the Reynolds stresses due to pressure fluctuations.

The unclosed dissipation rate tensor in (2.105) is defined by

$$\varepsilon_{ij} \equiv 2\nu \left\langle \frac{\partial u_i}{\partial x_k} \frac{\partial u_j}{\partial x_k} \right\rangle, \tag{2.113}$$

and describes how velocity fluctuations are dissipated near the Kolmogorov scale. For low-Reynolds-number turbulent flows, the off-diagonal components of ε_{ij} can be significant. However, for locally isotropic flows, the dissipation term simplifies to

$$\varepsilon_{ij} = \frac{2}{3}\varepsilon \delta_{ij}. \tag{2.114}$$

Thus, only the normal Reynolds stresses ($i = j$) are directly dissipated in a high-Reynolds-number turbulent flow. The shear stresses ($i \neq j$), on the other hand, are dissipated indirectly, i.e., the pressure–rate-of-strain tensor first transfers their energy to the normal stresses, where it can be dissipated directly. Without this redistribution of energy, the shear stresses would grow unbounded in a simple shear flow due to the unbalanced production term $\mathcal{P}_{12}$ given by (2.108). This fact is just one illustration of the key role played by $\mathcal{R}_{ij}$ in the Reynolds stress balance equation.

Summing (2.105) over the diagonal elements of the Reynolds stresses results in the transport equation for the turbulent kinetic energy:

$$\frac{\partial k}{\partial t} + \langle U_k \rangle \frac{\partial k}{\partial x_k} + \frac{1}{2}\frac{\partial \langle u_j u_j u_i \rangle}{\partial x_i} + \frac{1}{\rho}\frac{\partial \langle u_i p \rangle}{\partial x_i} = \mathcal{P} + \nu \nabla^2 k - \tilde{\varepsilon}, \tag{2.115}$$

where the turbulent kinetic energy production term is given by

$$\mathcal{P} \equiv -\langle u_i u_j \rangle \frac{\partial \langle U_i \rangle}{\partial x_j}, \tag{2.116}$$

and

$$\tilde{\varepsilon} \equiv \varepsilon - \nu \frac{\partial^2 \langle u_i u_j \rangle}{\partial x_i \partial x_j} \tag{2.117}$$

is the *pseudo dissipation rate*. At high Reynolds numbers, $\tilde{\varepsilon}$ is approximately equal to the unclosed dissipation rate ε. Note that in a simple shear flow, where $\partial \langle U_1 \rangle / \partial x_2$ is the only non-zero mean velocity gradient, the production term $\mathcal{P}$ involves the Reynolds shear stress $\langle u_1 u_2 \rangle$, and thus is unclosed. (Compare this with the Reynolds stress balance equation where the production term is closed.) Likewise, the triple correlation $\langle u_j u_j u_i \rangle$ and the pressure–velocity correlation $\langle u_i p \rangle$, which describe spatial transport of turbulent kinetic energy, are unclosed. On the other hand, the molecular transport term, $\nu \nabla^2 k$, will be negligible in high-Reynolds-number turbulent flows.

2.2.4 Turbulent dissipation rate

As discussed in Section 2.1, in high-Reynolds-number turbulent flows the scalar dissipation rate is equal to the rate of energy transfer through the inertial range of the turbulence energy spectrum. The usual modeling approach is thus to use a transport equation for the transfer rate instead of the detailed balance equation for the dissipation rate derived from (1.27). Nevertheless, in order to understand better the small-scale physical phenomena that determine ε, we will derive its transport equation starting from (2.99).

In order to simplify the notation, we will first denote the fluctuating velocity gradient by

$$g_{ji} \equiv \frac{\partial u_i}{\partial x_j}. \tag{2.118}$$

Differentiating both sides of (2.99) with respect to x_j then yields

$$\begin{aligned} \frac{\partial g_{ji}}{\partial t} + \langle U_k \rangle \frac{\partial g_{ji}}{\partial x_k} = & - u_k \frac{\partial g_{ji}}{\partial x_k} - g_{jk} g_{ki} + \nu \nabla^2 g_{ji} \\ & - g_{ki} \frac{\partial \langle U_k \rangle}{\partial x_j} - g_{jk} \frac{\partial \langle U_i \rangle}{\partial x_k} - u_k \frac{\partial^2 \langle U_i \rangle}{\partial x_j \partial x_k} \\ & - \frac{1}{\rho} \frac{\partial^2 p'}{\partial x_i \partial x_j} + \frac{\partial^2 \langle u_i u_k \rangle}{\partial x_k \partial x_j}. \end{aligned} \tag{2.119}$$

We next denote the random 'dissipation rate' by[18]

$$\epsilon \equiv \nu g_{ji} g_{ji}, \tag{2.120}$$

which is related to the turbulence dissipation rate by $\varepsilon = \langle \epsilon \rangle$. Multiplying (2.119) by $2g_{ji}$ then yields

$$\begin{aligned}\frac{\partial \epsilon}{\partial t} + \langle U_k \rangle \frac{\partial \epsilon}{\partial x_k} = & - u_k \frac{\partial \epsilon}{\partial x_k} - 2 g_{ji} g_{jk} g_{ki} + 2\nu^2 g_{ji} \nabla^2 g_{ji} \\ & - 2\nu g_{ji} g_{ki} \frac{\partial \langle U_k \rangle}{\partial x_j} - 2\nu g_{ji} g_{jk} \frac{\partial \langle U_i \rangle}{\partial x_k} \\ & - 2\nu u_k g_{ji} \frac{\partial^2 \langle U_i \rangle}{\partial x_j \partial x_k} - \frac{2\nu}{\rho} g_{ji} \frac{\partial^2 p'}{\partial x_i \partial x_j} + 2\nu g_{ji} \frac{\partial^2 \langle u_i u_k \rangle}{\partial x_k \partial x_j}.\end{aligned} \tag{2.121}$$

The first term on the right-hand side of (2.121) can be rewritten using the continuity equation as

$$-u_k \frac{\partial \epsilon}{\partial x_k} = -\frac{\partial u_k \epsilon}{\partial x_k}. \tag{2.122}$$

This term represents the spatial transport of ϵ by turbulent velocity fluctuations.

The third term on the right-hand side of (2.121) can be rewritten as the sum of two terms:

$$2\nu^2 g_{ji} \nabla^2 g_{ji} = \nu \nabla^2 \epsilon - 2\nu^2 \frac{\partial g_{ji}}{\partial x_k} \frac{\partial g_{ji}}{\partial x_k}. \tag{2.123}$$

The first term is the molecular transport of ϵ, while the second represents molecular dissipation of velocity gradients.

The pressure term in (2.121) can be rewritten using the continuity equation as

$$-\frac{2\nu}{\rho} g_{ji} \frac{\partial^2 p'}{\partial x_i \partial x_j} = -\frac{2\nu}{\rho} \frac{\partial}{\partial x_i} \left(g_{ji} \frac{\partial p'}{\partial x_j} \right). \tag{2.124}$$

The pressure term is thus responsible for spatial transport due to the fluctuating pressure field.

Reynolds averaging of (2.121) yields the final form for the dissipation-rate transport equation:

$$\frac{\partial \varepsilon}{\partial t} + \langle U_i \rangle \frac{\partial \varepsilon}{\partial x_i} = \nu \nabla^2 \varepsilon - \frac{\partial}{\partial x_i} \left[\langle u_i \epsilon \rangle + \mathcal{T}^{(p)}_{\varepsilon,i} \right] + \mathcal{S}_\varepsilon + \mathcal{C}_\varepsilon + \mathcal{V}_\varepsilon - \mathcal{D}_\varepsilon. \tag{2.125}$$

The first two terms on the right-hand side of this expression are the spatial transport terms. For homogeneous turbulence, these terms will be exactly zero. For inhomogeneous turbulence, the molecular transport term $\nu \nabla^2 \varepsilon$ will be negligible (order Re_L^{-1}). Spatial transport will thus be due to the unclosed velocity fluctuation term $\langle u_i \epsilon \rangle$, and the unclosed pressure transport term $\mathcal{T}^{(p)}_{\varepsilon,i}$ defined by

$$\mathcal{T}^{(p)}_{\varepsilon,i} \equiv \left\langle \frac{2\nu}{\rho} \frac{\partial u_i}{\partial x_j} \frac{\partial p}{\partial x_j} \right\rangle. \tag{2.126}$$

[18] Note a different symbol (ϵ) is used to denote the *random variable*, i.e., $\varepsilon \neq \epsilon$.

All of the remaining terms on the right-hand side of (2.125) are unclosed, and are defined as follows.

The mean-velocity-gradient production term $\mathcal{S}_\varepsilon$ is defined by

$$\mathcal{S}_\varepsilon \equiv -2\left(\left\langle \nu \frac{\partial u_j}{\partial x_i}\frac{\partial u_j}{\partial x_k}\right\rangle + \left\langle \nu \frac{\partial u_i}{\partial x_j}\frac{\partial u_k}{\partial x_j}\right\rangle\right)\frac{\partial \langle U_i\rangle}{\partial x_k}. \tag{2.127}$$

In homogeneous, locally isotropic turbulence the velocity-gradient correlation terms in (2.127) simplify to (Pope 2000)

$$\left\langle \nu \frac{\partial u_j}{\partial x_i}\frac{\partial u_j}{\partial x_k}\right\rangle = \left\langle \nu \frac{\partial u_i}{\partial x_j}\frac{\partial u_k}{\partial x_j}\right\rangle = \frac{1}{6}\varepsilon\delta_{ik}. \tag{2.128}$$

Thus, if the turbulence is locally isotropic, the mean-velocity-gradient production term reduces to

$$\mathcal{S}_\varepsilon = -\frac{2}{3}\varepsilon\delta_{ij}\frac{\partial \langle U_i\rangle}{\partial x_j} = 0, \tag{2.129}$$

where the final equality follows from applying the continuity equation for the mean velocity field, (2.89). At sufficiently high Reynolds numbers, $\mathcal{S}_\varepsilon$ should thus be negligible.

The mean-velocity-gradient curvature term $\mathcal{C}_\varepsilon$ is defined by

$$\mathcal{C}_\varepsilon \equiv -2\left\langle u_k \frac{\partial u_i}{\partial x_j}\right\rangle \nu \frac{\partial^2 \langle U_i\rangle}{\partial x_k \partial x_j}. \tag{2.130}$$

At high Reynolds numbers, the energy-containing scales of one velocity component and the dissipative scales of another velocity component will be uncorrelated. We can then write

$$2\left\langle u_k \frac{\partial u_i}{\partial x_j}\right\rangle = \frac{\partial \langle u_i^2\rangle}{\partial x_j}\delta_{ik}, \tag{2.131}$$

so that $\mathcal{C}_\varepsilon$ reduces to

$$\mathcal{C}_\varepsilon = -\nu \frac{\partial \langle u_i^2\rangle}{\partial x_j}\frac{\partial^2 \langle U_i\rangle}{\partial x_j \partial x_i}. \tag{2.132}$$

Applying the continuity equation for the mean velocity field, $\mathcal{C}_\varepsilon$ is exactly zero at high Reynolds numbers.

The remaining two terms in (2.125) are the vortex-stretching term $\mathcal{V}_\varepsilon$ defined by

$$\mathcal{V}_\varepsilon \equiv -2\left\langle \nu \frac{\partial u_i}{\partial x_j}\frac{\partial u_i}{\partial x_k}\frac{\partial u_k}{\partial x_j}\right\rangle, \tag{2.133}$$

and the gradient-dissipation term $\mathcal{D}_\varepsilon$ defined by

$$\mathcal{D}_\varepsilon \equiv 2\left\langle \left(\nu \frac{\partial^2 u_i}{\partial x_j \partial x_k}\right)\left(\nu \frac{\partial^2 u_i}{\partial x_j \partial x_k}\right)\right\rangle. \tag{2.134}$$

Both of these terms are large in high-Reynolds-number turbulent flows.

For example, the vortex-stretching term is a triple-correlation term that corresponds to the rate at which dissipation is created by spectral energy passing from the inertial range to the dissipative range of the energy spectrum (see (2.75), p. 43). Letting $\kappa_{\mathrm{DI}} \approx 0.1\kappa_\eta$ denote the wavenumber separating the inertial range and the dissipative range (Pope 2000), $\mathcal{V}_\varepsilon$ can be written in terms of the spectral energy transfer rate $\mathcal{T}_u(\kappa, t)$ as[19]

$$\mathcal{V}_\varepsilon = C_{\mathcal{V}1} \nu \kappa_{\mathrm{DI}}^2 \mathcal{T}_u(\kappa_{\mathrm{DI}}, t) + C_{\mathcal{V}2} \left(\frac{\varepsilon}{\nu} \right)^{1/2} \varepsilon, \tag{2.135}$$

where the first term on the right-hand side corresponds to the spectral energy flux into the dissipative range, the second term corresponds to vortex stretching in the dissipative range, and $C_{\mathcal{V}1}$ and $C_{\mathcal{V}2}$ are constants of order unity. Thus, using the fact that at spectral equilibrium $\mathcal{T}_u(\kappa_{\mathrm{DI}}) = \varepsilon$ (Pope 2000), we find that $\mathcal{V}_\varepsilon$ scales with Reynolds number as

$$\mathcal{V}_\varepsilon \sim C_{\mathcal{V}1} \frac{\nu}{\eta^2} \varepsilon + C_{\mathcal{V}2} \mathrm{Re}_L^{1/2} \frac{\varepsilon}{k} \varepsilon \sim \mathrm{Re}_L^{1/2} \frac{\varepsilon}{k} \varepsilon. \tag{2.136}$$

Likewise, because the gradient-dissipation term must balance the vortex-stretching term at spectral equilibrium, it also must scale with the Reynolds number as

$$\mathcal{D}_\varepsilon \sim \mathrm{Re}_L^{1/2} \frac{\varepsilon}{k} \varepsilon. \tag{2.137}$$

However, in general, we can write the gradient-dissipation term as the product of the dissipation rate and a characteristic gradient dissipation rate. The latter can be formed by dividing the fraction of the dissipation rate falling in the dissipation range, i.e.,[20]

$$\varepsilon_{\mathrm{D}}(t) \equiv 2 \int_{\kappa_{\mathrm{DI}}}^{\infty} \nu \kappa^2 E_u(\kappa, t) \, \mathrm{d}\kappa, \tag{2.138}$$

by the fraction of the turbulent kinetic energy in the dissipation range, i.e.,

$$k_{\mathrm{D}}(t) \equiv \int_{\kappa_{\mathrm{DI}}}^{\infty} E_u(\kappa, t) \, \mathrm{d}\kappa. \tag{2.139}$$

In terms of ε_{D} and k_{D}, the gradient-dissipation term can then be written as

$$\mathcal{D}_\varepsilon = C_{\mathcal{D}} \frac{\varepsilon_{\mathrm{D}}}{k_{\mathrm{D}}} \varepsilon_{\mathrm{D}}, \tag{2.140}$$

where $C_{\mathcal{D}}$ is a constant of order unity.[21]

At high Reynolds numbers, the small scales of the turbulent energy spectrum will be in dynamic equilibrium with the large scales, and we may use the model energy spectrum ((2.53), p. 39) to evaluate the right-hand sides of (2.138) and (2.139). By definition, κ_{DI} was chosen such that $\varepsilon_{\mathrm{D}} \approx \varepsilon$ so that the right-hand side of (2.138) can be approximated by ε. On the other hand, making the change of variables $\kappa \rightarrow \kappa\eta$ in (2.138) and evaluating the resultant integral using the model spectrum yields

$$k_{\mathrm{D}} \sim (\nu\varepsilon)^{1/2} = \mathrm{Re}_L^{-1/2} k. \tag{2.141}$$

Thus, employing these results in (2.140) again yields (2.137) at spectral equilibrium.

[19] Note that this expression could be used to approximate $\mathcal{V}_\varepsilon(t)$ for non-equilibrium turbulence. However, κ_{DI} would need to be redefined in terms of ν and the integral time scale τ_e (i.e., independent of ε).

[20] By definition of the dissipation range, $\varepsilon_{\mathrm{D}} \approx \varepsilon$ even when the turbulence is not in spectral equilibrium.

[21] Note that (2.140) will be valid even when the turbulent energy spectrum is not in spectral equilibrium. On the other hand, as shown next, (2.137) is strictly valid only at spectral equilibrium.

Table 2.4. *The turbulence statistics and unclosed quantities appearing in the transport equations for high-Reynolds-number inhomogeneous turbulent flows.*

Physical quantity	Transport equation	Unclosed terms
Mean velocity $\langle U_i \rangle$	(2.93), p. 47	$\langle u_i u_j \rangle$
Mean pressure $\langle p \rangle$	(2.98), p. 48	$\langle u_i u_j \rangle$
Reynolds stresses $\langle u_i u_j \rangle$	(2.105), p. 49	$\langle u_i u_j u_k \rangle$, Π_{ij}, ε
Turbulent kinetic energy k	(2.115), p. 51	$\langle u_i u_j \rangle$, $\langle u_i u_j u_k \rangle$, $\langle u_i p \rangle$, ε
Dissipation rate ε	(2.125), p. 52	$\langle u_i \epsilon \rangle$, $\mathcal{T}_{\varepsilon,i}^{(p)}$, $\mathcal{V}_\varepsilon$, $\mathcal{D}_\varepsilon$

For high-Reynolds-number homogeneous turbulent flows,[22] the right-hand side of the dissipation-rate transport equation thus reduces to the difference between two large terms:[23]

$$\frac{\mathrm{d}\varepsilon}{\mathrm{d}t} \approx \mathcal{V}_\varepsilon - \mathcal{D}_\varepsilon = C_{\mathcal{V}1} \mathrm{Re}_L^{1/2} \frac{1}{\tau_\mathrm{e}} \mathcal{T}_\mathrm{DI}(t) + C_{\mathcal{V}2} \left(\frac{\varepsilon}{\nu} \right)^{1/2} \varepsilon - C_\mathcal{D} \frac{\varepsilon^2}{k_\mathrm{D}}, \tag{2.142}$$

where $\mathcal{T}_\mathrm{DI}(t) \equiv \mathcal{T}_u(\kappa_\mathrm{DI}, t)$, Re_L is the integral-scale Reynolds number, and τ_e is the eddy turnover time defined by (2.49). Likewise, k_D is governed by

$$\frac{\mathrm{d}k_\mathrm{D}}{\mathrm{d}t} \approx \mathcal{T}_\mathrm{DI}(t) - \varepsilon. \tag{2.143}$$

Given initial conditions and $\mathcal{T}_\mathrm{DI}(t)$, these two ordinary differential equations could be solved to find $\varepsilon(t)$. But, inevitably,[24] for large Reynolds numbers one finds $\varepsilon(t) \approx \mathcal{T}_\mathrm{DI}(t)$. Hence, most engineering models for the dissipation rate are largely empirical fits that attempt to model the energy flux $\mathcal{T}_\mathrm{DI}(t)$ instead of the individual terms on the right-hand side of (2.125). We will look at the available models in Chapter 4.

For convenience, the turbulence statistics used in engineering calculations of inhomogeneous, high-Reynolds-number turbulent flows are summarized in Table 2.4 along with the unclosed terms that appear in their transport equations. Models for the unclosed terms are discussed in Chapter 4.

22 We are essentially assuming that the small scales are in dynamic equilibrium with the large scales. This may also hold in low-Reynolds-number turbulent flows. However, for low-Reynolds-number flows, one may need to account also for dissipation rate anisotropy by modeling all components in the dissipation-rate tensor ε_{ij}.

23 The difference appearing on the right-hand side is found experimentally to scale as ε^2/k.

24 The right-hand side of (2.142) scales as $\mathrm{Re}_L^{1/2}/\tau_\mathrm{e}$. In the 'usual' situation where the spectral flux is controlled by the rate at which energy enters the inertial range from the large scales, $\mathcal{T}_\mathrm{DI}(t)$ will vary on time scales of order τ_e. Thus, at large Reynolds numbers, $\mathcal{T}_\mathrm{DI}(t)$ will be in a quasi-steady state with respect to $\varepsilon(t)$.

3

Statistical description of turbulent mixing

The material contained in this chapter closely parallels the presentation in Chapter 2. In Section 3.1, we review the phenomenological description of turbulent mixing that is often employed in engineering models to relate the scalar mixing time to the turbulence time scales. In Section 3.2, the statistical description of homogeneous turbulent mixing is developed based on the one-point and two-point probability density function of the scalar field. In Section 3.3, the transport equations for one-point statistics used in engineering models of inhomogeneous scalar mixing are derived and simplified for high-Reynolds-number turbulent flows. Both inert and reacting scalars are considered. Finally, in Section 3.4, we consider the turbulent mixing of two inert scalars with *different* molecular diffusion coefficients. The latter is often referred to as *differential diffusion*, and is known to affect pollutant formation in gas-phase turbulent reacting flows (Bilger 1982; Bilger and Dibble 1982; Kerstein *et al.* 1995; Kronenburg and Bilger 1997; Nilsen and Kosály 1997; Nilsen and Kosály 1998).

3.1 Phenomenology of turbulent mixing

As seen in Chapter 2 for turbulent flow, the length-scale information needed to describe a homogeneous scalar field is contained in the *scalar energy spectrum* $E_\phi(\kappa, t)$, which we will look at in some detail in Section 3.2. However, in order to gain valuable intuition into the essential physics of scalar mixing, we will look first at the relevant length scales of a turbulent scalar field, and we develop a simple phenomenological model valid for fully developed, statistically stationary turbulent flow. Readers interested in the detailed structure of the scalar fields in turbulent flow should have a look at the remarkable experimental data reported in Dahm *et al.* (1991), Buch and Dahm (1996) and Buch and Dahm (1998).

3.1.1 Length scales of turbulent mixing

Two important length scales for describing turbulent mixing of an inert scalar are the scalar integral scale L_ϕ, and the Batchelor scale λ_B. The latter is defined in terms of the Kolmogorov scale η and the Schmidt number by

$$\lambda_B \equiv \mathrm{Sc}^{-1/2}\eta, \tag{3.1}$$

where Sc is the ratio of the kinematic viscosity and the molecular diffusivity:

$$\mathrm{Sc} \equiv \frac{\nu}{\Gamma}. \tag{3.2}$$

The scalar integral scale characterizes the largest structures in the scalar field, and is primarily determined by two processes: (1) initial conditions – the scalar field can be initialized with a characteristic L_ϕ that is completely independent of the turbulence field, and (2) turbulent mixing – the energy-containing range of a turbulent flow will create 'scalar eddies' with a characteristic length scale L_ϕ that is approximately equal to L_u.

In chemical-process equipment, both initial conditions and turbulent mixing must be accounted for in models for turbulent scalar mixing. For example, near the feed pipe in a stirred-tank reactor the turbulence field (and thus L_u) is primarily determined by the motion of the impeller, not by the motion of the fluid exiting the feed pipe. On the other hand, the feed pipe diameter will be primarily responsible for determining L_ϕ. Thus, at the feed pipe, the scalar integral scale may be much smaller than the turbulence integral scale. This would then result in faster mixing as compared with the case where $L_\phi = L_u$. Nevertheless, as the fluid moves away from the feed pipe, the scalar integral scale will approach the turbulence integral scale,[1] and thus the mixing rate will decrease.

Obviously, a successful model for turbulent mixing must, at a minimum, be able to account for the time dependence of L_ϕ in homogeneous turbulence.[2] However, the situation is further complicated by the fact that large-scale gradients in the scalar mean and the turbulence fields also exist in the reactor. Turbulent mixing due to eddies in the energy-containing range will interact with the mean scalar gradient to produce concentration eddies with characteristic length scale L_u. A successful model for turbulent mixing in a stirred-tank reactor must thus also account for the inhomogeneous distribution of L_ϕ due to boundary conditions and turbulent mixing.

Like the Kolmogorov scale in a turbulent flow, the Batchelor scale characterizes the smallest scalar eddies wherein molecular diffusion is balanced by turbulent mixing.[3] In gas-phase flows, $\mathrm{Sc} \approx 1$, so that the smallest scales are of the same order of magnitude as the Kolmogorov scale, as illustrated in Fig. 3.1. In liquid-phase flows, $\mathrm{Sc} \gg 1$ so that the scalar field contains much more fine-scale structure than the velocity field, as

[1] In chemical-reaction engineering, this process is sometimes referred to as *mesomixing*.

[2] Because the integral scale is defined in terms of the energy spectrum, an appropriate starting point would be the scalar spectral transport equation given in Section 3.2.

[3] The Batchelor scale applies when $1 \leq \mathrm{Sc}$. For $\mathrm{Sc} \ll 1$, a diffusion scale defined by $\lambda_d = (\Gamma^3/\epsilon)^{1/4}$ applies.

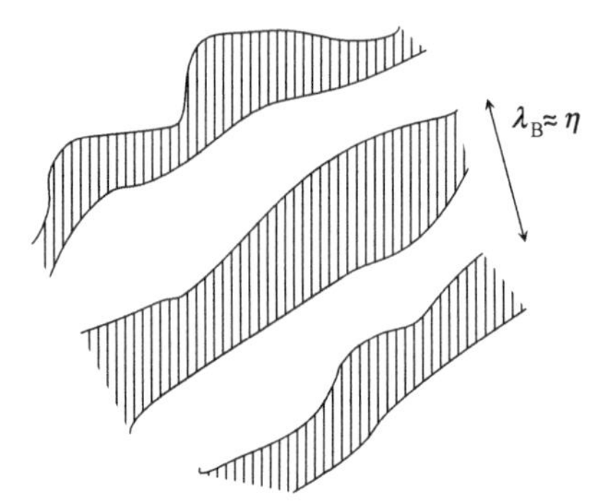

Figure 3.1. Sketch of Batchelor-scale scalar field in a gas-phase flow.

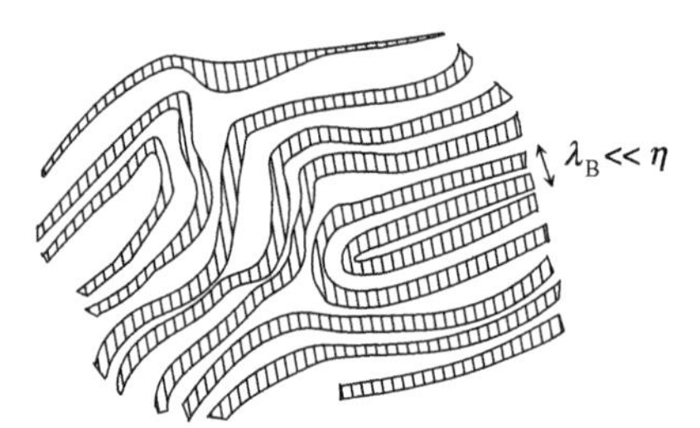

Figure 3.2. Sketch of Batchelor-scale scalar field in a liquid-phase flow.

illustrated in Fig. 3.2. For scalar eddies much larger than the Batchelor scale, molecular diffusion is negligible.[4] Thus, initially non-premixed scalar fields will remain segregated at scales larger than the Batchelor scale. This has important consequences for turbulent reacting flows because it implies that the chemical source term will be strongly coupled to turbulent mixing for many chemical reactions of practical importance. At high Reynolds numbers,[5] the small scales of the scalar field will be nearly isotropic and will evolve on a time scale that is much smaller than that of the large scales. Moreover, for a passive scalar, the characteristic time scales for mixing at length scales above the Batchelor scale will be determined solely by the turbulent flow. These observations lead to the following simple phenomenological model for estimating the turbulent mixing time in terms of the turbulent flow field statistics.

3.1.2 Phenomenological model for turbulent mixing

At scales larger than λ_B (or with Sc $\to \infty$), (1.28), p. 16, reduces to

$$\frac{D\phi}{Dt} = \frac{\partial \phi}{\partial t} + U_i \frac{\partial \phi}{\partial x_i} = 0, \tag{3.3}$$

4 Because it implies that (1.28) will be purely convective, this range of scalar wavenumbers is called the *convective range*.

5 In the discussion that follows, we will assume that $1 \leq$ Sc so that a high Reynolds number suffices to imply the existence of an inertial range for the turbulence *and* a convective range for the scalar.

Figure 3.3. Sketch of slab initial conditions at $t = 0$.

Figure 3.4. Sketch of scalar field with slab initial conditions at $0 < t$.

which implies that ϕ will be constant in fluid elements convected by the flow.[6] Thus, if the initial scalar field is segregated into two domains D and D^c as shown in Fig. 3.3, so that

$$\phi(\mathbf{x}, 0) = \begin{cases} 0 & \text{for } \mathbf{x} \in D \\ 1 & \text{for } \mathbf{x} \in D^c, \end{cases} \tag{3.4}$$

then the turbulence field will only change the *length-scale distribution*,[7] while keeping $\phi = 0$ or 1 at every point in the flow, as shown in Fig. 3.4. However, with non-zero diffusivity, the scalar field will eventually begin to 'move towards the mean,' as seen in Fig. 3.5.

In a fully developed turbulent flow, the rate at which the size of a scalar eddy of length l_ϕ decreases depends on its size relative to the turbulence integral scale L_u and the Kolmogorov scale η. For scalar eddies in the inertial sub-range ($\eta < l_\phi < L_u$), the scalar mixing rate can be approximated by the inverse of the spectral transfer time scale defined in (2.68), p. 42:[8]

$$\gamma(l_\phi) = \left(\frac{\varepsilon}{\nu}\right)^{1/2} \left(\frac{\eta}{l_\phi}\right)^{2/3} \quad \text{for} \quad \eta \leq l_\phi \leq L_u. \tag{3.5}$$

6 A Lagrangian description of the velocity field can be used to find the location $\mathbf{X}(t)$ of the fluid element at time $0 < t$ that started at $\mathbf{X}(0)$. In the Lagrangian description, (3.3) implies that the scalar field associated with the fluid element will remain unchanged, i.e., $\phi(\mathbf{X}(t), t) = \phi(\mathbf{X}(0), 0)$.

7 Or, equivalently, the turbulence field will change the scalar energy spectrum.

8 We have set the proportionality constant in (2.68) equal to unity. Applying the resultant formula at the Batchelor scale suggests that it may be closer to 0.5 (Batchelor 1959).

Figure 3.5. Sketch of scalar field with slab initial conditions and non-zero diffusivity at $0 \ll t$.

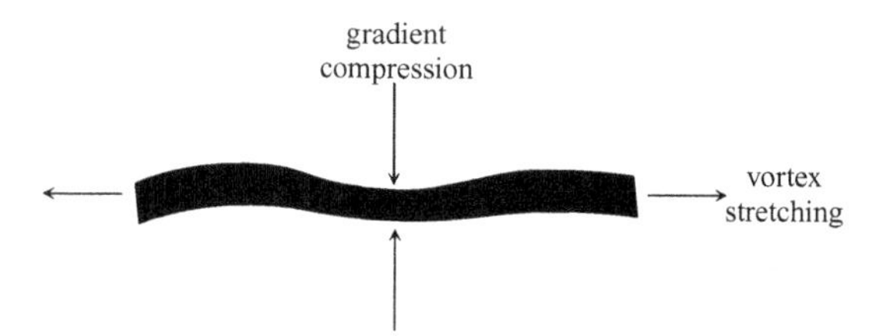

Figure 3.6. Sketch of vortex stretching at small scales.

Using the relationship between L_u and η given in Table 2.2, p. 36, the mixing rate at the velocity integral scale L_u found from (3.5) is approximately

$$\gamma(L_u) = \frac{\varepsilon}{k}, \tag{3.6}$$

while the mixing rate at the Kolmogorov scale is approximately

$$\gamma(\eta) = \left(\frac{\varepsilon}{\nu}\right)^{1/2} = \mathrm{Re}_L^{1/2}\gamma(L_u). \tag{3.7}$$

Thus, at high Reynolds numbers, Kolmogorov-scale mixing will be much faster than integral-scale mixing. Nevertheless, as shown below, the overall mixing time will be controlled by the slower process, i.e., integral-scale mixing.

For scalar eddies smaller than the Kolmogorov scale, the physics of scalar mixing changes. As illustrated in Fig. 3.6, vortex stretching causes the scalar field to become one-dimensional at a constant rate (Batchelor 1959). Thus, for $l_\phi \leq \eta$, the mixing rate can be approximated by

$$\gamma(l_\phi) = \left(\frac{\varepsilon}{\nu}\right)^{1/2} \quad \text{for} \quad l_\phi \leq \eta. \tag{3.8}$$

This process continues until l_ϕ reaches the Batchelor scale, where diffusion takes over and quickly destroys all scalar gradients. The scalar field is then completely mixed, i.e.,

$$\phi(\mathbf{x}, \infty) = \begin{cases} \langle\phi\rangle & \text{for } \mathbf{x} \in D \\ \langle\phi\rangle & \text{for } \mathbf{x} \in D^{\mathrm{c}}. \end{cases} \tag{3.9}$$

The total mixing time t_{mix} can thus be approximated by the time required to reduce $l_\phi(0) = L_\phi$ to $l_\phi(t_{\mathrm{mix}}) = \lambda_{\mathrm{B}}$.

The total mixing time can be computed by inserting the mixing rate $\gamma(l_\phi)$ into the following simple phenomenological model for the scalar length scale:

$$\frac{\mathrm{d}l_\phi}{\mathrm{d}t} = -\gamma(l_\phi)l_\phi. \tag{3.10}$$

Denoting by t_K the time it takes for l_ϕ to be reduced from L_ϕ to η, we find by solving (3.10) with (3.5) that

$$t_\mathrm{K} = \frac{3}{2}\left[\left(\frac{L_\phi}{\eta}\right)^{2/3} - 1\right]\tau_\eta. \tag{3.11}$$

Using the relationship between L_u and η from Table 2.2, p. 36, t_K can be rewritten as

$$t_\mathrm{K} = \frac{3}{2}\left(\frac{L_\phi}{L_u}\right)^{2/3}\tau_u - \frac{3}{2}\tau_\eta. \tag{3.12}$$

At high Reynolds numbers, $\tau_\eta \ll \tau_u$, and thus, if $L_\phi \sim L_u$, then

$$t_\mathrm{K} \approx \frac{3}{2}\left(\frac{L_\phi}{L_u}\right)^{2/3}\tau_u. \tag{3.13}$$

Likewise, letting t_B denote the time it takes to reduce l_ϕ from η to λ_B, we find by solving (3.10) with (3.8) that

$$t_\mathrm{B} = \frac{1}{2}\ln(\mathrm{Sc})\tau_\eta. \tag{3.14}$$

Thus, adding t_K and t_B, the total mixing time can be approximated by

$$t_\mathrm{mix} \approx \frac{3}{2}\left(\frac{L_\phi}{L_u}\right)^{2/3}\tau_u + \frac{1}{2}\ln(\mathrm{Sc})\tau_\eta. \tag{3.15}$$

From the form of this expression, we note the following.

(a) Unless $\mathrm{Sc} \gg 1$, the second term will be negligible. Thus, the Schmidt number must be fairly large before significant differences between the mixing rate in gas-phase and liquid-phase turbulent flows can be observed experimentally at high Reynolds numbers.[9]

(b) The ratio $L_\phi : L_u$ (scalar-to-velocity length-scale ratio) in (3.15) is a key parameter in scalar mixing. In decaying turbulence (i.e., without a source term for kinetic energy), the turbulence integral scale will increase with time. For these types of flows, the scalar mixing time will evolve with time in a non-trivial manner (Chasnov 1998).

(c) If the scalar eddies are initially generated by turbulent diffusivity acting on a mean scalar gradient, then $L_\phi \approx L_u$, and $t_\mathrm{mix} \approx \tau_u$ is determined entirely by the turbulent flow.

[9] This may not be true, however, if $l_\phi(0) \sim \eta$.

The mixing model (3.10) used to derive t_{mix} is intentionally simplistic. For example, it cannot account for the fact that the scalar field is composed of a wide distribution of length scales in a fully developed turbulent flow. In Section 3.2, we review the statistical description of homogeneous turbulent mixing, which parallels the statistical description of turbulence reviewed in Chapter 2. There it is shown that t_{mix} can be redefined in terms of the scalar variance and the scalar dissipation rate, which can be computed from the scalar energy spectrum. Corrsin (1964) has shown that the turbulent mixing time computed from a *fully developed* scalar energy spectrum has the same form as (3.15). This is again a manifestation of *spectral equilibrium* wherein the rate of scalar dissipation at small scales is controlled by the spectral transfer rate from large scales. Nonetheless, as discussed above for the stirred-tank reactor, turbulent scalar mixing often occurs in situations where we cannot assume that the scalar energy spectrum is in spectral equilibrium. In Chapter 4, we will look at models based on the scalar spectral transport equation that attempt to describe *non-equilibrium* turbulent scalar mixing.

3.2 Homogeneous turbulent mixing

A statistical description of homogeneous turbulent mixing can be developed using the same tools as in Section 2.1 for homogeneous turbulence. As discussed in Chapter 1, the principal advantage of using a PDF description of the scalar fields in a turbulent reacting flow is the fact that the chemical source term can be evaluated directly from the one-point composition PDF. Thus, we first discuss the one-point joint velocity, composition PDF, and related statistical functions needed for one-point models of turbulent reacting flows. Nevertheless, the problem of predicting the turbulent mixing time scale cannot be addressed at the one-point PDF level. This shortcoming of the one-point description leads us next to consider the two-point correlation function, and, equivalently in homogeneous turbulence, the scalar energy spectrum for inert scalars. Finally, as a starting point for describing non-equilibrium turbulent mixing, we look at the scalar spectral transport equation for inert scalars.

3.2.1 One-point velocity, composition PDF

Because the random velocity field $\mathbf{U}(\mathbf{x}, t)$ appears in (1.28), p. 16, a passive scalar field in a turbulent flow will be a random field that depends strongly on the velocity field (Warhaft 2000). Thus, turbulent scalar mixing can be described by a one-point joint velocity, composition PDF $f_{\mathbf{U},\phi}(\mathbf{V}, \psi; \mathbf{x}, t)$ defined by

$$f_{\mathbf{U},\phi}(\mathbf{V}, \psi; \mathbf{x}, t)\,\mathrm{d}\mathbf{V}\,\mathrm{d}\psi \equiv \mathrm{P}[\{\mathbf{V} \leq \mathbf{U}(\mathbf{x}, t) < \mathbf{V} + \mathrm{d}\mathbf{V}\} \cap \{\psi \leq \phi(\mathbf{x}, t) < \psi + \mathrm{d}\psi\}]. \tag{3.16}$$

Nevertheless, in many cases, a simpler description based on the one-point composition PDF $f_{\phi}(\psi; \mathbf{x}, t)$, defined by

$$f_{\phi}(\psi; \mathbf{x}, t)\,\mathrm{d}\psi \equiv \mathrm{P}[\psi \leq \phi(\mathbf{x}, t) < \psi + \mathrm{d}\psi], \tag{3.17}$$

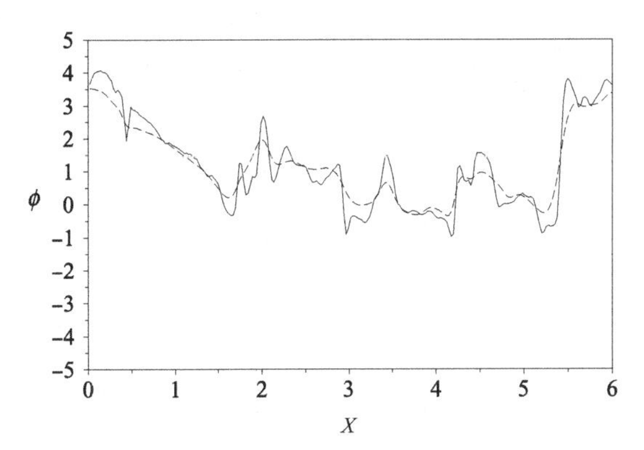

Figure 3.7. Two random scalar fields $\phi(\mathbf{x}, t^*)$ as a function of $x = x_1$ with fixed $t = t^*$. The scalar fields were extracted from DNS of isotropic turbulence ($R_\lambda = 140$, $\langle \mathbf{U} \rangle = \mathbf{0}$) with collinear uniform mean scalar gradients. Dashed line: Sc $= 1/8$; solid line: Sc $= 1$. The corresponding velocity field is shown in Fig. 2.1. (Courtesy of P. K. Yeung.)

can be successfully employed. The one-point composition PDF can be found from the one-point joint velocity, composition PDF by integrating over velocity sample space, i.e.,

$$f_\phi(\psi; \mathbf{x}, t) = \iiint_{-\infty}^{+\infty} f_{\mathbf{U},\phi}(\mathbf{V}, \psi; \mathbf{x}, t)\, \mathrm{d}\mathbf{V}. \tag{3.18}$$

We will make use of both PDFs in Chapters 6 and 7.

For scalar mixing in isotropic turbulence with a uniform mean scalar gradient, the scalar field $\phi(\mathbf{x}, t)$ will be homogeneous and nearly Gaussian with statistics that depend on the Schmidt number (Yeung 1998a). For fixed time $t = t^*$, the function $\phi(\mathbf{x}, t^*)$ varies randomly with respect to $\mathbf{x}$. This behavior is illustrated in Fig. 3.7 for two Schmidt numbers (Sc $= 1/8$ and 1). For fixed $\mathbf{x} = \mathbf{x}^*$, the Eulerian scalar field $\phi(\mathbf{x}^*, t)$ is a random process with respect to t, as shown in Fig. 3.8. Likewise, the Lagrangian scalar $\phi^+(t) \equiv \phi(\mathbf{x}^+(t), t)$, found by following a fluid particle, is illustrated in Fig. 3.9. (See Section 6.7 for more details. Additional samples can be found in Yeung (2001) and Yeung (2002).) Note that the Lagrangian scalars vary more slowly with time than their Eulerian counterpart shown in Fig. 3.8. On the other hand, the lower-Schmidt-number scalar field is smoother in Fig. 3.8, but varies more quickly in Fig. 3.9.[10] Nevertheless, the scalar fields for different Schmidt numbers remain highly correlated. This has important ramifications for Lagrangian PDF mixing models discussed in Section 6.7.

The scalar fields appearing in Figs. 3.7 to 3.9 were taken from the same DNS database as the velocities shown in Figs. 2.1 to 2.3. The one-point joint velocity, composition PDF found from any of these examples will be nearly Gaussian, even though the temporal and/or spatial variations are distinctly different in each case.[11] Due to the mean scalar

[10] The latter is difficult to discern from the figure, but appears clearly in the Lagrangian auto-correlation functions shown in Yeung (2002).

[11] A two-point two-time PDF would be required to describe such spatial and temporal variations.

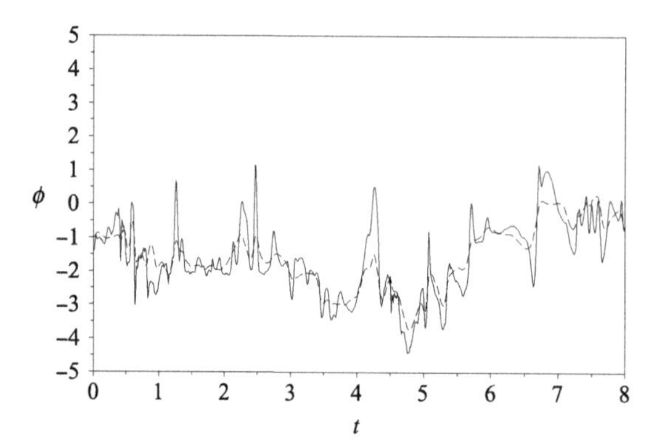

Figure 3.8. Two scalar random processes $\phi(\mathbf{x}^*, t)$ as a function of t with fixed $\mathbf{x} = \mathbf{x}^*$. The scalar fields were extracted from DNS of isotropic turbulence ($R_\lambda = 140$, $\langle \mathbf{U} \rangle = \mathbf{0}$) with collinear uniform mean scalar gradients. Dashed line: Sc $= 1/8$; solid line: Sc $= 1$. The corresponding velocity random process is shown in Fig. 2.2. (Courtesy of P. K. Yeung.)

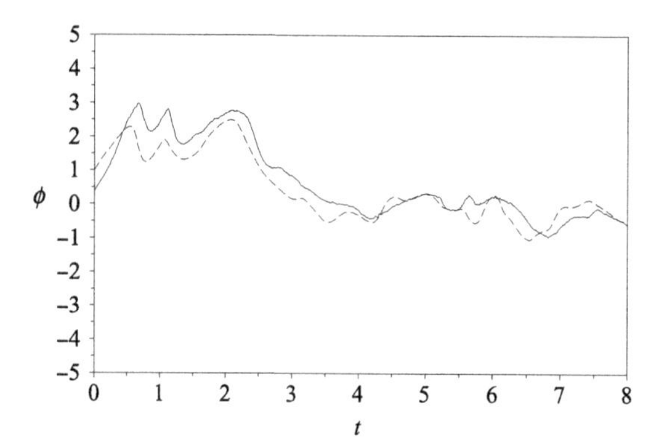

Figure 3.9. Lagrangian scalars $\phi^+(t)$ as a function of t. The scalar fields were extracted from DNS of isotropic turbulence ($R_\lambda = 140$, $\langle \mathbf{U} \rangle = \mathbf{0}$) with collinear uniform mean scalar gradients. Dashed line: Sc $= 1/8$. Solid line: Sc $= 1$. The corresponding Lagrangian velocity is shown in Fig. 2.3. (Courtesy of P. K. Yeung.)

gradients, the scalar fields are correlated with the velocity component only in the direction of the mean scalar gradient (i.e., $\langle u_1 \phi \rangle < 0$ and $\langle u_2 \phi \rangle = \langle u_3 \phi \rangle = 0$).

Unlike the PDF of the scalar fields shown in Figs. 3.7 to 3.9, the composition PDF is often far from Gaussian (Dopazo 1975; Dopazo 1979; Kollmann and Janicka 1982; Drake *et al.* 1986; Kosály 1989; Dimotakis and Miller 1990; Jayesh and Warhaft 1992; Jiang *et al.* 1992; Girimaji 1993; Cai *et al.* 1996; Jaberi *et al.* 1996; Warhaft 2000).[12] A typical situation is *binary mixing*, where initially the scalar field is completely segregated so that the homogeneous composition PDF is given by

$$f_\phi(\psi; 0) = p_0 \delta(\psi) + p_1 \delta(\psi - 1). \tag{3.19}$$

[12] Gaussian PDFs are found for homogeneous inert scalar mixing in the presence of a uniform mean scalar gradient. However, for turbulent reacting flows, the composition PDF is usually far from Gaussian due to the non-linear effects of chemical reactions.

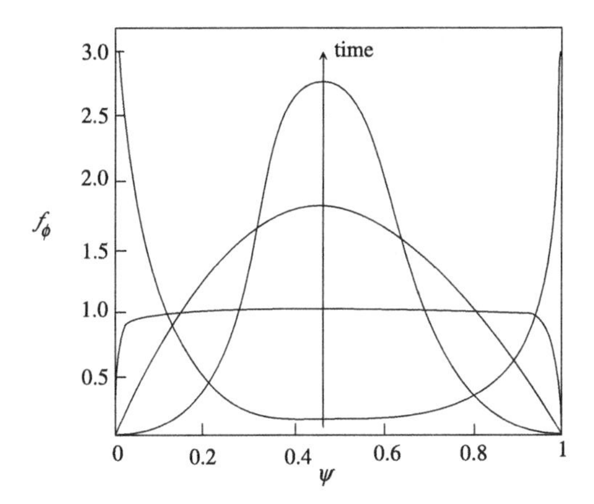

Figure 3.10. Evolution of the inert composition PDF for binary mixing.

The two constants p_0 and p_1 are equal to the volume fraction of the scalar field with value $\phi(\mathbf{x}, 0) = 0$ and $\phi(\mathbf{x}, 0) = 1$, respectively. By definition of the composition PDF, $p_0 + p_1 = 1$, and the initial scalar mean is given by

$$\langle \phi(\mathbf{x}, 0) \rangle = \int_{-\infty}^{+\infty} \psi f_\phi(\psi; 0) \, \mathrm{d}\psi = p_1. \tag{3.20}$$

Starting from these initial conditions, the composition PDF will evolve in a non-trivial manner due to turbulent mixing and molecular diffusion.[13] This process is illustrated in Fig. 3.10, where it can be seen that the shape of the composition PDF at early and intermediate times is far from Gaussian.[14] As discussed in Chapter 6, one of the principal challenges in transported PDF methods is to develop mixing models that can successfully describe the change in shape of the composition PDF due to molecular diffusion.

For homogeneous binary mixing of an inert scalar, the scalar mean will remain constant so that $\langle \phi(\mathbf{x}, t) \rangle = \langle \phi(\mathbf{x}, 0) \rangle = p_1$. Thus, the rate of scalar mixing can be quantified in terms of the scalar variance $\langle \phi'^2 \rangle$ defined by

$$\langle \phi'^2(\mathbf{x}, t) \rangle \equiv \int_{-\infty}^{+\infty} (\psi - \langle \phi \rangle)^2 f_\phi(\psi; t) \, \mathrm{d}\psi. \tag{3.21}$$

For binary mixing, the initial scalar variance is given by $\langle \phi'^2(\mathbf{x}, 0) \rangle = p_1(1 - p_1)$. Moreover, at large times, the composition PDF will collapse to the mean value so that $\langle \phi'^2(\mathbf{x}, \infty) \rangle = 0$ and

$$f_\phi(\psi; \infty) = \delta(\psi - p_1). \tag{3.22}$$

At intermediate times, the scalar variance will be a decreasing function of time. We can thus define the *intensity of segregation* – a measure of the departure of the scalar field

[13] Turbulent mixing is primarily responsible for fixing the rate at which the composition PDF evolves in time. Molecular diffusion, on the other hand, determines the shape of the composition PDF at different time instants.

[14] In fact, binary mixing of an inert scalar is well represented by a *beta PDF*.

Table 3.1. *The relationship between the chemical-reaction-engineering description of turbulent mixing and the one-point composition PDF description.*

For descriptions of the combinations given in the table, please see the text.

CRE description	Composition PDF description
Well macromixed and well micromixed	$f_\phi(\psi;\mathbf{x},t) = \delta(\psi - \langle\phi\rangle)$ $\langle\phi\rangle$ independent of $\mathbf{x}$ $\langle\phi'^2\rangle = 0$
Well macromixed and poorly micromixed	$f_\phi(\psi;\mathbf{x},t) = f_\phi(\psi;t)$ $\langle\phi\rangle$ independent of $\mathbf{x}$ $\langle\phi'^2\rangle > 0$ and independent of $\mathbf{x}$
Poorly macromixed and well micromixed	$f_\phi(\psi;\mathbf{x},t) = \delta(\psi - \langle\phi\rangle)$ $\langle\phi\rangle$ dependent on $\mathbf{x}$ $\langle\phi'^2\rangle = 0$
Poorly macromixed and poorly micromixed	$f_\phi(\psi;\mathbf{x},t)$ arbitrary $\langle\phi\rangle$ dependent on $\mathbf{x}$ $\langle\phi'^2\rangle > 0$ and dependent on $\mathbf{x}$

from molecular-scale homogeneity – in terms of the scalar variance by

$$I(t) \equiv \frac{\langle\phi'^2(\mathbf{x},t)\rangle}{\langle\phi'^2(\mathbf{x},0)\rangle}. \tag{3.23}$$

As shown below, the rate of change of $I(t)$ is determined from the length-scale distribution of the scalar field as characterized by the scalar energy spectrum.

The definition of the composition PDF and the intensity of segregation suffice for establishing the connection between the statistical description of scalar mixing and the chemical-reaction-engineering approach discussed in Section 1.2. The concept of macromixing is directly related to the dependence of $f_\phi(\psi;\mathbf{x},t)$ on $\mathbf{x}$. For example, a statistically homogeneous flow corresponds to a well macromixed reactor (i.e., the one-point composition PDF $f_\phi(\psi;t)$ is independent of $\mathbf{x}$). On the other hand, the concept of micromixing is directly related to $\langle\phi'^2\rangle$. For example, a well micromixed reactor corresponds to a composition PDF with $\langle\phi'^2\rangle = 0$. The four possible combinations of macromixing and micromixing are given in Table 3.1 along with the corresponding form of the composition PDF and the scalar mean and variance.

The first combination – well macromixed and well micromixed – is just the CSTR model for a stirred reactor wherein the scalar is assumed to be constant at every point in the reactor. The second combination – well macromixed and poorly micromixed – corresponds to a statistically homogeneous flow and is often assumed when deriving CRE micromixing models. The third combination – poorly macromixed and well micromixed – is often assumed in CFD models but, strictly speaking, cannot exist in real flows. In the presence of a mean scalar gradient, turbulent mixing will generate scalar fluctuations so that the scalar variance will be non-zero. The fourth combination – poorly macromixed and poorly micromixed – is the general case, and can be handled using transported PDF methods. Some authors further sub-divide the poorly micromixed case into two sub-cases: poorly mesomixed and poorly micromixed. The former corresponds to homogeneous flows with $L_\phi(0) \sim L_u$, while the latter corresponds to homogeneous flows with $L_\phi(0) \sim \eta$. Since $L_\phi(t)$ will generally increase with time so that initially poorly micromixed scalars will gradually become poorly mesomixed, we will not make the distinction between the two sub-cases in this work. However, as discussed in Section 3.1, the mixing time in the two sub-cases can be very different.

In Section 3.3, we will use (3.16) with the Navier–Stokes equation and the scalar transport equation to derive one-point transport equations for selected scalar statistics. As seen in Chapter 1, for turbulent reacting flows one of the most important statistics is the mean chemical source term, which is defined in terms of the one-point joint composition PDF $f_\phi(\psi; \mathbf{x}, t)$ by

$$\langle \mathbf{S}(\mathbf{x}, t) \rangle \equiv \int_{-\infty}^{+\infty} \cdots \int \mathbf{S}(\psi) f_\phi(\psi; \mathbf{x}, t) \, \mathrm{d}\psi. \tag{3.24}$$

Thus, in Chapter 6, the transport equations for $f_\phi(\psi; \mathbf{x}, t)$ and the one-point joint velocity, composition PDF $f_{\mathbf{U},\phi}(\mathbf{V}, \psi; \mathbf{x}, t)$ are derived and discussed in detail. Nevertheless, the computational effort required to solve the PDF transport equations is often considered to be too large for practical applications. Therefore, in Chapter 5, we will look at alternative closures that attempt to replace $f_\phi(\psi; \mathbf{x}, t)$ in (3.24) by a simplified expression that can be evaluated based on one-point scalar statistics that are easier to compute.

3.2.2 Conditional velocity and scalar statistics

In one-point models for turbulent mixing, extensive use of conditional statistics is made when developing simplified models. For example, in the PDF transport equation for $f_\phi(\psi; \mathbf{x}, t)$, the expected value of the velocity fluctuations conditioned on the scalars appears and is defined by

$$\langle \mathbf{u} | \psi \rangle \equiv \langle \mathbf{U} | \phi = \psi \rangle - \langle \mathbf{U} \rangle. \tag{3.25}$$

The first term on the right-hand side of this expression can be calculated from the conditional PDF $f_{\mathbf{U}|\phi}(\mathbf{V}|\psi; \mathbf{x}, t)$ defined by

$$f_{\mathbf{U}|\phi}(\mathbf{V}|\psi; \mathbf{x}, t) \equiv \frac{f_{\mathbf{U},\phi}(\mathbf{V}, \psi; \mathbf{x}, t)}{f_\phi(\psi; \mathbf{x}, t)}. \tag{3.26}$$

Using this expression, the conditional expectation is given by

$$\langle \mathbf{U}(\mathbf{x}, t)|\boldsymbol{\psi}\rangle \equiv \iiint_{-\infty}^{+\infty} \mathbf{V} f_{\mathbf{U}|\boldsymbol{\phi}}(\mathbf{V}|\boldsymbol{\psi}; \mathbf{x}, t)\, \mathrm{d}\mathbf{V}. \tag{3.27}$$

Another conditional expectation that frequently occurs in closures for the chemical source term is the conditional mean of the composition variables given the *mixture fraction*. The latter, defined in Chapter 5, is an inert scalar formed by taking a linear combination of the components of $\boldsymbol{\phi}$:

$$\xi(\mathbf{x}, t) = \sum_{\alpha} a_\alpha \phi_\alpha(\mathbf{x}, t), \tag{3.28}$$

where the sum on the right-hand side is over all composition variables. Formally, the mixture-fraction PDF can be found from the joint composition PDF using

$$f_\xi(\zeta; \mathbf{x}, t) \equiv \int_{-\infty}^{+\infty} \cdots \int \delta\left(\zeta - \sum_{\alpha} a_\alpha \psi_\alpha\right) f_{\boldsymbol{\phi}}(\boldsymbol{\psi}; \mathbf{x}, t)\, \mathrm{d}\boldsymbol{\psi}. \tag{3.29}$$

Moreover, the joint composition, mixture-fraction PDF can be written as

$$f_{\boldsymbol{\phi},\xi}(\boldsymbol{\psi}, \zeta; \mathbf{x}, t) \equiv \delta\left(\zeta - \sum_{\alpha} a_\alpha \psi_\alpha\right) f_{\boldsymbol{\phi}}(\boldsymbol{\psi}; \mathbf{x}, t). \tag{3.30}$$

In terms of these PDFs, the conditional PDF of the composition variables given the mixture fraction is defined by

$$f_{\boldsymbol{\phi}|\xi}(\boldsymbol{\psi}|\zeta; \mathbf{x}, t) \equiv \frac{f_{\boldsymbol{\phi},\xi}(\boldsymbol{\psi}, \zeta; \mathbf{x}, t)}{f_\xi(\zeta; \mathbf{x}, t)}. \tag{3.31}$$

The conditional expected value of $\boldsymbol{\phi}$ conditioned on the mixture fraction is then found by integration:

$$\langle \boldsymbol{\phi}|\zeta\rangle \equiv \int_{-\infty}^{+\infty} \cdots \int \boldsymbol{\psi} f_{\boldsymbol{\phi}|\xi}(\boldsymbol{\psi}|\zeta; \mathbf{x}, t)\, \mathrm{d}\boldsymbol{\psi}. \tag{3.32}$$

The conditional expected value $\langle \boldsymbol{\phi}|\zeta\rangle$ can be estimated from experimental data for turbulent reacting flows. In a turbulent flame far from extinction, it has been found that $f_{\boldsymbol{\phi}|\xi}(\boldsymbol{\psi}|\zeta; \mathbf{x}, t)$ is often well approximated by

$$f_{\boldsymbol{\phi}|\xi}(\boldsymbol{\psi}|\zeta; \mathbf{x}, t) \approx \prod_{\alpha} \delta\left(\psi_\alpha - \langle \phi_\alpha|\zeta\rangle\right) \equiv \boldsymbol{\delta}\left(\boldsymbol{\psi} - \langle \boldsymbol{\phi}|\zeta\rangle\right). \tag{3.33}$$

This, in turn, implies that the conditional chemical source term is closed when written in terms of the conditional means:

$$\langle \mathbf{S}|\zeta\rangle = \mathbf{S}(\langle \boldsymbol{\phi}|\zeta\rangle). \tag{3.34}$$

In cases where this approximation is acceptable, it thus suffices to provide a model for $\langle \boldsymbol{\phi}|\zeta\rangle$ in place of a model for the joint composition PDF. Approaches based on this approximation are referred to as *conditional-moment closures* (CMCs), and are discussed in detail in Section 5.8.

In developing closures for the chemical source term and the PDF transport equation, we will also come across conditional moments of the derivatives of a field conditioned on the value of the field. For example, in conditional-moment closures, we must provide a functional form for the scalar dissipation rate conditioned on the mixture fraction, i.e.,

$$\langle \epsilon_\xi | \zeta \rangle \equiv \left\langle 2\Gamma \frac{\partial \xi}{\partial x_i} \frac{\partial \xi}{\partial x_i} \right| \xi = \zeta \left. \vphantom{\frac{\partial \xi}{\partial x_i}} \right\rangle .$$

Conditional moments of this type cannot be evaluated using the one-point PDF of the mixture fraction alone (O'Brien and Jiang 1991). In order to understand better the underlying closure problem, it is sometimes helpful to introduce a new random field, i.e.,[15]

$$Z(\mathbf{x}, t) \equiv 2\Gamma \frac{\partial \xi}{\partial x_i} \frac{\partial \xi}{\partial x_i}, \tag{3.35}$$

and the one-point joint PDF of $Z(\mathbf{x}, t)$ and $\xi(\mathbf{x}, t)$ defined by

$$f_{Z,\xi}(z, \zeta; \mathbf{x}, t) \, \mathrm{d}z \, \mathrm{d}\zeta \equiv \mathrm{P}[\{z \leq Z(\mathbf{x}, t) < z + \mathrm{d}z\} \cap \{\zeta \leq \xi(\mathbf{x}, t) < \zeta + \mathrm{d}\zeta\}]. \tag{3.36}$$

As discussed in Chapter 1, given a sample mixture-fraction *field*[16] $\{\xi(\mathbf{x}, t) : \mathbf{x} \in L^3\}$, this joint PDF could be estimated and plotted in terms of z and ζ as a two-dimensional contour plot.

Mathematically, the conditional expected value $\langle Z | \zeta \rangle$ can be found by integration starting from the conditional PDF $f_{Z|\xi}(z|\zeta; \mathbf{x}, t)$:

$$\langle Z | \zeta \rangle \equiv \int_{-\infty}^{+\infty} z f_{Z|\xi}(z|\zeta; \mathbf{x}, t) \, \mathrm{d}z = \int_{-\infty}^{+\infty} z \frac{f_{Z,\xi}(z, \zeta; \mathbf{x}, t)}{f_\xi(\zeta; \mathbf{x}, t)} \, \mathrm{d}z. \tag{3.37}$$

The need to add new random variables defined in terms of *derivatives* of the random fields is simply a manifestation of the lack of two-point information. While it is possible to develop a two-point PDF approach, inevitably it will suffer from the lack of three-point information. Moreover, the two-point PDF approach will be computationally intractable for practical applications. A less ambitious approach that will still provide the length-scale information missing in the one-point PDF can be formulated in terms of the scalar spatial correlation function and scalar energy spectrum described next.

3.2.3 Spatial correlation functions

As mentioned above, the one-point PDF description does not provide the length-scale information needed to predict the decay rate of the scalar variance. For this purpose, a

[15] The transport equation for $Z(\mathbf{x}, t)$ is derived in Section 3.3 and, as expected, is unclosed at the one-point level due to the appearance of new terms involving the spatial derivatives of $Z(\mathbf{x}, t)$.

[16] If the complete field is available, then it can easily be differentiated to find the gradient field, and hence $\{Z(\mathbf{x}, t) : \mathbf{x} \in L^3\}$.

two-point composition PDF, or (for a homogeneous scalar field) a scalar spatial correlation function provides the relevant information. The scalar spatial correlation function in homogeneous turbulence can be written in terms of the fluctuating scalar field:

$$\phi'(\mathbf{x}, t) \equiv \phi(\mathbf{x}, t) - \langle \phi(\mathbf{x}, t) \rangle \tag{3.38}$$

as

$$R_\phi(\mathbf{r}, t) \equiv \langle \phi'(\mathbf{x}, t)\phi'(\mathbf{x}^*, t) \rangle, \tag{3.39}$$

where $\mathbf{r} = \mathbf{x}^* - \mathbf{x}$. Other spatial correlation functions of interest in scalar mixing studies are the scalar-velocity cross-correlation function defined by

$$R_{\phi i}(\mathbf{r}, t) \equiv \langle \phi'(\mathbf{x}, t) u_i(\mathbf{x}^*, t) \rangle, \tag{3.40}$$

and the scalar cross-correlation function defined by

$$R_{\alpha\beta}(\mathbf{r}, t) \equiv \langle {\phi'}_\alpha(\mathbf{x}, t) {\phi'}_\beta(\mathbf{x}^*, t) \rangle. \tag{3.41}$$

Note that evaluating the correlation functions at $\mathbf{r} = 0$ yields the corresponding one-point statistics. For example, $R_{\alpha\beta}(0, t)$ is equal to the scalar covariance $\langle {\phi'}_\alpha {\phi'}_\beta \rangle$.

Like the velocity spatial correlation function discussed in Section 2.1, the scalar spatial correlation function provides length-scale information about the underlying scalar field. For a homogeneous, isotropic scalar field, the spatial correlation function will depend only on $r \equiv |\mathbf{r}|$, i.e., $R_\phi(r, t)$. The scalar integral scale L_ϕ and the scalar Taylor microscale λ_ϕ can then be computed based on the normalized scalar spatial correlation function f_ϕ defined by

$$f_\phi(r, t) \equiv \frac{R_\phi(r, t)}{R_\phi(0, t)}. \tag{3.42}$$

In terms of this function, the scalar integral scale is defined by

$$L_\phi(t) \equiv \int_0^\infty f_\phi(r, t)\, \mathrm{d}r, \tag{3.43}$$

and the scalar Taylor microscale is defined by

$$\lambda_\phi(t) \equiv \left(-\frac{1}{2} \frac{\partial^2 f_\phi}{\partial r^2}(0, t) \right)^{-1/2}. \tag{3.44}$$

Moreover, like the relationship between the turbulence Taylor microscale and the dissipation rate ε, $\lambda_\phi(t)$ is related to the scalar variance decay rate by[17]

$$\frac{\mathrm{d}\langle \phi'^2 \rangle}{\mathrm{d}t} = -\frac{12\Gamma}{\lambda_\phi^2} \langle \phi'^2 \rangle = -\varepsilon_\phi, \tag{3.45}$$

[17] The scalar Taylor microscale is thus related to the mixing time t_{mix}, i.e., $t_{\mathrm{mix}} = \lambda_\phi^2/6\Gamma$.

where ε_ϕ is the *scalar dissipation rate* defined by

$$\varepsilon_\phi \equiv \left\langle 2\Gamma \frac{\partial \phi'}{\partial x_i} \frac{\partial \phi'}{\partial x_i} \right\rangle. \tag{3.46}$$

In general, the scalar Taylor microscale will be a function of the Schmidt number. However, for fully developed turbulent flows,[18] $L_\phi \sim L_{11}$ and $\lambda_\phi \sim \text{Sc}^{-1/2}\lambda_g$. Thus, a model for non-equilibrium scalar mixing could be formulated in terms of a dynamic model for $\lambda_\phi(t)$. But, due to the difficulties associated with working in terms of the scalar spatial correlation function, a simpler approach is to work with the scalar energy spectrum defined next.

3.2.4 Scalar energy spectrum

For homogeneous scalar fields, the *scalar spectrum* $\Phi_\phi(\boldsymbol{\kappa}, t)$ is related to the scalar spatial correlation function defined in (3.39) through the following Fourier transform pair (see Lesieur (1997) for details):

$$\Phi_\phi(\boldsymbol{\kappa}, t) = \frac{1}{(2\pi)^3} \iiint_{-\infty}^{+\infty} R_\phi(\mathbf{r}, t)\, \mathrm{e}^{-\mathrm{i}\boldsymbol{\kappa}\cdot\mathbf{r}}\, \mathrm{d}\mathbf{r}, \tag{3.47}$$

$$R_\phi(\mathbf{r}, t) = \iiint_{-\infty}^{+\infty} \Phi_\phi(\boldsymbol{\kappa}, t)\, \mathrm{e}^{\mathrm{i}\boldsymbol{\kappa}\cdot\mathbf{r}}\, \mathrm{d}\boldsymbol{\kappa}. \tag{3.48}$$

Similar Fourier transform pairs relate the spatial correlation functions defined in (3.40) and (3.41) to corresponding cospectra $\Phi_{\phi i}(\boldsymbol{\kappa}, t)$ and $\Phi_{\alpha\beta}(\boldsymbol{\kappa}, t)$, respectively.

Note that from its definition, the scalar spatial correlation function is related to the scalar variance by

$$R_\phi(0, t) = \langle \phi'^2 \rangle(t) = \iiint_{-\infty}^{+\infty} \Phi_\phi(\boldsymbol{\kappa}, t)\, \mathrm{d}\boldsymbol{\kappa}. \tag{3.49}$$

Thus, $\Phi_\phi(\boldsymbol{\kappa}, t)\, \mathrm{d}\boldsymbol{\kappa}$ roughly corresponds to the 'amount' of scalar variance located at point $\boldsymbol{\kappa}$ in wavenumber space at time t. Similar statements can be made concerning the relationship between $\Phi_{\phi i}$ and the scalar flux $\langle u_i \phi \rangle$, and between $\Phi_{\alpha\beta}$ and the covariance $\langle \phi'_\alpha \phi'_\beta \rangle$.

The *scalar energy spectrum* $E_\phi(\kappa, t)$ is defined in terms of the scalar spectrum by integrating out all directional information:

$$E_\phi(\kappa, t) = \iiint_{-\infty}^{+\infty} \Phi_\phi(\boldsymbol{\kappa}, t) \delta(\kappa - |\boldsymbol{\kappa}|)\, \mathrm{d}\boldsymbol{\kappa}, \tag{3.50}$$

[18] 'Fully developed' is again meant to imply that both the velocity and scalar energy spectra are fully developed.

where the integral is over shells in wavenumber space located at a distance κ from the origin. The *scalar flux energy spectrum* $E_{\phi i}(\kappa, t)$ and the *scalar covariance energy spectrum* $E_{\alpha\beta}(\kappa, t)$ are defined in a similar manner.

For an isotropic scalar field, the scalar spectrum is related to the scalar energy spectrum by

$$\Phi_\phi(\boldsymbol{\kappa}, t) = \frac{E_\phi(\kappa, t)}{4\pi\kappa^2}. \tag{3.51}$$

At high Reynolds numbers, we can again expect the small scales of the scalar field to be nearly isotropic. In the classical picture of turbulent mixing, one speaks of scalar eddies produced at large scales, with a distinct directional orientation, that lose their directional preference as they cascade down to small scales where they are dissipated by molecular diffusion.

By definition, the scalar variance can be found directly from the scalar energy spectrum by integrating over wavenumber space:

$$\langle \phi'^2 \rangle(t) = \int_0^\infty E_\phi(\kappa, t)\, \mathrm{d}\kappa. \tag{3.52}$$

Thus, $E_\phi(\kappa, t)\, \mathrm{d}\kappa$ represents the 'amount' of scalar variance located at wavenumber κ. For isotropic turbulence, the scalar integral length scale L_ϕ is related to the scalar energy spectrum by

$$L_\phi(t) = \frac{\pi}{2\langle \phi'^2 \rangle} \int_0^\infty \frac{E_\phi(\kappa, t)}{\kappa}\, \mathrm{d}\kappa. \tag{3.53}$$

Similar relationships exist for the scalar flux energy spectrum and the scalar covariance energy spectrum.

Likewise, the scalar dissipation rate is related to the scalar energy spectrum by

$$\varepsilon_\phi(t) = \int_0^\infty 2\Gamma\kappa^2 E_\phi(\kappa, t)\, \mathrm{d}\kappa = \int_0^\infty D_\phi(\kappa, t)\, \mathrm{d}\kappa, \tag{3.54}$$

where $D_\phi(\kappa, t)$ is the *scalar dissipation energy spectrum*. The scalar energy spectrum thus determines the scalar mixing time:

$$\tau_\phi(t) \equiv \frac{2\langle \phi'^2 \rangle}{\varepsilon_\phi}. \tag{3.55}$$

If the scalar mixing time is constant, then the classical exponential-decay result is recovered for the intensity of segregation defined by (3.23), p. 66:

$$I(t) = \mathrm{e}^{-2t/\tau_\phi}. \tag{3.56}$$

In general, τ_ϕ must be computed from a dynamic model for $E_\phi(\kappa, t)$. However, for fully developed scalar fields (equilibrium mixing), the mixing time can be approximated from a model spectrum for $E_\phi(\kappa)$.

3.2.5 Model scalar spectrum

As in Section 2.1 for the turbulent energy spectrum, a model scalar energy spectrum can be developed to describe $E_\phi(\kappa)$. However, one must account for the effect of the Schmidt number. For Sc $\ll 1$, the scalar-dissipation wavenumbers, defined by[19]

$$\kappa_{c1} \equiv \mathrm{Sc}^{3/4}\kappa_{\mathrm{DI}} \tag{3.57}$$

and

$$\kappa_{c2} \equiv \mathrm{Sc}^{3/4}\kappa_\eta, \tag{3.58}$$

fall in the inertial range of the turbulent energy spectrum. On the other hand, for Sc $\gg$ 1, the Batchelor wavenumber $\kappa_{\mathrm{B}} = \mathrm{Sc}^{1/2}\kappa_\eta$ will be much larger than the Kolmogorov wavenumber κ_η.

In either case, if the Reynolds number is sufficiently large, there will exist a range of wavenumbers ($\kappa \ll \kappa_{c1}$ and κ_{DI}) for which neither the molecular viscosity nor the molecular diffusivity will be effective. In this range, the stationary scalar energy spectrum should have the same form as $E_u(\kappa)$. Thus, as for the turbulent energy spectrum, the low-wavenumber range will be referred to as the *scalar energy-containing range* and is bounded above by κ_{EI} given by (2.56), p. 40. This range is followed by the *inertial-convective sub-range* (Obukhov 1949; Corrsin 1951b). The model scalar energy spectrum for these sub-ranges has the form

$$E_\phi(\kappa) = C_{\mathrm{OC}}\varepsilon_\phi\varepsilon^{-1/3}\kappa^{-5/3} f_{\mathrm{L}}(\kappa L_u) \quad \text{for} \quad \kappa \ll \min(\kappa_{c1}, \kappa_{\mathrm{DI}}), \tag{3.59}$$

where $C_{\mathrm{OC}} \approx 2/3$ is the Obukhov–Corrsin constant, and f_{L} is the low-wavenumber cut-off function appearing in the model spectrum for turbulent energy ((2.54), p. 39).

Note that as Re_L goes to infinity with Sc constant, both the turbulent energy spectrum and the scalar energy spectrum will be dominated by the energy-containing and inertial/inertial-convective sub-ranges. Thus, in this limit, the characteristic time scale for scalar variance dissipation defined by (3.55) becomes

$$\tau_\phi = \frac{2C_{\mathrm{OC}}}{C}\tau_u \approx \tau_u, \tag{3.60}$$

so that the scalar time scale will be nearly equal to the turbulence time scale and independent of the Schmidt number.

The form of the scalar energy spectrum for larger wavenumbers will depend on the Schmidt number. Considering first the case where Sc $\ll 1$, the range of wavenumbers between κ_{c2} and κ_{DI} is referred to as the *inertial-diffusive sub-range* (Batchelor *et al.* 1959). Note that this range can exist only for Schmidt numbers less than $\mathrm{Sc}_{\mathrm{id}}$, where

$$\mathrm{Sc}_{\mathrm{id}} \equiv \left(\frac{\kappa_{\mathrm{DI}}}{\kappa_\eta}\right)^{4/3}. \tag{3.61}$$

[19] We have defined two diffusion cut-off wavenumbers in terms of κ_{DI} and κ_η in order to be consistent with the model turbulent energy spectrum introduced in Chapter 2.

The form of the scalar energy spectrum in the inertial-diffusive sub-range can be found starting from the Navier–Stokes equation (see McComb (1990) for details) to be

$$E_\phi(\kappa) \propto \varepsilon_\phi \varepsilon^{2/3}\Gamma^{-3}\kappa^{-17/3} \quad \text{for} \quad \kappa_{\text{c}2} \ll \kappa \ll \kappa_{\text{DI}}. \tag{3.62}$$

For all values of Sc and wavenumbers larger than κ_η, the scalar energy spectrum drops off quickly in the *viscous-diffusive sub-range*, and can be approximated by (see McComb (1990) for details)

$$E_\phi(\kappa) \propto \varepsilon_\phi \varepsilon^{-1/2}\nu^{1/2}\kappa^{-1} \exp[-c_{\text{d}}\text{Sc}^{-1}(\kappa\eta)^2] \quad \text{for} \quad \kappa_\eta \ll \kappa. \tag{3.63}$$

This expression was derived originally by Batchelor (1959) under the assumption that the correlation time of the Kolmogorov-scale strain rate is large compared with the Kolmogorov time scale. Alternatively, Kraichnan (1968) derived a model spectrum of the form

$$E_\phi(\kappa) \propto \varepsilon_\phi \varepsilon^{-1/2}\nu^{1/2}\kappa^{-1}(1 + c_{\text{d}}\text{Sc}^{-1/2}\kappa\eta) \exp[-c_{\text{d}}\text{Sc}^{-1/2}\kappa\eta] \text{ for } \kappa_\eta \ll \kappa \tag{3.64}$$

by assuming that the correlation time is small compared with the Kolmogorov time scale (Kraichnan 1974). DNS data for passive scalar mixing (Bogucki *et al.* 1997; Yeung *et al.* 2002) with $\text{Sc} > 1$ show good agreement with (3.64).

A model scalar energy spectrum can be developed by combining the various theoretical spectra introduced above with appropriately defined cut-off functions and exponents:

$$E_\phi(\kappa) = C_{\text{OC}}\varepsilon_\phi \varepsilon^{-3/4}\nu^{5/4}(\kappa\eta)^{-\beta(\kappa\eta)} f_{\text{L}}(\kappa L_u) f_{\text{B}}(\kappa\eta), \tag{3.65}$$

where the scaling exponent is defined by

$$\beta(\kappa\eta) \equiv 1 + \frac{2}{3}\left[7 - 6 f_{\text{D}}(\kappa\eta)\right] f_\eta(\kappa\eta). \tag{3.66}$$

For small κ, this exponent starts at $5/3$ and then varies depending on the values of the cut-off functions. For large κ, the scaling exponent approaches unity. In between, the value of the scaling exponent depends on Sc, and it reaches $17/3$ when $\text{Sc} \ll 1$.

The cut-off functions $f_{\text{L}}(\kappa L_u)$ and $f_\eta(\kappa\eta)$ are taken from the model energy spectrum ((2.54) and (2.55)). The diffusive-scale cut-off function $f_{\text{D}}(\kappa\eta)$ is defined by

$$f_{\text{D}}(\kappa\eta) \equiv \left[1 + c_{\text{D}}\text{Sc}^{-d(\kappa\eta)/2}\kappa\eta\right] \exp\left[-c_{\text{D}}\text{Sc}^{-d(\kappa\eta)/2}\kappa\eta\right], \tag{3.67}$$

with $c_{\text{D}} = 2.59$, and the diffusive exponent is given by

$$d(\kappa\eta) \equiv \frac{1}{2} + \frac{1}{4} f_\eta(\kappa\eta). \tag{3.68}$$

The diffusive exponent is $3/4$ in the inertial range and $1/2$ in the viscous range. The values of c_{D} and the exponent in (3.67) have been chosen to reproduce roughly the spectral 'bump' observed in compensated spectra for $0.1 \leq \text{Sc} \leq 1$ (Yeung *et al.* 2002).

The Batchelor-scale cut-off function $f_{\text{B}}(\kappa\eta)$ can be defined by (*Batchelor*)

$$f_{\text{B}}(\kappa\eta) \equiv \exp\left[-c_{\text{d}}\text{Sc}^{-2d(\kappa\eta)}(\kappa\eta)^2\right], \tag{3.69}$$

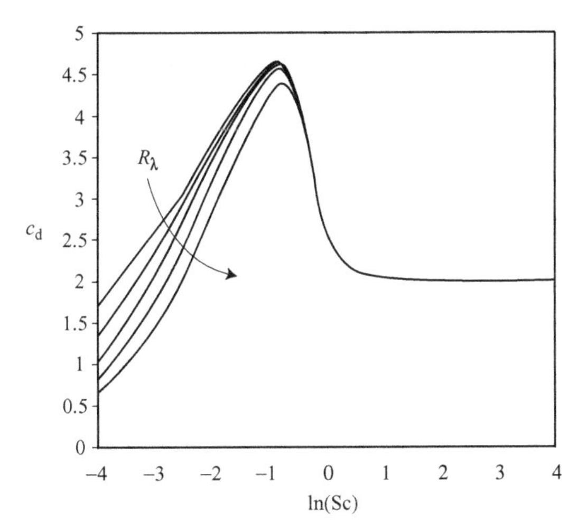

Figure 3.11. The scalar-dissipation constant c_d found with the Kraichnan cut-off function as a function of Schmidt number at various Reynolds numbers: $R_\lambda = 50$, 100, 200, 400, and 800. The arrow indicates the direction of increasing Reynolds number.

or by (*Kraichnan*)

$$f_B(\kappa\eta) \equiv \left[1 + c_d \mathrm{Sc}^{-d(\kappa\eta)}\kappa\eta\right] \exp\left[-c_d \mathrm{Sc}^{-d(\kappa\eta)}\kappa\eta\right]. \tag{3.70}$$

However, DNS data for Schmidt numbers near unity suggest that (3.70) provides the best model for the scalar-dissipation range (Yeung *et al.* 2002).

The scalar-dissipation constant c_d appearing in (3.69) and (3.70) is fixed by forcing the integral of the scalar-dissipation spectrum to satisfy (3.54):[20]

$$\int_0^\infty (\kappa\eta)^{2-\beta(\kappa\eta)} f_L(\kappa L_u) f_B(\kappa\eta)\, \mathrm{d}(\kappa\eta) = \frac{\mathrm{Sc}}{2C_{\mathrm{OC}}}. \tag{3.71}$$

The dependence of c_d on the Reynolds and Schmidt numbers found using (3.70) is shown in Fig. 3.11. Note that for $\mathrm{Sc} > 1$ (and $C_{\mathrm{OC}} = 2/3$), $c_d = 2$ for all values of Re and Sc. On the other hand, c_d is strongly dependent on the Schmidt number when $\mathrm{Sc} < 1$. Also note that, for large Sc, (3.71) yields $c_d = (6C_{\mathrm{OC}})^{1/2}$. Thus, the limiting value of c_d will depend on the choice of C_{OC}.

Example scalar energy spectra for a range of Schmidt numbers are plotted for $R_\lambda = 500$ in Fig. 3.12 using (3.70) for f_B. The model velocity spectrum is included in the figure for comparison. For low Schmidt numbers, the scalar spectrum falls off much faster than the velocity spectrum. For moderate Schmidt numbers, inertial-range $(-5/3)$ scaling is observed over a wide range of wavenumbers. For high Schmidt numbers, viscous-convective (-1) scaling is evident.

The behavior of the scalar spectra for Schmidt numbers near $\mathrm{Sc} = 1$ is distinctly different than the velocity spectrum. This can be most clearly seen by plotting compensated

[20] Another alternative is to fix c_d and to adjust C_{OC} to satisfy (3.71). For large Sc, either method will yield the same Sc- and Re-independent result. However, for small Sc, the Sc dependence is very strong.

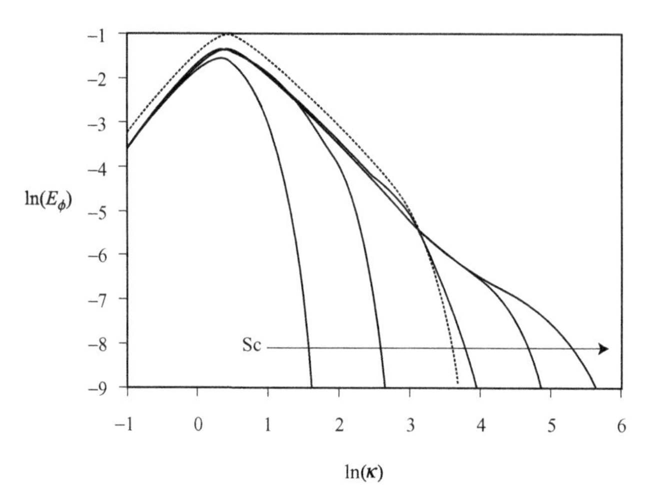

Figure 3.12. Model scalar energy spectra at $R_\lambda = 500$ normalized by the integral scales. The velocity energy spectrum is shown as a dotted line for comparison. The Schmidt numbers range from $\mathrm{Sc} = 10^{-4}$ to $\mathrm{Sc} = 10^4$ in powers of 10^2.

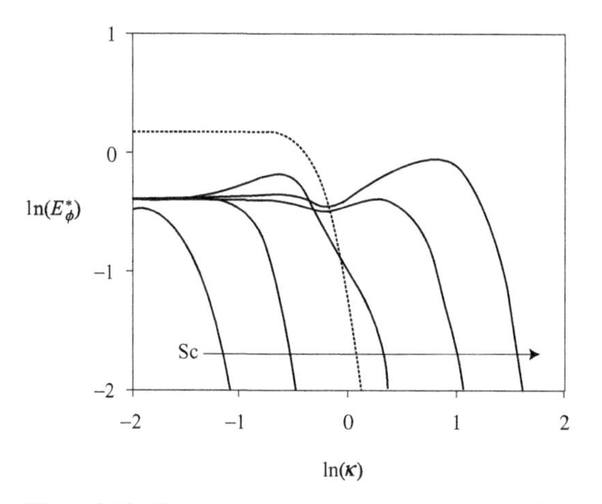

Figure 3.13. Compensated scalar energy spectra E^*_ϕ at $R_\lambda = 500$ normalized by the Kolmogorov scales. The compensated velocity energy spectrum is shown as a dotted line for comparison. The Schmidt numbers range from $\mathrm{Sc} = 10^{-2}$ to $\mathrm{Sc} = 10^2$ in powers of 10.

energy spectra, defined by $E^*(\kappa) = (\kappa\eta)^{5/3}E(\kappa)$. Compensated spectra for $R_\lambda = 500$ and Sc in the range $(0.01, 100)$ are shown in Fig. 3.13. For $\mathrm{Sc} = 1$, a spectral 'bump' is clearly evident. As noted above, the height and location of this bump are controlled by c_D.

Having defined the model scalar energy spectrum, it can now be used to compute the scalar mixing time τ_ϕ as a function of Sc and R_λ. In the turbulent mixing literature, the scalar mixing time is usually reported in a dimensionless form referred to as the *mechanical-to-scalar time-scale ratio* R defined by

$$R \equiv \frac{k\varepsilon_\phi}{\varepsilon\langle\phi'^2\rangle} = \frac{2\tau_u}{\tau_\phi}. \tag{3.72}$$

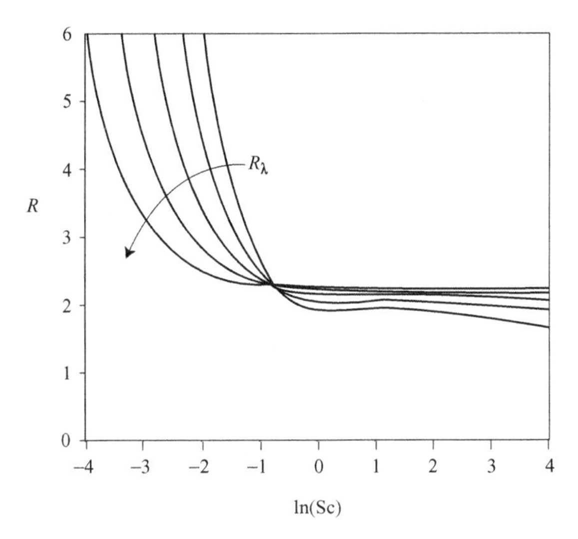

Figure 3.14. The mechanical-to-scalar time-scale ratio for a fully developed scalar energy spectrum as a function of the Schmidt number at various Reynolds numbers: $R_\lambda = 50, 100, 200, 400,$ and 800. The arrow indicates the direction of increasing Reynolds number.

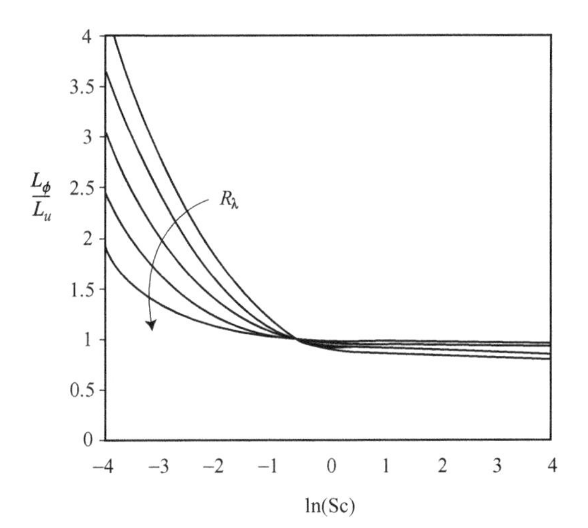

Figure 3.15. The scalar-to-velocity length-scale ratio for a fully developed scalar energy spectrum as a function of the Schmidt number at various Reynolds numbers: $R_\lambda = 50, 100, 200, 400,$ and 800. The arrow indicates the direction of increasing Reynolds number.

In Fig. 3.14, the mechanical-to-scalar time-scale ratio computed from the model scalar energy spectrum is plotted as a function of the Schmidt number at various Reynolds numbers. Consistent with (3.15), p. 61, for $1 \ll \mathrm{Sc}$ the mechanical-to-scalar time-scale ratio decreases with increasing Schmidt number as ln(Sc). Likewise, the scalar integral scale L_ϕ can be computed from the model spectrum. The ratio $L_\phi : L_u$ is plotted in Fig. 3.15, where it can be seen that it approaches unity at high Reynolds numbers.

Chasnov (1994) has carried out detailed studies of inert-scalar mixing at moderate Reynolds numbers using direct numerical simulations. He found that for decaying scalar fields the scalar spectrum at low wavenumbers is dependent on the initial scalar spectrum, and that this sensitivity is reflected in the mechanical-to-scalar time-scale ratio. Likewise, R is found to depend on both the Reynolds number and the Schmidt number in a non-trivial manner for decaying velocity and/or scalar fields (Chasnov 1991; Chasnov 1998).

The model scalar energy spectrum was derived for the limiting case of a fully developed scalar spectrum. As mentioned at the end of Section 3.1, in many applications the scalar energy spectrum cannot be assumed to be in spectral equilibrium. This implies that the mechanical-to-scalar time-scale ratio will depend on how the scalar spectrum was initialized, i.e., on $E_\phi(\kappa, 0)$. In order to compute R for non-equilibrium scalar mixing, we can make use of models based on the scalar spectral transport equation described below.

3.2.6 Scalar spectral transport

Like the turbulent energy spectrum discussed in Section 2.1, a transport equation can be derived for the scalar energy spectrum $E_\phi(\kappa, t)$ starting from (1.27) and (1.28) for an inert scalar (see McComb (1990) or Lesieur (1997) for details). The resulting equation is[21]

$$\frac{\partial E_\phi}{\partial t} = T_\phi(\kappa, t) - 2\Gamma\kappa^2 E_\phi(\kappa, t), \tag{3.73}$$

where $T_\phi(\kappa, t)$ is the unclosed *scalar spectral transfer function* involving three-point correlations between two values of the scalar field and one component of the velocity field. Like $T_u(\kappa, t)$, the scalar spectral transfer function is conservative:

$$\int_0^\infty T_\phi(\kappa, t)\, \mathrm{d}\kappa = 0. \tag{3.74}$$

Thus, the final term in (3.73) is responsible for scalar dissipation due to molecular diffusion at wavenumbers near κ_B.

The scalar spectral transport equation can be easily extended to the cospectrum of two inert scalars $E_{\alpha\beta}(\kappa, t)$. The resulting equation is

$$\frac{\partial E_{\alpha\beta}}{\partial t} = T_{\alpha\beta}(\kappa, t) - (\Gamma_\alpha + \Gamma_\beta)\kappa^2 E_{\alpha\beta}(\kappa, t), \tag{3.75}$$

where $T_{\alpha\beta}(\kappa, t)$ is the unclosed *cospectral transfer function* involving three-point correlations between one value of each scalar field and one component of the velocity field. Like $T_\phi(\kappa, t)$, the cospectral transfer is also conservative. Comparing (3.73) and (3.75), we can note that the 'effective' molecular diffusivity of the cospectrum is $(\Gamma_\alpha + \Gamma_\beta)/2$. This has important ramifications on differential diffusion effects, as will be seen in Section 3.4.

[21] This expression applies to the case where there is no mean scalar gradient. Adding a uniform mean scalar gradient generates an additional source term on the right-hand side involving the scalar-flux energy spectrum.

In order to understand better the physics of scalar spectral transport, it will again be useful to introduce the *scalar spectral energy transfer rates* $\mathcal{T}_\phi$ and $\mathcal{T}_{\alpha\beta}$ defined by

$$\mathcal{T}_\phi(\kappa, t) \equiv -\int_0^\kappa T_\phi(s, t)\,\mathrm{d}s \tag{3.76}$$

and

$$\mathcal{T}_{\alpha\beta}(\kappa, t) \equiv -\int_0^\kappa T_{\alpha\beta}(s, t)\,\mathrm{d}s. \tag{3.77}$$

In a fully developed turbulent flow,[22] the scalar spectral transfer rate in the inertial-convective sub-range is equal to the scalar dissipation rate, i.e., $\mathcal{T}_\phi(\kappa) = \varepsilon_\phi$ for $\kappa_{\mathrm{EI}} \leq \kappa \leq \kappa_{\mathrm{DI}}$. Likewise, when $\mathrm{Sc} \gg 1$, so that a viscous-convective sub-range exists, the scalar transfer rate will equal the scalar dissipation rate for wavenumbers in the viscous-convective sub-range, i.e., $\mathcal{T}_\phi(\kappa) = \varepsilon_\phi$ for $\kappa_{\mathrm{DI}} \leq \kappa \ll \kappa_{\mathrm{B}}$.

For a passive scalar, the turbulent flow will be unaffected by the presence of the scalar. This implies that for wavenumbers above the scalar dissipation range, the characteristic time scale for scalar spectral transport should be equal to that for velocity spectral transport τ_{st} defined by (2.67), p. 42. Thus, by equating the scalar and velocity spectral transport time scales, we have[23]

$$\frac{\kappa E_\phi(\kappa, t)}{\mathcal{T}_\phi(\kappa, t)} = \tau_{\mathrm{st}}(\kappa, t). \tag{3.78}$$

At spectral equilibrium, this expression can be rewritten as

$$E_\phi(\kappa) = \varepsilon_\phi \tau_{\mathrm{st}}(\kappa)\kappa^{-1} \quad \text{for } \kappa_{\mathrm{EI}} \leq \kappa \ll \kappa_{\mathrm{B}}. \tag{3.79}$$

For wavenumbers in the inertial range, $\tau_{\mathrm{st}}(\kappa)$ is given by (2.68), p. 42, and yields

$$E_\phi(\kappa) \propto \varepsilon_\phi \varepsilon^{-1/2}\kappa^{-5/3} \quad \text{for } \kappa_{\mathrm{EI}} \leq \kappa \leq \kappa_{\mathrm{DI}} \tag{3.80}$$

as expected for the inertial-convective sub-range. Likewise, setting $\tau_{\mathrm{st}}(\kappa) = \tau_\eta$ for $\kappa_\eta \leq \kappa \ll \kappa_{\mathrm{B}}$ yields the scalar spectrum for the viscous-convective sub-range. In general, models for $\mathcal{T}_\phi(\kappa, t)$ will be linear in $E_\phi(\kappa, t)$ and use a characteristic scalar spectral transport time scale defined in terms of the velocity spectrum (e.g., τ_{st}).

Following the approach used to derive (2.75), p. 43, the scalar spectral transport equation can also be used to generate a spectral model for the scalar dissipation rate for the case $1 \leq \mathrm{Sc}$.[24] Multiplying (3.73) by $2\Gamma\kappa^2$ yields the spectral transport equation for $D_\phi(\kappa, t)$:

$$\frac{\partial D_\phi}{\partial t} = 2\Gamma\kappa^2 T_\phi(\kappa, t) - 2\Gamma\kappa^2 D_\phi(\kappa, t). \tag{3.81}$$

[22] As before, 'fully developed' is taken to mean that both the velocity and scalar energy spectra are fully developed.

[23] It is tempting to try to use this expression as a simple *model* for $\mathcal{T}_\phi(\kappa, t)$ to close (3.75). However, the resultant transport equation is purely convective, and thus does not predict the correct time evolution for $E_\phi(\kappa, t)$.

[24] For $\mathrm{Sc} < 1$, a similar expression can be derived by taking $\kappa_{\mathrm{D}} = \kappa_{\mathrm{C}1}$. However, because $\kappa_{\mathrm{C}1} < \kappa_{\mathrm{DI}}$ lies in the inertial range, the characteristic spectral transport time for wavenumbers greater than $\kappa_{\mathrm{C}1}$ cannot be taken as constant.

Integration over all wavenumbers – with the same assumptions and manipulations that led to (2.75) – results in the following expression for ε_ϕ:

$$\frac{\mathrm{d}\varepsilon_\phi}{\mathrm{d}t} \approx 2\Gamma\kappa_\mathrm{D}^2 \mathcal{T}_\phi(\kappa_\mathrm{D}, t) + 2C_\mathrm{s}^\phi \left(\frac{\varepsilon}{\nu}\right)^{1/2} \varepsilon_\phi - \mathcal{D}_\varepsilon^\phi, \tag{3.82}$$

where the scalar-dissipation-dissipation rate $\mathcal{D}_\varepsilon^\phi$ is defined by

$$\mathcal{D}_\varepsilon^\phi = 2\int_0^\infty \Gamma\kappa^2 D_\phi(\kappa, t)\,\mathrm{d}\kappa, \tag{3.83}$$

and the wavenumber separating the viscous-convective sub-range and the viscous-diffusive sub-range is denoted by $\kappa_\mathrm{D} = \mathrm{Sc}^{1/2}\kappa_\mathrm{DI}$.[25]

Equation (3.82) illustrates the importance of the scalar spectral energy transfer rate in determining the scalar dissipation rate in high-Reynolds-number turbulent flows. Indeed, near spectral equilibrium, $\mathcal{T}_\phi(\kappa_\mathrm{D}, t)$ (like $\mathcal{T}_u(\kappa_\mathrm{DI}, t)$) will vary on time scales of the order of the eddy turnover time τ_e, while the characteristic time scale of (3.82) is $\tau_\eta \ll \tau_\mathrm{e}$. This implies that $\varepsilon_\phi(t)$ will be determined primarily by the 'slow' variations in $\mathcal{T}_\phi(\kappa_\mathrm{D}, t)$.

A spectral model similar to (3.82) can be derived from (3.75) for the *joint scalar dissipation rate* $\varepsilon_{\alpha\beta}$ defined by (3.139), p. 90. We will use these models in Section 3.4 to understand the importance of spectral transport in determining differential-diffusion effects. As we shall see in the next section, the spectral interpretation of scalar energy transport has important ramifications on the transport equations for one-point scalar statistics for inhomogeneous turbulent mixing.

3.3 Inhomogeneous turbulent mixing

Most of the results presented thus far in this chapter are strictly valid only for homogeneous turbulent mixing. However, turbulent mixing encountered in chemical-process equipment is almost always *inhomogeneous*. In order to model inhomogeneous turbulent mixing, transport equations are needed to describe how the various scalar quantities are transported from one location to another in the flow domain. In this section, transport equations for the most frequently employed scalar statistics are derived from the fundamental governing equations ((1.27) and (1.28)). As seen in Section 2.2, the non-linearities in the governing equations introduce unclosed terms that must be modeled. As discussed in Chapters 4 and 5, reacting-scalar transport models are fitted first to data for homogeneous inert-scalar mixing, and then validated for reacting flows. Nevertheless, as is the case with turbulence models, the model parameters must often be adjusted to simulate correctly complex flows of industrial interest.

Owing to the complexity of multi-point descriptions, almost all scalar transport models for complex flows are based on one-point statistics. As shown in Section 2.1, one-point turbulence statistics are found by integrating over the velocity sample space. Likewise,

[25] Note that $\Gamma\kappa_\mathrm{D}^2 = \nu\kappa_\mathrm{DI}^2$.

joint scalar, velocity statistics can be computed from the one-point joint velocity, composition PDF. For example, the scalar flux is defined by

$$\langle \mathbf{U}(\mathbf{x}, t)\phi(\mathbf{x}, t)\rangle \equiv \int_{-\infty}^{+\infty} \iiint_{-\infty}^{+\infty} \mathbf{V}\psi f_{\mathbf{U},\phi}(\mathbf{V}, \psi; \mathbf{x}, t) \, \mathrm{d}\mathbf{V} \, \mathrm{d}\psi. \tag{3.84}$$

Furthermore, space and time derivatives of mean quantities can be easily related to space and time derivatives of $f_{\mathbf{U},\phi}(\mathbf{V}, \psi; \mathbf{x}, t)$. For example, starting from (3.84), the time derivative of the scalar flux is given by

$$\frac{\partial \langle \mathbf{U}\phi\rangle}{\partial t} \equiv \int_{-\infty}^{+\infty} \iiint_{-\infty}^{+\infty} \mathbf{V}\psi \frac{\partial f_{\mathbf{U},\phi}}{\partial t}(\mathbf{V}, \psi; \mathbf{x}, t) \, \mathrm{d}\mathbf{V} \, \mathrm{d}\psi. \tag{3.85}$$

In general, given $f_{\mathbf{U},\phi}(\mathbf{V}, \psi; \mathbf{x}, t)$, transport equations for one-point statistics can be easily derived. This is the approach used in *transported PDF methods* as discussed in Chapter 6. In this section, as in Section 2.2, we will employ Reynolds averaging to derive the one-point transport equations for turbulent reacting flows.

3.3.1 Scalar mean

Starting with the scalar transport equation ((1.28), p. 16), Reynolds averaging leads to the transport equation for the scalar means:

$$\frac{\partial \langle \phi_\alpha \rangle}{\partial t} + \frac{\partial \langle U_j \phi_\alpha \rangle}{\partial x_j} = \Gamma_\alpha \nabla^2 \langle \phi_\alpha \rangle + \langle S_\alpha(\boldsymbol{\phi})\rangle. \tag{3.86}$$

The second term on the left-hand side of this expression can be further decomposed into two terms:

$$\frac{\partial \langle U_j \phi_\alpha \rangle}{\partial x_j} = \langle U_j \rangle \frac{\partial \langle \phi_\alpha \rangle}{\partial x_j} + \frac{\partial \langle u_j \phi_\alpha \rangle}{\partial x_j}. \tag{3.87}$$

The final form for the scalar mean transport equation in a turbulent reacting flow is given by

$$\frac{\partial \langle \phi_\alpha \rangle}{\partial t} + \langle U_j \rangle \frac{\partial \langle \phi_\alpha \rangle}{\partial x_j} = \Gamma_\alpha \nabla^2 \langle \phi_\alpha \rangle - \frac{\partial \langle u_j \phi_\alpha \rangle}{\partial x_j} + \langle S_\alpha(\boldsymbol{\phi})\rangle. \tag{3.88}$$

The first term on the right-hand side of this expression is the molecular transport term that scales as $\mathrm{Sc}_\alpha^{-1}\mathrm{Re}_L^{-1}$. Thus, at high Reynolds numbers,[26] it can be neglected. The two new unclosed terms in (3.88) are the scalar flux $\langle u_j \phi_\alpha \rangle$, and the mean chemical source term $\langle S_\alpha(\boldsymbol{\phi})\rangle$. For chemical reacting flows, the modeling of $\langle S_\alpha(\boldsymbol{\phi})\rangle$ is of greatest concern, and we discuss this aspect in detail in Chapter 5.

[26] The reader will recognize $\mathrm{Sc}\mathrm{Re}_L$ as the Péclet number. Here we will assume throughout that the Schmidt number is order unity or greater. Thus, high Reynolds number will imply high Péclet number.

For turbulent mixing of an inert scalar ϕ, the mean scalar transport equation reduces to

$$\frac{\partial \langle \phi \rangle}{\partial t} + \langle U_j \rangle \frac{\partial \langle \phi \rangle}{\partial x_j} = \Gamma \nabla^2 \langle \phi \rangle - \frac{\partial \langle u_j \phi \rangle}{\partial x_j}. \tag{3.89}$$

Thus, the closure problem reduces to finding an appropriate expression for the scalar flux $\langle u_j \phi \rangle$. In high-Reynolds-number turbulent flows, the molecular transport term is again negligible. Thus, the scalar-flux term is responsible for the rapid mixing observed in turbulent flows.

3.3.2 Scalar flux

Like the Reynolds stresses, the scalar flux obeys a transport equation that can be derived from the Navier–Stokes and scalar transport equations. We will first derive the transport equation for the scalar flux of an inert scalar from (2.99), p. 48, and the governing equation for inert-scalar fluctuations. The latter is found by subtracting (3.89) from (1.28) (p. 16), and is given by

$$\frac{\partial \phi'}{\partial t} + \langle U_k \rangle \frac{\partial \phi'}{\partial x_k} = \Gamma \nabla^2 \phi' - u_j \frac{\partial \phi'}{\partial x_j} - u_j \frac{\partial \langle \phi \rangle}{\partial x_j} + \frac{\partial \langle u_j \phi \rangle}{\partial x_j}. \tag{3.90}$$

The derivation of the scalar-flux transport equation proceeds in exactly the same manner as with the Reynolds stresses. We first multiply (2.99), p. 48, by ϕ' and (3.90) by u_i. By adding the resulting two expressions, we find

$$\begin{aligned}
\frac{\partial u_i \phi'}{\partial t} + \langle U_j \rangle \frac{\partial u_i \phi'}{\partial x_j} &= \Gamma u_i \nabla^2 \phi' + \nu \phi' \nabla^2 u_i - \phi' \frac{1}{\rho} \frac{\partial p'}{\partial x_i} \\
&\quad - u_i u_j \frac{\partial \phi'}{\partial x_j} - \phi' u_j \frac{\partial u_i}{\partial x_j} \\
&\quad - u_i u_j \frac{\partial \langle \phi \rangle}{\partial x_j} - \phi' u_j \frac{\partial \langle U_i \rangle}{\partial x_j} \\
&\quad + u_i \frac{\partial \langle u_j \phi \rangle}{\partial x_j} + \phi' \frac{\partial \langle u_j u_i \rangle}{\partial x_j}.
\end{aligned} \tag{3.91}$$

This expression can be further simplified by making use of the following identities:

$$\frac{\partial}{\partial x_j}\left(u_i \frac{\partial \phi'}{\partial x_j} \right) = u_i \nabla^2 \phi' + \frac{\partial u_i}{\partial x_j} \frac{\partial \phi'}{\partial x_j}, \tag{3.92}$$

$$\frac{\partial}{\partial x_j}\left(\phi' \frac{\partial u_i}{\partial x_j} \right) = \phi' \nabla^2 u_i + \frac{\partial u_i}{\partial x_j} \frac{\partial \phi'}{\partial x_j}, \tag{3.93}$$

and

$$\frac{\partial u_j u_i \phi'}{\partial x_j} = u_i \frac{\partial u_j \phi'}{\partial x_j} + \phi' \frac{\partial u_j u_i}{\partial x_j}. \tag{3.94}$$

Reynolds averaging of (3.91) then yields the transport equation for the inert-scalar flux:[27]

$$\frac{\partial \langle u_i \phi \rangle}{\partial t} + \langle U_j \rangle \frac{\partial \langle u_i \phi \rangle}{\partial x_j} = \frac{\partial}{\partial x_j} \left(\mathcal{T}_{ij}^{\phi} - \langle u_j u_i \phi \rangle \right) + \mathcal{P}_i^{\phi} + \Pi_i^{\phi} - \varepsilon_i^{\phi}. \tag{3.95}$$

Note that if the Reynolds stresses are known, then the production term $\mathcal{P}_i^{\phi}$ defined by

$$\mathcal{P}_i^{\phi} \equiv - \langle u_i u_j \rangle \frac{\partial \langle \phi \rangle}{\partial x_j} - \langle u_j \phi \rangle \frac{\partial \langle U_i \rangle}{\partial x_j} \tag{3.96}$$

will be closed. Thus, the scalar-flux transport equation is most useful in conjunction with turbulence models that directly solve for the Reynolds stresses. Note also that scalar flux can be generated either by turbulent fluctuations acting on the mean scalar gradient, or by a mean velocity gradient. All other terms on the right-hand side of (3.95) are unclosed and are defined as follows.

The triple-correlation term $\langle u_j u_i \phi' \rangle$ and the molecular-transport term $\mathcal{T}_{ij}^{\phi}$, defined by

$$\mathcal{T}_{ij}^{\phi} \equiv \nu \left\langle \phi' \frac{\partial u_i}{\partial x_j} \right\rangle + \Gamma \left\langle u_i \frac{\partial \phi'}{\partial x_j} \right\rangle, \tag{3.97}$$

are responsible for spatial transport of scalar flux. For homogeneous flows, these terms make no contribution to (3.95). In high-Reynolds-number flows, the molecular-transport term will be negligible. Thus, the triple-correlation term is primarily responsible for scalar-flux transport in high-Reynolds-number turbulent flows.

The pressure-scrambling term Π_i^{ϕ}, defined by

$$\Pi_i^{\phi} \equiv - \left\langle \frac{\phi'}{\rho} \frac{\partial p}{\partial x_i} \right\rangle, \tag{3.98}$$

is, like the related term in the Reynolds stress transport equation, a dominant term in the scalar-flux transport equation. This term can be decomposed into a pressure-diffusion term $\langle p \phi' \rangle$ and a pressure-scalar-gradient term $\mathcal{R}_i^{\phi}$:

$$\Pi_i^{\phi} = \mathcal{R}_i^{\phi} - \frac{1}{\rho} \frac{\partial \langle p \phi' \rangle}{\partial x_i}, \tag{3.99}$$

where

$$\mathcal{R}_i^{\phi} \equiv \left\langle \frac{p'}{\rho} \frac{\partial \phi}{\partial x_i} \right\rangle. \tag{3.100}$$

As we shall see in Chapter 4, models for $\mathcal{R}_i^{\phi}$ have much in common with those used for $\mathcal{R}_{ij}$ in the Reynolds stress transport equation. Indeed, as shown using transported PDF methods in Chapter 6, the model for $\mathcal{R}_{ij}$ uniquely determines the model for $\mathcal{R}_i^{\phi}$.

[27] Note that $\langle u_i \phi \rangle = \langle u_i \phi' \rangle + \langle u_i \rangle \langle \phi \rangle = \langle u_i \phi' \rangle$.

The final term on the right-hand side of (3.95) is the scalar-flux dissipation ε_i^ϕ defined by

$$\varepsilon_i^\phi \equiv (\nu + \Gamma)\left\langle \frac{\partial u_i}{\partial x_j}\frac{\partial \phi'}{\partial x_j}\right\rangle. \tag{3.101}$$

In locally isotropic turbulence, the fluctuating velocity gradient and scalar gradient will be uncorrelated, and ε_i^ϕ will be null. Thus, at sufficiently high Reynolds number, the scalar-flux dissipation is negligible.

The transport equation for the scalar flux of a reacting scalar ϕ_α can be derived following the same steps as those leading up to (3.95). The final expression reduces to

$$\frac{\partial \langle u_i\phi_\alpha\rangle}{\partial t} + \langle U_j\rangle\frac{\partial \langle u_i\phi_\alpha\rangle}{\partial x_j} = \frac{\partial}{\partial x_j}\left(\mathcal{T}_{ij}^\alpha - \langle u_j u_i {\phi'}_\alpha\rangle\right) + \mathcal{P}_i^\alpha + \Pi_i^\alpha - \varepsilon_i^\alpha + \langle u_i S_\alpha(\phi)\rangle, \tag{3.102}$$

where $\mathcal{T}_{ij}^\alpha$, $\mathcal{P}_i^\alpha$, Π_i^α, and ε_i^α are found from the corresponding terms in the inert-scalar-flux transport equation by replacing ϕ with ϕ_α. The new unclosed term in (3.102) is $\langle u_i S_\alpha\rangle$, which describes the correlation between the fluctuating velocity and chemical source term. Note that this term can lead to strong coupling between the scalar fields due to chemical interactions.

Only for an isothermal, first-order reaction where $S_\alpha = -k_1\phi_\alpha$ will the chemical source term in (3.102) be closed, i.e., $\langle u_i S_\alpha(\phi)\rangle = -k_1\langle u_i\phi_\alpha\rangle$. Indeed, for more complex chemistry, closure of the chemical source term in the scalar-flux transport equation is a major challenge. However, note that, unlike the scalar-flux dissipation term, which involves the correlation between gradients (and hence two-point statistical information), the chemical source term is given in terms of $\mathbf{u}(\mathbf{x}, t)$ and $\phi(\mathbf{x}, t)$. Thus, given the one-point joint velocity, composition PDF $f_{\mathbf{U},\phi}(\mathbf{V}, \psi; \mathbf{x}, t)$, the chemical source term is closed, and can be computed from

$$\langle u_i S_\alpha\rangle \equiv \int_{-\infty}^{+\infty}\cdots\int_{-\infty}^{+\infty}\iiint (V_i - \langle U_i\rangle) S_\alpha(\psi) f_{\mathbf{U},\phi}(\mathbf{V}, \psi; \mathbf{x}, t)\,\mathrm{d}\mathbf{V}\,\mathrm{d}\psi. \tag{3.103}$$

This is the approach taken in transported PDF methods, as discussed in detail in Chapter 6. Most of the other closure methods discussed in Chapter 5 require knowledge of the scalar variance, which can be found from a transport equation as shown next.

3.3.3 Scalar variance

The transport equation for the variance of an inert scalar $\langle \phi'^2\rangle$ can be easily derived by multiplying (3.90) by $2\phi'$, and Reynolds averaging the resultant expression. This process leads to an unclosed term of the form $2\phi'\nabla^2\phi'$, which can be rewritten as

$$2\langle \phi'\nabla^2\phi'\rangle = \nabla^2\langle \phi'^2\rangle - 2\left\langle \frac{\partial \phi'}{\partial x_i}\frac{\partial \phi'}{\partial x_i}\right\rangle. \tag{3.104}$$

The inert-scalar-variance transport equation can then be written as

$$\frac{\partial \langle \phi'^2 \rangle}{\partial t} + \langle U_j \rangle \frac{\partial \langle \phi'^2 \rangle}{\partial x_j} = \Gamma \nabla^2 \langle \phi'^2 \rangle - \frac{\partial \langle u_j \phi'^2 \rangle}{\partial x_j} + \mathcal{P}_\phi - \varepsilon_\phi. \tag{3.105}$$

The molecular-transport term $\Gamma \nabla^2 \langle \phi'^2 \rangle$ will be negligible at high Reynolds number. The scalar-variance-production term $\mathcal{P}_\phi$ is defined by

$$\mathcal{P}_\phi \equiv -2 \langle u_j \phi \rangle \frac{\partial \langle \phi \rangle}{\partial x_j}. \tag{3.106}$$

Thus, (3.105) has three unclosed terms: the scalar flux $\langle u_j \phi \rangle$, the scalar variance flux $\langle u_j \phi'^2 \rangle$, and the scalar dissipation rate ε_ϕ defined by

$$\varepsilon_\phi \equiv 2\Gamma \left\langle \frac{\partial \phi'}{\partial x_i} \frac{\partial \phi'}{\partial x_i} \right\rangle. \tag{3.107}$$

For homogeneous scalar mixing, the mean scalar gradient can be either null or constant, i.e., of the form:

$$\frac{\partial \langle \phi \rangle}{\partial x_i} = \beta_i.$$

Furthermore, in stationary isotropic turbulence the scalar flux is related to the mean scalar gradient by

$$\langle u_i \phi \rangle = -D_{\mathrm{T}} \frac{k^2}{\varepsilon} \beta_i = -\Gamma_{\mathrm{T}} \beta_i, \tag{3.108}$$

where D_{T} is the 'turbulent diffusivity' constant. Under these conditions, (3.105) reduces to

$$\frac{\mathrm{d} \langle \phi'^2 \rangle}{\mathrm{d}t} = 2\Gamma_{\mathrm{T}} \beta_i \beta_i - \varepsilon_\phi. \tag{3.109}$$

The first term on the right-hand side is always positive and represents a source of scalar variance due to turbulent mixing in the presence of a mean scalar gradient. On the other hand, the second term is always negative and represents molecular dissipation of scalar variance near the Batchelor scale. In the chemical-reaction-engineering literature, the first term has been referred to as *mesomixing* to emphasize the fact that it occurs primarily due to turbulent fluctuations in the energy-containing range of the velocity spectrum. Likewise, the second term is referred to as *micromixing* because it is a small-scale term controlled by scalar gradient correlations (see (3.107)). For inhomogeneous scalar mixing, the mean velocity and spatial transport terms in (3.105) will also be important.[28] The scalar-variance transport equation thus provides a systematic and complete representation of inert-scalar mixing in turbulent flows.

As discussed in Chapter 4, the modeling of the scalar dissipation rate in (3.105) is challenging due to the need to describe both equilibrium and non-equilibrium spectral

[28] As noted in Chapter 1, these processes are referred to as *macromixing* in the chemical-reaction-engineering literature.

transport. One starting point is the transport equation for the scalar dissipation rate, which we look at next.

3.3.4 Scalar dissipation rate

The transport equation for the scalar dissipation rate of an inert scalar can be derived starting from (3.90). We begin by defining the fluctuating scalar gradient as

$$g_i \equiv \frac{\partial \phi'}{\partial x_i}. \tag{3.110}$$

Differentiating both sides of (3.90) with respect to x_i, and multiplying the resultant expression by $2g_i$, then yields

$$\begin{aligned}\frac{\partial g_i g_i}{\partial t} + \langle U_j \rangle \frac{\partial g_i g_i}{\partial x_j} = 2\Gamma g_i \nabla^2 g_i - u_j \frac{\partial g_i g_i}{\partial x_j} - 2g_j g_i \frac{\partial u_j}{\partial x_i} \\ - 2g_j g_i \frac{\partial \langle U_j \rangle}{\partial x_i} - 2g_i \frac{\partial u_j}{\partial x_i} \frac{\partial \langle \phi \rangle}{\partial x_j} \\ - 2u_j g_i \frac{\partial^2 \langle \phi \rangle}{\partial x_j \partial x_i} + 2g_i \frac{\partial^2 \langle u_j \phi \rangle}{\partial x_i \partial x_j}.\end{aligned} \tag{3.111}$$

Next we define the fluctuating scalar dissipation rate ϵ_ϕ by

$$\epsilon_\phi \equiv 2\Gamma g_i g_i. \tag{3.112}$$

We can then rewrite the first term on the right-hand side of (3.111) as follows:

$$2\Gamma g_i \nabla^2 g_i = \frac{1}{2} \nabla^2 \epsilon_\phi - 2\Gamma \frac{\partial^2 \phi'}{\partial x_i \partial x_j} \frac{\partial^2 \phi'}{\partial x_i \partial x_j}. \tag{3.113}$$

Note that in the turbulent mixing literature, the scalar dissipation rate is often defined without the factor of two in (3.112). Likewise, in the combustion literature, the symbol χ_ϕ is used in place of ϵ_ϕ in (3.112). In this book, we will consistently use ϵ_ϕ to denote the fluctuating scalar dissipation rate, and $\varepsilon_\phi = \langle \epsilon_\phi \rangle$ to denote the scalar dissipation rate.

Multiplying (3.111) by 2Γ and Reynolds averaging yields the final form for the scalar-dissipation-rate transport equation:

$$\frac{\partial \varepsilon_\phi}{\partial t} + \langle U_j \rangle \frac{\partial \varepsilon_\phi}{\partial x_j} = \Gamma \nabla^2 \varepsilon_\phi - \frac{\partial \langle u_j \epsilon_\phi \rangle}{\partial x_j} + \mathcal{S}_\varepsilon^\phi + \mathcal{G}_\varepsilon^\phi + \mathcal{C}_\varepsilon^\phi + \mathcal{V}_\varepsilon^\phi - \mathcal{D}_\varepsilon^\phi. \tag{3.114}$$

The first two terms on the right-hand side of this expression are responsible for spatial transport of scalar dissipation. In high-Reynolds-number turbulent flows, the scalar-dissipation flux $\langle u_j \epsilon_\phi \rangle$ is the dominant term. The other terms on the right-hand side are similar to the corresponding terms in the dissipation transport equation ((2.125), p. 52), and are defined as follows.

The mean-velocity-gradient term $\mathcal{S}_\varepsilon^\phi$ is defined by

$$\mathcal{S}_\varepsilon^\phi \equiv -4\Gamma \left\langle \frac{\partial \phi'}{\partial x_i} \frac{\partial \phi'}{\partial x_j} \right\rangle \frac{\partial \langle U_j \rangle}{\partial x_i}. \tag{3.115}$$

If the small scales of the scalar field are locally isotropic, then

$$4\Gamma\left\langle\frac{\partial\phi'}{\partial x_i}\frac{\partial\phi'}{\partial x_j}\right\rangle = \frac{2}{3}\varepsilon_\phi\delta_{ij}. \tag{3.116}$$

Applying this expression in (3.115), and using the continuity equation for the mean velocity, yields $\mathcal{S}_\varepsilon^\phi = 0$. Thus, in high-Reynolds-number flows,[29] $\mathcal{S}_\varepsilon^\phi$ will be negligible.

The mean-scalar-gradient term $\mathcal{G}_\varepsilon^\phi$ is defined by

$$\mathcal{G}_\varepsilon^\phi \equiv -4\Gamma\left\langle\frac{\partial\phi'}{\partial x_i}\frac{\partial u_j}{\partial x_i}\right\rangle\frac{\partial\langle\phi\rangle}{\partial x_j}. \tag{3.117}$$

At high Reynolds numbers, the scalar gradient and the velocity gradient will be uncorrelated due to local isotropy. Thus, $\mathcal{G}_\varepsilon^\phi$ will be negligible. Likewise, the mean-scalar-curvature term $\mathcal{C}_\varepsilon^\phi$, defined by

$$\mathcal{C}_\varepsilon^\phi \equiv -4\left\langle u_j\frac{\partial\phi'}{\partial x_i}\right\rangle\Gamma\frac{\partial^2\langle\phi\rangle}{\partial x_j\partial x_i}, \tag{3.118}$$

will also be negligible at high Reynolds numbers due to the presence of Γ on the right-hand side.

The remaining two terms in (3.114) will be large in high-Reynolds-number turbulent flows. The vortex-stretching term $\mathcal{V}_\varepsilon^\phi$ is defined by

$$\mathcal{V}_\varepsilon^\phi \equiv -4\Gamma\left\langle\frac{\partial\phi'}{\partial x_j}\frac{\partial u_j}{\partial x_i}\frac{\partial\phi'}{\partial x_i}\right\rangle. \tag{3.119}$$

The right-hand side can be rewritten in terms of the fluctuating rate-of-strain tensor s_{ij} ((2.50), p. 38):

$$\mathcal{V}_\varepsilon^\phi = -4\Gamma\left\langle\frac{\partial\phi'}{\partial x_j}s_{ij}\frac{\partial\phi'}{\partial x_i}\right\rangle. \tag{3.120}$$

The second-order fluctuating rate-of-strain tensor is real and symmetric. Thus, its three eigenvalues are real and, due to continuity, sum to zero. The latter implies that one eigenvalue (α) is always positive, and one eigenvalue (γ) is always negative. In the turbulence literature (Pope 2000), γ is referred to as the most compressive strain rate.

Direct numerical simulation studies of turbulent mixing (e.g., Ashurst *et al.* 1987) have shown that the fluctuating scalar gradient is nearly always aligned with the eigenvector of the most compressive strain rate. For a fully developed scalar spectrum, the vortex-stretching term can be expressed as

$$\mathcal{V}_\varepsilon^\phi = C_\mathcal{V}^\phi|\gamma|\varepsilon_\phi, \tag{3.121}$$

where $C_\mathcal{V}^\phi$ is a constant of order unity. Moreover, because it describes the small scales of the turbulent flow,[30] γ will be proportional to $1/\tau_\eta$. Thus, for a fully developed scalar

[29] In addition to the Reynolds number, local isotropy for the scalar field will depend on the Schmidt number: Sc must be large enough to allow for a inertial-convective sub-range in the scalar energy spectrum.

[30] Strictly speaking, this scaling is valid for $1 \leq \mathrm{Sc}$. For smaller Schmidt numbers, an inertial-range time scale can be used.

spectrum, the vortex-stretching term scales with Reynolds number as

$$\mathcal{V}_\varepsilon^\phi \sim \mathrm{Re}_L^{1/2} \frac{\varepsilon}{k} \varepsilon_\phi. \tag{3.122}$$

We extend this model to non-fully developed scalar spectra in (3.130) below.

The gradient-dissipation term $\mathcal{D}_\varepsilon^\phi$ is defined by

$$\mathcal{D}_\varepsilon^\phi \equiv 4 \left\langle \left(\Gamma \frac{\partial^2 \phi'}{\partial x_i \partial x_j} \right) \left(\Gamma \frac{\partial^2 \phi'}{\partial x_i \partial x_j} \right) \right\rangle. \tag{3.123}$$

Like its counterpart $\mathcal{D}_\varepsilon$ in the turbulence-dissipation transport equation, $\mathcal{D}_\varepsilon^\phi$ can be written as

$$\mathcal{D}_\varepsilon^\phi = C_\mathcal{D}^\phi \frac{\varepsilon_\phi}{\langle \phi'^2 \rangle_\mathrm{D}} \varepsilon_\phi, \tag{3.124}$$

where the fraction of the scalar variance in the scalar dissipation range $\langle \phi'^2 \rangle_\mathrm{D}$ is defined by

$$\langle \phi'^2 \rangle_\mathrm{D} \equiv \int_{\kappa_\mathrm{D}}^\infty E_\phi(\kappa, t)\, \mathrm{d}\kappa. \tag{3.125}$$

The scalar-dissipation wavenumber κ_D is defined in terms of κ_DI by $\kappa_\mathrm{D} = \mathrm{Sc}^{1/2} \kappa_\mathrm{DI}$.

Like the fraction of the turbulent kinetic energy in the dissipation range k_D ((2.139), p. 54), for a fully developed scalar spectrum the fraction of scalar variance in the scalar dissipation range scales with Reynolds number as

$$\langle \phi'^2 \rangle_\mathrm{D} \sim \mathrm{Re}_L^{-1/2} \langle \phi'^2 \rangle. \tag{3.126}$$

It then follows under the same conditions that the gradient-dissipation term scales as

$$\mathcal{D}_\varepsilon^\phi \sim \mathrm{Re}_L^{1/2} \frac{\varepsilon_\phi}{\langle \phi'^2 \rangle} \varepsilon_\phi. \tag{3.127}$$

Combining (3.122) and (3.127), the scalar dissipation rate for homogeneous turbulent mixing can be expressed as[31]

$$\frac{\mathrm{d}\varepsilon_\phi}{\mathrm{d}t} \approx \mathcal{V}_\varepsilon^\phi - \mathcal{D}_\varepsilon^\phi = C_\mathcal{V}^\phi \mathrm{Re}_L^{1/2} \frac{\varepsilon}{k} \varepsilon_\phi - C_\mathcal{D}^\phi \mathrm{Re}_L^{1/2} \frac{\varepsilon_\phi}{\langle \phi'^2 \rangle} \varepsilon_\phi. \tag{3.128}$$

For large Reynolds numbers, the right-hand side of this expression will be large, thereby forcing the scalar dissipation rate to attain a stationary solution quickly. Thus, for a fully developed scalar spectrum, the scalar mixing rate is related to the turbulent frequency by

$$\frac{\varepsilon_\phi}{\langle \phi'^2 \rangle} = C_\phi \frac{\varepsilon}{k}, \tag{3.129}$$

where $C_\phi = C_\mathcal{V}^\phi / C_\mathcal{D}^\phi$ is a constant of order unity. As noted earlier, this is the 'equilibrium' closure model for the scalar dissipation rate in CFD calculations of high-Reynolds-number turbulent mixing.

When the scalar spectrum is not fully developed, the vortex-stretching term $\mathcal{V}_\varepsilon^\phi$ will depend on the scalar spectral energy transfer rate evaluated at the scalar-dissipation wavenumber $\mathcal{T}_\phi(\kappa_\mathrm{D}, t)$.[32] Like the vortex-stretching term $\mathcal{V}_\varepsilon$ appearing in the transport

[31] We have also assumed that the turbulence is homogeneous and fully developed.

[32] For a fully developed scalar spectrum in stationary, homogeneous turbulence, $\mathcal{T}_\phi(\kappa_\mathrm{D}) = \varepsilon_\phi$.

Table 3.2. *The scalar statistics and unclosed quantities appearing in the transport equations for inhomogeneous turbulent mixing of an inert scalar.*

Physical quantity	Transport equation	Unclosed terms
Scalar mean $\langle\phi\rangle$	(3.89), p. 82	$\langle u_i\phi\rangle$
Scalar flux $\langle u_i\phi\rangle$	(3.90), p. 82	$\langle u_iu_j\rangle$, $\langle u_iu_j\phi'\rangle$, Π_i^ϕ
Scalar variance $\langle\phi'^2\rangle$	(3.105), p. 85	$\langle u_i\phi\rangle$, $\langle u_i\phi'^2\rangle$, ε_ϕ
Scalar dissipation rate ε_ϕ	(3.114), p. 86	$\langle u_i\epsilon_\phi\rangle$, $\mathcal{V}_\varepsilon^\phi$, $\mathcal{D}_\varepsilon^\phi$

equation for the turbulence dissipation rate ((2.135), p. 54), $\mathcal{V}_\varepsilon^\phi$ can be expressed as

$$\mathcal{V}_\varepsilon^\phi = C_{\mathcal{V}1}^\phi \mathrm{Re}_L^{1/2}\frac{1}{\tau_\mathrm{e}}\mathcal{T}_\phi(\kappa_\mathrm{D}, t) + C_{\mathcal{V}2}^\phi\left(\frac{\epsilon}{\nu}\right)^{1/2}\varepsilon_\phi, \tag{3.130}$$

where the first term on the right-hand side corresponds to the rate of transfer of scalar spectral energy into the scalar dissipation range, and the second term describes vortex stretching in the scalar dissipation range.[33]

Using (3.130), the transport equation for the scalar dissipation rate in high-Reynolds-number homogeneous turbulence becomes

$$\frac{\mathrm{d}\varepsilon_\phi}{\mathrm{d}t} \approx C_{\mathcal{V}1}^\phi \mathrm{Re}_L^{1/2}\frac{1}{\tau_\mathrm{e}}\mathcal{T}_\phi(\kappa_\mathrm{D}, t) + C_{\mathcal{V}2}^\phi\left(\frac{\varepsilon}{\nu}\right)^{1/2}\varepsilon_\phi - C_\mathcal{D}^\phi\frac{\varepsilon_\phi}{\langle\phi'^2\rangle_\mathrm{D}}\varepsilon_\phi, \tag{3.131}$$

where the fraction of the scalar variance in the scalar dissipation range is found from

$$\frac{\mathrm{d}\langle\phi'^2\rangle_\mathrm{D}}{\mathrm{d}t} \approx \mathcal{T}_\phi(\kappa_\mathrm{D}, t) - \varepsilon_\phi. \tag{3.132}$$

The scalar spectral energy transfer rate $\mathcal{T}_\phi(\kappa_\mathrm{D}, t)$ will vary on time scales proportional to the eddy turnover time τ_e. At high Reynolds numbers, (3.131) and (3.132) quickly attain a quasi-steady state wherein

$$\varepsilon_\phi = \mathcal{T}_\phi(\kappa_\mathrm{D}, t) \tag{3.133}$$

and

$$\langle\phi'^2\rangle_\mathrm{D} \propto \tau_\mathrm{e}\mathcal{T}_\phi(\kappa_\mathrm{D}, t)\mathrm{Re}_L^{-1/2}. \tag{3.134}$$

Thus, like the turbulence dissipation rate, the scalar dissipation rate of an inert scalar is primarily determined by the rate at which spectral energy enters the scalar dissipation range. Most engineering models for the scalar dissipation rate attempt to describe $\mathcal{T}_\phi(\kappa_\mathrm{D}, t)$ in terms of one-point turbulence statistics. We look at some of these models in Chapter 4.

The scalar statistics used in engineering calculations of high-Reynolds-number turbulent mixing of an inert scalar are summarized in Table 3.2 along with the unclosed terms that appear in their transport equations. In Chapter 4, we will discuss methods for modeling the unclosed terms in the RANS transport equations.

[33] Again, the use of the Kolmogorov time scale in the second term implies that $1 \leq \mathrm{Sc}$.

3.3.5 Scalar covariance

As discussed in Chapter 5, the complexity of the chemical source term restricts the applicability of closures based on second- and higher-order moments of the scalars. Nevertheless, it is instructive to derive the scalar covariance equation for two scalars ϕ_α and ϕ_β with molecular-diffusion coefficients Γ_α and Γ_β, respectively. Starting from (1.28), p. 16, the transport equation for $\langle \phi'_\alpha \phi'_\beta \rangle$ can be found following the same steps that were used for the Reynolds stresses. This process yields[34]

$$\begin{aligned}\frac{\partial \phi'_\alpha \phi'_\beta}{\partial t} + \langle U_i \rangle \frac{\partial \phi'_\alpha \phi'_\beta}{\partial x_i} = {} & \Gamma_\alpha \phi'_\beta \nabla^2 \phi'_\alpha + \Gamma_\beta \phi'_\alpha \nabla^2 \phi'_\beta - \frac{\partial u_i \phi'_\alpha \phi'_\beta}{\partial x_i} \\ & + \phi'_\beta u_i \frac{\partial \langle \phi_\alpha \rangle}{\partial x_i} + \phi'_\alpha u_i \frac{\partial \langle \phi_\beta \rangle}{\partial x_i} \\ & + \phi'_\beta \frac{\partial \langle u_i \phi_\alpha \rangle}{\partial x_i} + \phi'_\alpha \frac{\partial \langle u_i \phi_\beta \rangle}{\partial x_i} \\ & + \phi'_\beta S'_\alpha(\boldsymbol{\phi}) + \phi'_\alpha S'_\beta(\boldsymbol{\phi}),\end{aligned} \tag{3.135}$$

where the fluctuating chemical source term is defined by $\mathbf{S}'(\boldsymbol{\phi}) \equiv \mathbf{S}(\boldsymbol{\phi}) - \langle \mathbf{S}(\boldsymbol{\phi}) \rangle$.

Reynolds averaging of (3.135) leads to terms of the form $\langle \phi'_\beta \nabla^2 \phi'_\alpha \rangle$, which can be rewritten as in (3.92). The final covariance transport equation can be expressed as

$$\frac{\partial \langle \phi'_\alpha \phi'_\beta \rangle}{\partial t} + \langle U_i \rangle \frac{\partial \langle \phi'_\alpha \phi'_\beta \rangle}{\partial x_i} = \frac{\partial}{\partial x_i} \left(\mathcal{T}_i^{\alpha\beta} - \langle u_i \phi'_\alpha \phi'_\beta \rangle \right) + \mathcal{P}_{\alpha\beta} - \gamma_{\alpha\beta} \varepsilon_{\alpha\beta} + S_{\alpha\beta}. \tag{3.136}$$

The covariance-production term $\mathcal{P}_{\alpha\beta}$ is defined by

$$\mathcal{P}_{\alpha\beta} \equiv - \langle u_i \phi_\alpha \rangle \frac{\partial \langle \phi_\beta \rangle}{\partial x_i} - \langle u_i \phi_\beta \rangle \frac{\partial \langle \phi_\alpha \rangle}{\partial x_i}, \tag{3.137}$$

and will be closed if the scalar fluxes are known. However, all other terms on the right-hand side of (3.136) will be unclosed and are defined as follows.

Spatial transport of scalar covariance is described by the triple-correlation term $\langle u_i \phi'_\alpha \phi'_\beta \rangle$, and the molecular-transport term $\mathcal{T}_i^{\alpha\beta}$ defined by

$$\mathcal{T}_i^{\alpha\beta} \equiv \Gamma_\alpha \left\langle \phi'_\beta \frac{\partial \phi'_\alpha}{\partial x_i} \right\rangle + \Gamma_\beta \left\langle \phi'_\alpha \frac{\partial \phi'_\beta}{\partial x_i} \right\rangle. \tag{3.138}$$

In high-Reynolds-number turbulent flows, $\mathcal{T}_i^{\alpha\beta}$ is negligible, and thus the triple-correlation term is largely responsible for spatial transport of scalar covariance.

The 'dissipation' term[35] in (3.136) is written as the product of the joint scalar dissipation rate $\varepsilon_{\alpha\beta}$ defined by

$$\varepsilon_{\alpha\beta} \equiv 2(\Gamma_\alpha \Gamma_\beta)^{1/2} \left\langle \frac{\partial \phi'_\alpha}{\partial x_i} \frac{\partial \phi'_\beta}{\partial x_i} \right\rangle, \tag{3.139}$$

[34] As usual, no summation is implied with respect to Greek-letter indices.

[35] When $\alpha \neq \beta$, $\varepsilon_{\alpha\beta}$ may be negative. This generally occurs, for example, when the covariance $\langle \phi'_\alpha \phi'_\beta \rangle$ is negative.

and the arithmetic-to-geometric-diffusivity ratio $\gamma_{\alpha\beta}$ defined by[36]

$$\gamma_{\alpha\beta} \equiv \frac{\Gamma_\alpha + \Gamma_\beta}{2(\Gamma_\alpha \Gamma_\beta)^{1/2}} = \frac{1}{2}\left(\frac{\Gamma_\alpha}{\Gamma_\beta}\right)^{1/2} + \frac{1}{2}\left(\frac{\Gamma_\beta}{\Gamma_\alpha}\right)^{1/2}. \tag{3.140}$$

Unlike the turbulence dissipation rate tensor, which is isotropic at high Reynolds number, the joint scalar dissipation rate tensor is usually highly anisotropic. Indeed, when $\Gamma_\alpha = \Gamma_\beta$, it is often the case for inert scalars that $\varepsilon_{\alpha\beta} = \varepsilon_{\alpha\alpha} = \varepsilon_\phi$, so that the joint scalar dissipation rate tensor is singular.

The physics associated with the joint scalar dissipation rate is further complicated by two additional factors:

(i) differential-diffusion effects that occur when $\Gamma_\alpha \neq \Gamma_\beta$;
(ii) gradient coupling caused by small-scale structure formation due to chemical reactions.

The first factor occurs even in homogeneous flows with two inert scalars, and is discussed in Section 3.4. The second factor is present in nearly all turbulent reacting flows with moderately fast chemistry. As discussed in Chapter 4, modeling the joint scalar dissipation rate is challenging due to the need to include all important physical processes. One starting point is its transport equation, which we derive below.

The final term in (3.136) is the covariance chemical source term $S_{\alpha\beta}$ defined by

$$S_{\alpha\beta} \equiv \langle \phi'_\beta S_\alpha(\boldsymbol{\phi})\rangle + \langle \phi'_\alpha S_\beta(\boldsymbol{\phi})\rangle. \tag{3.141}$$

For complex chemical source terms, this expression generates new unclosed terms that are particularly difficult to model. Even for an isothermal, second-order reaction with

$$S_1 = -k_1\phi_1\phi_2, \tag{3.142}$$

$$S_2 = -k_1\phi_1\phi_2, \tag{3.143}$$

triple-correlation terms of the form $\langle \phi'^2_1 \phi'_2\rangle$ and $\langle \phi'_1 \phi'^2_2\rangle$ are generated in the covariance chemical source term:

$$S_{12} = -k_1\left(\langle \phi'^2_1 \phi'_2\rangle + 2\langle \phi'^2_1\rangle\langle\phi_2\rangle + 2\langle\phi_1\rangle\langle\phi'^2_2\rangle + \langle \phi'_1\phi'^2_2\rangle\right). \tag{3.144}$$

In Chapter 5, we will review models referred to as *moment methods*, which attempt to close the chemical source term by expressing the unclosed higher-order moments in terms of lower-order moments. However, in general, such models are of limited applicability. On the other hand, transported PDF methods (discussed in Chapter 6) treat the chemical source term exactly.

[36] Note that $\gamma_{\alpha\beta}$ depends only on the ratio of the molecular diffusivities. Moreover, its value is the same for either ratio, i.e., $\Gamma_\alpha/\Gamma_\beta$ or $\Gamma_\beta/\Gamma_\alpha$.

3.3.6 Joint scalar dissipation rate

The transport equation for the joint scalar dissipation rate can be derived following the same steps used to derive (3.110). We begin by defining the fluctuating scalar gradients as

$$g_{i\alpha} \equiv \frac{\partial \phi'_\alpha}{\partial x_i}. \tag{3.145}$$

We will also need to define the Jacobian matrix of the chemical source term $\mathbf{J}$ with components

$$J_{\alpha\beta}(\boldsymbol{\phi}) \equiv \frac{\partial S_\beta}{\partial \phi_\alpha}(\boldsymbol{\phi}). \tag{3.146}$$

For example, for the second-order reaction in (3.142) the Jacobian matrix is given by

$$\mathbf{J} = -k_1 \begin{bmatrix} \phi_2 & \phi_2 \\ \phi_1 & \phi_1 \end{bmatrix}. \tag{3.147}$$

Differentiating both sides of (1.28) with respect to x_i yields the transport equation for $g_{i\alpha}$:

$$\begin{aligned} \frac{\partial g_{i\alpha}}{\partial t} + \langle U_j \rangle \frac{\partial g_{i\alpha}}{\partial x_j} &= \Gamma_\alpha \nabla^2 g_{i\alpha} - g_{j\alpha} \frac{\partial u_j}{\partial x_i} - u_j \frac{\partial g_{i\alpha}}{\partial x_j} - g_{j\alpha} \frac{\partial \langle U_j \rangle}{\partial x_i} \\ &\quad - \frac{\partial u_j}{\partial x_i} \frac{\partial \langle \phi_\alpha \rangle}{\partial x_j} - u_j \frac{\partial^2 \langle \phi_\alpha \rangle}{\partial x_j \partial x_i} + \frac{\partial^2 \langle u_j \phi_\alpha \rangle}{\partial x_j \partial x_i} + \sum_\gamma J_{\gamma\alpha} g_{i\gamma}, \end{aligned} \tag{3.148}$$

where the summation in the final term is over all scalars. Likewise, the transport equation for $g_{i\beta}$ can be found by replacing α with β in (3.148).

The derivation of the transport equation for $g_{i\alpha} g_{i\beta}$ is analogous to that used to derive the transport equation for the scalar covariance. The resultant expression is

$$\begin{aligned} \frac{\partial g_{i\alpha} g_{i\beta}}{\partial t} + \langle U_j \rangle \frac{\partial g_{i\alpha} g_{i\beta}}{\partial x_j} &= \Gamma_\alpha g_{i\beta} \nabla^2 g_{i\alpha} + \Gamma_\beta g_{i\alpha} \nabla^2 g_{i\beta} - u_j \frac{\partial g_{i\alpha} g_{i\beta}}{\partial x_j} \\ &\quad - (g_{j\alpha} g_{i\beta} + g_{i\alpha} g_{j\beta}) \frac{\partial u_j}{\partial x_i} \\ &\quad - (g_{j\alpha} g_{i\beta} + g_{i\alpha} g_{j\beta}) \frac{\partial \langle U_j \rangle}{\partial x_i} \\ &\quad - g_{i\alpha} \frac{\partial u_j}{\partial x_i} \frac{\partial \langle \phi_\beta \rangle}{\partial x_j} - g_{i\beta} \frac{\partial u_j}{\partial x_i} \frac{\partial \langle \phi_\alpha \rangle}{\partial x_j} \\ &\quad - u_j g_{i\alpha} \frac{\partial^2 \langle \phi_\beta \rangle}{\partial x_j \partial x_i} - u_j g_{i\beta} \frac{\partial^2 \langle \phi_\alpha \rangle}{\partial x_j \partial x_i} \\ &\quad + g_{i\alpha} \frac{\partial^2 \langle u_j \phi_\beta \rangle}{\partial x_i \partial x_j} + g_{i\beta} \frac{\partial^2 \langle u_j \phi_\alpha \rangle}{\partial x_i \partial x_j} \\ &\quad + \sum_\gamma J_{\gamma\alpha} g_{i\beta} g_{i\gamma} + \sum_\gamma J_{\gamma\beta} g_{i\alpha} g_{i\gamma}. \end{aligned} \tag{3.149}$$

We next define the fluctuating joint scalar dissipation rate $\epsilon_{\alpha\beta}$ by

$$\epsilon_{\alpha\beta} \equiv 2(\Gamma_\alpha \Gamma_\beta)^{1/2} g_{i\alpha} g_{i\beta}, \tag{3.150}$$

from which the joint scalar dissipation rate is defined by $\varepsilon_{\alpha\beta} = \langle \epsilon_{\alpha\beta} \rangle$. Multiplying (3.149) by $2(\Gamma_\alpha \Gamma_\beta)^{1/2}$ and Reynolds averaging the resultant expression yields the final form for the joint scalar dissipation rate transport equation:

$$\begin{aligned} \frac{\partial \varepsilon_{\alpha\beta}}{\partial t} + \langle U_j \rangle \frac{\partial \varepsilon_{\alpha\beta}}{\partial x_j} = {} & \frac{\partial}{\partial x_j} \left(\mathcal{T}_{\varepsilon j}^{\alpha\beta} - \langle u_j \epsilon_{\alpha\beta} \rangle \right) \\ & + \mathcal{S}_\varepsilon^{\alpha\beta} + \mathcal{G}_\varepsilon^{\alpha\beta} + \mathcal{C}_\varepsilon^{\alpha\beta} + \mathcal{V}_\varepsilon^{\alpha\beta} - \gamma_{\alpha\beta} \mathcal{D}_\varepsilon^{\alpha\beta} + S_\varepsilon^{\alpha\beta}. \end{aligned} \tag{3.151}$$

The joint-dissipation-flux term $\langle u_j \epsilon_{\alpha\beta} \rangle$ and the molecular-transport term $\mathcal{T}_{\varepsilon j}^{\alpha\beta}$, defined by

$$\mathcal{T}_{\varepsilon j}^{\alpha\beta} \equiv 2\Gamma_\alpha \left\langle (\Gamma_\alpha \Gamma_\beta)^{1/2} \frac{\partial \phi'_\beta}{\partial x_i} \frac{\partial^2 \phi'_\alpha}{\partial x_j \partial x_i} \right\rangle + 2\Gamma_\beta \left\langle (\Gamma_\alpha \Gamma_\beta)^{1/2} \frac{\partial \phi'_\alpha}{\partial x_i} \frac{\partial^2 \phi'_\beta}{\partial x_j \partial x_i} \right\rangle, \tag{3.152}$$

are responsible for spatial transport of the joint scalar dissipation. As usual, in high-Reynolds-number turbulent flows, $\langle u_j \epsilon_{\alpha\beta} \rangle$ is the dominant spatial transport term. The other terms on the right-hand side of (3.151) are similar to the corresponding terms in the dissipation transport equation ((2.125), p. 52), and are defined as follows.

The mean-velocity-gradient term $\mathcal{S}_\varepsilon^{\alpha\beta}$ is defined by

$$\mathcal{S}_\varepsilon^{\alpha\beta} \equiv -2(\Gamma_\alpha \Gamma_\beta)^{1/2} \left(\left\langle \frac{\partial \phi'_\alpha}{\partial x_i} \frac{\partial \phi'_\beta}{\partial x_j} \right\rangle + \left\langle \frac{\partial \phi'_\beta}{\partial x_i} \frac{\partial \phi'_\alpha}{\partial x_j} \right\rangle \right) \frac{\partial \langle U_j \rangle}{\partial x_i}. \tag{3.153}$$

If the small scales of the scalar field are locally isotropic, then

$$2(\Gamma_\alpha \Gamma_\beta)^{1/2} \left\langle \frac{\partial \phi'_\alpha}{\partial x_i} \frac{\partial \phi'_\beta}{\partial x_j} \right\rangle = \frac{1}{3} \varepsilon_{\alpha\beta} \delta_{ij}. \tag{3.154}$$

Combining this expression with (3.153), and using the continuity equation for $\langle \mathbf{U} \rangle$, yields $\mathcal{S}_\varepsilon^{\alpha\beta} = 0$. Note that local isotropy does not exclude the possibility that scalar gradient vectors are collinear as is often the case when $\Gamma_\alpha = \Gamma_\beta$.

The mean-scalar-gradient term $\mathcal{G}_\varepsilon^{\alpha\beta}$ is defined by

$$\mathcal{G}_\varepsilon^{\alpha\beta} \equiv -2(\Gamma_\alpha \Gamma_\beta)^{1/2} \left(\left\langle \frac{\partial \phi'_\alpha}{\partial x_i} \frac{\partial u_j}{\partial x_i} \right\rangle \frac{\partial \langle \phi_\beta \rangle}{\partial x_j} + \left\langle \frac{\partial \phi'_\beta}{\partial x_i} \frac{\partial u_j}{\partial x_i} \right\rangle \frac{\partial \langle \phi_\alpha \rangle}{\partial x_j} \right). \tag{3.155}$$

At high Reynolds numbers, the scalar gradient and the velocity gradient will be uncorrelated due to local isotropy. Thus, $\mathcal{G}_\varepsilon^{\alpha\beta}$ will be negligible. Furthermore, the mean-scalar-curvature term $\mathcal{C}_\varepsilon^{\alpha\beta}$, defined by

$$\mathcal{C}_\varepsilon^{\alpha\beta} \equiv -2(\Gamma_\alpha \Gamma_\beta)^{1/2} \left(\left\langle u_j \frac{\partial \phi'_\alpha}{\partial x_i} \right\rangle \frac{\partial^2 \langle \phi_\beta \rangle}{\partial x_j \partial x_i} + \left\langle u_j \frac{\partial \phi'_\beta}{\partial x_i} \right\rangle \frac{\partial^2 \langle \phi_\alpha \rangle}{\partial x_j \partial x_i} \right), \tag{3.156}$$

will also be negligible at high Reynolds numbers.

As for the scalar dissipation rate, the remaining terms in (3.151) will be large in high-Reynolds-number turbulent flows. The vortex-stretching term $\mathcal{V}_{\varepsilon}^{\alpha\beta}$ is defined by

$$\mathcal{V}_{\varepsilon}^{\alpha\beta} \equiv -4(\Gamma_\alpha \Gamma_\beta)^{1/2} \left\langle \frac{\partial \phi'_\alpha}{\partial x_j} s_{ij} \frac{\partial \phi'_\beta}{\partial x_i} \right\rangle. \tag{3.157}$$

The physics of this term can be understood using the scalar cospectrum $E_{\alpha\beta}(\kappa, t)$, which, similar to the scalar spectrum, represents the fraction of the scalar covariance at wavenumber κ, i.e.,

$$\langle \phi'_\alpha \phi'_\beta \rangle = \int_0^\infty E_{\alpha\beta}(\kappa, t) \, \mathrm{d}\kappa. \tag{3.158}$$

In homogeneous turbulence, spectral transport can be quantified by the scalar cospectral energy transfer rate $\mathcal{T}_{\alpha\beta}(\kappa, t)$. We can also define the wavenumber that separates the viscous-convective and the viscous-diffusive sub-ranges $\kappa_{\mathrm{D}}^{\alpha\beta}$ by introducing the arithmetic-mean molecular diffusivity $\Gamma_{\alpha\beta}$ defined by

$$\Gamma_{\alpha\beta} \equiv \frac{\Gamma_\alpha + \Gamma_\beta}{2}, \tag{3.159}$$

and a Schmidt number $\mathrm{Sc}_{\alpha\beta}$ defined by

$$\mathrm{Sc}_{\alpha\beta} \equiv \frac{\nu}{\Gamma_{\alpha\beta}}. \tag{3.160}$$

We then have $\kappa_{\mathrm{D}}^{\alpha\beta} = \mathrm{Sc}_{\alpha\beta}^{1/2} \kappa_{\mathrm{DI}}$, and at spectral equilibrium $\mathcal{T}_{\alpha\beta}(\kappa_{\mathrm{D}}^{\alpha\beta}, t) = \gamma_{\alpha\beta} \varepsilon_{\alpha\beta}$.

As in (3.130), the vortex-stretching term can be expressed as

$$\mathcal{V}_{\varepsilon}^{\alpha\beta} = C_{\mathcal{V}1}^{\alpha\beta} \mathrm{Re}_L^{1/2} \frac{1}{\gamma_{\alpha\beta} \tau_{\mathrm{e}}} \mathcal{T}_{\alpha\beta}\left(\kappa_{\mathrm{D}}^{\alpha\beta}, t\right) + C_{\mathcal{V}2}^{\alpha\beta} \left(\frac{\epsilon}{\nu}\right)^{1/2} \varepsilon_{\alpha\beta}, \tag{3.161}$$

where $C_{\mathcal{V}1}^{\alpha\beta}$ and $C_{\mathcal{V}2}^{\alpha\beta}$ are constants of order unity. Thus, the vortex-stretching term describes the rate at which cospectral energy arrives in the covariance-dissipation range.

The next term in (3.151) is the joint-gradient-dissipation term $\mathcal{D}_{\varepsilon}^{\alpha\beta}$ defined by

$$\mathcal{D}_{\varepsilon}^{\alpha\beta} \equiv 4 \left\langle \left(\Gamma_\alpha \frac{\partial^2 \phi'_\alpha}{\partial x_i \partial x_j} \right) \left(\Gamma_\beta \frac{\partial^2 \phi'_\beta}{\partial x_i \partial x_j} \right) \right\rangle. \tag{3.162}$$

Like (3.123), the joint-gradient-dissipation term can be written as

$$\mathcal{D}_{\varepsilon}^{\alpha\beta} = C_{\mathcal{D}}^{\alpha\beta} \gamma_{\alpha\beta} \frac{\varepsilon_{\alpha\beta}}{\langle \phi'_\alpha \phi'_\beta \rangle_{\mathrm{D}}} \varepsilon_{\alpha\beta}, \tag{3.163}$$

where the fraction of the covariance in the covariance-dissipation range $\langle \phi'_\alpha \phi'_\beta \rangle_{\mathrm{D}}$ is defined by

$$\langle \phi'_\alpha \phi'_\beta \rangle_{\mathrm{D}} = \int_{\kappa_{\mathrm{D}}^{\alpha\beta}}^\infty E_{\alpha\beta}(\kappa, t) \, \mathrm{d}\kappa. \tag{3.164}$$

Note that in (3.163) the characteristic time scale for scalar-covariance dissipation is $\varepsilon_{\alpha\beta} / \langle \phi'_\alpha \phi'_\beta \rangle_{\mathrm{D}}$. We show in Chapter 4 that $\varepsilon_{\alpha\beta}$ and $\langle \phi'_\alpha \phi'_\beta \rangle_{\mathrm{D}}$ will always have the same sign so that this time scale is well defined, even when the integral-scale covariance $\langle \phi'_\alpha \phi'_\beta \rangle$ has the opposite sign.

For inert-scalar mixing in homogeneous turbulence, $\langle \phi'_\alpha \phi'_\beta \rangle_{\mathrm{D}}$ can be found from[37]

$$\frac{\mathrm{d}\langle \phi'_\alpha \phi'_\beta \rangle_{\mathrm{D}}}{\mathrm{d}t} \approx \mathcal{T}_{\alpha\beta}\left(\kappa_{\mathrm{D}}^{\alpha\beta}, t\right) - \gamma_{\alpha\beta}\varepsilon_{\alpha\beta}. \tag{3.165}$$

Likewise, the joint dissipation rate can be approximated by

$$\frac{\mathrm{d}\varepsilon_{\alpha\beta}}{\mathrm{d}t} \approx C_{\mathrm{V}1}^{\alpha\beta} \mathrm{Re}_L^{1/2} \frac{1}{\gamma_{\alpha\beta}\tau_{\mathrm{e}}} \mathcal{T}_{\alpha\beta}\left(\kappa_{\mathrm{D}}^{\alpha\beta}, t\right) + C_{\mathrm{V}2}^{\alpha\beta}\left(\frac{\epsilon}{\nu}\right)^{1/2} \varepsilon_{\alpha\beta} - C_D^{\alpha\beta} \gamma_{\alpha\beta} \frac{\varepsilon_{\alpha\beta}}{\langle \phi'_\alpha \phi'_\beta \rangle_{\mathrm{D}}} \varepsilon_{\alpha\beta}. \tag{3.166}$$

At high Reynolds number, we again find from (3.165) and (3.166) that the joint scalar dissipation rate is proportional to the cospectral energy transfer rate, i.e.,

$$\varepsilon_{\alpha\beta} = \frac{1}{\gamma_{\alpha\beta}} \mathcal{T}_{\alpha\beta}\left(\kappa_{\mathrm{D}}^{\alpha\beta}, t\right), \tag{3.167}$$

and that $\langle \phi'_\alpha \phi'_\beta \rangle_{\mathrm{D}}$ is given by

$$\langle \phi'_\alpha \phi'_\beta \rangle_{\mathrm{D}} \propto \tau_e \mathcal{T}_{\alpha\beta}\left(\kappa_{\mathrm{D}}^{\alpha\beta}, t\right) \mathrm{Re}_L^{-1/2}. \tag{3.168}$$

Engineering models for the joint dissipation rate of inert scalars must thus provide a description of the cospectral energy transfer rate. We will look at such a model in Chapter 4.

The final term in (3.151) is the joint-scalar-dissipation chemical source term $S_\varepsilon^{\alpha\beta}$ defined by

$$S_\varepsilon^{\alpha\beta} = \sum_\gamma \left(\left(\frac{\Gamma_\alpha}{\Gamma_\gamma}\right)^{1/2} \langle J_{\gamma\alpha}(\boldsymbol{\phi}) \epsilon_{\beta\gamma} \rangle + \left(\frac{\Gamma_\beta}{\Gamma_\gamma}\right)^{1/2} \langle J_{\gamma\beta}(\boldsymbol{\phi}) \epsilon_{\alpha\gamma} \rangle \right). \tag{3.169}$$

For non-linear chemical reactions, this term leads to new unclosed terms that are difficult to model. For example, even the isothermal second-order reaction, (3.142), where the joint dissipation chemical source term is given by

$$S_\varepsilon^{12} = -k_1 \left(\langle \phi_2 \epsilon_{12} \rangle + \left(\frac{\Gamma_1}{\Gamma_2}\right)^{1/2} \langle \phi_1 \epsilon_{22} \rangle + \left(\frac{\Gamma_2}{\Gamma_1}\right)^{1/2} \langle \phi_2 \epsilon_{11} \rangle + \langle \phi_1 \epsilon_{12} \rangle \right), \tag{3.170}$$

generates four new unclosed terms.

In some applications, separation of scales can be invoked to argue that $\boldsymbol{\phi}$ and $\epsilon_{\alpha\beta}$ are uncorrelated so that

$$\langle J_{\gamma\alpha}(\boldsymbol{\phi}) \epsilon_{\beta\gamma} \rangle = \langle J_{\gamma\alpha}(\boldsymbol{\phi}) \rangle \varepsilon_{\beta\gamma}. \tag{3.171}$$

However, because fast chemical reactions often occur over very small scales, strong coupling may exist between the chemical-source-term Jacobian and the joint scalar dissipation rate. These difficulties render computational approaches based on solving the joint scalar

[37] In principle, the same equation could be used for a reacting scalar. However, one would need to know the spectral distribution of the covariance chemical source term $S_{\alpha\beta}$, (3.141), in order to add the corresponding covariance-dissipation-range chemical source term to (3.165).

Table 3.3. *The one-point scalar statistics and unclosed quantities appearing in the transport equations for inhomogeneous turbulent mixing of multiple reacting scalars at high Reynolds numbers.*

Physical quantity	Transport equation	Unclosed terms
Scalar mean $\langle \phi_\alpha \rangle$	(3.88), p. 81	$\langle u_i \phi_\alpha \rangle$, $\langle S_\alpha \rangle$
Scalar flux $\langle u_i \phi_\alpha \rangle$	(3.102), p. 84	$\langle u_i u_j \rangle$, $\langle u_i u_j \phi'_\alpha \rangle$, Π_i^α, $\langle u_i S_\alpha \rangle$
Covariance $\langle \phi'_\alpha \phi'_\beta \rangle$	(3.136), p. 90	$\langle u_i \phi_\alpha \rangle$, $\langle u_i \phi_\beta \rangle$, $\langle u_i \phi'_\alpha \phi'_\beta \rangle$, $\varepsilon_{\alpha\beta}$, $S_{\alpha\beta}$
Joint dissipation rate $\varepsilon_{\alpha\beta}$	(3.151), p. 93	$\langle u_i \epsilon_{\alpha\beta} \rangle$, $\mathcal{V}_\varepsilon^{\alpha\beta}$, $\mathcal{D}_\varepsilon^{\alpha\beta}$, $\mathcal{S}_\varepsilon^{\alpha\beta}$

dissipation rate transport equation for multiple reacting scalars intractable. An alternative approach based on solving for the joint PDF of ϕ and $\epsilon_{\alpha\beta}$ is discussed in Section 6.10.

The one-point scalar statistics used in engineering calculations of high-Reynolds-number turbulent mixing of reacting scalars is summarized in Table 3.3 along with the unclosed terms that appear in their transport equations. In Chapter 4, we will discuss methods for modeling the unclosed terms in the RANS transport equations.

3.4 Differential diffusion

Differential diffusion occurs when the molecular diffusivities of the scalar fields are not the same. For the simplest case of two inert scalars, this implies $\Gamma_1 \neq \Gamma_2$ and $\gamma_{12} > 1$ (see (3.140)). In homogeneous turbulence, one effect of differential diffusion is to de-correlate the scalars. This occurs first at the diffusive scales, and then 'backscatters' to larger scales until the energy-containing scales de-correlate. Thus, one of the principal difficulties of modeling differential diffusion is the need to account for this length-scale dependence.

Scalar correlation at the diffusive scales can be measured by the scalar-gradient correlation function:

$$g_{\alpha\beta} \equiv \frac{\langle (\nabla \phi'_\alpha) \cdot (\nabla \phi'_\beta) \rangle}{(\langle |\nabla \phi'_\alpha|^2 \rangle \langle |\nabla \phi'_\beta|^2 \rangle)^{1/2}} = \frac{\varepsilon_{\alpha\beta}}{(\varepsilon_{\alpha\alpha} \varepsilon_{\beta\beta})^{1/2}}. \tag{3.172}$$

Likewise, scalar correlation at the energy-containing scales is measured by the scalar correlation function:

$$\rho_{\alpha\beta} \equiv \frac{\langle \phi'_\alpha \phi'_\beta \rangle}{\left(\langle \phi'^2_\alpha \rangle \langle \phi'^2_\beta \rangle \right)^{1/2}}. \tag{3.173}$$

The Reynolds-number dependence of differential-diffusion effects on $g_{\alpha\beta}$ is distinctly different than on $\rho_{\alpha\beta}$, and can be best understood by looking at scalars in homogeneous, stationary turbulence with and without uniform mean scalar gradients.

3.4.1 Homogeneous turbulence

In homogeneous turbulence, the governing equations for the scalar covariance, (3.137), and the joint scalar dissipation rate, (3.166), reduce, respectively, to

$$\frac{\mathrm{d}\langle\phi'_\alpha\phi'_\beta\rangle}{\mathrm{d}t} = -\langle u_i\phi_\alpha\rangle G_{i\beta} - \langle u_i\phi_\beta\rangle G_{i\alpha} - \gamma_{\alpha\beta}\varepsilon_{\alpha\beta} \tag{3.174}$$

and[38]

$$\frac{\mathrm{d}\varepsilon_{\alpha\beta}}{\mathrm{d}t} = C_\mathrm{B}\left(\frac{\epsilon}{\nu}\right)^{1/2}\varepsilon_{\alpha\beta} - C_\mathrm{d}\gamma_{\alpha\beta}\frac{\varepsilon_{\alpha\beta}}{\langle\phi'_\alpha\phi'_\beta\rangle_\mathrm{D}}\varepsilon_{\alpha\beta}, \tag{3.175}$$

wherein the uniform scalar gradients,

$$G_{i\alpha} \equiv \frac{\partial\langle\phi_\alpha\rangle}{\partial x_i}, \tag{3.176}$$

are constant. Note that in (3.175) we have used the same constants C_B (gradient amplification) and C_d (molecular dissipation) for all components of the joint dissipation rate. This is consistent with the DNS results of Vedula (2001) and Vedula *et al.* (2001). As a result, the only direct effect of differential diffusion in (3.175) comes from $\gamma_{\alpha\beta} > 1$.

In order to proceed, we will need to close (3.174) by using a gradient-diffusion model for the scalar fluxes:

$$\langle u_i\phi_\alpha\rangle = -\Gamma_\mathrm{T}G_{i\alpha}. \tag{3.177}$$

At high Reynolds number and for Schmidt numbers near unity or larger, we are justified in assuming that Γ_T is nearly independent of Schmidt number. We will also need a closure for $\langle\phi'_\alpha\phi'_\beta\rangle_\mathrm{D}$ in (3.175). In general, the dissipation-range variance scales as $\mathrm{Re}_1^{-1} \equiv \mathrm{Re}_L^{-1/2}$ (Fox and Yeung 1999; Vedula 2001). We will thus model the covariances by

$$\langle\phi'_\alpha\phi'_\beta\rangle_\mathrm{D} = \frac{C_{\alpha\beta}}{\mathrm{Re}_1}\langle\phi'_\alpha\phi'_\beta\rangle, \tag{3.178}$$

where $C_{\alpha\beta}$ depends, in general, on Sc_α and Sc_β and is a weak function of Re_1. (For example, when $\mathrm{Re}_1 = 1$, $C_{\alpha\beta} = 1$.) However, as the Reynolds number increases, $C_{\alpha\beta}$ will approach an Sc-independent value (Fox 1999).

Using (3.177) and (3.178), the governing equations reduce to

$$\frac{\mathrm{d}\langle\phi'_\alpha\phi'_\beta\rangle}{\mathrm{d}t} = 2\Gamma_\mathrm{T}G_{i\alpha}G_{i\beta} - \gamma_{\alpha\beta}\varepsilon_{\alpha\beta} \tag{3.179}$$

and

$$\frac{\mathrm{d}\varepsilon_{\alpha\beta}}{\mathrm{d}t} = C_\mathrm{B}\left(\frac{\epsilon}{\nu}\right)^{1/2}\varepsilon_{\alpha\beta} - \left(\frac{\mathrm{Re}_1 C_\mathrm{d}}{C_{\alpha\beta}}\right)\gamma_{\alpha\beta}\frac{\varepsilon_{\alpha\beta}}{\langle\phi'_\alpha\phi'_\beta\rangle}\varepsilon_{\alpha\beta}. \tag{3.180}$$

We will now use these equations to find expressions for the correlation coefficients for cases with and without mean scalar gradients.

[38] Because the cospectral energy transfer term is combined with the gradient-amplification term, this expression can only be valid for 'equilibrium' scalar spectra. A more complete model is discussed in Section 4.7.

3.4.2 Mean scalar gradients

With mean scalar gradients, the scalar covariances and joint dissipation rates attain steady-state values found by setting the right-hand sides of (3.179) and (3.180) equal to zero. This yields

$$\varepsilon_{\alpha\beta} = 2\Gamma_{\mathrm{T}} \frac{G_{i\alpha} G_{i\beta}}{\gamma_{\alpha\beta}} \tag{3.181}$$

and

$$\langle \phi'_{\alpha} \phi'_{\beta} \rangle = \left(\frac{k C_{\mathrm{d}}}{\varepsilon C_{\mathrm{B}}} \right) \left(\frac{\gamma_{\alpha\beta} \varepsilon_{\alpha\beta}}{C_{\alpha\beta}} \right). \tag{3.182}$$

The correlation coefficients are thus

$$g_{\alpha\beta} = \frac{1}{\gamma_{\alpha\beta}} \frac{G_{i\alpha} G_{i\beta}}{|G_{\alpha}||G_{\beta}|} = \frac{\cos(\theta_{\alpha\beta})}{\gamma_{\alpha\beta}}, \tag{3.183}$$

where $\theta_{\alpha\beta}$ is the angle between the mean scalar gradients, and

$$\rho_{\alpha\beta} = \left(\frac{(C_{\alpha\alpha} C_{\beta\beta})^{1/2}}{C_{\alpha\beta}} \right) \gamma_{\alpha\beta} g_{\alpha\beta} = \left(\frac{(C_{\alpha\alpha} C_{\beta\beta})^{1/2}}{C_{\alpha\beta}} \right) \cos(\theta_{\alpha\beta}). \tag{3.184}$$

For perfectly aligned mean scalar gradients, $\cos(\theta_{\alpha\beta}) = 1$. Note also that in order for (3.184) to hold, $0 < C_{\alpha\alpha} C_{\beta\beta} \leq C_{\alpha\beta}^2$, where the equality holds when the Schmidt numbers are equal. This condition has important ramifications when developing models for $\langle \phi'_{\alpha} \phi'_{\beta} \rangle_{\mathrm{D}}$, as discussed in Section 4.7.

Because $\gamma_{\alpha\beta}$ depends only on the Schmidt numbers of the two scalars, we can deduce from (3.183) that $g_{\alpha\beta} \neq \cos(\theta_{\alpha\beta})$ for infinite Reynolds number. On the other hand, $\rho_{\alpha\beta} = \cos(\theta_{\alpha\beta})$ for infinite Reynolds number. Thus, we can conclude that the scalar gradients will be de-correlated by differential diffusion for any value of the Reynolds number, even though the scalars themselves will show less and less de-correlation as the Reynolds number increases. Analogous trends are seen with DNS (Yeung and Moseley 1995; Yeung 1998a; Yeung *et al.* 2000), and should also be observed with reacting scalars. Yeung (1998a) has also studied the correlation coefficient as a function of the wavenumber κ for the case $\cos(\theta_{\alpha\beta}) = 1$, and has shown that $\rho_{\alpha\beta}(\kappa)$ is near unity for low wavenumbers and near $g_{\alpha\beta}$ at high wavenumbers. Moreover, as the Reynolds number increases, the slope of $\rho_{\alpha\beta}(\kappa)$ with κ decreases, leading to a larger value for $\rho_{\alpha\beta}$.

3.4.3 Decaying scalars

In the absence of mean scalar gradients, the scalar covariances and joint dissipation rates will decay towards zero. For this case, it is convenient to work with the governing equations for $g_{\alpha\beta}$ and $\rho_{\alpha\beta}$ directly. These expressions can be derived from (3.179) and (3.180):

$$\frac{\mathrm{d}\rho_{\alpha\beta}}{\mathrm{d}t} = -\frac{1}{2} \left(2 \frac{\gamma_{\alpha\beta} \varepsilon_{\alpha\beta}}{\langle \phi'_{\alpha} \phi'_{\beta} \rangle} - \frac{\varepsilon_{\alpha\alpha}}{\langle \phi'^2_{\alpha} \rangle} - \frac{\varepsilon_{\beta\beta}}{\langle \phi'^2_{\beta} \rangle} \right) \tag{3.185}$$

and

$$\frac{\mathrm{d}g_{\alpha\beta}}{\mathrm{d}t} = -\frac{\mathrm{Re}_1 C_\mathrm{d}}{2}\left(\frac{2}{C_{\alpha\beta}}\frac{\gamma_{\alpha\beta}\varepsilon_{\alpha\beta}}{\langle\phi'_\alpha\phi'_\beta\rangle} - \frac{1}{C_{\alpha\alpha}}\frac{\varepsilon_{\alpha\alpha}}{\langle\phi'^2_\alpha\rangle} - \frac{1}{C_{\beta\beta}}\frac{\varepsilon_{\beta\beta}}{\langle\phi'^2_\beta\rangle}\right). \tag{3.186}$$

In addition, we will need the governing equations for the scalar time scales:

$$\frac{\mathrm{d}}{\mathrm{d}t}\left(\frac{\gamma_{\alpha\beta}\varepsilon_{\alpha\beta}}{\langle\phi'_\alpha\phi'_\beta\rangle}\right) = \left(\frac{\varepsilon}{\nu}\right)^{1/2}\left[C_\mathrm{B} - \left(C_\mathrm{d} - \frac{C_{\alpha\beta}}{\mathrm{Re}_1}\right)\frac{k\gamma_{\alpha\beta}\varepsilon_{\alpha\beta}}{\varepsilon C_{\alpha\beta}\langle\phi'_\alpha\phi'_\beta\rangle}\right]\frac{\gamma_{\alpha\beta}\varepsilon_{\alpha\beta}}{\langle\phi'_\alpha\phi'_\beta\rangle}, \tag{3.187}$$

where $\mathrm{Re}_1 C_\mathrm{d} > C_{\alpha\beta}$. We will now look at the high-Reynolds-number behavior of (3.185)–(3.187).

The stable steady-state solution to (3.187) can be expressed as

$$\frac{k\gamma_{\alpha\beta}\varepsilon_{\alpha\beta}}{\varepsilon C_{\alpha\beta}\langle\phi'_\alpha\phi'_\beta\rangle} = \frac{C_\mathrm{B}}{C_\mathrm{d} - C_{\alpha\beta}/\mathrm{Re}_1}. \tag{3.188}$$

Thus, in the limit of large (but finite) Reynolds numbers,

$$\frac{\gamma_{\alpha\beta}\varepsilon_{\alpha\beta}}{C_{\alpha\beta}\langle\phi'_\alpha\phi'_\beta\rangle} = \frac{\varepsilon C_\mathrm{B}}{kC_\mathrm{d}}\left[1 + \frac{C_{\alpha\beta}}{C_\mathrm{d}}\mathrm{Re}_1^{-1} + \mathcal{O}\left(\mathrm{Re}_1^{-2}\right)\right]. \tag{3.189}$$

Note that the first term in the series on the right-hand side of this expression is independent of Schmidt number.

Using (3.189) in (3.185) and (3.186), and keeping only the terms of leading order in Re_1, yields

$$\frac{\mathrm{d}\rho_{\alpha\beta}}{\mathrm{d}t} = -\frac{\varepsilon C_\mathrm{B}}{2kC_\mathrm{d}}(2C_{\alpha\beta} - C_{\alpha\alpha} - C_{\beta\beta}) \tag{3.190}$$

and

$$\frac{\mathrm{d}g_{\alpha\beta}}{\mathrm{d}t} = \frac{\mathrm{d}\rho_{\alpha\beta}}{\mathrm{d}t}. \tag{3.191}$$

Thus, at high Reynolds numbers, the correlation functions both decay at the same rate, which is proportional to the turbulence integral time scale $\tau_u = k/\varepsilon$.

Note that in order for the correlation functions to decay, a strong condition on $C_{\alpha\beta}$ is required, namely $0 < C_{\alpha\alpha} + C_{\beta\beta} \leq 2C_{\alpha\beta}$. On the other hand, if only the weak condition $0 < C_{\alpha\alpha}C_{\beta\beta} \leq C^2_{\alpha\beta}$ is satisfied, then the correlation functions can increase with time. Thus, in order to determine if the correlation functions decay to zero, a model is needed for $C_{\alpha\beta}$, or equivalently for $\langle\phi'_\alpha\phi'_\beta\rangle_\mathrm{D}$. Such a model is discussed in Section 4.7. For now, we will simply note that, as the Reynolds number increases, $C_{\alpha\beta}$ approaches a Sc-independent value. Hence, the difference term on the right-hand side of (3.190) will become very small, resulting in a reduction in the rate of de-correlation at high Reynolds numbers. Similar behavior has been observed in DNS (Yeung and Pope 1993). If, in addition, the turbulence is decaying so that $\tau_u(t)$ grows with time, the behavior of the correlation functions will be even more complex.

4

Models for turbulent transport

This chapter is devoted to methods for describing the turbulent transport of passive scalars. The basic transport equations resulting from Reynolds averaging have been derived in earlier chapters and contain unclosed terms that must be modeled. Thus the available models for these terms are the primary focus of this chapter. However, to begin the discussion, we first review transport models based on the direct numerical simulation of the Navier–Stokes equation, and other models that do not require one-point closures. The presentation of turbulent transport models in this chapter is not intended to be comprehensive. Instead, the emphasis is on the differences between particular classes of models, and how they relate to models for turbulent reacting flow. A more detailed discussion of turbulent-flow models can be found in Pope (2000). For practical advice on choosing appropriate models for particular flows, the reader may wish to consult Wilcox (1993).

4.1 Direct numerical simulation

Direct numerical simulation (DNS) involves a full numerical solution of the Navier–Stokes equations without closures (Rogallo and Moin 1984; Givi 1989; Moin and Mahesh 1998). A detailed introduction to the numerical methods used for DNS can be found in Ferziger and Perić (2002). The principal advantage of DNS is that it provides extremely detailed information about the flow. For example, the instantaneous pressure at any point in the flow can be extracted from DNS, but is nearly impossible to measure experimentally. Likewise, Lagrangian statistics can be obtained for any flow quantity and used to develop new turbulence models based on Lagrangian PDF methods (Yeung 2002). The application of DNS to inhomogeneous turbulent flows is limited to simple 'canonical' flows at relatively modest Reynolds numbers. Thus we will limit our discussion here to the simplest application of DNS: homogeneous turbulent flow in a periodic domain.

4.1.1 Homogeneous turbulence

For homogeneous turbulent flows (no walls, periodic boundary conditions, zero mean velocity), pseudo-spectral methods are usually employed due to their relatively high accuracy. In order to simulate the Navier–Stokes equation,

$$\frac{\partial \mathbf{U}}{\partial t} + \mathbf{U} \cdot \nabla \mathbf{U} = \nu \nabla^2 \mathbf{U} - \frac{1}{\rho} \nabla p, \tag{4.1}$$

we can use the Fourier-transformed velocity vector $\hat{\mathbf{U}}_{\boldsymbol{\kappa}}$.[1] This vector is related to the velocity vector through the discrete Fourier transform,

$$\mathbf{U}(\mathbf{x}, t) = \sum_{\boldsymbol{\kappa}} \hat{\mathbf{U}}_{\boldsymbol{\kappa}}(t) \, \mathrm{e}^{\mathrm{i}\boldsymbol{\kappa} \cdot \mathbf{x}}, \tag{4.2}$$

where $\boldsymbol{\kappa}$ is the wavenumber vector. The summation in (4.2) extends over a cube with volume K^3 centered at the origin in wavenumber space. The maximum wavenumber that can be resolved on the DNS grid is thus equal to K.

Using (4.2), the transformed Navier–Stokes equation becomes[2]

$$\frac{\mathrm{d}\hat{\mathbf{U}}_{\boldsymbol{\kappa}}}{\mathrm{d}t} = -\nu \kappa^2 \hat{\mathbf{U}}_{\boldsymbol{\kappa}} - \mathcal{F}\{\mathbf{U} \cdot \nabla \mathbf{U}\} + \left(\frac{\boldsymbol{\kappa} \cdot \mathcal{F}\{\mathbf{U} \cdot \nabla \mathbf{U}\}}{\kappa^2} \right) \boldsymbol{\kappa}, \tag{4.3}$$

where $\mathcal{F}\{\cdot\}$ is the discrete Fourier transform of the (non-linear) term in the argument, e.g.,

$$\mathcal{F}\{\mathbf{U}\} \Rightarrow \hat{\mathbf{U}}_{\boldsymbol{\kappa}}. \tag{4.4}$$

Continuity requires that $\boldsymbol{\kappa}$ and $\hat{\mathbf{U}}_{\boldsymbol{\kappa}}$ be orthogonal:

$$\nabla \cdot \mathbf{U} = 0 \qquad \Longrightarrow \qquad \boldsymbol{\kappa} \cdot \hat{\mathbf{U}}_{\boldsymbol{\kappa}} = 0. \tag{4.5}$$

At any fixed time t, the velocity field can be recovered by an inverse Fourier transform:

$$\mathcal{F}^{-1}\{\hat{\mathbf{U}}_{\boldsymbol{\kappa}}\} \Rightarrow \mathbf{U}. \tag{4.6}$$

DNS of homogeneous turbulence thus involves the solution of a large system of ordinary differential equations (ODEs; see (4.3)) that are coupled through the convective and pressure terms (i.e., the terms involving $\mathcal{F}$).

The time step required for accurate solutions of (4.3) is limited by the need to resolve the shortest time scales in the flow. In Chapter 3, we saw that the smallest eddies in a homogeneous turbulent flow can be characterized by the Kolmogorov length and time scales. Thus, the time step h must satisfy[3]

$$h \sim \left(\frac{\nu}{\epsilon} \right)^{1/2} = \text{Kolmogorov time scale.} \tag{4.7}$$

[1] The independent variables on which $\hat{\mathbf{U}}_{\boldsymbol{\kappa}}$ depends are $\boldsymbol{\kappa}$ and t. The principal advantage of using this formulation is that spatial derivatives become summations over wavenumber space. The resulting numerical solutions have higher accuracy compared with finite-difference methods using the same number of grid points.

[2] Alternatively, the right-hand side can be written in terms of a projection tensor and a convolution (Pope 2000). The form given here is used in the *pseudo-spectral* method.

[3] See Pope (2000) for an alternative estimate of the time-step scaling based on the Courant number. The overall conclusion, however, remains the same: DNS is computationally prohibitive for high Reynolds numbers.

Since the flow must be simulated for several integral time scales, the total number of time steps required will increase with Reynolds number as

$$N_{\text{dns}} \sim \frac{\tau_u}{\tau_\eta} \sim \text{Re}_L^{1/2}. \tag{4.8}$$

Likewise, good spatial accuracy requires that the maximum wavenumber (K) be inversely proportional to the Kolmogorov length scale:

$$K \sim \frac{1}{\eta} = \left(\frac{\epsilon}{\nu^3}\right)^{1/4} = \frac{1}{L_u}\text{Re}_L^{3/4}. \tag{4.9}$$

In a three-dimensional simulation, a physical-space cube of volume $V \sim L_u^3$ (i.e., L_u in each direction) is employed. Thus, the total number of ODEs that must be solved scales as

$$VK^3 \sim \text{Re}_L^{9/4} \tag{4.10}$$

for a fully resolved homogeneous turbulent flow.

Given the number of time steps and the number of equations, the total CPU time, *CPU*, for DNS of a homogeneous turbulent flow will scale as

$$CPU \sim N_{\text{dns}}(VK^3) \sim \text{Re}_L^{11/4}, \tag{4.11}$$

and thus increasing the Reynolds number by an order of magnitude would require 562 times more computing resources! Moreover, because all of the variables must be stored in computer memory, DNS simulations require a large amount of computer RAM. On large multi-processor computers, direct numerical simulations with $K = 1024$ ($\text{Re}_L \approx 50\,000$) are possible. Nevertheless, the high cost of DNS limits its application to the study of fundamental physical phenomena in idealized flows. For example, DNS is extremely valuable for studying the behavior of the unclosed terms in the Reynolds-averaged Navier–Stokes equation. Indeed, many of the recent advances in turbulence modeling have been made possible by the availability of DNS of canonical flows (Eswaran and Pope 1988; Nomura and Elgobashi 1992; Overholt and Pope 1996; Moin and Mahesh 1998; Yeung 2002).

4.1.2 Reacting flow

For DNS of turbulent reacting flow, it is also necessary to solve for the scalar fields, and thus the computational cost can increase considerably as compared with DNS of turbulent flow (Givi and McMurtry 1988; Leonard and Hill 1988; Jou and Riley 1989; Leonard and Hill 1991; Leonard and Hill 1992; Mell *et al.* 1994; Vervisch and Poinsot 1998; Poinsot and Veynante 2001). For this reason, most direct numerical simulations of reacting flow are done at rather low Reynolds numbers (well below the threshold where high-Reynolds-number scaling rules can be applied). Owing to the computational expense (see, for example, Cook and Riley (1996)), very few studies of inhomogeneous turbulent reacting flow have been reported; and even these are limited to fairly small Damköhler numbers in order to obtain fully resolved scalar fields.

As a simple example, consider the isothermal one-step reaction

$$\mathrm{A} + \mathrm{B} \xrightarrow{k_1} \mathrm{P}$$

in a homogeneous turbulent flow with passive scalars. For this reaction, one must simulate the transport equations for reactants A and B, e.g.,

$$\frac{\partial \mathrm{A}}{\partial t} + \mathbf{U} \cdot \nabla \mathrm{A} = \Gamma_{\mathrm{A}} \nabla^2 \mathrm{A} - k_1 \mathrm{AB}. \tag{4.12}$$

As for the velocity field, the (discrete) Fourier-transformed scalar field $\hat{\mathrm{A}}_{\kappa}$ can be defined by

$$\mathrm{A}(\mathbf{x}, t) = \sum_{\kappa} \hat{\mathrm{A}}_{\kappa}(t)\, \mathrm{e}^{\mathrm{i}\kappa\cdot\mathbf{x}}. \tag{4.13}$$

The transformed scalar transport equation then becomes

$$\frac{\mathrm{d}\hat{\mathrm{A}}_{\kappa}}{\mathrm{d}t} = -\Gamma_{\mathrm{A}} \kappa^2 \hat{\mathrm{A}}_{\kappa} - \mathcal{F}\{\mathbf{U} \cdot \nabla \mathrm{A} + k_1 \mathrm{AB}\}. \tag{4.14}$$

In the pseudo-spectral method, the non-linear convective and reaction terms in (4.14) are evaluated in physical space and then transformed back to wavenumber space:

$$\begin{pmatrix} \hat{\mathbf{U}}_{\kappa} \\ \hat{\mathrm{A}}_{\kappa} \\ \hat{\mathrm{B}}_{\kappa} \end{pmatrix} \xrightarrow{\mathcal{F}^{-1}} \begin{pmatrix} \mathbf{U}(\mathbf{x}, t) \\ \mathrm{A}(\mathbf{x}, t) \\ \mathrm{B}(\mathbf{x}, t) \end{pmatrix} \longrightarrow \begin{pmatrix} \mathbf{U} \cdot \nabla \mathrm{A} \\ k_1 \mathrm{AB} \end{pmatrix} \xrightarrow{\mathcal{F}} \mathcal{F}\{\mathbf{U} \cdot \nabla \mathrm{A} + k_1 \mathrm{AB}\}. \tag{4.15}$$

The Fourier and inverse Fourier transforms couple all of the coefficients $(\hat{\mathbf{U}}_{\kappa}, \hat{\mathrm{A}}_{\kappa}, \hat{\mathrm{B}}_{\kappa})$, and typically represent the most expensive part of the computation. Care must also be taken to ensure that the scalar field remains bounded in composition space when applying the inverse Fourier transform.

When the Schmidt number is greater than unity, addition of a scalar transport equation places a new requirement on the maximum wavenumber K. For $\mathrm{Sc} \geq 1$, the smallest characteristic length scale of the scalar field is the Batchelor scale, λ_{B}. Thus, the maximum wavenumber will scale with Reynolds and Schmidt number as

$$K \sim \frac{1}{\lambda_{\mathrm{B}}} = \frac{1}{L_u} \mathrm{Sc}^{1/2} \mathrm{Re}_L^{3/4}, \tag{4.16}$$

so that the total number of Fourier coefficients scales as

$$V K^3 \sim \mathrm{Sc}^{3/2} \mathrm{Re}_L^{9/4}. \tag{4.17}$$

Thus, due to limitations on the available computer memory, DNS of homogeneous turbulent reacting flows has been limited to $\mathrm{Sc} \sim 1$ (i.e., gas-phase reactions). Moreover, because explicit ODE solvers (e.g., Runge-Kutta) are usually employed for time stepping, numerical stability puts an upper limit on reaction rate k_1. Although more complex chemistry can be implemented in a DNS, the range of reaction rates is severely limited by resolution and stability requirements.

In general, liquid-phase reactions ($\mathrm{Sc} \gg 1$) and fast chemistry are beyond the range of DNS. The treatment of inhomogeneous flows (e.g., a chemical reactor) adds further restrictions. Thus, although DNS is a valuable tool for studying fundamentals,[4] it is not a useful tool for chemical-reactor modeling. Nonetheless, much can be learned about scalar transport in turbulent flows from DNS. For example, valuable information about the effect of molecular diffusion on the joint scalar PDF can be easily extracted from a DNS simulation and used to validate the micromixing closures needed in other scalar transport models.

4.2 Large-eddy simulation

As discussed in Pope (2000), at high Reynolds numbers over 99 percent of the computational expense of DNS is used to resolve the dissipation range of the turbulent energy spectrum. However, as discussed in Chapter 2, the energy-containing scales determine most of the flow-dependent transport properties. Thus, in this sense, DNS 'wastes' most of the computational effort on resolving scales that are not very important for determining second-order quantities such as the Reynolds stresses and the scalar flux. Large-eddy simulation (LES), on the other hand, attempts to overcome this limitation by resolving only the largest (flow-dependent) scales. At high enough Reynolds number, it is hoped that the small scales will be flow-independent, and thus that they can be successfully modeled by an appropriate sub-grid-scale (SGS) model. Because they are fully three-dimensional and time-dependent, large-eddy simulations are still relatively expensive compared with RANS models. Moreover, because the small scales are not resolved, a closure is needed for SGS micromixing and chemical reactions. Nevertheless, the increased accuracy of the description of the energy-containing scales makes LES an attractive method for simulating turbulent transport in complex flows (Akselvoll and Moin 1996; Calmet and Magnaudet 1997; Desjardin and Frankel 1998; Pitsch and Steiner 2000; Wall *et al.* 2000; Hughes *et al.* 2001a; Hughes *et al.* 2001b).

4.2.1 Filtered Navier–Stokes equation

A detailed description of LES filtering is beyond the scope of this book (see, for example, Meneveau and Katz (2000) or Pope (2000)). However, the basic idea can be understood by considering a so-called 'sharp-spectral' filter in wavenumber space. For this filter, a cut-off frequency κ_c in the inertial range of the turbulent energy spectrum is chosen (see Fig. 4.1), and a low-pass filter is applied to the Navier–Stokes equation to separate the

4 See, for example, Eswaran and Pope (1988), Swaminathan and Bilger (1997), Vervisch and Poinsot (1998), and Poinsot and Veynante (2001).

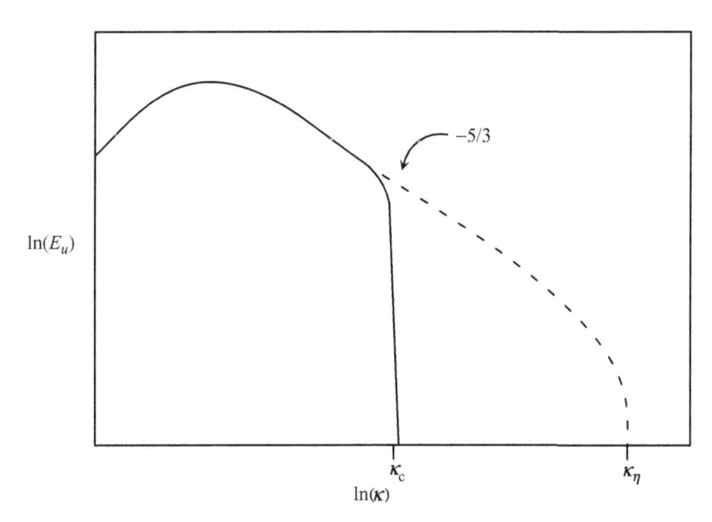

Figure 4.1. Sketch of LES energy spectrum with the sharp-spectral filter. Note that all information about length scales near the Kolmogorov scale is lost after filtering.

resolved scales from the sub-grid scales. In homogeneous turbulence, the filtered velocity field is related to the Fourier coefficients (i.e., in (4.2)) of the unfiltered velocity field by

$$\overline{\mathbf{U}}(\mathbf{x}, t) = \sum_{|\boldsymbol{\kappa}| \leq \kappa_c} \hat{\mathbf{U}}_{\boldsymbol{\kappa}}(t)\, e^{i\boldsymbol{\kappa}\cdot\mathbf{x}}, \tag{4.18}$$

where the sum is over a sphere of radius κ_c in wavenumber space. In physical space, the filtering operation is written as

$$\overline{\mathbf{U}}(\mathbf{x}, t) = \int_{-\infty}^{+\infty} \mathbf{U}(\mathbf{r}, t) G(\mathbf{r} - \mathbf{x})\, d\mathbf{r}, \tag{4.19}$$

where G is a *filtering function* with the bandwidth (Pope 2000)

$$\Delta = \frac{\pi}{\kappa_c}. \tag{4.20}$$

The filtering process removes all SGS fluctuations so that $\overline{\mathbf{U}}(\mathbf{x}, t)$ is considerably 'smoother' than $\mathbf{U}(\mathbf{x}, t)$. As a general rule, the bandwidth should be at least twice the smallest resolved scale in an LES calculation. However, it is not unusual to see LES simulations reported where the bandwidth is nearly equal to the grid spacing.

It can be easily shown (Pope 2000) that filtering and space/time derivatives commute in homogeneous flows. The filtered Navier–Stokes equation can thus be expressed as

$$\frac{\partial \overline{U}_i}{\partial t} + \overline{U}_j \frac{\partial \overline{U}_i}{\partial x_j} + \frac{\partial \tau^r_{ij}}{\partial x_j} = -\frac{1}{\rho}\frac{\partial \overline{p}}{\partial x_i} + \nu \nabla^2 \overline{U}_i, \tag{4.21}$$

where the *anisotropic residual stress tensor* is defined by

$$\tau^r_{ij} \equiv \overline{U_i U_j} - \overline{U}_i \overline{U}_j - \frac{1}{3}(\overline{U_i U_i} - \overline{U}_i \overline{U}_i), \tag{4.22}$$

and the *modified filtered pressure* is defined by

$$\overline{p} \equiv \int_{-\infty}^{+\infty} p(\mathbf{r}, t) G(\mathbf{r} - \mathbf{x}) \, \mathrm{d}\mathbf{r} + \frac{1}{3}(\overline{U_i U_i} - \overline{U}_i \overline{U}_i). \tag{4.23}$$

The divergence of (4.21) yields a Poisson equation for $\overline{p}$. However, the residual stress tensor τ_{ij}^{r} is unknown because it involves unresolved SGS terms (i.e., $\overline{U_i U_j}$). Closure of the residual stress tensor is thus a major challenge in LES modeling of turbulent flows.

The form of (4.21) is very similar to that of (2.93), p. 47, for the mean velocity $\langle \mathbf{U} \rangle$ found by Reynolds averaging. However, unlike the Reynolds stresses, the residual stresses depend on how the filter function G is defined (Pope 2000). Perhaps the simplest model for the residual stress tensor was proposed by Smagorinsky (1963):

$$\tau_{ij}^{\mathrm{r}} = -2(C_{\mathrm{S}}\Delta)^2 |\overline{S}| \overline{S}_{ij}, \tag{4.24}$$

where the filtered strain rate is defined by

$$\overline{S}_{ij} \equiv \frac{1}{2} \left(\frac{\partial \overline{U}_i}{\partial x_j} + \frac{\partial \overline{U}_j}{\partial x_i} \right), \tag{4.25}$$

and the characteristic filtered strain rate is given by

$$|\overline{S}| \equiv (2\overline{S}_{ij}\overline{S}_{ij})^{1/2}. \tag{4.26}$$

The Smagorinsky constant $C_{\mathrm{S}} \approx 0.17$ holds at high Reynolds number if the cut-off wavenumber κ_{c} falls in the inertial sub-range (Pope 2000). However, the same value is often used for LES at relatively low Reynolds number where no clear inertial sub-range can exist.

For inhomogeneous flows, $C_{\mathrm{S}} = 0.17$ has been shown to be too large. For example, for LES of turbulent channel flow a value in the range 0.065–0.10 yields better agreement with experimental data. The development of 'improved' SGS models that are applicable to more complex flows is an active area of research (Germano *et al.* 1991; Meneveau *et al.* 1996; Langford and Moser 1999). (See Meneveau and Katz (2000) and Pope (2000) for a detailed discussion of current models.) To date, no completely general model with a wide range of applicability is available and, due to the complexity of turbulence, perhaps none should be expected. Nevertheless, since the computational cost of LES is much smaller than that required for DNS, LES opens up the possibility of studying turbulent flows of industrial interest in more detail than is possible with RANS models.

4.2.2 LES velocity PDF

The filtering process used in LES results in a loss of information about the SGS velocity field. For homogeneous turbulence and the sharp-spectral filter, the residual velocity field[5]

[5] The residual velocity field is defined by $\mathbf{u}' = \mathbf{U} - \overline{\mathbf{U}}$.

can be expressed in terms of the discrete Fourier coefficients in (4.2) by

$$\mathbf{u}'(\mathbf{x}, t) = \sum_{|\boldsymbol{\kappa}|>\kappa_c} \hat{\mathbf{U}}_{\boldsymbol{\kappa}}(t)\, e^{i\boldsymbol{\kappa}\cdot\mathbf{x}}, \tag{4.27}$$

where the sum is now over all wavenumbers outside a sphere of radius κ_c. Note that, subject to the constraints of continuity, the Fourier coefficients on the right-hand side of (4.27) can be varied randomly without changing the filtered velocity $\overline{\mathbf{U}}$.[6] Thus, $\overline{\mathbf{U}}$ and $\mathbf{u}'$ can be treated as separate random fields. It is then possible to define a conditional one-point PDF for $\mathbf{u}'$ that describes the residual velocity fluctuations conditioned on the event that all Fourier coefficients in (4.18) are fixed (i.e., given that the random field $\overline{\mathbf{U}} = \overline{\mathbf{U}}^*$).[7] The sample space filtered velocity field $\overline{\mathbf{U}}^*$ is then assumed to be known, or can be found by solving the filtered Navier–Stokes equation. Unlike with the one-point velocity PDF defined in Chapter 2, the reader should note that the LES velocity PDF is defined in terms of the entire field $\overline{\mathbf{U}}^*$ (i.e., it depends on the value of the filtered velocity at every point in the flow). Thus, in theory, models for the LES velocity PDF should include non-local effects such as integrals and/or spatial derivatives of $\overline{\mathbf{U}}^*$.

Denoting the LES velocity PDF by $f_{\mathbf{U}|\overline{\mathbf{U}}}(\mathbf{V}|\overline{\mathbf{U}}^*; t)$,[8] the mean and covariance with respect to the LES velocity PDF are defined, respectively, by

$$\begin{aligned}\langle \mathbf{U}|\overline{\mathbf{U}}^*\rangle &\equiv \iiint_{-\infty}^{+\infty} \mathbf{V} f_{\mathbf{U}|\overline{\mathbf{U}}}(\mathbf{V}|\overline{\mathbf{U}}^*)\, d\mathbf{V} \\ &= \overline{\mathbf{U}}^*(\mathbf{x}, t) + \langle \mathbf{u}'|\overline{\mathbf{U}}^*\rangle \\ &= \overline{\mathbf{U}}^*(\mathbf{x}, t),\end{aligned} \tag{4.28}$$

and[9]

$$\begin{aligned}\langle u_i' u_j'|\overline{\mathbf{U}}^*\rangle &\equiv \iiint_{-\infty}^{+\infty} (V_i - \overline{U}_i)(V_j - \overline{U}_j) f_{\mathbf{U}|\overline{\mathbf{U}}}(\mathbf{V}|\overline{\mathbf{U}}^*)\, d\mathbf{V} \\ &= \langle U_i U_j|\overline{\mathbf{U}}^*\rangle - \overline{U}_i^*(\mathbf{x}, t)\overline{U}_j^*(\mathbf{x}, t).\end{aligned} \tag{4.29}$$

Note that we have made use of the fact that for a homogeneous flow with an isotropic filter $\langle \mathbf{u}'|\overline{\mathbf{U}}^*\rangle = \mathbf{0}$. More generally, the conditional expected value of the residual velocity field will depend on the filter choice.

Compared with the Reynolds stresses $\langle u_i u_j\rangle$, less is known concerning the statistical properties of the residual Reynolds stresses $\langle u_i' u_j'|\overline{\mathbf{U}}^*\rangle$. However, because they represent

[6] More precisely, the Fourier coefficients in (4.27) can be replaced by random variables with the following properties: $\boldsymbol{\kappa} \cdot \hat{\mathbf{U}}_{\boldsymbol{\kappa}} = 0$ and $\langle \hat{\mathbf{U}}_{\boldsymbol{\kappa}}\rangle = \mathbf{0}$ for all $\boldsymbol{\kappa}$ such that $|\boldsymbol{\kappa}| > \kappa_c$. An 'energy-conserving' scheme would also require that the expected value of the residual kinetic energy be the same for all choices of the random variable.

[7] The LES velocity PDF is a conditional PDF that can be defined in the usual manner by starting from the joint PDF for the discrete Fourier coefficients $\hat{\mathbf{U}}_{\boldsymbol{\kappa}}$.

[8] The velocity field is assumed to be statistically homogeneous. Thus, the LES velocity PDF does not depend explicitly on $\mathbf{x}$, only implicitly through $\overline{\mathbf{U}}^*(\mathbf{x}, t)$.

[9] The residual velocity covariance should not be confused with the Reynolds stresses. Indeed, most of the contribution to the Reynolds stresses comes from the filtered velocity field. Thus, in general, $\langle u_i' u_j'|\overline{\mathbf{U}}^*\rangle \ll \langle u_i u_j\rangle$.

the smallest scales of the flow, it can be expected that they will be more nearly isotropic, and thus simpler to model, than the Reynolds stresses.

An alternative method for describing the SGS velocity fluctuations is to define the filter density function (FDF) first proposed by Pope (1990):

$$h_{\mathbf{U}}(\mathbf{V};\mathbf{x},t) = \int_{-\infty}^{+\infty} \delta\left[\mathbf{U}(\mathbf{r},t) - \mathbf{V}\right] G(\mathbf{r}-\mathbf{x})\, \mathrm{d}\mathbf{r}. \tag{4.30}$$

Note that $h_{\mathbf{U}}$ operates on the random field $\mathbf{U}(\mathbf{r}, t)$ and (for fixed parameters $\mathbf{V}$, $\mathbf{x}$, and t) produces a real number. Thus, unlike the LES velocity PDF described above, the FDF is in fact a *random variable* (i.e., its value is different for each realization of the random field) defined on the ensemble of all realizations of the turbulent flow. In contrast, the LES velocity PDF is a true conditional PDF defined on the sub-ensemble of all realizations of the turbulent flow *that have the same filtered velocity field*. Hence, the filtering function enters into the definition of $f_{\mathbf{U}|\overline{\mathbf{U}}}(\mathbf{V}|\overline{\mathbf{U}}^*)$ only through the specification of the members of the sub-ensemble.

The LES velocity PDF and the FDF can also be defined for inhomogeneous flows, and for other filters (Gao and O'Brien 1993; Gao and O'Brien 1994). However, it is important to recognize that the statistical properties embodied in the LES velocity PDF and the FDF will most certainly depend on the choice of the filter. For example, in contrast to the LES velocity PDF, the FDF approach requires a 'positive' filter (Verman *et al.* 1994) in order to ensure that $h_{\mathbf{U}}$ is non-negative. Thus, since the sharp-spectral filter used in (4.27) is not a positive filter (Pope 2000), it should not be used to define the FDF. The functional form of the LES velocity PDF will also depend on the filter function. Thus, even at high Reynolds numbers, the LES velocity PDF for fully developed homogeneous turbulence may not become Gaussian, or even self-similar. However, when deviations from Gaussian behavior are small, it is possible to develop transported PDF models for the SGS velocity field (using either approach) based on linear stochastic differential equations, and we will explore this topic in more detail in Chapter 6. The methods described in Chapter 7 can then be used to simulate the effects of SGS velocity fluctuations on scalar transport.

4.2.3 Scalar transport

The filtered transport equation for an inert, passive scalar has the form

$$\frac{\partial \overline{\phi}}{\partial t} + \overline{U}_j \frac{\partial \overline{\phi}}{\partial x_j} + \frac{\partial \tau^{\mathrm{r}}_{\phi i}}{\partial x_i} = \Gamma \nabla^2 \overline{\phi}, \tag{4.31}$$

where the *residual scalar flux* is defined by

$$\tau^{\mathrm{r}}_{\phi i} \equiv \overline{U_i \phi} - \overline{U}_i \overline{\phi}. \tag{4.32}$$

By analogy with the Smagorinsky model, the SGS scalar flux can be modeled using a gradient-diffusion model (Eidson 1985):

$$\tau^{\mathrm{r}}_{\phi i} = -\Gamma_{\mathrm{sgs}} \frac{\partial \overline{\phi}}{\partial x_i}, \tag{4.33}$$

where the SGS turbulent diffusion coefficient is defined by

$$\Gamma_{\mathrm{sgs}} \equiv \frac{2(C_{\mathrm{S}}\Delta)^2}{\mathrm{Sc}_{\mathrm{sgs}}} |\overline{S}|. \tag{4.34}$$

The sub-grid-scale turbulent Schmidt number has a value of $\mathrm{Sc}_{\mathrm{sgs}} \approx 0.4$ (Pitsch and Steiner 2000), and controls the magnitude of the SGS turbulent diffusion. Note that due to the filtering process, the filtered scalar field will be considerably 'smoother' than the original field. For high-Schmidt-number scalars, the molecular diffusion coefficient (Γ) will be much smaller than the SGS diffusivity, and can thus usually be neglected.

As with the LES velocity PDF, a conditional PDF for the residual scalar field can be developed in terms of the LES composition PDF, denoted by $f_{\phi|\overline{\mathbf{U}},\overline{\phi}}(\psi|\overline{\mathbf{U}}^*, \overline{\phi}^*; \mathbf{x}, t)$.[10] For a homogeneous scalar field with an isotropic filter, the conditional expected value of the scalar will have the property $\langle \phi | \overline{\mathbf{U}}, \overline{\phi}^* \rangle = \overline{\phi}^*$. Moreover, a transport equation can be derived for the residual scalar variance defined by[11]

$$\langle \phi'^2 | \overline{\mathbf{U}}^*, \overline{\phi}^* \rangle \equiv \int_{-\infty}^{+\infty} (\psi - \overline{\phi})^2 f_{\phi|\overline{\mathbf{U}},\overline{\phi}}(\psi|\overline{\mathbf{U}}^*, \overline{\phi}^*) \, \mathrm{d}\psi. \tag{4.35}$$

For high-Schmidt-number scalars, the residual scalar variance can be significant, and a model for the LES composition PDF is needed to describe the residual scalar fluctuations. For example, as discussed in Section 5.3 for the mixture-fraction vector, a presumed PDF can be used to model the LES mixture-fraction PDF. Alternatively, for reacting flows, an LES composition PDF can be modeled using transported PDF methods, and we will look at this possibility in more detail in Chapter 6. In any case, it is important to understand that all SGS information concerning scalar fluctuations is lost due to filtering when LES is used to describe scalar transport (Tong 2001; Wang and Tong 2002; Rajagopalan and Tong 2003). If this information is needed, for example, to close the chemical source term, an SGS closure model must be used to close the LES scalar transport equation.

4.2.4 Reacting flow

For turbulent reacting flows, LES introduces an additional closure problem due to filtering of the chemical source term (Cook and Riley 1994; Cook *et al.* 1997; Jiménez *et al.* 1997; Cook and Riley 1998; Desjardin and Frankel 1998; Wall *et al.* 2000). For the one-step

[10] Alternatively, an LES joint velocity, composition PDF can be defined where both ϕ and $\mathbf{U}$ are random variables: $f_{\mathbf{U},\phi|\overline{\mathbf{U}},\overline{\phi}}(\mathbf{V}, \psi|\overline{\mathbf{U}}^*, \overline{\phi}^*; \mathbf{x}, t)$. In either case, the sample space fields $\overline{\mathbf{U}}^*$ and $\overline{\phi}^*$ are assumed to be known.

[11] In this definition, $\phi' = \phi - \overline{\phi}$ is the residual scalar field.

reaction appearing in (4.12), the filtered chemical source term for species A becomes

$$\overline{S}_{\mathrm{A}} = -k_1 \overline{\mathrm{AB}}, \tag{4.36}$$

and depends on the unknown filtered scalar product $\overline{\mathrm{AB}}$. Because chemical reactions occur at the unresolved scales, there is no easy (or generally applicable) way to close this term. Thus, unlike DNS, reacting-flow LES requires closure models (like the ones discussed in Chapter 5) to handle the chemical source term (Poinsot and Veynante 2001).

Generally speaking, any of the chemical-source-term closures that have been developed for RANS models can be applied to reacting-flow LES by making appropriate modifications. For example, methods based on the presumed mixture-fraction PDF can be applied by replacing the mixture-fraction mean $\langle \xi \rangle$ and variance $\langle \xi'^2 \rangle$ in (5.142), p. 175, by $\overline{\xi}^*$ and $\langle \xi'^2 | \overline{\mathbf{U}}^*, \overline{\xi}^* \rangle$, respectively. The LES transport equation for the filtered mixture fraction $\overline{\xi}^*$ has the same form as (4.30). The residual mixture-fraction variance $\langle \xi'^2 | \overline{\mathbf{U}}^*, \overline{\xi}^* \rangle$ can be found either by deriving and solving an LES transport equation similar to (3.105), p. 85, or by postulating a 'similarity' closure that depends only on $\overline{\xi}$. In any case, it is important to realize that LES of turbulent reacting flows does not avoid the closure problems associated with the chemical source term. Rational closure models that depend on the chemical time scales of the flow (see Section 5.1) must be implemented in order to produce accurate LES predictions for reacting flows. However, due to the relatively high cost of LES, it can be expected that closures based on presumed PDF methods will find the widest application. (See Branley and Jones (2001) for an example application using a beta PDF.) In Section 5.10, we will look at a class of closures based on multi-environment presumed PDF models that can be easily implemented in reacting-flow LES codes.

4.3 Linear-eddy model

We have seen that for high Reynolds and/or Schmidt numbers, DNS of complex flows is intractable due to the large computational requirements. LES, on the other hand, requires closures to describe molecular mixing and chemical reactions. One possibility for overcoming these limitations is to use a simple model for turbulent transport coupled with a detailed model for the scalar fields. The linear-eddy model (LEM) (Kerstein 1988) accomplishes this objective by introducing a one-dimensional approximation for the scalar fields. In the resulting model, the computational requirements for treating reactions and molecular diffusion without closures are significantly reduced as compared with DNS. The LEM has been successfully applied to study a number of turbulent mixing problems (Kerstein 1989; Kerstein 1990; Kerstein 1991a; Kerstein 1991b; Kerstein 1992).

The modeling ideas used in the LEM have recently been extended to include a one-dimensional description of the turbulence (Kerstein 1999a). This one-dimensional turbulence (ODT) model has been applied to shear-driven (Kerstein 1999a; Kerstein and Dreeben 2000) and buoyancy-driven flows (Kerstein 1999b), as well as to simple reacting shear flows (Echekki *et al.* 2001; Hewson and Kerstein 2001; Kerstein 2002). Since the

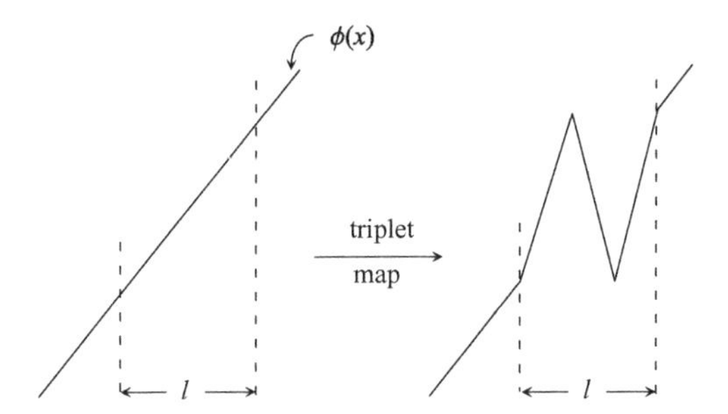

Figure 4.2. Sketch of triplet map.

basic ideas behind the LEM and ODT model are very similar, we will restrict our attention to the LEM and its extension to inhomogeneous flows.

4.3.1 Homogeneous flows

In the LEM, a one-dimensional, reaction-diffusion equation of the form[12]

$$\frac{\partial \phi}{\partial t} = \mathbf{\Gamma} \frac{\partial^2 \phi}{\partial x^2} + \mathbf{S}(\phi) \tag{4.37}$$

is solved over the entire range of length scales present in a turbulent scalar field. The computational grid spacing (Δx) is chosen such that $\Delta x \sim \lambda_{\mathrm{B}}$, where λ_{B} is the Batchelor length scale of the least-diffusive scalar. The size of the computational domain (L) is set such that $L_u \sim L$. In terms of Re_L and Sc, the LEM requires a spatial resolution with a total number of grid points that scales as $\mathrm{Sc}^{1/2}\mathrm{Re}_L^{3/4}$. Thus, for high Reynolds and Schmidt numbers, the computational requirements are large compared with RANS models, but still tractable compared with DNS.[13] The reaction-diffusion equation is solved numerically over *finite* time intervals interspersed with random 'turbulent-advection' events as described next.

In the LEM, turbulence is modeled by a random rearrangement process that compresses the scalar field locally to simulate the reduction in length scales that results from turbulent mixing. For example, with the triplet map, defined schematically in Fig. 4.2, a random length scale l is selected at a random point in the computational domain, and the scalar field is then compressed by a factor of three.[14] The PDF for l,

$$f(l) = \frac{5}{3} \frac{l^{-8/3}}{\left(\eta^{-5/3} - L_u^{-5/3}\right)} \quad \text{for} \quad \eta \leq l \leq L_u, \tag{4.38}$$

[12] $\mathbf{\Gamma}$ denotes the diagonal molecular diffusion matrix, which is assumed to be constant. However, in principle, any detailed molecular diffusion model could be used.

[13] Perhaps even more important is the fact that LEM does not require a numerical solution to the Navier–Stokes equation. Indeed, even a three-dimensional diffusion equation is generally less computationally demanding than the Poisson equation needed to find the pressure field.

[14] Although the triplet map has been employed in most applications, other maps could be used with similar results.

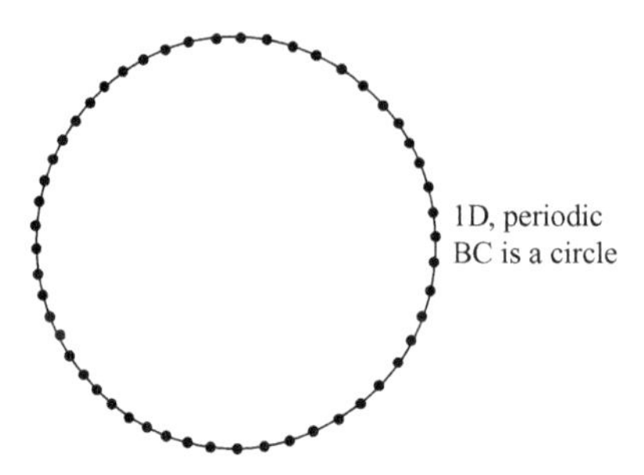

Figure 4.3. Sketch of computational domain used in the LEM for homogeneous flows. The circumference of the circular domain scales as $L \sim L_u$. Thus, the number of grid points required for fully resolved simulations scales as $\mathrm{Sc}^{1/2}\mathrm{Re}_L^{3/4}$.

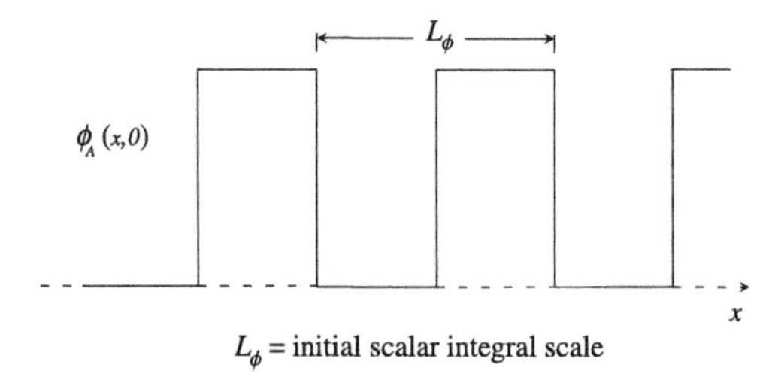

Figure 4.4. Sketch of initial scalar field used in the LEM for homogeneous flows.

and the frequency of random mixing events (to leading order at high Re),

$$\lambda = \frac{54}{5} \frac{\Gamma_\mathrm{T}}{L_u^3} \left(\frac{L_u}{\eta} \right)^{5/3} = \frac{54}{5} \frac{\epsilon}{k} \frac{1}{L_u} \mathrm{Re}_L^{5/4}, \tag{4.39}$$

where Γ_T is the local turbulent diffusivity, have been chosen to mimic inertial-range scaling. During the simulation, points x are randomly chosen in the domain $[0, L]$ according to a Poisson process with frequency λL. Then, with probability $f(l)$, an 'eddy event' of length l is selected at random, and the triplet map is applied at point x.

During the time intervals between random eddy events, (4.37) is solved numerically using the scalar fields that result from the random rearrangement process as initial conditions. A standard one-dimensional parabolic equation solver with periodic boundary conditions (BCs) is employed for this step. The computational domain is illustrated in Fig. 4.3. For a homogeneous scalar field, the evolution of $\phi(x, t)$ will depend on the characteristic length scale of the initial scalar fields. Typical initial conditions are shown in Fig. 4.4. Note that no model is required for the scalar dissipation rate because Batchelor-scale diffusion is treated exactly. Note also that this implies that the reaction-diffusion equation will be very stiff relative to RANS models, and thus will require considerably more computing resources.

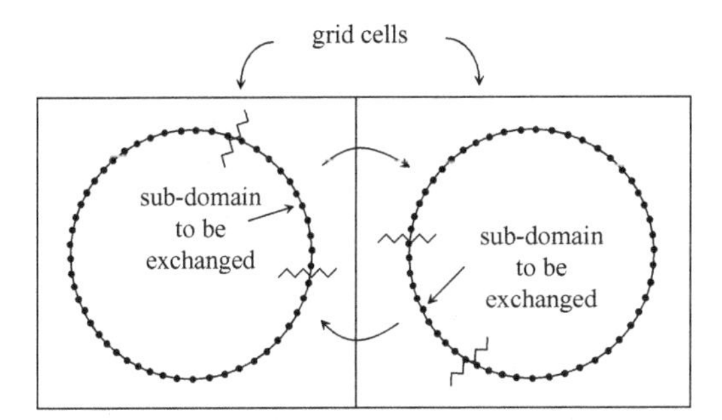

Figure 4.5. Sketch of how LEM can be applied to an inhomogeneous flow. At fixed time intervals, sub-domains from neighboring grid cells are exchanged to mimic advection and turbulent diffusion.

4.3.2 Inhomogeneous flows

For inhomogeneous flows, the LEM can be employed as a sub-grid-scale model inside each CFD cell as shown in Fig. 4.5. Turbulent diffusion and convection is then simulated by exchanging sub-domains between neighboring computational cells. The length of each sub-domain will be approximately equal to the local turbulence integral scale (i.e., L_u), and thus will contain many grid points from the LEM computational domain. For LES, the length of the sub-domain will be roughly the grid spacing. Attempts at implementing this algorithm have shown it to be computationally feasible, but very expensive when accurate results are desired. Indeed, inside each CFD grid cell, the LEM computational domain can require up to 10 000 grid points at moderate Reynolds numbers in order to solve accurately for the strong coupling between the chemistry and molecular diffusion in (4.37) (Caillau 1994).

Another method for using the LEM to model inhomogeneous flows is to let each LEM computational domain represent a Lagrangian fluid particle (Pit 1993). The transported PDF transport models described in Chapter 6 can then be used to model the effects of turbulent advection on each fluid particle, and a simple turbulent mixing model can be invoked to exchange LEM sub-domains between 'neighboring' fluid particles. However, the computational expense of such a detailed model would be at least an order of magnitude larger than that required for transported PDF simulations (see Chapter 7). Thus, 'Lagrangian LEM' will most likely remain computationally intractable for some time to come. Note, also, that Lagrangian LEM could be used for LES of turbulent reacting flows.

For high Reynolds and Schmidt numbers, LEM contains too much detail concerning the dissipation range of the scalar spectrum. Indeed, given the large differences in the length and time scales that are resolved, it is usually possible to treat a turbulent-reacting-flow problem more efficiently by invoking a suitably defined sub-grid-scale closure for the chemical source term. As with DNS, the LEM does not require a closure for chemical reactions and diffusion. Thus, it is very useful for studying the physics of scalar mixing (e.g., McMurtry *et al.* 1993), and for using these results to develop new sub-grid-scale molecular mixing models. The LEM has also been employed to evaluate sub-grid-scale turbulent combustion models (e.g., Desjardin and Frankel 1996) with limited success.

4.4 RANS turbulence models

In Section 2.2, the Reynolds-averaged Navier–Stokes (RANS) equations were derived. The resulting transport equations and unclosed terms are summarized in Table 2.4. In this section, the most widely used closures are reviewed. However, due to the large number of models that have been proposed, no attempt at completeness will be made. The reader interested in further background information and an in-depth discussion of the advantages and limitations of RANS turbulence models can consult any number of textbooks and review papers devoted to the topic. In this section, we will follow most closely the presentation by Pope (2000).

RANS turbulence models are the workhorse of CFD applications for complex flow geometries. Moreover, due to the relatively high cost of LES, this situation is not expected to change in the near future. For turbulent reacting flows, the additional cost of dealing with complex chemistry will extend the life of RANS models even further. For this reason, the chemical-source-term closures discussed in Chapter 5 have all been formulated with RANS turbulence models in mind. The focus of this section will thus be on RANS turbulence models based on the turbulent viscosity hypothesis and on second-order models for the Reynolds stresses.

The presentation in this section has been intentionally kept short, as our primary objective is to present the 'standard' form of each model so that the reader can refer to them in later chapters. Nevertheless, it is extremely important for the reader to realize that the quality of a reacting-flow simulation will in no small part depend on the performance of the turbulence model. The latter will depend on a number of issues[15] that are outside the scope of this text, but that should not be neglected when applying RANS models to complex flows.

4.4.1 Turbulent-viscosity-based models

As shown in Chapter 2, the transport equation for the Reynolds-averaged velocity is given by

$$\frac{\partial \langle U_i \rangle}{\partial t} + \langle U_j \rangle \frac{\partial \langle U_i \rangle}{\partial x_j} + \frac{\partial}{\partial x_j} \langle u_j u_i \rangle = -\frac{1}{\rho} \frac{\partial \langle p \rangle}{\partial x_i} + \nu \nabla^2 \langle U_i \rangle. \tag{4.40}$$

In order to use this equation for CFD simulations, the unclosed term involving the Reynolds stresses ($\langle u_i u_j \rangle$) must be modeled. Turbulent-viscosity-based models rely on the following

[15] For example, boundary conditions, Reynolds and Schmidt number effects, applicability of a model to a particular flow, etc.

hypothesis:

$$\langle u_i u_j \rangle = \frac{2}{3} k \delta_{ij} - \nu_T \left(\frac{\partial \langle U_i \rangle}{\partial x_j} + \frac{\partial \langle U_j \rangle}{\partial x_i} \right), \tag{4.41}$$

where ν_T is the so-called 'turbulent viscosity.' A discussion of the implicit and explicit assumptions that underlie this model can be found in Durbin *et al.* (1994) and Pope (2000). Here it suffices to note that (4.41) shifts the problem of modeling the six components of the Reynolds stress tensor to the 'simpler' problem of modeling the scalar field ν_T.

The simplest models for the turbulent viscosity require no additional transport equations. For example, the assumption that ν_T is constant can be applied to only a small class of turbulent flows, but is manifestly the simplest to apply. The next level – mixing length models – relates the turbulent viscosity to the mean rate of strain by introducing a characteristic 'mixing length.' The applicability of this model is again extremely limited, and thus we shall not discuss it further. Readers interested in more background information on these so-called 'algebraic' models for the turbulent viscosity should consult Pope (2000).

The next level of turbulence models introduces a transport equation to describe the variation of the turbulent viscosity throughout the flow domain. The simplest models in this category are the so-called 'one-equation' models wherein the turbulent viscosity is modeled by

$$\nu_T(\mathbf{x}, t) = l_m (k(\mathbf{x}, t))^{1/2}, \tag{4.42}$$

where k is the turbulent kinetic energy, and l_m is the mixing length, which is assumed to be known. The model is closed by solving the transport equation for k ((2.115), p. 51); however, models are required for the unclosed terms in (2.115). The 'standard' closures are as follows:

$$\frac{1}{2} \langle u_j u_j u_i \rangle + \frac{1}{\rho} \langle u_i p \rangle - \nu \nabla^2 k - \nu \frac{\partial^2 \langle u_i u_j \rangle}{\partial x_i \partial x_j} = - \frac{\nu_T}{\sigma_k} \frac{\partial k}{\partial x_i} \tag{4.43}$$

and

$$\varepsilon = C_D \frac{k^{3/2}}{l_m}, \tag{4.44}$$

where $\sigma_k = 1.0$ and C_D are model constants. The transport equation for k then becomes

$$\frac{\partial k}{\partial t} + \langle \mathbf{U} \rangle \cdot \nabla k = \nabla \cdot \left(\frac{\nu_T}{\sigma_k} \nabla k \right) + \mathcal{P} - \varepsilon, \tag{4.45}$$

where $\mathcal{P}$ is the turbulent kinetic energy production term given by (2.116), p. 51. The principal weakness of the one-equation model is that the mixing length must be fixed by the user. Indeed, for complex turbulent flows, it is unreasonable to expect that the mixing length will be the same at every point in the flow.

At the next level of complexity, a second transport equation is introduced, which effectively removes the need to fix the mixing length. The most widely used 'two-equation' model is the k–ε model wherein a transport equation for the turbulent dissipation rate is formulated. Given $k(\mathbf{x}, t)$ and $\varepsilon(\mathbf{x}, t)$, the turbulent viscosity is specified by

$$\nu_{\mathrm{T}} = C_{\mu} \frac{k^2}{\varepsilon}, \tag{4.46}$$

where $C_{\mu} = 0.09$ is a new model constant. As discussed in Chapter 2, the transport equation for ε is not based on a term-by-term closure of (2.125) on p. 52. Instead, it is modeled by analogy to the transport equation for the turbulent kinetic energy. The 'standard' model has the form:

$$\frac{\partial \varepsilon}{\partial t} + \langle \mathbf{U} \rangle \cdot \nabla \varepsilon = \nabla \cdot \left(\frac{\nu_{\mathrm{T}}}{\sigma_{\varepsilon}} \nabla \varepsilon \right) + \frac{\varepsilon}{k} \left(C_{\varepsilon 1} \mathcal{P} - C_{\varepsilon 2} \varepsilon \right), \tag{4.47}$$

where the model 'constants' have been fitted to experimental data for canonical flows: $\sigma_{\varepsilon} = 1.3$, $C_{\varepsilon 1} = 1.44$, and $C_{\varepsilon 2} = 1.92$. However, it must be recognized that different values will be needed to model specific flows accurately (e.g., shear flow, turbulent jets, etc.), and thus the 'standard' values represent a compromise chosen to give the 'best overall' results (Pope 2000).

Other two-equation models have been developed using various combinations of k and ε to derive alternative transport equations to replace (4.47). The most popular is, perhaps, the k–ω model (see Wilcox (1993) for a detailed discussion of its advantages), where the turbulence frequency $\omega \equiv \varepsilon / k$ is used in place of ε. The 'standard' transport equation for $\omega(\mathbf{x}, t)$ is

$$\frac{\partial \omega}{\partial t} + \langle \mathbf{U} \rangle \cdot \nabla \omega = \nabla \cdot \left(\frac{\nu_{\mathrm{T}}}{\sigma_{\omega}} \nabla \omega \right) + C_{\omega 1} \frac{\omega}{k} \mathcal{P} - C_{\omega 2} \omega^2, \tag{4.48}$$

where $\sigma_{\omega} = 1.3$, $C_{\omega 1} = 0.44$, and $C_{\omega 2} = 0.92$. Note that (4.48) does not follow directly by combining (4.45) and (4.47). Normally, such a disagreement would be unacceptable; however, given the uncertainty inherent in the derivation of (4.47), in this case it can be overlooked. One noteworthy advantage of the k–ω model over the k–ϵ model is its treatment of the near-wall region in boundary-layer flows (Wilcox 1993), especially for low-Reynolds-number flows. Indeed, the k–ω equation can be applied well into the viscous sub-layer, while the k–ϵ model (with wall functions) requires the first grid point away from the wall to lie in the log layer.

Turbulence models based on the turbulent-viscosity hypothesis have been applied to the greatest range of turbulent-flow problems, and are computationally the most economical. Nevertheless, while it can be acceptably accurate for simple flows (e.g., two-dimensional thin shear flows without significant streamline curvature or mean pressure gradients), the k–ε model is known to be quite inaccurate for others (i.e., the predicted mean velocity is qualitatively incorrect). Corrections based on non-linear turbulent-viscosity models have proven to be useful for predicting secondary flows; however, their accuracy for more complex flows with strong mean streamline curvature and/or rapid changes in the mean velocity has not been demonstrated (Pope 2000). For this reason, models based on solving a transport equation for the Reynolds stresses have received considerable attention. We look at these models next.

4.4.2 Reynolds-stress transport equation

The second-order modeling approach starts with the transport equation for the Reynolds stresses ((2.105), p. 49):

$$\frac{\partial \langle u_i u_j \rangle}{\partial t} + \langle U_k \rangle \frac{\partial \langle u_i u_j \rangle}{\partial x_k} + \frac{\partial \langle u_k u_i u_j \rangle}{\partial x_k} = \mathcal{P}_{ij} + \Pi_{ij} + \nu \nabla^2 \langle u_i u_j \rangle - \varepsilon_{ij}, \tag{4.49}$$

where the production term $\mathcal{P}_{ij}$ ((2.106), p. 49) is closed. For inhomogeneous flows, the velocity–pressure-gradient tensor Π_{ij} can be decomposed into two terms by defining the following functions (Pope 2000):

$$\text{pressure transport:} \quad \mathcal{T}_{kij}^{(\mathrm{p})} \equiv \frac{2\langle u_k p \rangle}{3\rho} \delta_{ij}, \tag{4.50}$$

$$\text{anisotropic rate-of-strain tensor:} \quad \mathcal{R}_{ij}^{(\mathrm{a})} \equiv \Pi_{ij} - \frac{1}{3} \Pi_{kk} \delta_{ij}. \tag{4.51}$$

The Reynolds-stress transport equation then becomes

$$\frac{\partial \langle u_i u_j \rangle}{\partial t} + \langle U_k \rangle \frac{\partial \langle u_i u_j \rangle}{\partial x_k} + \frac{\partial}{\partial x_k} \left(\langle u_k u_i u_j \rangle + \mathcal{T}_{kij}^{(\mathrm{p})} \right) = \nu \nabla^2 \langle u_i u_j \rangle + \mathcal{P}_{ij} + \mathcal{R}_{ij}^{(\mathrm{a})} - \varepsilon_{ij}. \tag{4.52}$$

Models are required to close the flux terms on the left-hand side, and the final two terms on the right-hand side of (4.52).

Experience with applying the Reynolds-stress model (RSM) to complex flows has shown that the most critical term in (4.52) to model precisely is the anisotropic rate-of-strain tensor $\mathcal{R}_{ij}^{(\mathrm{a})}$ (Pope 2000). Relatively simple models are thus usually employed for the other unclosed terms. For example, the dissipation term is often assumed to be isotropic:

$$\varepsilon_{ij} = \frac{2}{3} \varepsilon \delta_{ij}. \tag{4.53}$$

However, at low Reynolds number (e.g., the near-wall region), the dissipation tensor can be highly anisotropic and other models are required (Pope 2000). For example, a simple extension of (4.53) yields (Rotta 1951)

$$\varepsilon_{ij} = \frac{\langle u_i u_j \rangle}{k} \varepsilon. \tag{4.54}$$

More elaborate models have been proposed to improve the near-wall treatment, and these are discussed in Pope (2000).

The flux terms are usually modeled by invoking a gradient-diffusion hypothesis. For example, one of the simplest models is

$$\langle u_k u_i u_j \rangle + \mathcal{T}_{kij}^{(\mathrm{p})} = -C_{\mathrm{s}} \frac{k}{\varepsilon} \langle u_k u_l \rangle \frac{\partial \langle u_i u_j \rangle}{\partial x_l}, \tag{4.55}$$

where $C_{\mathrm{s}} = 0.22$. A more general form has also been suggested in the literature:

$$\langle u_k u_i u_j \rangle + \mathcal{T}_{kij}^{(\mathrm{p})} = -C_{\mathrm{s}} \frac{k}{\varepsilon} \left(\langle u_i u_l \rangle \frac{\partial \langle u_j u_k \rangle}{\partial x_l} + \langle u_j u_l \rangle \frac{\partial \langle u_k u_i \rangle}{\partial x_l} + \langle u_k u_l \rangle \frac{\partial \langle u_i u_j \rangle}{\partial x_l} \right). \tag{4.56}$$

However, comparison with data (Launder 1996) indicates that the additional complexity does not yield much improvement over (4.55).

In (4.54) and (4.55), k is found directly from $\langle u_i u_j \rangle$, but ε requires a separate transport equation that is found by a slight modification of (4.47):

$$\frac{\partial \varepsilon}{\partial t} + \langle \mathbf{U} \rangle \cdot \nabla \varepsilon = \frac{\partial}{\partial x_i} \left(C_\varepsilon \frac{k}{\varepsilon} \langle u_i u_j \rangle \frac{\partial \varepsilon}{\partial x_j} \right) + \frac{\varepsilon}{k} \left(C_{\varepsilon 1} \mathcal{P} - C_{\varepsilon 2} \varepsilon \right), \tag{4.57}$$

where $C_\varepsilon = 0.15$ and $\mathcal{P}$ is found directly from (2.116) on p. 51 using the Reynolds stresses. A logical extension of the 'spectral equilibrium' model for ε would be to consider non-equilibrium spectral transport from large to small scales. Such models are an active area of current research (e.g., Schiestel 1987 and Hanjalić *et al.* 1997). Low-Reynolds-number models for ε typically add new terms to correct for the viscous sub-layer near walls, and adjust the model coefficients to include a dependency on Re_L. These models are still *ad hoc* in the sense that there is little physical justification – instead models are validated and tuned for particular flows.

Returning to (4.52), it should be noted that many Reynolds-stress models have been proposed in the literature, which differ principally by the closure used for the anisotropic rate-of-strain tensor. Nevertheless, almost all closures can be written as (Pope 2000)

$$\mathcal{R}_{ij}^{(\mathrm{a})} = \varepsilon \sum_{n=1}^{8} f^{(n)} \mathcal{T}_{ij}^{(n)}, \tag{4.58}$$

where $f^{(n)}$ are scalar coefficients, and $\mathcal{T}_{ij}^{(n)}$ are non-dimensional, symmetric, deviatoric tensors. The latter are defined by

$$\mathcal{T}_{ij}^{(1)} \equiv b_{ij}, \tag{4.59}$$

$$\mathcal{T}_{ij}^{(2)} \equiv b_{ij}^2 - \frac{1}{3} b_{kk}^2 \delta_{ij}, \tag{4.60}$$

$$\mathcal{T}_{ij}^{(3)} \equiv \hat{S}_{ij}, \tag{4.61}$$

$$\mathcal{T}_{ij}^{(4)} \equiv \hat{S}_{ik} b_{kj} + b_{ik} \hat{S}_{kj} - \frac{2}{3} \hat{S}_{kl} b_{lk} \delta_{ij}, \tag{4.62}$$

$$\mathcal{T}_{ij}^{(5)} \equiv \hat{\Omega}_{ik} b_{kj} - b_{ik} \hat{\Omega}_{kj}, \tag{4.63}$$

$$\mathcal{T}_{ij}^{(6)} \equiv \hat{S}_{ik} b_{kj}^2 + b_{ik}^2 \hat{S}_{kj} - \frac{2}{3} \hat{S}_{kl} b_{lk}^2 \delta_{ij}, \tag{4.64}$$

$$\mathcal{T}_{ij}^{(7)} \equiv \hat{\Omega}_{ik} b_{kj}^2 - b_{ik}^2 \hat{\Omega}_{kj}, \tag{4.65}$$

and

$$\mathcal{T}_{ij}^{(8)} \equiv b_{ik} \hat{S}_{kl} b_{lj} - \frac{1}{3} \hat{S}_{kl} b_{lk} \delta_{ij}. \tag{4.66}$$

In these expressions, the anisotropy tensor is defined by

$$b_{ij} \equiv \frac{\langle u_i u_j \rangle}{2k} - \frac{1}{3} \delta_{ij}, \tag{4.67}$$

Table 4.1. *The scalar coefficients $f^{(n)}$ for four different Reynolds-stress models.* The SSG model is generally considered to yield the best results for a fairly wide range of flows. However, all models are known to yield poor results in specific cases. For example, certain wall-bounded flows are particularly problematic.

	LLR-IP[a]	LRR-IQ[a]	JM[a]	SSG[a]
$f^{(1)}$	$-2C_1$	$-2C_1$	$-2C_1$	$-C_1 - C_1^* \frac{\mathcal{P}}{\varepsilon}$
$f^{(2)}$	0	0	0	C_2
$f^{(3)}$	$\frac{4}{3}C_2$	$\frac{4}{5}$	$2(C_4 - C_2)$	$C_3 - C_3^*(b_{ij}b_{ij})^{1/2}$
$f^{(4)}$	$2C_2$	$\frac{6}{11}(2 + 3C_2)$	$-3C_2$	C_4
$f^{(5)}$	$2C_2$	$\frac{2}{11}(10 - 7C_2)$	$3C_2 + 4C_3$	C_5
$f^{(6)}$	0	0	0	0
$f^{(7)}$	0	0	0	0
$f^{(8)}$	0	0	0	0
	$C_1 = 1.8$	$C_1 = 1.5$	$C_1 = 1.5$	$C_1 = 3.4$, $C_1^* = 1.8$
	$C_2 = 0.6$	$C_2 = 0.4$	$C_2 = -0.53$	$C_2 = 4.2$
			$C_3 = 0.67$	$C_3 = 0.8$, $C_3^* = 1.3$
			$C_4 = -0.12$	$C_4 = 1.25$
				$C_5 = 0.4$

[a] Defined in Pope (2000).

and is exactly zero in isotropic turbulence. The other terms appearing in the definition of $\mathcal{T}_{ij}^{(n)}$ are the normalized mean rate of strain,

$$\hat{S}_{ij} \equiv \frac{k}{2\varepsilon}\left(\frac{\partial \langle U_i \rangle}{\partial x_j} + \frac{\partial \langle U_j \rangle}{\partial x_i}\right), \tag{4.68}$$

and the normalized rate of rotation,

$$\hat{\Omega}_{ij} \equiv \frac{k}{2\varepsilon}\left(\frac{\partial \langle U_i \rangle}{\partial x_j} - \frac{\partial \langle U_j \rangle}{\partial x_i}\right). \tag{4.69}$$

The various models differ by the choice of the scalar coefficients $f^{(n)}$. Four examples of widely used models (described in Pope 2000) are shown in Table 4.1.

It is important to be aware of the fact that most turbulence models have been developed for and validated against turbulent shear flows parallel to walls. Models for separated and recirculating flows have received much less attention. The turbulent impinging jet shown in Fig. 4.6 is a good example of a flow used to test the shortcomings of widely employed turbulence models. The unique characteristics of this flow include the following.[16]

(1) Near the centerline turbulence is created by normal straining as opposed to shear.
(2) RMS velocity fluctuations normal to the wall are larger than those parallel to the wall.

[16] See Cooper *et al.* (1993a) and Cooper *et al.* (1993b) for further details.

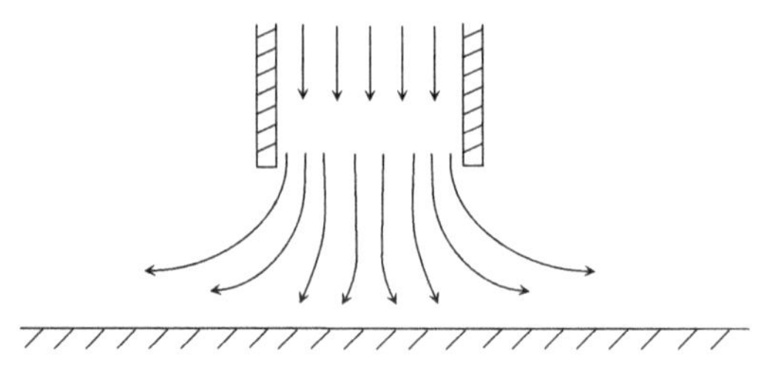

Figure 4.6. Sketch of impinging flow.

(3) Length scales near the wall are strongly influenced by the jet and depend only weakly on the distance to the wall.
(4) Convective transport towards the wall is important.

Owing to these unique characteristics, additional wall-reflection terms are required in the pressure redistribution model in order to obtain satisfactory agreement with data for the impinging jet flow. A detailed discussion of RANS models that employ a more physically realistic description of the pressure fluctuations in the near-wall region can be found in Pope (2000). However, one obvious shortcoming of current wall models is that they typically depend explicitly on the unit normal to the wall, which makes it very difficult to apply them to complex geometries.

In summary, a number of different RANS turbulence models are available for CFD simulation of turbulent flows. None, however, are capable of modeling all flows with reasonable accuracy, and in many cases application of the 'wrong' model can result in large errors in the predictions of even the mean velocity field (Wilcox 1993). Since the accuracy of the turbulence model will have a strong influence on models for scalar transport, it is important to understand the properties and limitations of the turbulence model being used and, if possible, to validate the flow predictions against experimental data. This will be particularly important for turbulent reacting flows wherein the sub-grid-scale mixing model will depend on the turbulence integral time scale ($\tau_u = k/\varepsilon$). Accurate predictions for τ_u in the reaction zone will be crucial for obtaining accurate predictions of the effect of mixing on chemical reactions. Unfortunately, given the inherent uncertainty in existing models and the difficulties of experimental measurements for ε, it is often difficult to assess the quality of the predictions for τ_u in complex flows.

4.5 RANS models for scalar mixing

The transport equation for the mean of an inert scalar was derived in Section 3.3:

$$\frac{\partial \langle \phi \rangle}{\partial t} + \langle U_i \rangle \frac{\partial \langle \phi \rangle}{\partial x_i} = \Gamma \nabla^2 \langle \phi \rangle - \frac{\partial \langle u_i \phi \rangle}{\partial x_i}. \tag{4.70}$$

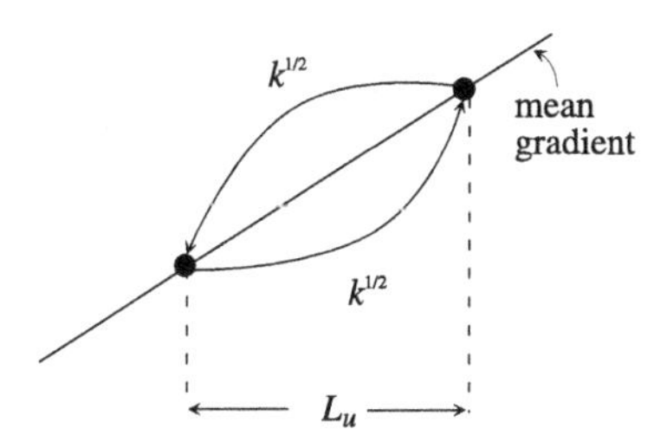

Figure 4.7. Sketch of physical mechanism responsible for gradient transport in homogeneous turbulence with a mean scalar gradient. Velocity fluctuations of characteristic size $k^{1/2}$ transport fluid particles up and down the mean scalar gradient over a distance with characteristic length L_u. The net result is a scalar flux down the mean scalar gradient.

The last term on the right-hand side is unclosed and represents scalar transport due to velocity fluctuations. The turbulent scalar flux $\langle u_i \phi \rangle$ varies on length scales on the order of the turbulence integral scales L_u, and hence is independent of molecular properties (i.e., ν and Γ).[17] In a CFD calculation, this implies that the grid size needed to resolve (4.70) must be proportional to the integral scale, and not the Batchelor scale as required in DNS. In this section, we look at two types of models for the scalar flux. The first is an extension of turbulent-viscosity-based models to describe the scalar field, while the second is a second-order model that is used in conjunction with Reynolds-stress models.

4.5.1 Turbulent-diffusivity-based models

In order to understand the physical basis for turbulent-diffusivity-based models for the scalar flux, we first consider a homogeneous turbulent flow with zero mean velocity gradient[18] and a uniform mean scalar gradient (Taylor 1921). In this flow, velocity fluctuations of characteristic size

$$u = |\mathbf{u}| \propto k^{1/2} \tag{4.71}$$

transport 'fluid particles' of size L_u randomly up and down the scalar gradient as shown in Fig. 4.7. The scalar gradient between two points x_1 and x_2 separated by a turbulence integral length scale in the direction of the mean scalar gradient can be approximated by

$$\frac{\phi(x_2) - \phi(x_1)}{L_u} \approx \frac{\mathrm{d}\langle \phi \rangle}{\mathrm{d}x}. \tag{4.72}$$

The random displacement of fluid particles along the scalar gradient creates a 'diffusive' flux proportional to the characteristic velocity and the characteristic concentration

[17] This is strictly true only at high Reynolds numbers, but most CFD codes do not account for molecular effects seen at low to moderate Reynolds numbers.

[18] Flows with shear do not exhibit simple gradient transport. See, for example, the homogeneous shear flow results reported by Rogers *et al.* (1986).

difference:

$$\begin{aligned}\langle u\phi\rangle &\propto k^{1/2}[\phi(x_2)-\phi(x_1)] \\ &\propto -L_u k^{1/2}\frac{\mathrm{d}\langle\phi\rangle}{\mathrm{d}x} \\ &\propto -\frac{k^2}{\varepsilon}\frac{\mathrm{d}\langle\phi\rangle}{\mathrm{d}x}.\end{aligned} \tag{4.73}$$

The scalar flux is then proportional to a 'turbulent-diffusion' coefficient:

$$\Gamma_\mathrm{T} \propto \frac{k^2}{\varepsilon} \tag{4.74}$$

multiplied by the mean scalar gradient. Note that the turbulent diffusivity is completely independent of the molecular-diffusion coefficient Γ. Indeed, Γ_T is generally several orders of magnitude larger than Γ. Note also that this argument implies that Γ_T will be the same for every chemical species.[19] Thus, at high Reynolds numbers, Γ_T will be the same for both gases and liquids (Pope 1998).

The turbulent diffusivity defined by (4.74) is proportional to the turbulent viscosity defined by (4.46). Turbulent-diffusivity-based models for the scalar flux extend this idea to arbitrary mean scalar gradients. The 'standard' gradient-diffusion model has the form

$$\langle u_i\phi\rangle = -\frac{\nu_\mathrm{T}}{\mathrm{Sc_T}}\frac{\partial\langle\phi\rangle}{\partial x_i}, \tag{4.75}$$

where $\mathrm{Sc_T} \approx 0.7$ is the turbulent Schmidt number that relates Γ_T to ν_T:

$$\Gamma_\mathrm{T} \equiv \frac{\nu_\mathrm{T}}{\mathrm{Sc_T}}. \tag{4.76}$$

While employed in most CFD codes, (4.75) is strictly valid only for isotropic turbulence (Biferale *et al.* 1995). Even for the simple case of a homogeneous turbulent shear flow (i.e., a uniform mean velocity gradient and a uniform mean scalar gradient), the gradient-diffusion model fails (Pope 2000) due to the fact that the scalar flux is not aligned with the mean scalar gradient. A slight modification of (4.75) leads to the modified gradient-diffusion model:[20]

$$\langle u_i\phi\rangle = -\frac{k}{\mathrm{Sc_T}\varepsilon}\langle u_i u_j\rangle\frac{\partial\langle\phi\rangle}{\partial x_j}, \tag{4.77}$$

which is often used in conjunction with a Reynolds-stress model for the turbulent flow. Although without a sound theoretical basis, the modified gradient-diffusion model does allow for a misalignment (albeit inaccurate!) between the scalar flux and the mean scalar gradient. More accurate models can be found by working directly with the scalar-flux transport equation.

[19] Again, this is strictly only true at sufficiently high Reynolds number where differential-diffusion effects become negligible.

[20] This type of model is usually referred to as an algebraic scalar-flux model. Similar models for the Reynolds-stress tensor are referred to as algebraic second-moment (ASM) closures. They can be derived from the scalar-flux transport equation by ignoring time-dependent and spatial-transport terms.

4.5.2 Scalar-flux transport equation

Like the Reynolds stresses, the scalar flux obeys a transport equation that was derived in Section 3.3:

$$\frac{\partial \langle u_i \phi \rangle}{\partial t} + \langle U_j \rangle \frac{\partial \langle u_i \phi \rangle}{\partial x_j} = \frac{\partial}{\partial x_j} \left(\mathcal{T}_{ij}^{\phi} - \langle u_i u_j \phi' \rangle - \frac{1}{\rho} \langle p \phi' \rangle \delta_{ij} \right) + \mathcal{P}_i^{\phi} + \mathcal{R}_i^{(\phi)} - \varepsilon_i^{\phi}, \tag{4.78}$$

where the production term $\mathcal{P}_i^{\phi}$ is closed and represents the source of scalar flux produced by the mean scalar gradient and by the mean velocity gradient. The pressure-scalar-gradient term $\mathcal{R}_i^{(\phi)}$ is unclosed and, like the related term in the Reynolds-stress transport equation, is the central focus of model development. The flux terms $\langle u_i u_j \phi' \rangle$ and $\langle p \phi' \rangle$ are responsible for spatial transport of scalar flux. On the other hand, the molecular-transport term $\mathcal{T}_{ij}^{\phi}$ will be small at high Reynolds numbers, and is usually neglected. Likewise, the scalar-flux dissipation term ε_i^{ϕ} will be null if local (small-scale) isotropy prevails.

In analogy to (4.55), the turbulent flux terms in (4.78) are usually modeled by invoking a gradient-diffusion hypothesis:[21]

$$\langle u_i u_j \phi' \rangle - \frac{1}{\rho} \langle p \phi' \rangle \delta_{ij} = -C_{\mathrm{s}}^{\phi} \frac{k}{\varepsilon} \langle u_j u_k \rangle \frac{\partial \langle u_i \phi \rangle}{\partial x_k}, \tag{4.79}$$

where $C_{\mathrm{s}}^{\phi} = 0.22$. However, in analogy to (4.47), a more general 'symmetric' form can also be used:

$$\langle u_i u_j \phi' \rangle + \frac{1}{\rho} \langle p \phi' \rangle \delta_{ij} = -C_{\mathrm{s}}^{\phi} \frac{k}{\varepsilon} \left(\langle u_i u_k \rangle \frac{\partial \langle u_j \phi \rangle}{\partial x_k} + \langle u_j u_k \rangle \frac{\partial \langle u_i \phi \rangle}{\partial x_k} + \langle u_k \phi \rangle \frac{\partial \langle u_i u_j \rangle}{\partial x_k} \right). \tag{4.80}$$

Again, however, comparison with data (Launder 1996) indicates that the additional complexity does not yield much improvement over (4.79).

For passive scalars, the model for $\mathcal{R}_i^{(\phi)}$ cannot be chosen independently from the model for the anisotropic rate-of-strain tensor ($\mathcal{R}_{ij}^{(\mathrm{a})}$) used for the Reynolds stresses (Pope 1994b). Nevertheless, many authors have proposed inconsistent models by postulating terms for $\mathcal{R}_i^{(\phi)}$ without using the corresponding terms in $\mathcal{R}_{ij}^{(\mathrm{a})}$ (or vice versa). One way to avoid this is to derive the scalar-flux transport equation directly from the joint velocity, composition PDF transport equation, as is done in Chapter 6. The scalar-flux coefficient tensor corresponding to the Reynolds-stress models in Table 4.1 that results from following this approach can be found in Pope (1994b). A consistent second-moment modeling approach would use the RSM from Table 4.1 with the corresponding scalar-flux model. However, these models have not been as extensively validated as the simple model given below.

[21] It can be noted that this model is not symmetric with respect to interchanging i and j, although the unclosed term is symmetric.

As for the models derived from the PDF transport equation, nearly all widely used models for $\mathcal{R}_i^{(\phi)}$ can be expressed as

$$\mathcal{R}_i^{(\phi)} = G_{ij} \langle u_j \phi \rangle, \tag{4.81}$$

where the scalar-flux coefficient tensor G_{ij} is a function of known quantities:

$$G_{ij} \left(\langle u_i u_j \rangle, \frac{\partial \langle U_i \rangle}{\partial x_j}, \epsilon \right). \tag{4.82}$$

At high Reynolds numbers, the modeled scalar-flux transport equation thus reduces (using (4.78)) to

$$\begin{aligned} \frac{\partial \langle u_i \phi \rangle}{\partial t} + \langle U_j \rangle \frac{\partial \langle u_i \phi \rangle}{\partial x_j} = {} & \frac{\partial}{\partial x_j} \left(C_{\mathrm{s}}^{\phi} \frac{k}{\varepsilon} \langle u_j u_k \rangle \frac{\partial \langle u_i \phi \rangle}{\partial x_k} \right) \\ & - \langle u_i u_j \rangle \frac{\partial \langle \phi \rangle}{\partial x_j} - \langle u_j \phi \rangle \frac{\partial \langle U_i \rangle}{\partial x_j} + G_{ij} \langle u_j \phi \rangle, \end{aligned} \tag{4.83}$$

and must be solved in conjunction with a Reynolds-stress model. A relatively simple model for G_{ij} is often employed:

$$G_{ij} \equiv -C_1 \frac{\varepsilon}{k} \delta_{ij} - C_2 \frac{\varepsilon}{k} b_{ij} + C_3 \frac{\partial \langle U_i \rangle}{\partial x_j}, \tag{4.84}$$

and has been found to yield good agreement with data for simple flows (Launder 1996).

Note that the last two terms on the right-hand of (4.83) can be combined to define

$$M_{ij} \equiv G_{ij} - \frac{\partial \langle U_i \rangle}{\partial x_j}. \tag{4.85}$$

If one then neglects the accumulation and transport terms in (4.83),[22] an algebraic second-moment (ASM) model for the scalar flux results:

$$M_{ij} \langle u_j \phi \rangle = \langle u_i u_j \rangle \frac{\partial \langle \phi \rangle}{\partial x_j}, \tag{4.86}$$

and relates the scalar flux to the mean scalar gradient. In general, the second-order tensor M_{ij} will be non-symmetric, and the resulting scalar-flux model cannot be expressed in a simple gradient-diffusion form. Nevertheless, a widely employed simplification is to assume that M_{ij} is isotropic:

$$M_{ij} \propto -\frac{\epsilon}{k} \delta_{ij}, \tag{4.87}$$

so that the scalar flux can be expressed as in (4.77).

Similar techniques have been employed to derive ASM models for turbulent reacting flows (Adumitroaie *et al.* 1997). It can be noted from (3.102) on p. 84 that the chemical source term will affect the scalar flux. For example, for a scalar involved in a first-order

[22] This assumption is strictly valid only for stationary, homogeneous turbulence.

reaction, (4.83) is closed:

$$\frac{\partial \langle u_i\phi\rangle}{\partial t} + \langle U_j\rangle \frac{\partial \langle u_i\phi\rangle}{\partial x_j} = \frac{\partial}{\partial x_j}\left(C_{\mathrm{s}}^{\phi}\frac{k}{\varepsilon}\langle u_j u_k\rangle \frac{\partial \langle u_i\phi\rangle}{\partial x_k}\right) - \langle u_i u_j\rangle \frac{\partial \langle \phi\rangle}{\partial x_j} - \langle u_j\phi\rangle \frac{\partial \langle U_i\rangle}{\partial x_j} + G_{ij}\langle u_j\phi\rangle - k_1\langle u_i\phi\rangle. \tag{4.88}$$

The resulting isotropic ASM then becomes

$$\langle u_i\phi\rangle \propto -\left(\frac{\tau_u}{1+\tau_u k_1}\right)\langle u_i u_j\rangle \frac{\partial \langle \phi\rangle}{\partial x_j}. \tag{4.89}$$

Thus, the ASM scalar flux in a first-order reacting flow will decrease with increasing reaction rate. For higher-order reactions, the chemical source term in (3.102) will be unclosed, and its net effect on the scalar flux will be complex. For this reason, transported PDF methods offer a distinct advantage: terms involving the chemical source term are closed so that its effect on the scalar flux is treated exactly. We look at these methods in Chapter 6.

4.5.3 Scalar-variance transport equation

For fast, non-premixed reactions,[23] the reaction rate is controlled by molecular diffusion.[24] The reaction rate should thus be largest at the scales characteristic of molecular diffusion. For slow reactions, on the other hand, mixing is complete all the way down to the molecular scale before much reaction can occur. Thus the reaction occurs in a premixed mode at all length scales, and is dominated by scales near L_ϕ since they represent the bulk of the scalar field. Because the characteristic length scale of the chemical source terms depends on the rates of the chemical reactions, $\langle \mathbf{S}(\phi)\rangle$ has no simple closure in terms of moments of the scalar fields. This is the major distinction (and difficulty) between turbulent mixing of inert tracers and turbulent reacting flows. For very fast, non-premixed reactions, the reaction is completely controlled by the rate of turbulent mixing. In this limit, the rate of mixing of an inert scalar can often be quantified by the scalar variance $\langle \phi'^2\rangle$ of an inert scalar (i.e., mixture fraction).

Starting with the scalar transport equation, a transport equation for the inert-scalar variance was derived in Section 3.3 ((3.105), p. 85):

$$\frac{\partial \langle \phi'^2\rangle}{\partial t} + \langle U_i\rangle \frac{\partial \langle \phi'^2\rangle}{\partial x_i} + \frac{\partial \langle u_i\phi'^2\rangle}{\partial x_i} = \Gamma \nabla^2 \langle \phi'^2\rangle + \mathcal{P}_\phi - \varepsilon_\phi. \tag{4.90}$$

Assuming that the scalar flux $\langle u_i\phi\rangle$ has been closed,[25] this equation has two unclosed terms. The first is the scalar-variance flux $\langle u_i\phi'^2\rangle$, which is usually closed by invoking a

[23] The standard example is A + B → P, where the reaction time scale is much shorter than the Kolmogorov time scale.

[24] In other words, reaction can only occur once A and B have diffused together so that they coexist at the same spatial location.

[25] It is imperative that the same closure for the scalar flux be used in (4.70) to find the scalar mean and the scalar-variance-production term $\mathcal{P}_\phi$.

gradient-diffusion model:

$$\langle u_i \phi'^2 \rangle = -\Gamma_{\mathrm{T}} \frac{\partial \langle \phi'^2 \rangle}{\partial x_i}. \tag{4.91}$$

However, if a Reynolds-stress model is used to describe the turbulence, a modified gradient-diffusion model can be employed:

$$\langle u_i \phi'^2 \rangle = -\frac{k}{\mathrm{Sc}_{\mathrm{T}} \varepsilon} \langle u_i u_j \rangle \frac{\partial \langle \phi'^2 \rangle}{\partial x_j}. \tag{4.92}$$

The second unclosed term is the scalar dissipation rate ε_ϕ. The most widely used closure for this term is the 'equilibrium' model (Spalding 1971; Béguier *et al.* 1978):

$$\varepsilon_\phi = C_\phi \frac{\varepsilon}{k} \langle \phi'^2 \rangle. \tag{4.93}$$

As its name implies, this closure is valid in fully developed turbulence when the dissipation scales are in spectral equilibrium with the energy-containing scales. In fact, as noted at the end of Section 3.2, (4.93) is actually a model for the scalar spectral energy transfer rate $\mathcal{T}_\phi$ through the inertial-convective sub-range in homogeneous turbulence. For inhomogeneous turbulence, it can be expected that ε_ϕ will depend on the degree of turbulent anisotropy (i.e., the Reynolds stresses) and the mean shear rate, since these will surely affect the energy transfer rate from large to small scales (Durbin 1982; Elgobashi and Launder 1983; Dowling 1991; Dowling 1992). These effects can be (partially) modeled by introducing a transport equation for the scalar dissipation rate.

4.5.4 Scalar-dissipation transport equation

A number of different authors have proposed transport models for the scalar dissipation rate in the same general 'scale-similarity' form as (4.47):

$$\frac{\partial \varepsilon_\phi}{\partial t} + \langle \mathbf{U} \rangle \cdot \nabla \varepsilon_\phi = \nabla \cdot \left(\frac{\nu_{\mathrm{T}}}{\sigma_{\varepsilon_\phi}} \nabla \varepsilon_\phi \right) + \left(C_{\mathrm{P1}} \frac{1}{\langle \phi'^2 \rangle} \mathcal{P}_\phi + C_{\mathrm{P2}} \frac{1}{k} \mathcal{P} - C_{\mathrm{D1}} \frac{\varepsilon_\phi}{\langle \phi'^2 \rangle} - C_{\mathrm{D2}} \frac{\varepsilon}{k} \right) \varepsilon_\phi. \tag{4.94}$$

The model constant C_{P1} controls the production due to mean scalar gradients, while C_{P2} controls the effect of mean velocity gradients. C_{D1} and C_{D2} control the molecular dissipation rate, and the corresponding terms differ only by the definition of the characteristic dissipation time scale. For non-premixed inflow conditions, the inlet boundary conditions for (4.94) are simply $\varepsilon_\phi = 0$. Sanders and Gökalp (1998) have investigated the predictions of available models of this form for axisymmetric turbulent jets; they have found that $C_{\mathrm{P1}} = C_{\mathrm{D1}} = 1.0$, $C_{\mathrm{D1}} = 0.725$, and $C_{\mathrm{D2}} = 0.95$ yield good agreement with the experimental data. Nevertheless, they also found that the general features of scalar dissipation in round turbulent jets were well predicted by all models. The effect of small-scale anisotropy of the scalar-gradient fluctuations has also been added to (4.94) by Gonzalez (2000). The predictive ability of this type of model for more complex turbulent flows has not been extensively investigated due to the lack of experimental data for ε_ϕ. However, it can be expected that 'scale-similarity' models of this form will be inadequate for describing non-equilibrium scalar fields resulting, for example, from non-equilibrium inlet flow conditions.

4.6 Non-equilibrium models for scalar dissipation

In many reacting flows, the reactants are introduced into the reactor with an integral scale L_ϕ that is significantly different from the turbulence integral scale L_u. For example, in a CSTR, L_u is determined primarily by the actions of the impeller. However, L_ϕ is fixed by the feed tube diameter and feed flow rate. Thus, near the feed point the scalar energy spectrum will not be in equilibrium with the velocity spectrum. A relaxation period of duration on the order of τ_u is required before equilibrium is attained. In a reacting flow, because the relaxation period is relatively long, most of the fast chemical reactions can occur before the equilibrium model, (4.93), is applicable.

Non-equilibrium models have been proposed to account for the 'redistribution' of scalar energy in wavenumber space. For example, Baldyga (1989) derived an *ad hoc* multi-scale model to describe the forward cascade of scalar energy from large to small scales, and has applied it to several reacting flows (e.g., Baldyga and Henczka 1995; Baldyga and Henczka 1997), However, this model lacks a specific expression for the scalar dissipation rate, and does not include 'backscatter' from small to large scales. Thus it is unable to describe non-equilibrium relaxation (like that shown in Fig. 4.13).

In a similar manner, the spectral relaxation (SR) model (Fox 1995) accounts for non-equilibrium effects by dividing the scalar variance into a finite number of wavenumber bands, each with its own characteristic scalar spectral energy transfer rate. These rates follow directly (Fox 1999) from the scalar spectral energy transfer function ((3.73), p. 78), and include both the 'forward cascade' and 'backscatter.' A sketch of the spectral partition is shown in Fig. 4.8 for the case where the Schmidt number is greater than unity. In terms of a simple model, one can represent the scalar energy spectrum by a series of back-mixed CSTRs with known flow rates (see Fig. 4.9). Because the bulk of the scalar dissipation occurs at the smallest scales, only the final stage contributes to the scalar-variance decay. The total number of stages depends explicitly on $\text{Re}_1 \equiv \text{Re}_L^{1/2}$ and Sc. In general, the number of stages in the inertial-convective sub-range increases as $\log(\text{Re}_1)$ (Fox 1995), so that even four stages suffice to describe fairly high Reynolds numbers. Likewise, for $1 < \text{Sc}$, the number of stages in the viscous-convective sub-range increases as $\log(\text{Sc})$.

Despite the explicit dependence on Reynolds number, in its present form the model does not describe low-Reynolds-number effects on the *steady-state* mechanical-to-scalar time-scale ratio (R defined by (3.72), p. 76). In order to include such effects, they would need to be incorporated in the scalar spectral energy transfer rates. In the original model, the spectral energy transfer rates were chosen such that $R(t) \to R_0 = 2$ for $\text{Sc} = 1$ and $\mathcal{P}_\phi = 0$ in stationary turbulence. In the version outlined below, we allow R_0 to be a model parameter. DNS data for $90 \lesssim R_\lambda$ suggest that R_0 is nearly constant. However, for lower

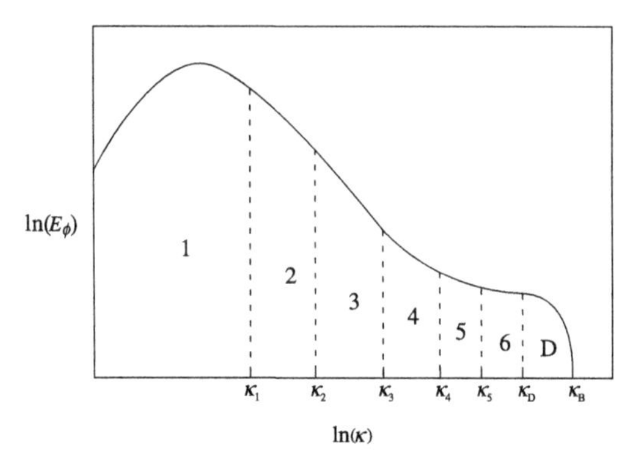

Figure 4.8. Sketch of wavenumber bands in the spectral relaxation (SR) model. The scalar-dissipation wavenumber κ_D lies one decade below the Batchelor-scale wavenumber κ_B. All scalar dissipation is assumed to occur in wavenumber band $[\kappa_D, \infty)$. Wavenumber band $[0, \kappa_1)$ denotes the energy-containing scales. The inertial-convective sub-range falls in wavenumber bands $[\kappa_1, \kappa_3)$, while wavenumber bands $[\kappa_3, \kappa_D)$ contain the viscous-convective sub-range.

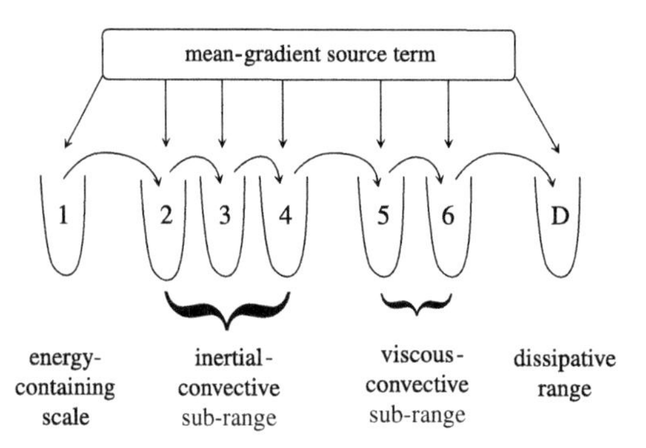

Figure 4.9. Sketch of CSTR representation of the SR model for $1 < \mathrm{Sc}$. Each wavenumber band is assumed to be 'well mixed' in the sense that it can be represented by a single variable $\langle \phi'^2 \rangle_n$. Scalar energy 'cascades' from large scales to the dissipative range where it is destroyed. 'Backscatter' also occurs in the opposite direction, and ensures that any arbitrary initial spectrum will eventually attain a self-similar equilibrium form. In the presence of a mean scalar gradient, scalar energy is added to the system by the scalar-flux energy spectrum. The fraction of this energy that falls in a particular wavenumber band is determined by forcing the self-similar spectrum for $\mathrm{Sc} = 1$ to be the same for all values of the mean-gradient source term.

Reynolds numbers, its value is significantly smaller than the high-Reynolds-number limit. Despite its inability to capture low-Reynolds-number effects on the steady-state scalar dissipation rate, the SR model does account for Reynolds-number and Schmidt-number effects on the *dynamic* behavior of $R(t)$.

4.6.1 Spectral relaxation model

As an illustrative example, we will consider the SR model equations for $\mathrm{Sc} \lesssim 1$ and $\mathrm{Re}_\lambda < 100$. (A general derivation of the model is given in Appendix A.) For this range of

Reynolds numbers, the inertial-convective sub-range contains two stages. The wavenumber bands are defined by the cut-off wavenumbers:[26]

$$\kappa_0 \equiv 0, \tag{4.95}$$

$$\kappa_1 \equiv \mathrm{Re}_1^{-3/2} \kappa_\eta, \tag{4.96}$$

$$\kappa_2 \equiv \left(\frac{3}{C_u \mathrm{Re}_1 + 2} \right)^{3/2} \kappa_u, \tag{4.97}$$

$$\kappa_3 \equiv \mathrm{Sc}^{1/2} \kappa_u, \tag{4.98}$$

where the Kolmogorov wavenumber is defined by[27]

$$\kappa_\eta \equiv \frac{1}{\eta} = \frac{\mathrm{Re}_1^{3/2}}{L_u}, \tag{4.99}$$

and the velocity-dissipation wavenumber is defined by[28]

$$\kappa_u \equiv C_u^{3/2} \kappa_\eta \tag{4.100}$$

with $C_u = (0.1)^{2/3} = 0.2154$. For $\mathrm{Sc} \lesssim 1$, the scalar-dissipation wavenumber is proportional to the velocity-dissipation wavenumber:[29]

$$\kappa_{\mathrm{D}} \equiv \kappa_3 = \mathrm{Sc}^{1/2} \kappa_u. \tag{4.101}$$

The definitions of the cut-off wavenumbers for higher Reynolds numbers and for $1 < \mathrm{Sc}$ can be found in Fox (1997).

In the SR model, the scalar energy in each wavenumber band is defined by[30]

$$\langle \phi'^2 \rangle_n(t) \equiv \int_{\kappa_{n-1}}^{\kappa_n} E_\phi(\kappa, t)\, \mathrm{d}\kappa. \tag{4.102}$$

The model equation for the scalar energies can then be derived from (3.73) on p. 78:

$$\frac{\mathrm{d}\langle \phi'^2 \rangle_1}{\mathrm{d}t} = \mathcal{T}_1 + \gamma_1 \mathcal{P}_\phi, \tag{4.103}$$

$$\frac{\mathrm{d}\langle \phi'^2 \rangle_2}{\mathrm{d}t} = \mathcal{T}_2 + \gamma_2 \mathcal{P}_\phi, \tag{4.104}$$

$$\frac{\mathrm{d}\langle \phi'^2 \rangle_3}{\mathrm{d}t} = \mathcal{T}_3 + \gamma_3 \mathcal{P}_\phi, \tag{4.105}$$

[26] As described in Fox (1995), the wavenumber bands are chosen to be as large as possible, subject to the condition that the characteristic time scales decrease as the band numbers increase. This condition is needed to ensure that scalar energy does not 'pile up' at intermediate wavenumber bands. The rate-controlling step in equilibrium spectral decay is then the scalar spectral energy transfer rate ($\mathcal{T}_1$) from the lowest wavenumber band.

[27] This definition implies that κ_1 actually falls within the energy-containing range, and not at the top of the inertial range as suggested in Fig. 4.8. In fact, at high Reynolds numbers, the amount of scalar energy in the first wavenumber band approaches zero, suggesting that it represents more accurately the low-wavenumber spectrum and not the energy-containing range. Its placement, however, is dictated by the desire to have its spectral transfer rate equal to the steady-state scalar dissipation rate.

[28] κ_u is defined to be the same as κ_{DI} from the model energy spectrum in Chapter 2. Thus, it lies approximately one decade below the Kolmogorov wavenumber.

[29] For $1 < \mathrm{Sc}$, $\kappa_{\mathrm{D}} \equiv 0.1\mathrm{Sc}^{1/2}\kappa_{\mathrm{B}}$.

[30] For $n = \mathrm{D}$, the integral covers $\kappa = \kappa_{\mathrm{D}}$ to $\kappa = \infty$.

and

$$\frac{d\langle\phi'^2\rangle_D}{dt} = \mathcal{T}_D + \gamma_D \mathcal{P}_\phi - \varepsilon_\phi, \tag{4.106}$$

where the scalar spectral energy transfer rates are defined by

$$\mathcal{T}_n \equiv \int_{\kappa_{n-1}}^{\kappa_n} T_\phi(\kappa, t)\, d\kappa \tag{4.107}$$

and

$$\mathcal{T}_D \equiv \int_{\kappa_D}^{\infty} T_\phi(\kappa, t)\, d\kappa. \tag{4.108}$$

Note that by using (3.74) on p. 78, it is easily shown that $\mathcal{T}_1 + \mathcal{T}_2 + \mathcal{T}_3 + \mathcal{T}_D = 0$ (i.e., spectral energy transfer is conservative).

The terms involving γ_n in the SR model equations correspond to the fraction of the scalar-variance production that falls into a particular wavenumber band. In principle, γ_n could be found from the scalar-flux spectrum (Fox 1999). Instead, it is convenient to use a 'self-similarity' hypothesis that states that for Sc = 1 at spectral equilibrium the fraction of scalar variance that lies in a particular wavenumber band will be independent of $\mathcal{P}_\phi$. Applying this hypothesis to (4.103)–(4.106) yields:[31]

$$\gamma_1 = (1-b)(1-a)(1-a/2), \tag{4.109}$$

$$\gamma_2 = a(1-b)(1-a/2), \tag{4.110}$$

$$\gamma_3 = a(1-b)/2, \tag{4.111}$$

$$\gamma_D = b, \tag{4.112}$$

where[32]

$$a \equiv 1 - \frac{1}{C_u \mathrm{Re}_1} \tag{4.113}$$

and[33]

$$b \equiv \frac{2C_d}{1 + C_B \mathrm{Re}_1/R_0 + [(1 + C_B \mathrm{Re}_1/R_0)^2 - 4C_d C_D \mathrm{Re}_1/R_0]^{1/2}}. \tag{4.114}$$

In the definition of b, R_0 is the equilibrium mechanical-to-scalar time-scale ratio found with Sc = 1 and $\mathcal{P}_\phi = 0$.[34] The parameters C_D, C_B, and C_d appear in the SR model for the scalar dissipation rate discussed below. Note that, by definition, $\gamma_1 + \gamma_2 + \gamma_3 + \gamma_D = 1$.

31 These relations originally appeared in Fox (1999) with a different definition for b and for three inertial-range stages. By comparing with the earlier definitions, the reader will note a pattern that can be used to find γ_n for any arbitrary number of stages.

32 The condition that a must be positive limits the applicability of the model to $1 < C_u\mathrm{Re}_1$ or $12 < R_\lambda$. This corresponds to $\kappa_1 = \kappa_u = 0.1\kappa_\eta$ so that scalar energy is transferred directly from the lowest-wavenumber band to the dissipative range. However, at such low Reynolds numbers, the spectral transfer rates used in the model cannot be expected to be accurate. In particular, the value of R_0 would need to account for low-Reynolds-number effects.

33 This definition for b is different than in Fox (1999). In the current version of the model, C_D no longer depends on C_s and C_d, and thus the steady-state value of $\langle\phi'^2\rangle_D$ has a more complicated dependence on model parameters.

34 As noted earlier, in the original model $R_0 = 2$.

The scalar variance is found by summation of the spectral energies:

$$\langle \phi'^2 \rangle = \sum_{n=1}^{3} \langle \phi'^2 \rangle_n + \langle \phi'^2 \rangle_D. \tag{4.115}$$

Thus the SR model yields the standard scalar-variance transport equation for homogeneous flow:

$$\frac{d\langle \phi'^2 \rangle}{dt} = \mathcal{P}_\phi - \varepsilon_\phi. \tag{4.116}$$

The only other equation needed to close the model is an expression for the scalar dissipation rate, which is found by starting from (3.81) on p. 79 (also see Appendix A):[35]

$$\frac{d\varepsilon_\phi}{dt} = \gamma_D \mathcal{P}_\phi \frac{\varepsilon_\phi}{\langle \phi'^2 \rangle_D} + C_D \left(\frac{\varepsilon}{\nu}\right)^{1/2} \mathcal{T}_\varepsilon + C_s \left(\frac{\varepsilon}{\nu}\right)^{1/2} \varepsilon_\phi - C_d \frac{\varepsilon_\phi}{\langle \phi'^2 \rangle_D} \varepsilon_\phi, \tag{4.117}$$

where

$$C_D \equiv 2\Gamma \kappa_D^2 \left(\frac{\nu}{\varepsilon}\right)^{1/2} = 0.02, \tag{4.118}$$

$C_s = C_B - C_D$, $C_B = 1$, and $C_d = 3$ (Fox 1995).[36] Note that at spectral equilibrium, $\mathcal{P}_\phi = \varepsilon_\phi$, $\mathcal{T}_\varepsilon = \mathcal{T}_D = \varepsilon_\phi(1 - \gamma_D)$, and (with $Sc = 1$) $R = R_0$. The right-hand side of (4.117) then yields (4.114). Also, it is important to recall that unlike (4.94), which models the *flux of scalar energy* into the dissipation range, (4.117) is a true small-scale model for ε_ϕ. For this reason, integral-scale terms involving the mean scalar gradients and the mean shear rate do not appear in (4.117). Instead, these effects must be accounted for in the model for the spectral transfer rates.

As discussed in Section 3.3, the terms on the right-hand side of (4.117) have specific physical interpretations in terms of the transport equation for the scalar dissipation rate ((3.114), p. 86). The first term corresponds to production by the mean scalar gradient. The second term corresponds to scalar dissipation generated by the flux of scalar energy into the scalar dissipation range. The third term corresponds to vortex stretching (or gradient amplification) due to coupling with the velocity field in the dissipation range. The last term corresponds to molecular dissipation. Note that all the terms on the right-hand side – except the mean-gradient-production term – scale as Re_l. At high Reynolds numbers a dynamic equilibrium will result wherein the mechanical-to-scalar time-scale ratio $R(t)$ is related to the spectral transfer rate by

$$\frac{C_d}{Re_l} \frac{\langle \phi'^2 \rangle}{\langle \phi'^2 \rangle_D} R^2(t) - C_s R(t) - C_D \frac{k}{\varepsilon \langle \phi'^2 \rangle} \mathcal{T}_\varepsilon(t) = 0. \tag{4.119}$$

The dynamic behavior of $R(t)$ will thus be controlled by $\mathcal{T}_\varepsilon(t)$, which in turn will be controlled by the rate of spectral transfer from large to small scales.

[35] In the original model (Fox 1999), the definition of ε_ϕ differed by a factor of two. Likewise, the parameters C_s and C_D have been changed by the same factor.

[36] DNS data suggest that $C_B \approx 0.6$. The effect of changing C_B on the predicted dynamical behavior is, however, small.

4.6.2 Spectral transfer rates

The spectral transfer rates are assumed to be local in wavenumber space:[37]

$$\mathcal{T}_n(\langle \phi'^2 \rangle_{n-1}, \langle \phi'^2 \rangle_n, \langle \phi'^2 \rangle_{n+1}). \tag{4.120}$$

In principle, the forward and backward transfer rates can be computed directly from DNS (see Appendix A). However, they are more easily computed by assuming idealized forms for the scalar energy spectrum (Fox 1995). In the general formulation (Fox 1999), they include both a forward cascade (α) and backscatter (β):

$$\mathcal{T}_1 = -(\alpha_{12} + \beta_{12})\langle \phi'^2 \rangle_1 + \beta_{21}\langle \phi'^2 \rangle_2, \tag{4.121}$$

$$\begin{aligned} \mathcal{T}_2 = {} & (\alpha_{12} + \beta_{12})\langle \phi'^2 \rangle_1 - (\alpha_{23} + \beta_{23})\langle \phi'^2 \rangle_2 \\ & - \beta_{21}\langle \phi'^2 \rangle_2 + \beta_{32}\langle \phi'^2 \rangle_3, \end{aligned} \tag{4.122}$$

$$\begin{aligned} \mathcal{T}_3 = {} & (\alpha_{23} + \beta_{23})\langle \phi'^2 \rangle_2 - (\alpha_{3\mathrm{D}} + \beta_{3\mathrm{D}})\langle \phi'^2 \rangle_3 \\ & - \beta_{32}\langle \phi'^2 \rangle_3 + \beta_{\mathrm{D}3}\langle \phi'^2 \rangle_{\mathrm{D}}, \end{aligned} \tag{4.123}$$

$$\mathcal{T}_{\mathrm{D}} = (\alpha_{3\mathrm{D}} + \beta_{3\mathrm{D}})\langle \phi'^2 \rangle_3 - \beta_{\mathrm{D}3}\langle \phi'^2 \rangle_{\mathrm{D}}, \tag{4.124}$$

and

$$\mathcal{T}_{\varepsilon} = (\alpha_{3\mathrm{D}} + \beta_{3\mathrm{D}})\langle \phi'^2 \rangle_3 - \beta_{\varepsilon}\varepsilon_{\phi}. \tag{4.125}$$

The forward cascade parameters can be determined from the fully developed *isotropic* velocity spectrum (Fox 1995):[38]

$$\alpha_{12} = R_0 \frac{\varepsilon}{k}, \qquad \alpha_{23} = \frac{\alpha_{12}}{a}, \qquad \alpha_{3\mathrm{D}} = \frac{2C_u \mathrm{Re}_1 \alpha_{12}}{aC_u \mathrm{Re}_1 + 3(1 - \mathrm{Sc}^{-1/3})}, \tag{4.126}$$

$$\beta_{12} = c_b \alpha_{12}, \qquad \beta_{23} = c_b \alpha_{23}, \qquad \beta_{3\mathrm{D}} = c_b \alpha_{3\mathrm{D}}, \tag{4.127}$$

where $c_{\mathrm{b}} = 1$ controls the relative amount of backscatter. (Note that in order to account for *anisotropic* turbulence, it would be necessary to modify the cascade parameters to include terms that depend on the mean velocity gradients.) The backscatter parameters are then chosen such that the equilibrium spectrum is the same for all values of c_{b} (Fox 1999):

$$\beta_{21} = c_{\mathrm{b}}(\alpha_{23} - \alpha_{12}), \qquad \beta_{32} = c_{\mathrm{b}}(\alpha_{3\mathrm{D}} - \alpha_{12}), \tag{4.128}$$

$$\beta_{\mathrm{D}3} = c_{\mathrm{b}}\alpha_{12}\mathrm{Sc}^{1/2}(1 - b)/b, \qquad \beta_{\varepsilon} = c_{\mathrm{b}}\mathrm{Sc}^{1/2}(1 - b). \tag{4.129}$$

The Schmidt-number dependence of $\beta_{\mathrm{D}3}$ is a result of the definition of κ_{D}, and has been verified using DNS (Fox and Yeung 1999).

The initial conditions for $\langle \phi'^2 \rangle_n$ and ε_ϕ are determined by the initial scalar spectrum. Examples of SR model predictions with four different initial conditions are shown in

[37] This is consistent with DNS data for passive scalar mixing in isotropic turbulence (Yeung 1996).

[38] From the definition of $\alpha_{3\mathrm{D}}$, it can be seen that the minimum Schmidt number that can be used corresponds to $\kappa_2 = \kappa_{\mathrm{D}}$.

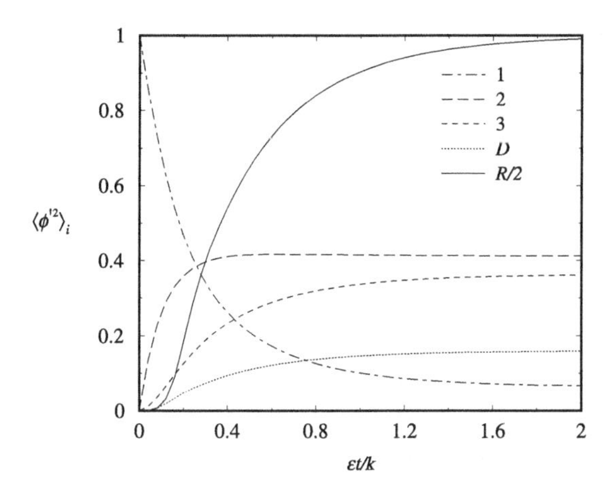

Figure 4.10. Predictions of the SR model for $\mathrm{Re}_\lambda = 90$ and $\mathrm{Sc} = 1$ for homogeneous scalar mixing in stationary turbulence. For these initial conditions, all scalar energy is in the first wavenumber band. Curves 1–3 and D correspond to the fraction of scalar energy in each wavenumber band: $\langle\phi'^2\rangle_n/\langle\phi'^2\rangle$. Note the relatively long transient period needed for $R(t)$ to approach its asymptotic value of $R_0 = 2$.

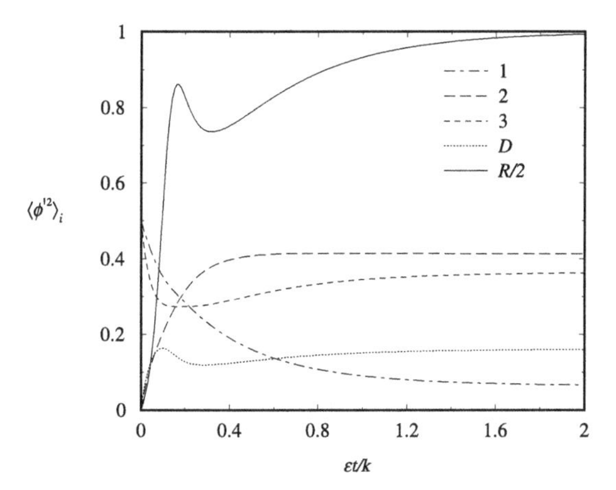

Figure 4.11. Predictions of the SR model for $\mathrm{Re}_\lambda = 90$ and $\mathrm{Sc} = 1$ for homogeneous scalar mixing. For these initial conditions, the scalar energy is equally divided between the first and third wavenumber bands. Curves 1–3 and D correspond to the fraction of scalar energy in each wavenumber band: $\langle\phi'^2\rangle_n/\langle\phi'^2\rangle$.

Figs. 4.10–4.13. In stationary turbulence (k and ε constant), the SR model attains a dynamical steady state wherein the variables $\langle\phi'^2\rangle_n$ and ε_ϕ reach their equilibrium values. In this limit (for $\mathrm{Sc} = 1$),

$$R = \frac{k\varepsilon_\phi}{\varepsilon\langle\phi'^2\rangle} = R_0 \tag{4.130}$$

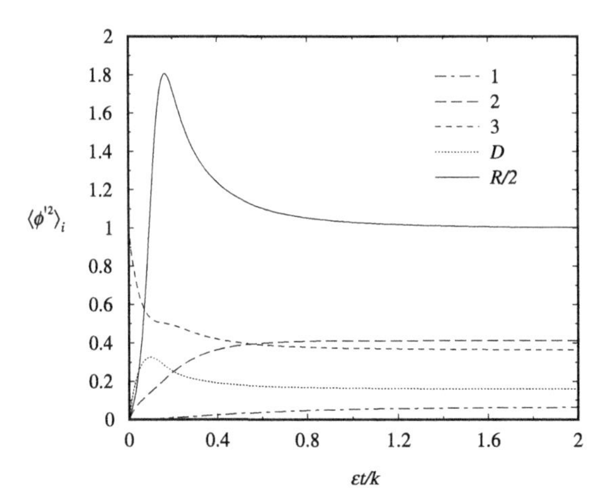

Figure 4.12. Predictions of the SR model for $\mathrm{Re}_\lambda = 90$ and $\mathrm{Sc} = 1$ for homogeneous scalar mixing. For these initial conditions, all scalar energy is in the third wavenumber band. Curves 1–3 and D correspond to the fraction of scalar energy in each wavenumber band: $\langle\phi'^2\rangle_n/\langle\phi'^2\rangle$. Note that the forward cascade from 3 to 4 is larger than backscatter from 3 to 2. Thus the scalar dissipation rate rises suddenly before dropping back to the steady-state value.

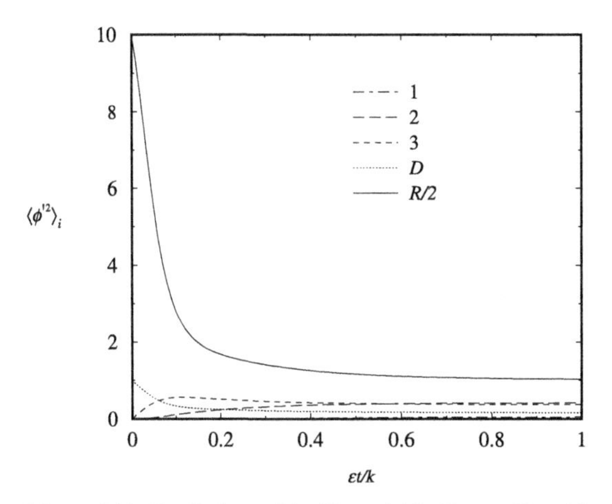

Figure 4.13. Predictions of the SR model for $\mathrm{Re}_\lambda = 90$ and $\mathrm{Sc} = 1$ for homogeneous scalar mixing with $\mathcal{P}_\phi = 0$. For these initial conditions, all scalar energy is in the dissipative range. Curves 1–3 and D correspond to the fraction of scalar energy in each wavenumber band: $\langle\phi'^2\rangle_n/\langle\phi'^2\rangle$. Note that backscatter plays a very important role in this case as it is the only mechanism for transferring scalar energy from the dissipative range to wavenumber bands 1–3.

as predicted by classical scalar mixing arguments. However, as seen in Fig. 4.10, the transient period lasts relatively long (on the order of $2\tau_u$), and thus the rates of fast chemical reactions will depend strongly on the initial conditions. In contrast, if the scalar integral scale is initially very small (as in Fig. 4.13), the transient period is very short, and the scalar variance falls rapidly due to the large value of $R(0)$.

4.6.3 Extensions of the SR model

The SR model can be extended to inhomogeneous flows (Tsai and Fox 1996a), and a Lagrangian PDF version (LSR model) has been developed and validated against DNS data (Fox 1997; Vedula *et al.* 2001). We will return to the LSR model in Section 6.10.

4.7 Models for differential diffusion

At high Reynolds numbers, the arguments developed in Chapter 3 indicate that the scalar dissipation rate should depend weakly on the Schmidt number Sc, and for fixed Sc this dependence should diminish as the Reynolds number increases. For chemically reacting flows, one is also interested in effects due to differential diffusion, i.e., when the molecular-diffusion coefficients for the scalars are different. For differential diffusion between two scalars, the Schmidt numbers[39] Sc_1 and Sc_2 will be unequal and, as discussed in Section 3.4, this will cause the scalars to de-correlate. DNS studies have shown the degree of de-correlation is strongly wavenumber-dependent, i.e., high wavenumbers de-correlate more quickly than low wavenumbers (Yeung 1998a). For long times, the degree of de-correlation is thus strongly dependent on backscatter from small to large scales.

At present, there exists no completely general RANS model for differential diffusion. Note, however, that because it solves (4.37) directly, the linear-eddy model discussed in Section 4.3 can describe differential diffusion (Kerstein 1990; Kerstein *et al.* 1995). Likewise, the laminar flamelet model discussed in Section 5.7 can be applied to describe differential diffusion in flames (Pitsch and Peters 1998). Here, in order to understand the underlying physics, we will restrict our attention to a multi-variate version of the SR model for inert scalars (Fox 1999).

4.7.1 Multi-variate SR model

The SR model introduced in Section 4.6 describes length-scale effects and contains an explicit dependence on Sc. In this section, we extend the SR model to describe differential diffusion (Fox 1999). The key extension is the inclusion of a model for the scalar covariance $\langle \phi'_\alpha \phi'_\beta \rangle$ and the joint scalar dissipation rate $\varepsilon_{\alpha\beta}$. In homogeneous turbulence, the covariance transport equation is given by (3.179), p. 97. Consistent with the cospectrum transport equation ((3.75), p. 78), we will define the molecular diffusivity of the covariance as

$$\Gamma_{\alpha\beta} \equiv \frac{\Gamma_\alpha + \Gamma_\beta}{2}. \tag{4.131}$$

A corresponding Schmidt number is defined by $\mathrm{Sc}_{\alpha\beta} \equiv \nu / \Gamma_{\alpha\beta}$. Then, when defining the SR model for the covariance, we will simply replace Sc by $\mathrm{Sc}_{\alpha\beta}$ (e.g., in (4.98),

[39] If one of the scalars is enthalpy, the Schmidt number is replaced by the Prandtl number. The ratio of the Schmidt number to the Prandtl number is the Lewis number – a measure of the relative diffusivity of chemical species and enthalpy. Non-unity Lewis number effects can be important in combustion.

(4.126), and (4.129)). The SR model for differential diffusion thus requires three sets of equations, one for each possible $\mathrm{Sc}_{\alpha\beta}$, $\alpha, \beta = 1, 2$. For simplicity, we will again assume that the number of wavenumber bands needed for both scalars (i.e., $\mathrm{Sc}_{\alpha\alpha}$ and $\mathrm{Sc}_{\beta\beta}$) is three. Note that only the third cut-off wavenumber, κ_3, (4.99), depends on the Schmidt number.

Like the scalar variance, (4.102), in the multi-variate SR model the scalar covariance is divided into finite wavenumber bands: $\langle \phi'_\alpha \phi'_\beta \rangle_n$, where (see (4.115))

$$\langle \phi'_\alpha \phi'_\beta \rangle = \sum_{n=1}^{3} \langle \phi'_\alpha \phi'_\beta \rangle_n + \langle \phi'_\alpha \phi'_\beta \rangle_\mathrm{D}. \tag{4.132}$$

These variables are governed by exactly the same model equations (e.g., (4.103)) as the scalar variances (inter-scale transfer at scales larger than the dissipation scale thus conserves scalar correlation), except for the dissipation range (e.g., (4.106)), where

$$\frac{\mathrm{d}\langle \phi'_\alpha \phi'_\beta \rangle_\mathrm{D}}{\mathrm{d}t} = \mathcal{T}_\mathrm{D} + \gamma_\mathrm{D} \mathcal{P}_{\alpha\beta} - \gamma_{\alpha\beta} \varepsilon_{\alpha\beta}. \tag{4.133}$$

Because the parameters $\alpha_{3\mathrm{D}}$, $\beta_{3\mathrm{D}}$, and $\beta_{\mathrm{D}3}$ appearing in $\mathcal{T}_\mathrm{D}$ depend on the Schmidt number, the dissipation-range spectral transfer rates will be different for each covariance component. $\mathcal{P}_{\alpha\beta}$ is the covariance-production term defined by (3.137) on p. 90.

De-correlation at small scales is generated by the molecular-diffusion term in the model for $\varepsilon_{\alpha\beta}$:

$$\frac{\mathrm{d}\varepsilon_{\alpha\beta}}{\mathrm{d}t} = \mathcal{P}_\varepsilon + C_\mathrm{D} \left(\frac{\varepsilon}{\nu}\right)^{1/2} \mathcal{T}_\varepsilon + C_\mathrm{s} \left(\frac{\varepsilon}{\nu}\right)^{1/2} \varepsilon_{\alpha\beta} - C_\mathrm{d} \frac{\gamma_{\alpha\beta}\varepsilon_{\alpha\beta}}{\langle \phi'_\alpha \phi'_\beta \rangle_\mathrm{D}} \varepsilon_{\alpha\beta}, \tag{4.134}$$

wherein the covariance-dissipation-production term $\mathcal{P}_\varepsilon$ is defined by

$$\mathcal{P}_\varepsilon \equiv -\frac{\gamma_\mathrm{D}}{\gamma_{\alpha\beta}} \left(\frac{\varepsilon_\alpha}{\langle \phi'^2_\alpha \rangle_\mathrm{D}} \langle u_i \phi_\alpha \rangle \frac{\partial \langle \phi_\beta \rangle}{\partial x_i} + \frac{\varepsilon_\beta}{\langle \phi'^2_\beta \rangle_\mathrm{D}} \langle u_i \phi_\beta \rangle \frac{\partial \langle \phi_\alpha \rangle}{\partial x_i} \right). \tag{4.135}$$

For differential diffusion, $\gamma_{\alpha\beta} \geq 1$. Thus, the dissipation transfer rate,

$$\mathcal{T}_\varepsilon = \gamma_{\alpha\beta}^{-1} (\alpha_{3\mathrm{D}} + \beta_{3\mathrm{D}}) \langle \phi'_\alpha \phi'_\beta \rangle_3 - \beta_\varepsilon \varepsilon_\phi, \tag{4.136}$$

will depend on the Schmidt number through $\gamma_{\alpha\beta}$ and β_ε, and will be different for each component of $\varepsilon_{\alpha\beta}$. Note that the model constants $C_\mathrm{D} = 0.02$, $C_\mathrm{s} = 0.98$, and $C_\mathrm{d} = 3$ are the same for all components of $\varepsilon_{\alpha\beta}$. The factor $\gamma_{\alpha\beta}^{-1}$ in (4.136) results from the fact that $\varepsilon_{\alpha\beta}$ is defined using $(\Gamma_\alpha \Gamma_\beta)^{1/2}$ instead of $\Gamma_{\alpha\beta}$. The resulting equation (4.134) is then consistent with (3.166) on p. 95.

In summary, the multi-variate SR model is found by applying the uni-variate SR model to each component of $\langle \phi'_\alpha \phi'_\beta \rangle_n$ and $\varepsilon_{\alpha\beta}$. For the case where $\Gamma_\alpha = \Gamma_\beta$, the model equations for all components will be identical. The model predictions then depend only on the initial conditions, which need not be identical for each component. In order to see how differential

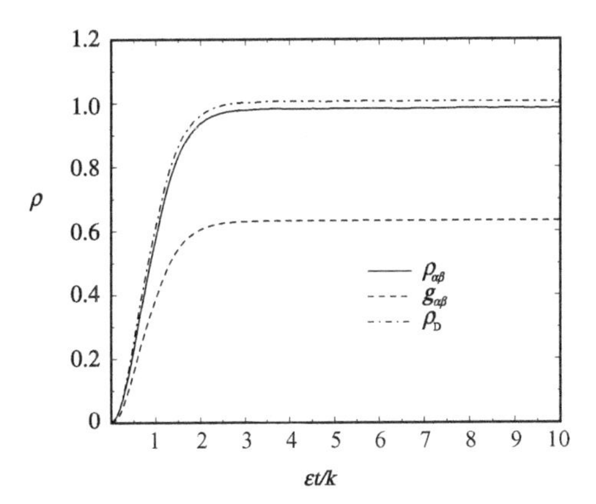

Figure 4.14. Predictions of the multi-variate SR model for $\mathrm{Re}_\lambda = 90$ and $\mathrm{Sc} = (1, 1/8)$ with collinear mean scalar gradients and no backscatter ($c_\mathrm{b} = 0$). For these initial conditions, the scalars are uncorrelated: $\rho_{\alpha\beta}(0) = g_{\alpha\beta}(0) = 0$. The correlation coefficient for the dissipation range, ρ_D, is included for comparison with $\rho_{\alpha\beta}$.

diffusion affects the model predictions, we will look next at the two cases introduced in Section 3.4.

4.7.2 Mean scalar gradients

The case of uniform mean scalar gradients was introduced in Section 3.4, where $G_{i\alpha}$ (see (3.176)) denotes the ith component of the gradient of $\langle\phi_\alpha\rangle$. In this section, we will assume that the mean scalar gradients are collinear so that $G_{i\alpha}G_{i\beta} = G_{i\alpha}G_{i\alpha} = G_{i\beta}G_{i\beta} \equiv G^2$. The scalar covariance production term then reduces to $\mathcal{P}_{\alpha\beta} = 2\Gamma_\mathrm{T}G^2$. In the absence of differential diffusion, the two scalars will become perfectly correlated in all wavenumber bands, i.e., $\langle\phi'_\alpha\phi'_\beta\rangle_n = \langle\phi'^2_\alpha\rangle_n = \langle\phi'^2_\beta\rangle_n$.

In order to illustrate how the multi-variate SR model works, we consider a case with constant $\mathrm{Re}_\lambda = 90$ and Schmidt number pair $\mathrm{Sc} = (1, 1/8)$. If we assume that the scalar fields are initially uncorrelated (i.e., $\rho_{\alpha\beta}(0) = 0$), then the model can be used to predict the transient behavior of the correlation coefficients (e.g., $\rho_{\alpha\beta}(t)$). Plots of the correlation coefficients without ($c_\mathrm{b} = 0$) and with backscatter ($c_\mathrm{b} = 1$) are shown in Figs. 4.14 and 4.15, respectively. As expected from (3.183), the scalar-gradient correlation coefficient $g_{\alpha\beta}(t)$ approaches $1/\gamma_{\alpha\beta} = 0.629$ for large t in both figures. On the other hand, the steady-state value of scalar correlation $\rho_{\alpha\beta}$ depends on the value of c_b. For the case with no backscatter, the effects of differential diffusion are confined to the small scales (i.e., $\langle\phi'_\alpha\phi'_\beta\rangle_3$ and $\langle\phi'_\alpha\phi'_\beta\rangle_\mathrm{D}$) and, because these scales contain a relatively small amount of the scalar energy, the steady-state value of $\rho_{\alpha\beta}$ is close to unity. In contrast, for the case with backscatter, de-correlation is transported back to the large scales, resulting in a lower steady-state value for $\rho_{\alpha\beta}$.

In Section 3.4, the analysis of the uniform mean-scalar-gradient case led to a relation for $\rho_{\alpha\beta}$ in (3.184) that involves $C_{\alpha\beta}$ (defined by (3.178)). The dissipation equation (4.134) in

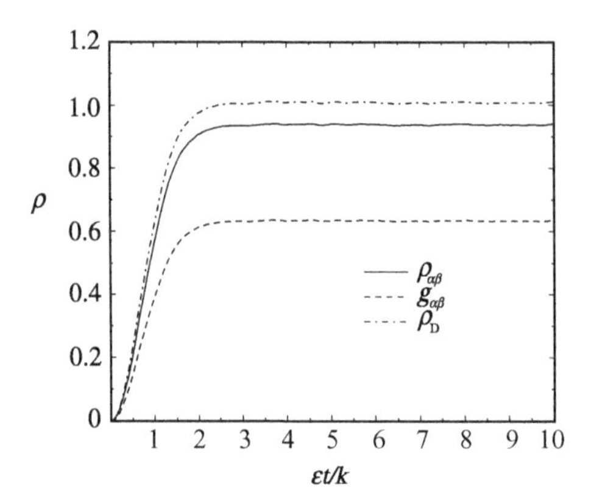

Figure 4.15. Predictions of the multi-variate SR model for $\text{Re}_\lambda = 90$ and $\text{Sc} = (1, 1/8)$ with collinear mean scalar gradients and backscatter ($c_\text{b} = 1$). For these initial conditions, the scalars are uncorrelated: $\rho_{\alpha\beta}(0) = g_{\alpha\beta}(0) = 0$. The correlation coefficient for the dissipation range, ρ_D, is included for comparison with $\rho_{\alpha\beta}$.

the multi-variate SR model has a form similar to (3.180). Thus, since $C_{\alpha\beta}$ is proportional to $\langle \phi'^2_\alpha \rangle_\text{D}$, it is of interest to consider the dissipation-range correlation function:

$$\rho_\text{D} \equiv \frac{\langle \phi'_\alpha \phi'_\beta \rangle_\text{D}}{\left(\langle \phi'^2_\alpha \rangle_\text{D} \langle \phi'^2_\beta \rangle_\text{D} \right)^{1/2}} = \frac{C_{\alpha\beta}}{(C_{\alpha\alpha} C_{\beta\beta})^{1/2}}. \tag{4.137}$$

Note that, unlike $\rho_{\alpha\beta}$, $|\rho_\text{D}|$ is not bounded by unity. This is because the dissipation-range cut-off wavenumber, (4.101), used to define $\langle \phi'_\alpha \phi'_\beta \rangle_\text{D}$ depends on $\text{Sc}_{\alpha\beta}$, which is different for each scalar pair (i.e., $\text{Sc}_{\alpha\beta} \neq \text{Sc}_{\alpha\alpha} \neq \text{Sc}_{\beta\beta}$). Indeed, for the mean-scalar-gradient case, (3.184) implies that at steady state $\rho_\text{D} > 1$. As shown in Figs. 4.14 and 4.15, the multi-variate SR model also predicts that ρ_D will approach a steady-state value greater than unity. However, due to the spectral transfer terms in the SR model, $\rho_{\alpha\beta} \neq 1/\rho_\text{D}$ as predicted by (3.184).

4.7.3 Decaying scalars

We will next consider the case of decaying scalars where $\mathcal{P}_{\alpha\beta} = 0$. For this case, it is convenient to assume that the scalars are initially perfectly correlated so that $\rho_{\alpha\beta}(0) = g_{\alpha\beta}(0) = \rho_\text{D}(0) = 1$. The multi-variate SR model can then be used to describe how the scalars de-correlate with time. We will again consider a case with constant $\text{Re}_\lambda = 90$ and Schmidt number pair $\text{Sc} = (1, 1/8)$.

Typical model predictions without and with backscatter are shown in Figs. 4.16 and 4.17, respectively. It can be noted that for decaying scalars the effect of backscatter on de-correlation is dramatic. For the case without backscatter (Fig. 4.16), after a short transient period the correlation coefficients all approach steady-state values. In contrast, when backscatter is included (Fig. 4.17), the correlation coefficients slowly approach zero. The rate of long-time de-correlation in the multi-variate SR model is thus proportional to the backscatter constant c_b.

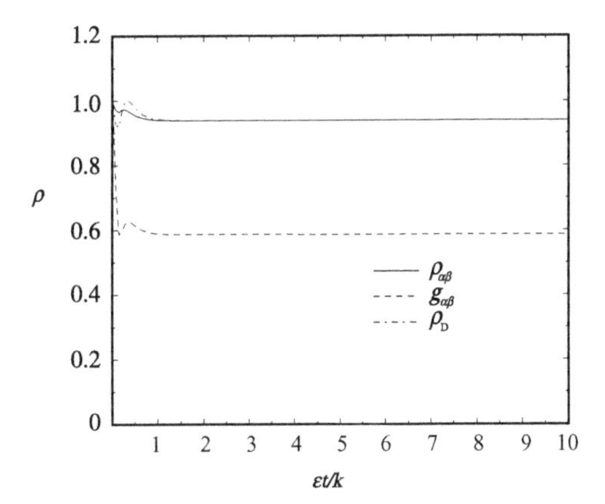

Figure 4.16. Predictions of the multi-variate SR model for $\mathrm{Re}_\lambda = 90$ and $\mathrm{Sc} = (1, 1/8)$ for decaying scalars with no backscatter ($c_\mathrm{b} = 0$). For these initial conditions, the scalars are perfectly correlated: $\rho_{\alpha\beta}(0) = g_{\alpha\beta}(0) = 1$. The correlation coefficient for the dissipation range, ρ_D, is included for comparison with $\rho_{\alpha\beta}$.

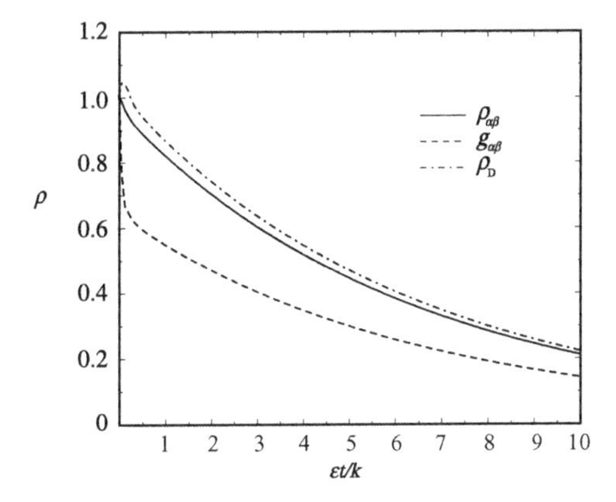

Figure 4.17. Predictions of the multi-variate SR model for $\mathrm{Re}_\lambda = 90$ and $\mathrm{Sc} = (1, 1/8)$ for decaying scalars with backscatter ($c_\mathrm{b} = 1$). For these initial conditions, the scalars are perfectly correlated: $\rho_{\alpha\beta}(0) = g_{\alpha\beta}(0) = 1$. The correlation coefficient for the dissipation range, ρ_D, is included for comparison with $\rho_{\alpha\beta}$.

The analysis in Section 3.4 for decaying scalars revealed that for large Reynolds number $g_{\alpha\beta}(t) \propto \rho_{\alpha\beta}(t)$ (see (3.191)). This prediction is consistent with Figs. 4.16 and 4.17, where, for large t, $g_{\alpha\beta}(t) \approx \rho_{\alpha\beta}(t)/\gamma_{\alpha\beta}$. It can also be seen from Fig. 4.16 that, without backscatter, $\rho_\mathrm{D}(t) = \rho_{\alpha\beta}(t)$, which implies that $C_{\alpha\beta} = C_{\alpha\alpha} = C_{\beta\beta}$. Using this result in (3.190) on p. 99 leads to $\rho_{\alpha\beta}$ becoming independent of t, which is consistent with Fig. 4.16. On the other hand, with backscatter the model predicts that $\rho_\mathrm{D} > \rho_{\alpha\beta}$, leading to long-time de-correlation. The available DNS data (Yeung and Pope 1993) for decaying scalars also show long-time de-correlation of the scalars. As shown in Fox (1999), the value $c_\mathrm{b} = 1$ leads to de-correlation rates that are consistent with the DNS data. This sensitivity to backscatter makes decaying scalars a discriminating test case for any differential-diffusion model.

4.8 Transported PDF methods

The turbulence models discussed in this chapter attempt to model the flow using low-order moments of the velocity and scalar fields. An alternative approach is to model the one-point joint velocity, composition PDF directly. For reacting flows, this offers the significant advantage of avoiding a closure for the chemical source term. However, the numerical methods needed to solve for the PDF are very different than those used in 'standard' CFD codes. We will thus hold off the discussion of transported PDF methods until Chapters 6 and 7 after discussing closures for the chemical source term in Chapter 5 that can be used with RANS and LES models.

5

Closures for the chemical source term

In this chapter, we present the most widely used methods for closing the chemical source term in the Reynolds-averaged scalar transport equation. Although most of these methods were not originally formulated in terms of the joint composition PDF, we attempt to do so here in order to clarify the relationships between the various methods. A schematic of the closures discussed in this chapter is shown in Fig. 5.1. In general, a closure for the chemical source term must assume a particular form for the joint composition PDF. This can be done either directly (e.g., presumed PDF methods), or indirectly by breaking the joint composition PDF into parts (e.g., by conditioning on the mixture-fraction vector). In any case, the assumed form will be strongly dependent on the functional form of the chemical source term. In Section 5.1, we begin by reviewing the methods needed to render the chemical source term in the simplest possible form. As stated in Chapter 1, the treatment of *non-premixed* turbulent reacting flows is emphasized in this book. For these flows, it is often possible to define a *mixture-fraction vector*, and thus the necessary theory is covered in Section 5.3.

5.1 Overview of the closure problem

In this section, we first introduce the 'standard' form of the chemical source term for both *elementary* and *non-elementary* reactions. We then show how to transform the composition vector into *reacting* and *conserved* vectors based on the form of the reaction coefficient matrix. We conclude by looking at how the chemical source term is affected by Reynolds averaging, and define the *chemical time scales* based on the Jacobian of the chemical source term.

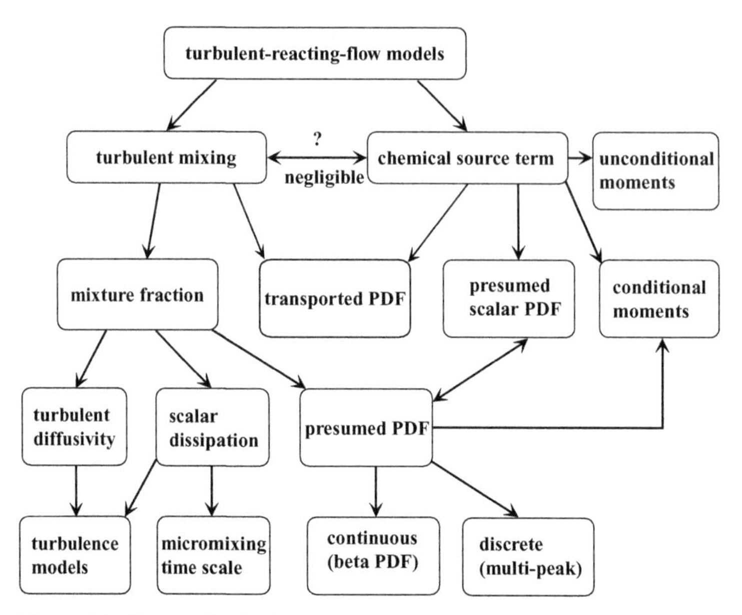

Figure 5.1. Closures for the chemical source term can be understood in terms of their relationship to the joint composition PDF. The simplest methods attempt to represent the joint PDF by its (lower-order) moments. At the next level, the joint PDF is expressed in terms of the product of the conditional joint PDF and the mixture-fraction PDF. The conditional joint PDF can then be approximated by invoking the 'fast-chemistry' or 'flamelet' limits, by modeling the conditional means of the compositions, or by assuming a functional form for the PDF. Similarly, it is also possible to assume a functional form for the joint composition PDF. The 'best' method to employ depends strongly on the functional form of the chemical source term and its characteristic time scales.

5.1.1 Chemical source term

As noted in Chapter 1, the chemical source term $\mathbf{S}(\phi)$ in (1.28) on p. 16 is the principal cause of closure difficulties when modeling turbulent reacting flows. The 'standard' expression[1] for the chemical source term begins with a set of I elementary reactions (Hill 1977):[2]

$$\sum_{\beta=1}^{K} \upsilon_{\beta i}^{\mathrm{f}} \mathrm{A}_{\beta} \underset{k_i^{\mathrm{r}}}{\overset{k_i^{\mathrm{f}}}{\rightleftharpoons}} \sum_{\beta=1}^{K} \upsilon_{\beta i}^{\mathrm{r}} \mathrm{A}_{\beta} \quad \text{for } i \in 1, \ldots, I, \tag{5.1}$$

involving K chemical species A_β with *integer* stoichiometric coefficients $\upsilon_{\beta i}^{\mathrm{f}}$ and $\upsilon_{\beta i}^{\mathrm{r}}$, and forward and reverse rate constants k_i^{f} and k_i^{r}. Note, however, that (5.1) can also be

[1] The chemical-source-term closures described in this chapter do not require that the chemical source term be derived from a set of elementary reactions as done here. Indeed, closure difficulties will result whenever $\mathbf{S}(\phi)$ is a non-linear function of ϕ.

[2] In elementary reactions, *chemical elements* are conserved. For example, in

$$2\mathrm{H_2} + \mathrm{O_2} \rightleftharpoons 2\mathrm{H_2O}$$

the numbers of H atoms (4) and O atoms (2) are the same on both sides of the reaction.

employed for non-elementary reactions if we allow the coefficient matrices $\boldsymbol{\upsilon}^{\mathrm{f}}$ and $\boldsymbol{\upsilon}^{\mathrm{r}}$ to have non-integer components.[3]

The *composition vector* $\boldsymbol{\phi}$ of length $N_{\mathrm{s}} = K + 1$ is made up of the *molar concentration vector* $\mathbf{c}$ (mol/cm^3) for the K chemical species and the temperature T (K):[4]

$$\boldsymbol{\phi} = \begin{bmatrix} \mathbf{c} \\ T \end{bmatrix}. \tag{5.2}$$

Based on (5.1), the chemical source term for the chemical species has the form[5]

$$\mathbf{S}_{\mathrm{c}}(\boldsymbol{\phi}) = \boldsymbol{\Upsilon}\mathbf{R}(\boldsymbol{\phi}) = \left[\sum_{i=1}^{I} \left(\upsilon^{\mathrm{r}}_{\alpha i} - \upsilon^{\mathrm{f}}_{\alpha i} \right) R_i(\boldsymbol{\phi}) \right], \tag{5.3}$$

where $\boldsymbol{\Upsilon}$ is the $K \times I$ *reaction coefficient matrix* and $\mathbf{R}(\boldsymbol{\phi})$ (mol/(cm^{3}s)) is the *reaction rate vector* with components $R_i(\boldsymbol{\phi})$ $(i \in 1, \ldots, I)$. Likewise, for a constant-pressure system, the chemical source term for temperature can be derived from an enthalpy balance and has the form

$$S_{\mathrm{T}}(\boldsymbol{\phi}) = -\frac{1}{\rho(\boldsymbol{\phi})c_p(\boldsymbol{\phi})}\mathbf{H}^{\mathrm{T}}(T)\mathbf{S}_{\mathrm{c}}(\boldsymbol{\phi}), \tag{5.4}$$

where ρ (g/cm^3) is the fluid density, c_p (J/(g K)) is the specific heat of the fluid, and the components of $\mathbf{H}(T)$ (J/mol) are the specific enthalpies of the chemical species $H_k(T)$ $(k \in 1, \ldots, K)$. Thus, the chemical source term for the composition vector can be written as

$$\mathbf{S}(\boldsymbol{\phi}) = \begin{bmatrix} \mathbf{S}_{\mathrm{c}}(\boldsymbol{\phi}) \\ S_{\mathrm{T}}(\boldsymbol{\phi}) \end{bmatrix} = \begin{bmatrix} \mathbf{I} \\ -\frac{1}{\rho(\boldsymbol{\phi})c_p(\boldsymbol{\phi})}\mathbf{H}^{\mathrm{T}}(T) \end{bmatrix} \boldsymbol{\Upsilon}\mathbf{R}(\boldsymbol{\phi}). \tag{5.5}$$

Note that if the fluid properties are independent of temperature (i.e., if $c_p(\boldsymbol{\phi})$ and $\mathbf{H}(T)$ are constant), then the chemical source term will depend on $\boldsymbol{\phi}$ only through $\mathbf{R}(\boldsymbol{\phi})$.

The *reaction rate functions* $R_i(\boldsymbol{\phi})$ $(i \in 1, \ldots, I)$ appearing in (5.5) usually have the following form[6] (no summation is implied with respect to i in these expressions) (Hill 1977):

$$R_i(\boldsymbol{\phi}) = k^{\mathrm{f}}_i(T) \prod_{\beta=1}^{K} c_\beta^{\upsilon^{\mathrm{f}}_{\beta i}} - k^{\mathrm{r}}_i(T) \prod_{\beta=1}^{K} c_\beta^{\upsilon^{\mathrm{r}}_{\beta i}}, \tag{5.6}$$

[3] We will denote the $(K \times I)$ forward and reverse coefficient matrices by $\boldsymbol{\upsilon}^{\mathrm{f}} = [\upsilon^{\mathrm{f}}_{\alpha i}]$ and $\boldsymbol{\upsilon}^{\mathrm{r}} = [\upsilon^{\mathrm{r}}_{\alpha i}]$, respectively.

[4] Alternatively, the specific enthalpy of the fluid can be used in place of the temperature. Likewise, the mass fractions or mole fractions can be used in place of the molar concentrations.

[5] We denote a matrix in terms of its components by $\mathbf{X} \equiv [x_{ij}]$.

[6] It will be implicitly assumed (as is almost always the case) that the components of $\mathbf{R}(\boldsymbol{\phi})$ are *linearly independent*, e.g., $R_i(\boldsymbol{\phi}) \neq \alpha R_j(\boldsymbol{\phi})$ for all $i \neq j$. If two components were linearly dependent, then the corresponding columns of $\boldsymbol{\Upsilon}$ could be replaced by a single column, and the linearly dependent component of $\mathbf{R}(\boldsymbol{\phi})$ removed. Linear independence implies that the Jacobian matrix formed from $\mathbf{R}(\boldsymbol{\phi})$ will be full rank for arbitrary choices of $\boldsymbol{\phi}$.

where the forward (k^{f}) and reverse (k^{r}) rate constants are given by[7]

$$k_i^{\mathrm{f}}(T) = A_i T^{\beta_i} \exp\left(-\frac{E_i}{RT}\right) \tag{5.7}$$

and

$$k_i^{\mathrm{r}}(T) = \frac{k_i^{\mathrm{f}}(T)}{K_i^{\mathrm{c}}(T)}. \tag{5.8}$$

The equilibrium constant K_i^{c} for reaction i is computed from thermodynamic considerations using Gibbs free energies. The Arrhenius form for the rate constants is written in terms of the pre-exponential factor A_i, the temperature exponent β_i, the activation energy E_i, and the universal gas constant R in the same units as the activation energy.

5.1.2 Elementary reactions

For elementary reactions (Hill 1977), the values of the stoichiometric coefficients are constrained by the fact that all chemical elements must be conserved in (5.1). Mathematically, this can be expressed in terms of an $E \times K$ *element matrix* $\mathbf{\Lambda}$ where E is the total number of chemical elements present in the reacting flow. Each column of $\mathbf{\Lambda}$ thus corresponds to a particular chemical species, and each row to a particular chemical element. As an example, consider a system containing $E = 2$ elements: O and H, and $K = 3$ species: H_2, O_2 and H_2O. The 2×3 element matrix for this system is

$$\mathbf{\Lambda} = \begin{bmatrix} 0 & 2 & 1 \\ 2 & 0 & 2 \end{bmatrix}. \tag{5.9}$$

Note that the element matrix contains only integer constants, and thus its space and time derivatives are null. Furthermore, for a well defined chemical system, $K \geq E$, and $\mathbf{\Lambda}$ will be full rank, i.e., $\mathrm{rank}(\mathbf{\Lambda}) = E$.

Chemical element conservation can be expressed in terms of $\mathbf{\Lambda}$ and $\mathbf{S}_{\mathrm{c}}$ by

$$\mathbf{\Lambda S}_{\mathrm{c}} = \left[\sum_{\alpha=1}^{K} \Lambda_{e\alpha} S_\alpha(\boldsymbol{\phi})\right] = \mathbf{0}. \tag{5.10}$$

However, since the chemical source term can be expressed in terms of $\boldsymbol{\Upsilon}$ and $\mathbf{R}$, the element conservation constraints can be rewritten as[8]

$$\mathbf{\Lambda \Upsilon} = \left[\sum_{\alpha=1}^{K} \Lambda_{e\alpha}\left(\upsilon_{\alpha i}^{\mathrm{r}} - \upsilon_{\alpha i}^{\mathrm{f}}\right)\right] = \mathbf{0}. \tag{5.11}$$

7 Chemical databases for gas-phase chemistry such as *Chemkin-II* utilize this format for the chemical source term.

8 Note that this constraint implies that the (I) columns of $\boldsymbol{\Upsilon}$ are *orthogonal* to the (E) rows of $\mathbf{\Lambda}$. Thus, since each column of $\boldsymbol{\Upsilon}$ represents one elementary reaction, the maximum number of *linearly independent* elementary reactions is equal to $K - E$, i.e., $N_\Upsilon \equiv \mathrm{rank}(\boldsymbol{\Upsilon}) \leq K - E$. For most chemical kinetic schemes, $N_\Upsilon = K - E$; however, this need not be the case.

Note that the right-hand side of this expression is an $E \times I$ null matrix, and thus element conservation must hold for any choice of $e \in 1, \dots, E$ and $i \in 1, \dots, I$. Moreover, since the element matrix is constant, (5.10) can be applied to the scalar transport equation ((1.28), p. 16) in order to eliminate the chemical source term in at least E of the K equations.[9] The chemically reacting flow problem can thus be described by only $K - E$ transport equations for the chemically reacting scalars, and E transport equations for non-reacting (*conserved*) scalars.[10]

As an example, consider again the chemical species whose element matrix is given by (5.9) and the reaction ($I = 1$)

$$2\mathrm{H_2} + \mathrm{O_2} \rightleftarrows 2\mathrm{H_2O}. \tag{5.12}$$

For simplicity, we will assume that the system is isothermal so that it can be described by the $K = 3$ species concentrations $\mathbf{c}$. The stoichiometric coefficients for this reaction are $\upsilon^{\mathrm{f}}_{11} = 2$, $\upsilon^{\mathrm{f}}_{21} = 1$, $\upsilon^{\mathrm{f}}_{31} = 0$, $\upsilon^{\mathrm{r}}_{11} = 0$, $\upsilon^{\mathrm{r}}_{21} = 0$, and $\upsilon^{\mathrm{r}}_{31} = 2$. Applying (5.9) and (5.5) then yields

$$\sum_{\alpha=1}^{3} \Lambda_{e\alpha} \upsilon^{\mathrm{f}}_{\alpha 1} = \sum_{\alpha=1}^{3} \Lambda_{e\alpha} \upsilon^{\mathrm{r}}_{\alpha 1} = \begin{cases} 2 & \text{if } e = 1 \\ 4 & \text{if } e = 2, \end{cases} \tag{5.13}$$

and thus $\mathbf{\Lambda \Upsilon} = \mathbf{0}$.[11] For this system, the scalar transport equation can be rewritten in terms of $K - E = 1$ reacting and $E = 2$ conserved scalars.

Formally, we are free to choose any linear combination of the three chemical species as the reacting scalar under the condition that the combination is linearly independent of the rows of $\mathbf{\Lambda}$.[12] Arbitrarily choosing c_3, a new scalar vector can be defined by the linear transformation[13]

$$\mathbf{c}^* \equiv \mathbf{Mc}, \tag{5.14}$$

where[14]

$$\mathbf{M} \equiv \begin{bmatrix} 0 & 2 & 1 \\ 2 & 0 & 2 \\ 0 & 0 & 1 \end{bmatrix}. \tag{5.15}$$

9 The total number of chemical source terms that can be eliminated is equal to $K - N_{\Upsilon}$.

10 In the special case where all chemical species have the same molecular diffusivity, only one transport equation is often required to describe the conserved scalars. The single conserved scalar can then be expressed in terms of the *mixture fraction.*

11 $\mathbf{\Upsilon}^{\mathrm{T}} = [-2 \quad -1 \quad 2]$.

12 In practice, the transformation matrix $\mathbf{M}$ can be found numerically from the *singular value decomposition* of $\mathbf{\Upsilon}$. The rank of $\mathbf{\Upsilon}$ is equal to the number of non-zero singular values, and the transformation matrix corresponds to the transpose of the premultiplier orthogonal matrix. This process is illustrated below for the non-elementary reaction case.

13 In the general case, another obvious choice would be to choose the transposes of $K - E$ linearly independent columns from $\mathbf{\Upsilon}$. However, if $N_{\Upsilon} < K - E$, then additional linearly independent rows must be added to make $\mathbf{M}$ full rank. These additional rows should be chosen to be orthogonal to the columns of $\mathbf{\Upsilon}$.

14 Note that the first two rows of $\mathbf{M}$ are just $\mathbf{\Lambda}$, and the third row multiplied by $\mathbf{c}$ is just c_3, i.e., $\mathbf{M}^{\mathrm{T}} = [\mathbf{\Lambda}^{\mathrm{T}} \;\; \mathbf{e}_3]$.

By definition, the transformation matrix $\mathbf{M}$ is full rank, and thus the inverse transformation is well defined, i.e., $\mathbf{c} = \mathbf{M}^{-1}\mathbf{c}^*$, where

$$\mathbf{M}^{-1} = \frac{1}{2}\begin{bmatrix} 0 & 1 & -2 \\ 1 & 0 & -1 \\ 0 & 0 & 2 \end{bmatrix}. \tag{5.16}$$

Applying the linear transformation to (1.28) on p. 16 then yields the transport equation for the transformed scalars:

$$\frac{\partial \mathbf{c}^*}{\partial t} + U_j \frac{\partial \mathbf{c}^*}{\partial x_j} = \mathbf{\Gamma}^* \frac{\partial^2 \mathbf{c}^*}{\partial x_j \partial x_j} + \mathbf{S}_c^*(\mathbf{c}^*), \tag{5.17}$$

where the transformed chemical source term is given by[15]

$$\mathbf{S}_c^*(\mathbf{c}^*) \equiv \mathbf{M}\mathbf{S}_c(\mathbf{M}^{-1}\mathbf{c}^*) = \begin{bmatrix} 0 \\ 0 \\ S_{c3}(\mathbf{M}^{-1}\mathbf{c}^*) \end{bmatrix}, \tag{5.18}$$

and the transformed molecular-diffusion-coefficient matrix is given by[16]

$$\mathbf{\Gamma}^* \equiv \mathbf{M}\,\mathbf{diag}(\Gamma_1, \Gamma_2, \Gamma_3)\mathbf{M}^{-1} = \begin{bmatrix} \Gamma_2 & 0 & \Gamma_3 - \Gamma_2 \\ 0 & \Gamma_1 & 2(\Gamma_3 - \Gamma_1) \\ 0 & 0 & \Gamma_3 \end{bmatrix}. \tag{5.19}$$

Note from (5.18) that only one of the three transformed scalars is reacting. Note also from (5.19) that the diffusion matrix is greatly simplified when $\Gamma_1 = \Gamma_2 = \Gamma_3$.[17]

5.1.3 Non-elementary reactions

The results presented above were discussed in terms of the special case of elementary reactions. However, if we relax the condition that the coefficients $\upsilon^{\mathrm{f}}_{\alpha i}$ and $\upsilon^{\mathrm{r}}_{\alpha i}$ must be integers, (5.1) is applicable to nearly all chemical reactions occurring in practical applications. In this general case, the element conservation constraints are no longer applicable. Nevertheless, all of the results presented thus far can be expressed in terms of the reaction coefficient matrix $\boldsymbol{\Upsilon}$, defined as before by

$$\boldsymbol{\Upsilon} \equiv \left[\left(\upsilon^{\mathrm{r}}_{\alpha i} - \upsilon^{\mathrm{f}}_{\alpha i}\right)\right]. \tag{5.20}$$

[15] Note that the number of null components in $\mathbf{S}_c^*$ is equal to $N = K - N_\Upsilon$.

[16] $\mathbf{diag}(x_{11}, \dots, x_{nn})$ denotes the $n \times n$ diagonal matrix with diagonal components x_{ii}.

[17] In this case, if the boundary and initial conditions allow it, either c_1^* or c_2^* can be used to define the mixture fraction. The number of conserved scalar transport equations that must be solved then reduces to one. In general, depending on the initial conditions, it may be possible to reduce the number of conserved scalar transport equations that must be solved to $\min(M_1, M_2)$ where $M_1 = K - N_\Upsilon$ and $M_2 =$ number of feed streams $- 1$. In many practical applications of turbulent reacting flows, $M_1 = E$ and $M_2 = 1$, and one can assume that the molecular-diffusion coefficients are equal; thus, only one conserved scalar transport equation (i.e., the *mixture fraction*) is required to describe the flow.

The key quantity that determines the number of reacting and conserved scalars is the rank of $\boldsymbol{\Upsilon}$. For the non-elementary reaction case, $N_\Upsilon \equiv \mathrm{rank}(\boldsymbol{\Upsilon}) \leq K$.[18] The number of reacting scalars will then be equal to N_Υ, and the number of conserved scalars will be equal to $N \equiv K - N_\Upsilon$.

In order to illustrate the non-elementary reaction case, consider the two-step ($I = 2$) reaction

$$\begin{aligned} \mathrm{A} + \mathrm{B} &\xrightarrow{k_1} \mathrm{R} \\ \mathrm{B} + \mathrm{R} &\xrightarrow{k_2} \mathrm{S} \end{aligned} \tag{5.21}$$

with $K = 4$ 'chemical species': A, B, R, and S. For simplicity, we will again assume that the reactions are isothermal so that the system is described by molar concentration vector $\mathbf{c}$. The reaction coefficient matrices for this reaction are given by

$$\boldsymbol{\upsilon}^{\mathrm{f}} = \begin{bmatrix} 1 & 0 \\ 1 & 1 \\ 0 & 1 \\ 0 & 0 \end{bmatrix}, \quad \boldsymbol{\upsilon}^{\mathrm{r}} = \begin{bmatrix} 0 & 0 \\ 0 & 0 \\ 1 & 0 \\ 0 & 1 \end{bmatrix}, \quad \boldsymbol{\Upsilon} = \begin{bmatrix} -1 & 0 \\ -1 & -1 \\ 1 & -1 \\ 0 & 1 \end{bmatrix}, \tag{5.22}$$

and the reaction rate function is given by

$$\mathbf{R}(\mathbf{c}) = \begin{bmatrix} k_1 c_1 c_2 \\ k_2 c_2 c_3 \end{bmatrix}. \tag{5.23}$$

Note that $N_\Upsilon = 2$. Thus, by applying an appropriate linear transformation, it should be possible to rewrite the scalar transport equation in terms of two reacting and two conserved scalars.

In order to find a linear transformation matrix to simplify the scalar transport equation, we will make use of the *singular value decomposition* (SVD) of $\boldsymbol{\Upsilon}$:

$$\boldsymbol{\Upsilon} = \mathbf{U}_{\mathrm{sv}} \boldsymbol{\Sigma}_{\mathrm{sv}} \mathbf{V}_{\mathrm{sv}}^{\mathrm{T}}, \tag{5.24}$$

where the $(K \times K)$ matrix $\mathbf{U}_{\mathrm{sv}}$ and the $(I \times I)$ matrix $\mathbf{V}_{\mathrm{sv}}$ are orthogonal matrices, and the $(K \times I)$ diagonal matrix $\boldsymbol{\Sigma}_{\mathrm{sv}}$ has the non-negative *singular values* as diagonal elements:

$$\boldsymbol{\Sigma}_{\mathrm{sv}} \equiv \mathbf{diag}\left(\sigma_1, \dots, \sigma_{\min(I,K)}\right). \tag{5.25}$$

Note that only $N_\Upsilon \leq \min(I, K)$ singular values will be non-zero. The columns of $\mathbf{U}_{\mathrm{sv}}$ corresponding to the non-zero singular values span the reacting-scalar sub-space, and the columns corresponding to the zero singular values span the conserved sub-space. The desired linear transformation matrix is thus

$$\mathbf{M} = \mathbf{U}_{\mathrm{sv}}^{\mathrm{T}}. \tag{5.26}$$

[18] It is again implicitly assumed that the components of $\mathbf{R}(\boldsymbol{\phi})$ are *linearly independent*. If two components were linearly dependent, then the corresponding columns of $\boldsymbol{\Upsilon}$ could be replaced by a single column, and the linearly dependent component of $\mathbf{R}(\boldsymbol{\phi})$ could be removed.

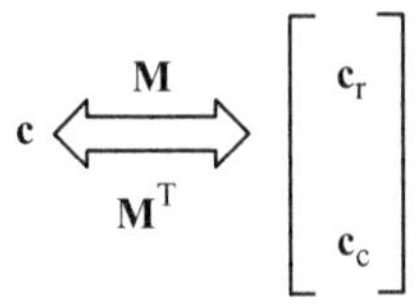

$$\mathbf{c} \underset{\mathbf{M}^{\mathrm{T}}}{\overset{\mathbf{M}}{\Longleftrightarrow}} \begin{bmatrix} \mathbf{c}_{\mathrm{r}} \\ \\ \mathbf{c}_{\mathrm{c}} \end{bmatrix}$$

Figure 5.2. The composition vector $\mathbf{c}^*$ can be partitioned by a linear transformation into two parts: $\mathbf{c}_{\mathrm{r}}$, a reacting-scalar vector of length N_Υ; and $\mathbf{c}_{\mathrm{c}}$, a conserved-scalar vector of length N. The linear transformation is independent of $\mathbf{x}$ and t, and is found from the singular value decomposition of the reaction coefficient matrix Υ.

The N_Υ rows of $\mathbf{M}$ that correspond to the non-zero singular values will yield the reacting scalars. The remaining N rows yield the conserved scalars.[19]

Applying SVD to $\boldsymbol{\Upsilon}$ in (5.22) yields[20]

$$\mathbf{M} = \begin{bmatrix} -0.5774 & -0.5774 & 0.5774 & 0 \\ 0 & -0.5774 & -0.5774 & 0.5774 \\ 0.7887 & -0.2887 & 0.5000 & 0.2113 \\ -0.2113 & 0.5000 & 0.2887 & 0.7887 \end{bmatrix} \tag{5.27}$$

and $\mathbf{M}^{-1} = \mathbf{M}^{\mathrm{T}}$. Hence, the transformed scalars are $\mathbf{c}^* = \mathbf{Mc}$, and the transformed chemical source term, (5.18), becomes

$$\mathbf{S}_{\mathrm{c}}^*(\mathbf{c}^*) = -1.7321 \begin{bmatrix} R_1(\mathbf{M}^{\mathrm{T}}\mathbf{c}^*) \\ R_2(\mathbf{M}^{\mathrm{T}}\mathbf{c}^*) \\ 0 \\ 0 \end{bmatrix}. \tag{5.28}$$

Note that because the columns of $\boldsymbol{\Upsilon}$ happen to be orthogonal, the linear transformation matrix results in a diagonal form for $\mathbf{M}\boldsymbol{\Upsilon}$. The reaction rate functions in the transformed chemical source term then act on each of the transformed scalars individually.[21]

As illustrated in Fig. 5.2, the SVD of $\boldsymbol{\Upsilon}$ can be used to partition $\mathbf{c}^*$ into reacting and conserved sub-spaces:

$$\mathbf{c}^* = \begin{bmatrix} \mathbf{c}_{\mathrm{r}} \\ \mathbf{c}_{\mathrm{c}} \end{bmatrix}. \tag{5.29}$$

Assuming (as is usually the case) that the non-zero singular values are arranged in order of decreasing magnitude:

$$\sigma_1 \geq \cdots \geq \sigma_{N_\Upsilon},$$

[19] In most SVD algorithms, the singular values are arranged in $\boldsymbol{\Sigma}_{\mathrm{sv}}$ in *descending order*. Thus, the first N_Υ rows of $\mathbf{M}$ yield the reacting scalars and the remaining N rows yield the conserved scalars.

[20] SVD algorithms are widely available in the form of linear algebra sub-routines. Here, we have employed the SVD routine in MATLAB. Note that the first two rows of $\mathbf{M}$ are just $0.5774\boldsymbol{\Upsilon}^{\mathrm{T}}$. In general, this would not be the case. However, the two rows happen to be orthogonal for this reaction scheme.

[21] Reaction scheme (5.21) is an example of 'simple' chemistry (i.e., $N_\Upsilon = I$). In Section 5.5 we show that for simple chemistry it is always possible to choose a linear transformation that puts the chemical source term in diagonal form. Obviously, when $N_\Upsilon < I$ this will not be possible since the number of reactions will be greater than the number of reacting scalars.

the reacting-scalar vector $\mathbf{c}_r$ of length N_Υ is defined by

$$\mathbf{c}_r \equiv \mathbf{M}_r \mathbf{c}, \tag{5.30}$$

where $\mathbf{M}_r$ is the $(N_\Upsilon \times K)$ matrix containing the *first* N_Υ rows of $\mathbf{M}$ defined by (5.26). Likewise, the conserved-scalar vector $\mathbf{c}_c$ of length N is defined by

$$\mathbf{c}_c \equiv \mathbf{M}_c \mathbf{c}, \tag{5.31}$$

where $\mathbf{M}_c$ is the $(N \times K)$ matrix containing the *last* N rows of $\mathbf{M}$. The transformed chemical source term for the chemical species can then be partitioned into two parts:

$$\mathbf{S}_c^*(\mathbf{c}^*) = \begin{bmatrix} \mathbf{S}_r(\mathbf{c}_r, \mathbf{c}_c) \\ \mathbf{0} \end{bmatrix}, \tag{5.32}$$

where

$$\mathbf{S}_r(\mathbf{c}_r, \mathbf{c}_c) = \mathbf{diag}\left(\sigma_1, \ldots, \sigma_{N_\Upsilon}\right) \mathbf{V}_r \mathbf{R} \left(\mathbf{M}^{\mathrm{T}} \begin{bmatrix} \mathbf{c}_r \\ \mathbf{c}_c \end{bmatrix} \right). \tag{5.33}$$

In this expression, $\mathbf{V}_r$ is the $(N_\Upsilon \times I)$ matrix containing the *first* N_Υ rows of $\mathbf{V}_{\mathrm{sv}}^{\mathrm{T}}$. Note that the reacting scalars $\mathbf{c}_r$ are coupled to the conserved scalars $\mathbf{c}_c$ through (5.33).

In the transformed scalar transport equation, if the molecular-diffusion coefficients of the scalars are not equal, the transformed molecular-diffusion-coefficient matrix $\mathbf{\Gamma}^* = \mathbf{M}\mathbf{diag}(\Gamma_1, \ldots, \Gamma_4)\mathbf{M}^{\mathrm{T}}$ will be quite complicated. For a *laminar* reacting flow, transformation of the scalar transport equation simplifies the chemical source term at the expense of greatly complicating the molecular-diffusion term. However, as discussed in Chapter 3, in a *turbulent* reacting flow the Reynolds-averaged molecular-diffusion terms are often controlled by turbulent mixing, and thus one can safely assume that the molecular-diffusion coefficients are all equal.[22] The transformation of the scalar transport equation then results in significant simplifications in the formulation of the chemical-source-term closure as discussed in detail in Section 5.3.

The transformation given above is completely general in the sense that it can be applied to any set of chemical reactions. However, in the case of 'simple' reactions where the number of reactions (I) is small compared with the number of species (K), further simplification of (5.33) is possible. In particular, if $N_\Upsilon = I$, then the matrix

$$\mathbf{V}^* \equiv \mathbf{diag}(\sigma_1, \ldots, \sigma_{N_\Upsilon})\mathbf{V}_r \tag{5.34}$$

will be square and non-singular. Thus, by defining a new reacting-scalar vector by

$$\mathbf{c}_r^* \equiv \mathbf{V}^{*-1}\mathbf{c}_r, \tag{5.35}$$

it is possible to put the transformed chemical source term in an 'uncoupled' form:

$$\mathbf{S}_r^*(\mathbf{c}_r^*, \mathbf{c}_c) = \mathbf{R} \left(\mathbf{M}^{*\mathrm{T}} \begin{bmatrix} \mathbf{c}_r^* \\ \mathbf{c}_c \end{bmatrix} \right), \tag{5.36}$$

22 As discussed in Section 3.4, differential-diffusion effects will decrease with increasing Reynolds number. A single molecular mixing time scale τ_ϕ can be employed for all scalars at high Reynolds numbers.

where

$$\mathbf{M}^* \equiv \begin{bmatrix} \mathbf{V}^{*\mathrm{T}} & \mathbf{0} \\ \mathbf{0} & \mathbf{I} \end{bmatrix} \mathbf{M}. \tag{5.37}$$

Note that in (5.36), each reaction rate function affects only one scalar. Thus, transport equations in this form can be easily manipulated in limiting cases where one reaction rate constant (k_i^{f}) is zero or infinite.[23]

For the more general case of non-isothermal systems, the SVD of $\boldsymbol{\Upsilon}$ can still be used to partition the chemical species into reacting and conserved sub-spaces. Thus, in addition to the dependence on $\mathbf{c}^*$, the transformed chemical source term for the chemical species, $\mathbf{S}_{\mathrm{c}}^*$, will also depend on T. The non-zero chemical source term for the temperature, S_{T}, must also be rewritten in terms of $\mathbf{c}^*$ in the transport equation for temperature.

The special case where the fluid properties are *independent of temperature* leads to a further reduction in the number of composition variables affected by chemical reactions. For this case, (5.5) becomes[24]

$$\mathbf{S}(\boldsymbol{\phi}) = \boldsymbol{\Upsilon}_{\mathrm{c}} \mathbf{R}(\boldsymbol{\phi}), \tag{5.38}$$

where the $N_{\mathrm{s}} \times I$ coefficient matrix $\boldsymbol{\Upsilon}_{\mathrm{c}}$, defined by

$$\boldsymbol{\Upsilon}_{\mathrm{c}} \equiv \begin{bmatrix} \mathbf{I} \\ -\frac{1}{\rho c_p} \mathbf{H}^{\mathrm{T}} \end{bmatrix} \boldsymbol{\Upsilon}, \tag{5.39}$$

is *constant*. In general, $\mathrm{rank}(\boldsymbol{\Upsilon}_{\mathrm{c}}) = N_{\Upsilon}$. Thus, as in the example above, the SVD of $\boldsymbol{\Upsilon}_{\mathrm{c}}$ can be used to define N_{Υ} reacting scalars and $(N_{\mathrm{s}} - N_{\Upsilon})$ conserved scalars. The fact that the fluid properties are constant implies that the transformation matrices will also be constant. The scalar transport equations for the chemical species *and* temperature can then be easily rewritten in terms of the transformed reacting/conserved scalars.

5.1.4 Reynolds-averaged chemical source term

The Reynolds-averaged scalar transport equation for chemical species A_α ((4.72), p. 121) contains a chemical source term of the form

$$\langle S_\alpha(\boldsymbol{\phi}) \rangle = \sum_{i=1}^{I} \left(\upsilon_{\alpha i}^{\mathrm{f}} - \upsilon_{\alpha i}^{\mathrm{r}} \right) \left(\left\langle k_i^{\mathrm{f}}(T) \prod_{\beta=1}^{K} c_\beta^{\upsilon_{\beta i}^{\mathrm{f}}} \right\rangle - \left\langle k_i^{\mathrm{r}}(T) \prod_{\beta=1}^{K} c_\beta^{\upsilon_{\beta i}^{\mathrm{r}}} \right\rangle \right), \tag{5.40}$$

which is non-linear in both the temperature T and the molar concentrations $\mathbf{c}$. The chemical source term is thus unclosed at the level of the first-order moments. Moreover, due to the exponential non-linearity in T, the chemical source term will be unclosed at any finite level of moments.[25]

[23] When $k_i^{\mathrm{f}} = 0$, $c_{\mathrm{r}i}^*$ will be a *conserved* scalar. On the other hand, when $k_i^{\mathrm{f}} \to \infty$, the reaction rate function must remain bounded. Thus, $R_i / k_i^{\mathrm{f}} \to 0$, which yields an algebraic equation that replaces the transport equation for $c_{\mathrm{r}i}^*$.

[24] For this case, the temperature can be treated as just another 'chemical species.' The reaction coefficients in the final row of $\boldsymbol{\Upsilon}_{\mathrm{c}}$ correspond to temperature and will be positive (negative) if the reaction is exothermic (endothermic).

[25] In other words, $\langle \mathrm{e}^x \rangle = 1 - \langle x \rangle + \frac{1}{2} \langle x^2 \rangle - \cdots + \frac{1}{n!} (-1)^n \langle x^n \rangle + \cdots$ involves moments of all orders.

The chemical-source-term closure problem occurs even for relatively simple isothermal reactions. For example, consider again the simple two-step reaction (5.21) where[26]

$$S_A(\mathbf{c}) = -k_1 c_A c_B, \tag{5.41}$$

$$S_B(\mathbf{c}) = -k_1 c_A c_B - k_2 c_B c_R, \tag{5.42}$$

and

$$S_R(\mathbf{c}) = k_1 c_A c_B - k_2 c_B c_R. \tag{5.43}$$

The Reynolds-averaged chemical source terms can be written in terms of the scalar means and covariances. For example,

$$\begin{aligned} \langle S_A(\mathbf{c})\rangle &= -k_1 \langle c_A c_B\rangle \\ &= -k_1 \langle c_A\rangle\langle c_B\rangle - k_1 \langle c'_A c'_B\rangle \end{aligned} \tag{5.44}$$

and

$$\begin{aligned} \langle S_B(\mathbf{c})\rangle &= k_1 \langle c_A c_B\rangle - k_2 \langle c_B c_R\rangle \\ &= k_1 \langle c_A\rangle\langle c_B\rangle + k_1 \langle c'_A c'_B\rangle - k_2 \langle c_B\rangle\langle c_R\rangle - k_2 \langle c'_B c'_R\rangle. \end{aligned} \tag{5.45}$$

However, the covariance terms will be complicated functions of the reaction rates and feed/flow conditions:

$$\langle c'_A c'_B\rangle = \mathrm{cov}_{A,B}(k_1, k_2, \text{feed conditions, flow conditions}) \tag{5.46}$$

and

$$\langle c'_B c'_R\rangle = \mathrm{cov}_{B,R}(k_1, k_2, \text{feed conditions, flow conditions}). \tag{5.47}$$

The closure problem thus reduces to finding general methods for modeling higher-order moments of the composition PDF that are valid over a wide range of chemical time scales.

5.1.5 Chemical time scales

The chemical time scales can be defined in terms of the *eigenvalues* of the *Jacobian matrix* of the chemical source term.[27] For example, for an isothermal system the $K \times K$ Jacobian matrix of the chemical source term is given by

$$\mathbf{J}(\mathbf{c}) \equiv \frac{\partial \mathbf{S}_c}{\partial \mathbf{c}}(\mathbf{c}) = \boldsymbol{\Upsilon}\mathbf{J}^R(\mathbf{c}). \tag{5.48}$$

The $I \times K$ components of $\mathbf{J}^R$ are defined by[28]

$$J^R_{\alpha,\beta} \equiv \frac{\partial R_\alpha}{\partial c_\beta}(\mathbf{c}), \tag{5.49}$$

26 For simplicity, we will use the fictitious chemical species as the subscripts whenever there is no risk of confusion. As shown above, this reaction can be rewritten in terms of two reacting scalars and thus two components for $\mathbf{S}^*$. The closure problem, however, cannot be eliminated by any linear transformation of the scalar variables.

27 Another valid choice for the chemical time scales would be the non-zero *singular values* resulting from the SVD of the Jacobian matrix.

28 By assumption, the components of $\mathbf{R}(\mathbf{c})$ are linearly independent for arbitrary choices of $\mathbf{c}$. Thus, for arbitrary choices of $\mathbf{c}$, the Jacobian matrix formed from $\mathbf{R}$ will be full rank: $\mathrm{rank}(\mathbf{J}^R) = \min(I, K)$; and $\mathrm{rank}(\mathbf{J}) = N_\Upsilon$. Nevertheless, for *particular* choices of $\mathbf{c}$ (e.g., $\mathbf{c} = \mathbf{0}$), $\mathbf{J}^R$ may be rank-deficient.

and (for non-linear reactions) will depend explicitly on $\mathbf{c}$ and the parameters appearing in the reaction rate functions in (5.6). The dependence on $\mathbf{c}$ implies that the eigenvalues of $\mathbf{J}$ will also depend on $\mathbf{c}$. Thus, the chemical time scales will vary depending on the location of $\mathbf{c}$ in composition space, and hence will vary with time and space throughout the flow field.[29] Nevertheless, given $\mathbf{c}$, the chemical-source-term Jacobian is well defined and the eigenvalues can be computed numerically.

Denoting the eigenvalues by μ_α ($\alpha \in 1, \dots, K$), the chemical time scales can be defined as[30]

$$\tau_\alpha \equiv \frac{1}{|\mu_\alpha|}, \qquad \alpha \in 1, \dots, K. \tag{5.50}$$

Consistent with the notion of a 'fast reaction,' large eigenvalues will yield very small chemical time scales. On the other hand, zero eigenvalues (corresponding to conserved scalars) will yield chemical time scales that are infinite. In a complex chemical kinetic scheme, the chemical time scales can range over several orders of magnitude. Closures for the chemical source term must take into account the chemical time scales that will be present in the reacting flow under consideration. In particular, the values of the chemical time scales relative to the time scales of the flow (see Fig. 5.3) must be properly accounted for when closing the chemical source term.

The chemical time scales τ_α and the mixing time scale τ_ϕ can be used to define the Damköhler number(s) $\mathrm{Da}_\alpha \equiv \tau_\phi/\tau_\alpha$. Note that fast reactions correspond to large Da, and slow reactions to small Da. In general, for reacting scalars one-point statistics such as the mean, scalar flux, and joint scalar dissipation rate will depend strongly on the value of Da. This Da dependence should be accounted for in closures for the chemical source term.

For chemically reacting flows defined in terms of elementary reactions, at least E of the eigenvalues appearing in (5.50) will be null. This fact is easily shown starting with (5.18) wherefrom the following relations can be derived:

$$\mathbf{J}^* \equiv \frac{\partial \mathbf{S}_c^*}{\partial \mathbf{c}^*} = \mathbf{M}\mathbf{J}\mathbf{M}^{-1} \quad \Rightarrow \quad \mathbf{J} = \mathbf{M}^{-1}\mathbf{J}^*\mathbf{M}. \tag{5.51}$$

Due to the conservation of elements, the rank of $\mathbf{J}^*$ will be less than or equal to $K - E$.[31] In general, $\mathrm{rank}(\mathbf{J}) = N_\Upsilon \leq K - E$, which implies that $N = K - N_\Upsilon$ eigenvalues of $\mathbf{J}$ are null. Moreover, since $\mathbf{M}$ is a similarity transformation, (5.51) implies that the eigenvalues of $\mathbf{J}$ and those of $\mathbf{J}^*$ are identical. We can thus limit the definition of the chemical time scales to include only the N_Υ finite τ_α found from (5.50). The other N components of the transformed composition vector correspond to conserved scalars for which no chemical-source-term closure is required. The same comments would apply if the N_Υ non-zero singular values of $\mathbf{J}$ were used to define the chemical time scales.

[29] In non-isothermal systems, the Jacobian is usually most sensitive to the value of temperature T. Indeed, for low temperatures, the components of the Jacobian are often nearly zero, while at high temperatures they can be extremely large.

[30] As noted below, the set of the chemical time scales can be restricted to include only those corresponding to non-zero eigenvalues. For a non-reacting flow, the set would thus be empty.

[31] For non-elementary reactions, it suffices to set $E = 0$ here and in the following discussion.

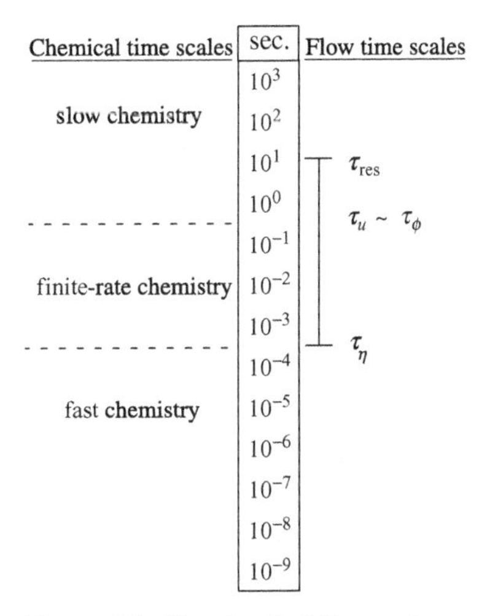

Figure 5.3. The chemical time scales must be considered in relation to the flow time scales occurring in a turbulent reacting flow. The flow time scales range from the Kolmogorov time scale τ_η, through the turbulence time scale τ_u, up to the mean residence time τ_{res}. In a turbulent reacting flow, the micromixing time τ_ϕ will usually lie between τ_η and τ_u. In a typical plant-scale chemical reactor, the range of flow time scales (which depends on the Reynolds number) is usually not more than three to four orders of magnitude. On the other hand, the chemical time scales in a non-isothermal reacting flow can easily range over more than ten orders of magnitude. The appropriate chemical-source-term closure will depend on how the two ranges of time scales overlap. The relative overlap of the time scales can be roughly divided into three cases: *slow chemistry* (chemical time scales all larger than τ_ϕ); *fast chemistry* (chemical time scales all smaller than τ_η); and *finite-rate chemistry*.

5.2 Moment closures

The simplest closure for the chemical source term is to assume that the joint composition PDF can be represented by its moments. In general, this assumption is of limited validity. Nevertheless, in this section we review methods based on moment closures in order to illustrate their limitations.

5.2.1 First-order moment closures

The limiting case where the chemical time scales are all large compared with the mixing time scale τ_ϕ, i.e., the *slow-chemistry limit*, can be treated by a simple first-order moment closure. In this limit, micromixing is fast enough that the composition variables can be approximated by their mean values (i.e., the first-order moments $\langle \phi \rangle$). We can then write, for example,

$$\left\langle k_i^{\mathrm{f}}(T) \prod_{\beta=1}^{K} c_\beta^{\upsilon_{\beta i}^{\mathrm{f}}} \right\rangle \approx k_i^{\mathrm{f}}(\langle T \rangle) \prod_{\beta=1}^{K} \langle c_\beta \rangle^{\upsilon_{\beta i}^{\mathrm{f}}}. \tag{5.52}$$

Or, to put it another way, the simplest first-order moment closure is to assume that all scalar covariances are zero:

$$\langle \phi'_\alpha \phi'_\beta \rangle = 0 \quad \text{for all } \alpha \text{ and } \beta, \tag{5.53}$$

so that $\langle \mathbf{S}(\phi) \rangle = \mathbf{S}(\langle \phi \rangle)$.

In most commercial CFD codes, (5.52) is the 'default' closure that is employed for modeling reacting flows. However, if any of the chemical time scales are small compared with the mixing time scale, this approximation will yield poor predictions and may lead to poor convergence! Physically, when the chemical time scales are small relative to the mixing time scale, the chemical reactions will be limited by sub-grid-scale mixing. The composition covariances (i.e., the second-order moments) will then be equal in magnitude to the first-order moments. This fact is most easily demonstrated for the one-step reaction

$$\mathrm{A} + \mathrm{B} \xrightarrow{k_1} \mathrm{P}. \tag{5.54}$$

When the rate constant k_1 is very large, chemical species A and B cannot coexist at the same spatial location. This implies that either $c_{\mathrm{A}} = 0$ or $c_{\mathrm{B}} = 0$ at every point in the flow. Thus, the expected value of their product will be null:

$$\langle c_{\mathrm{A}} c_{\mathrm{B}} \rangle = 0. \tag{5.55}$$

Rewriting $\langle c_{\mathrm{A}} c_{\mathrm{B}} \rangle$ in terms of the means and fluctuating components then yields

$$\langle c'_{\mathrm{A}} c'_{\mathrm{B}} \rangle = -\langle c_{\mathrm{A}} \rangle \langle c_{\mathrm{B}} \rangle. \tag{5.56}$$

Assuming that the covariance is null would thus lead to an overprediction of the mean reaction rate $\langle \mathbf{S} \rangle$.

For a one-step reaction, most commercial CFD codes provide a simple 'fix' for the fast-chemistry limit. This (first-order moment) method consists of simply slowing down the reaction rate whenever it is faster than the micromixing rate:

$$\langle S_A \rangle = \min \left(S_A(\langle \phi \rangle), C_1 \frac{\epsilon}{k} \right), \tag{5.57}$$

where C_1 is an empirical coefficient that must be adjusted to match experimental data. In principle, the same method can be extended to treat finite-rate chemistry in the form of (5.3) by assuming

$$\langle R_i(\phi) \rangle = \min \left(R_i(\langle \phi \rangle), C_i \frac{\epsilon}{k} \right), \tag{5.58}$$

where C_i ($i \in 1, \ldots, I$) are empirical constants. However, for mixing-sensitive reactions like the competitive-consecutive reaction:

$$\mathrm{A} + \mathrm{B} \xrightarrow{k_1} \mathrm{R} \quad \text{where} \quad \frac{1}{\tau_\phi} \ll k_1,$$

$$\mathrm{B} + \mathrm{R} \xrightarrow{k_2} \mathrm{S} \quad \text{where} \quad \frac{1}{\tau_\phi} \approx k_2,$$

the yield of the desired product (R), defined in terms of the concentrations at the reactor outlet $\mathbf{c}^\infty$ by

$$Y = \frac{c_\mathrm{R}^\infty}{c_\mathrm{R}^\infty + 2c_\mathrm{S}^\infty}, \tag{5.59}$$

will depend strongly on the choice of the empirical constants.[32] Indeed, they usually must be adjusted for each new set of flow conditions in order to obtain satisfactory agreement with experimental data. Moreover, because the yield is a 'global' (i.e., based on concentrations at the reactor outlet), as opposed to a 'local' measure of the reaction, values of (C_1, C_2) that produce the correct yield cannot be expected also to give an accurate representation of the local concentration fields. In general, simple first-order moment closures like (5.58) are inadequate for mixing-sensitive reactions dominated by finite-rate chemistry effects.

5.2.2 Higher-order moment closures

The failure of first-order moment closures for the treatment of mixing-sensitive reactions has led to the exploration of higher-order moment closures (Dutta and Tarbell 1989; Heeb and Brodkey 1990; Shenoy and Toor 1990). The simplest closures in this category attempt to relate the covariances of reactive scalars to the variance of the mixture fraction $\langle \xi'^2 \rangle$. The latter can be found by solving the inert-scalar-variance transport equation ((3.105), p. 85) along with the transport equation for $\langle \xi \rangle$. For example, for the one-step reaction in (5.54) the unknown scalar covariance can be approximated by

$$\langle c'_\mathrm{A} c'_\mathrm{B} \rangle = -\frac{\langle \xi'^2 \rangle}{\langle \xi \rangle (1 - \langle \xi \rangle)} \langle c_\mathrm{A} \rangle \langle c_\mathrm{B} \rangle. \tag{5.60}$$

The mean chemical source term (5.45) then becomes

$$\langle S_\mathrm{A} \rangle = -k_1 \left(1 - \frac{\langle \xi'^2 \rangle}{\langle \xi \rangle (1 - \langle \xi \rangle)} \right) \langle c_\mathrm{A} \rangle \langle c_\mathrm{B} \rangle. \tag{5.61}$$

In the limit where the mixture-fraction-variance dissipation rate is null, the mixture-fraction variance is related to the mean mixture fraction by

$$\langle \xi'^2 \rangle = \langle \xi \rangle (1 - \langle \xi \rangle) \quad \text{when } \varepsilon_\xi = 0. \tag{5.62}$$

[32] 'Rules of thumb' have been suggested for fixing the values of the empirical constants. For example, reactions involving chemical species entering through the feed streams use larger values of the constants than those involving chemical species generated by other reactions.

In this limit, the mixing time scale (τ_ϕ) is infinite and thus (5.61) predicts correctly that $\langle S_{\rm A} \rangle$ is null. In the other limit where τ_ϕ is null, the mixture-fraction variance will also be null, so that (5.61) again predicts the correct limiting value for the mean chemical source term. Between these two limits, the magnitude of $\langle S_{\rm A} \rangle$ will be decreased due to the finite rate of scalar mixing.

The extension of simple relationships such as (5.60) to multiple-step chemistry has proven to be elusive. For example, (5.60) implies that the covariance between two scalars is always negative.[33] However, this need not be the case. For example, if B were produced by another reaction between A and R:[34]

$$\begin{aligned} &\mathrm{A} + \mathrm{B} \xrightarrow{k_1} \mathrm{R} \\ &\mathrm{A} + \mathrm{R} \xrightarrow{k_2} 2\mathrm{B} + \mathrm{P}, \end{aligned}$$

then the sign of the covariance of A and B would depend not only on the relative values of k_1 and k_2, but also on the spatial location in the flow.

The other obvious higher-moment closure strategy that could be tried to close the chemical source term is to solve the transport equations for the scalar covariances ((3.136), p. 90). However, as shown for a second-order reaction in (3.144) on p. 91, this leads to a closure problem for the covariance chemical source term $S_{\alpha\beta}$ in the scalar-covariance transport equation. In general, non-linear chemical source terms will lead to difficult closure problems when moment methods are employed for finite-rate chemistry. Other closure schemes based on knowledge of the joint composition PDF, instead of just its lower-order moments, are more successful for finite-rate chemistry. On the other hand, in the equilibrium-chemistry limit, an accurate closure can be developed by using the mixture-fraction PDF. However, before discussing this limiting case, we will first develop a general definition of the mixture-fraction vector.

5.3 Mixture-fraction vector

For *non-premixed* turbulent reacting flows, it is often possible to define a *mixture-fraction vector* ($\boldsymbol{\xi}$) that can be employed to develop chemical-source-term closures that are much more successful than moment closures. In this section, we discuss a general method for finding the mixture-fraction vector (when it exists) for a given set of initial/inlet conditions. We also show that when a mixture-fraction vector exists, it is possible to transform the reacting-scalar vector into a *reaction-progress vector* ($\varphi_{\rm rp}$) that is null for all initial and inlet conditions. Thus, the turbulent reacting flow can be *most simply* described in terms of the reaction-progress vector and the mixture-fraction vector. These vectors are found using a constant-coefficient linear transformation matrix, (5.107), that depends only on the inlet/initial conditions and the reaction coefficient matrix. We conclude the section by

[33] The case where the covariance is always positive can be handled simply by changing the sign on the right-hand side.

[34] This set of reactions is *auto-catalytic*, i.e., production of B by the second step enhances the reaction rate of the first step.

introducing the joint PDF of the mixture-fraction vector, which is needed to compute the Reynolds-averaged chemical source term in closures based on the conditional composition PDF.

5.3.1 General formulation

In Section 5.1, we have seen (Fig. 5.2) that the molar concentration vector $\mathbf{c}$ can be transformed using the SVD of the reaction coefficient matrix $\boldsymbol{\Upsilon}$ into a vector $\mathbf{c}^*$ that has N_Υ reacting components $\mathbf{c}_\mathrm{r}$ and N conserved components $\mathbf{c}_\mathrm{c}$.[35] In the limit of equilibrium chemistry, the behavior of the N_Υ reacting scalars will be dominated by the transformed chemical source term $\mathbf{S}_\mathrm{c}^*$.[36] On the other hand, the behavior of the N conserved scalars will depend on the turbulent flow field and the inlet and initial conditions for the flow domain. However, they will be independent of the chemical reactions, which greatly simplifies the mathematical description.

At high Reynolds numbers, it is usually possible to assume that the mean scalar fields (e.g., $\langle \mathbf{c}_\mathrm{c} \rangle$) are independent of molecular-scale quantities such as the molecular-diffusion coefficients. In this case, it is usually safe to assume that all scalars have the same molecular diffusivity Γ. The conserved-scalar transport equation then simplifies to[37]

$$\frac{\partial \mathbf{c}_\mathrm{c}}{\partial t} + U_j \frac{\partial \mathbf{c}_\mathrm{c}}{\partial x_j} = \Gamma \frac{\partial^2 \mathbf{c}_\mathrm{c}}{\partial x_j \partial x_j}. \tag{5.63}$$

Note that (5.63) is linear, and thus $\mathbf{c}_\mathrm{c}$ will be uniquely determined by the initial and boundary conditions on the flow domain.

As shown in Fig. 5.4, the flow domain can be denoted by Ω with inlet streams at N_in boundaries denoted by $\partial\Omega_i$ $(i \in 1, \ldots, N_\mathrm{in})$. In many scalar mixing problems, the initial conditions in the flow domain are uniform, i.e., $\mathbf{c}_\mathrm{c}(\mathbf{x}, 0) = \mathbf{c}_\mathrm{c}^{(0)}$. Likewise, the scalar values at the inlet streams are often constant so that $\mathbf{c}_\mathrm{c}(\mathbf{x} \in \partial\Omega_i, t) = \mathbf{c}_\mathrm{c}^{(i)}$ for all $i \in 1, \ldots, N_\mathrm{in}$. Under these assumptions,[38] the *principle of linear superposition* leads to the following relationship:

$$\mathbf{c}_\mathrm{c}(\mathbf{x}, t) = \left[\mathbf{c}_\mathrm{c}^{(0)} \mathbf{c}_\mathrm{c}^{(1)} \quad \cdots \quad \mathbf{c}_\mathrm{c}^{(N_\mathrm{in})}\right] \boldsymbol{\alpha}(\mathbf{x}, t) = \boldsymbol{\Phi}_\mathrm{c} \boldsymbol{\alpha}(\mathbf{x}, t), \tag{5.64}$$

where $\boldsymbol{\Phi}_\mathrm{c}$ is an $N \times (N_\mathrm{in} + 1)$ matrix formed from the initial and inlet conditions. In (5.64), $\boldsymbol{\alpha}$ is a coefficient vector of length $N_\mathrm{in} + 1$ whose components are found by solving[39]

$$\frac{\partial \alpha_i}{\partial t} + U_j \frac{\partial \alpha_i}{\partial x_j} = \Gamma \frac{\partial^2 \alpha_i}{\partial x_j \partial x_j} \quad \text{for} \quad i \in 0, \ldots, N_\mathrm{in}, \tag{5.65}$$

[35] In order to simplify the notation, we will consider only the isothermal case in this section. The constant-fluid-property case will be identical when $\boldsymbol{\Upsilon}$ is replaced by $\boldsymbol{\Upsilon}_\mathrm{c}$. In the non-isothermal case, an additional transport equation will be needed to determine the temperature.

[36] Recall that the Jacobian of $\mathbf{S}_\mathrm{c}^*$ will generate N_Υ chemical time scales. In the equilibrium-chemistry limit, all N_Υ chemical times are assumed to be much smaller than the flow time scales.

[37] We will assume throughout this section that the composition vector, as well as the initial and boundary conditions, have been previously transformed into the reacting/conserved sub-spaces.

[38] The general case of non-uniform initial conditions and time-dependent boundary conditions can also be expressed analytically, and leads to convolution integrals in space and time. Nevertheless, for most practical applications, the assumptions of uniform initial conditions and constant boundary conditions are adequate.

[39] Note that, exceptionally, the components of $\boldsymbol{\alpha}$ are numbered starting from zero.

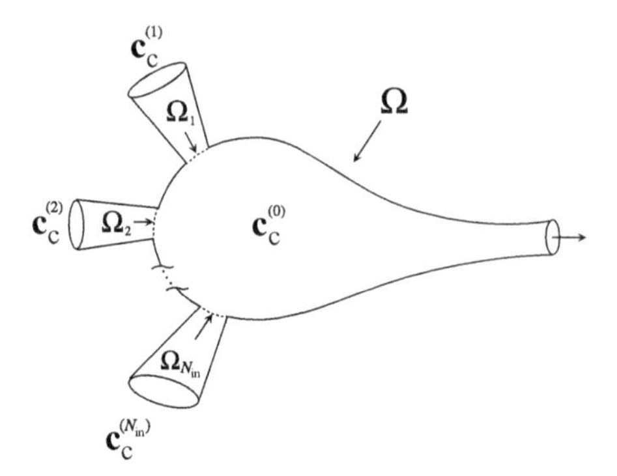

Figure 5.4. The flow domain Ω with inlet streams at N_{in} boundaries $\partial\Omega_i$ $(i \in 1, \ldots, N_{in})$. The initial conditions for the conserved scalars in the flow domain are assumed to be uniform and equal to $\mathbf{c}_c^{(0)}$. Likewise, the inlet conditions for the conserved scalars at the ith inlet are assumed to be constant and equal to $\mathbf{c}_c^{(i)}$.

with the following inlet/initial conditions:[40]

- $\alpha_0(\mathbf{x} \in \Omega, 0) = 1$, $\alpha_0(\mathbf{x} \in \partial\Omega_i, t) = 0$ for all $i \in 1, \ldots, N_{in}$;
- $\alpha_1(\mathbf{x} \in \Omega, 0) = 0$, $\alpha_1(\mathbf{x} \in \partial\Omega_1, t) = 1$ and $\alpha_1(\mathbf{x} \in \partial\Omega_i, t) = 0$ for $i \neq 1$;
- $\vdots$
- $\alpha_{N_{in}}(\mathbf{x} \in \Omega, 0) = 0$, $\alpha_{N_{in}}(\mathbf{x} \in \partial\Omega_{N_{in}}, t) = 1$ and $\alpha_{N_{in}}(\mathbf{x} \in \partial\Omega_i, t) = 0$ for $i \neq N_{in}$.

Note that the coefficient vector found from (5.65) is overdetermined. Indeed, by summing the component governing equations and the initial and boundary conditions, it is easily shown that

$$\sum_{i=0}^{N_{in}} \alpha_i(\mathbf{x}, t) = 1. \tag{5.66}$$

Thus, *at most* N_{in} transport equations must be solved to determine the coefficient vector. In order to eliminate one component of $\boldsymbol{\alpha}$, we can subtract any of the columns of $\boldsymbol{\Phi}_c$ to form a new conserved-scalar vector, e.g.,[41]

$$\boldsymbol{\varphi}_c \equiv \mathbf{c}_c - \mathbf{c}_c^{(0)}. \tag{5.67}$$

Applying the principle of linear superposition, $\boldsymbol{\varphi}_c$ can be written as[42]

$$\boldsymbol{\varphi}_c(\mathbf{x}, t) = \left[\boldsymbol{\varphi}_c^{(1)} \quad \cdots \quad \boldsymbol{\varphi}_c^{(N_{in})}\right]\boldsymbol{\alpha}(\mathbf{x}, t) = \boldsymbol{\Phi}_c^{(0)}\boldsymbol{\alpha}(\mathbf{x}, t), \tag{5.68}$$

where the coefficient vector $\boldsymbol{\alpha}$ is now length N_{in}, and the columns of $\boldsymbol{\Phi}_c^{(0)}$ are defined by

$$\boldsymbol{\varphi}_c^{(i)} \equiv \mathbf{c}_c^{(i)} - \mathbf{c}_c^{(0)} \quad \text{for} \quad i \in 1, \ldots, N_{in}. \tag{5.69}$$

[40] The boundary conditions are otherwise zero flux at the walls and outflow conditions at the outlet(s).

[41] We have chosen the initial conditions as the reference vector. For steady-state problems, $\alpha_0 = 0$ and any non-zero inlet vector can be chosen.

[42] The superscript on $\boldsymbol{\Phi}_c^{(0)}$ is used as a reminder that $\mathbf{c}^{(0)}$ was chosen as the reference vector.

$$\boldsymbol{\varphi}_c = (\mathbf{c}_c - \mathbf{c}_c^{(k)}) \underset{\mathbf{U}^{(k)}}{\overset{\mathbf{U}^{(k)\mathrm{T}}}{\Longleftrightarrow}} \begin{bmatrix} \boldsymbol{\varphi}_{cv} \\ \boldsymbol{\varphi}_{cc} \end{bmatrix} = \begin{bmatrix} \boldsymbol{\varphi}_{cv} \\ \mathbf{0} \end{bmatrix}$$

Figure 5.5. The conserved-scalar vector φ_c can be partitioned by a linear transformation into two parts: φ_{cv}, a conserved-variable vector of length $N_{\Phi_c^{(0)}}$; and $\varphi_{cc} = \mathbf{0}$, a conserved-constant vector of length $N - N_{\Phi_c^{(0)}}$. The linear transformation depends only on the matrix of initial/boundary conditions $\Phi_c^{(0)}$, and thus is independent of $\mathbf{x}$ and t.

The question of exactly how many coefficients are required to describe the conserved-scalar field does not end with (5.68), and is a significant one since it will determine the *minimum* number of transport equations that must be solved to describe the flow completely. From the form of (5.68), it can be seen that φ_c is a linear combination of the columns of $\mathbf{\Phi}_c^{(0)}$.[43] Thus, only a *linearly independent sub-set* of the columns is required to describe φ_c completely. The exact number is thus equal to $N_{\Phi_c^{(0)}} \equiv \mathrm{rank}(\mathbf{\Phi}_c^{(0)}) \leq \min(N, N_{\mathrm{in}})$.

As in Section 5.1 for the reacting- and conserved-scalar sub-spaces, φ_c can be further partitioned into two sub-spaces corresponding to $N_{\Phi_c^{(0)}}$ components that vary with space/time and $(N - N_{\Phi_c^{(0)}})$ components that are uniform/constant. This is illustrated in Fig. 5.5. The partitioning is most easily carried out using the SVD of $\mathbf{\Phi}_c^{(0)}$:

$$\mathbf{\Phi}_c^{(0)} = \mathbf{U}^{(0)}\mathbf{\Sigma}^{(0)}\mathbf{V}^{(0)\mathrm{T}}, \tag{5.70}$$

where the $(N \times N)$ matrix $\mathbf{U}^{(0)}$ and the $(N_{\mathrm{in}} \times N_{\mathrm{in}})$ matrix $\mathbf{V}^{(0)}$ are orthogonal, and the $(N \times N_{\mathrm{in}})$ diagonal matrix $\mathbf{\Sigma}^{(0)}$ has $N_{\Phi_c^{(0)}}$ non-zero singular values in descending order across the diagonal. The partitioned conserved-scalar vector can be expressed as

$$\varphi_c^{**}(\mathbf{x}, t) \equiv \begin{bmatrix} \varphi_{cv}(\mathbf{x}, t) \\ \varphi_{cc}(\mathbf{x}, t) \end{bmatrix} \equiv \mathbf{U}^{(0)\mathrm{T}}\varphi_c(\mathbf{x}, t), \tag{5.71}$$

where φ_{cv} is a vector of length $N_{\Phi_c^{(0)}}$ corresponding to the conserved-variable sub-space, and φ_{cc} is a vector of length $(N - N_{\Phi_c^{(0)}})$ corresponding to the conserved-constant sub-space.

Applying the linear transformation in (5.71) to (5.68) yields[44]

$$\begin{aligned} \varphi_c^{**}(\mathbf{x}, t) &= \mathbf{U}^{(0)\mathrm{T}}\mathbf{\Phi}_c^{(0)}\alpha(\mathbf{x}, t) \\ &= \mathbf{\Sigma}^{(0)}\mathbf{V}^{(0)\mathrm{T}}\alpha(\mathbf{x}, t) \\ &= \begin{bmatrix} \mathbf{\Phi}_{cv}^{(0)} \\ \mathbf{0} \end{bmatrix} \alpha(\mathbf{x}, t), \end{aligned} \tag{5.72}$$

[43] Using terminology from linear algebra, the columns of $\mathbf{\Phi}_c^{(0)}$ *span* the solution space of dimension $N_{\Phi_c^{(0)}} = \mathrm{rank}(\mathbf{\Phi}_c^{(0)})$. A linearly independent sub-set of $N_{\Phi_c^{(0)}}$ columns can be used to form a *basis*.

[44] The zero matrix in the final expression results from the $(N - N_{\Phi_c^{(0)}})$ zero singular values in $\mathbf{\Sigma}^{(0)}$. Thus, if $N = N_{\Phi_c^{(0)}}$, all conserved scalars are variable, and no zero matrix will be present.

$$\boldsymbol{\varphi} = \left(\mathbf{c} - \mathbf{c}^{(k)}\right) \underset{\mathbf{M}^{(k)\mathrm{T}}}{\overset{\mathbf{M}^{(k)}}{\Longleftrightarrow}} \begin{bmatrix} \boldsymbol{\varphi}_{\mathrm{r}} \\ \boldsymbol{\varphi}_{\mathrm{cv}} \\ \mathbf{0} \end{bmatrix} = \boldsymbol{\varphi}^{**}$$

Figure 5.6. The molar concentration vector $\mathbf{c}$ of length K can be partitioned by a linear transformation into three parts: φ_{r}, a reacting vector of length N_{Υ}; φ_{cv}, a conserved-variable vector of length $N_{\Phi_c^{(0)}}$; and $\mathbf{0}$, a null vector of length $K - N_{\Upsilon} - N_{\Phi_c^{(0)}}$. The linear transformation matrix $\mathbf{M}^{(i)}$ depends on the reference concentration vector $\mathbf{c}^{(i)}$ and the reaction coefficient matrix $\boldsymbol{\Upsilon}$.

where $\boldsymbol{\Phi}_{\mathrm{cv}}^{(0)}$ is a full rank, $(N_{\Phi_c^{(0)}} \times N_{\mathrm{in}})$ matrix. Thus, comparing with (5.71), we see that $\varphi_{\mathrm{cc}}(\mathbf{x}, t) = \mathbf{0}$ (as it must since all components are constant) and[45]

$$\varphi_{\mathrm{cv}}(\mathbf{x}, t) = \boldsymbol{\Phi}_{\mathrm{cv}}^{(0)} \boldsymbol{\alpha}(\mathbf{x}, t). \tag{5.73}$$

Also, by applying the linear transformation $\boldsymbol{\Phi}_{\mathrm{cv}}^{(0)}$ to (5.65), we can see that the conserved-variable scalar vector obeys a transformed scalar transport equation of the form

$$\frac{\partial \varphi_{\mathrm{cv}}}{\partial t} + U_j \frac{\partial \varphi_{\mathrm{cv}}}{\partial x_j} = \Gamma \frac{\partial^2 \varphi_{\mathrm{cv}}}{\partial x_j \partial x_j} \tag{5.74}$$

with initial condition $\varphi_{\mathrm{cv}}(\mathbf{x} \in \Omega, 0) = \mathbf{0}$ and boundary conditions $\varphi_{\mathrm{cv}}(\mathbf{x} \in \partial\Omega_i) = \boldsymbol{\Phi}_{\mathrm{cv}}^{(0)} \mathbf{e}_i$ $(i \in 1, \ldots, N_{\mathrm{in}})$.[46]

In summary, for a non-stationary turbulent reacting flow wherein all scalars can be assumed to have identical molecular diffusivities and the initial and boundary conditions are uniform/constant, the K-component molar concentration vector

$$\varphi(\mathbf{x}, t) = \mathbf{c}(\mathbf{x}, t) - \mathbf{c}^{(0)} \tag{5.75}$$

can be partitioned into three distinct parts: (1) N_{Υ} reacting scalars φ_{r}, (2) $N_{\Phi_c^{(0)}}$ conserved-variable scalars φ_{cv}, and (3) $(K - N_{\Upsilon} - N_{\Phi_c^{(0)}})$ conserved-constant scalars $\varphi_{\mathrm{cc}} = \mathbf{0}$. This process is illustrated in Fig. 5.6. The orthogonal linear transformation that yields the partitioned composition vector

$$\varphi^{**} \equiv \begin{bmatrix} \varphi_{\mathrm{r}} \\ \varphi_{\mathrm{cv}} \\ \mathbf{0} \end{bmatrix} = \mathbf{M}^{(0)} \left(\mathbf{c} - \mathbf{c}^{(0)}\right) \tag{5.76}$$

is defined by (5.26) and (5.71):[47]

$$\mathbf{M}^{(0)} \equiv \begin{bmatrix} \mathbf{I} & \mathbf{0} \\ \mathbf{0} & \mathbf{U}^{(0)\mathrm{T}} \end{bmatrix} \mathbf{U}_{\mathrm{sv}}^{\mathrm{T}}, \tag{5.77}$$

45 Since $N_{\Phi_c^{(0)}} \leq N_{\mathrm{in}}$, $\boldsymbol{\Phi}_{\mathrm{cv}}^{(0)}$ will never have more rows than it has columns. The columns of $\boldsymbol{\Phi}_{\mathrm{cv}}^{(0)}$ thus span the solution space of $\varphi_{\mathrm{cv}}(\mathbf{x}, t)$, but may be more than are required to form a basis.

46 $\mathbf{e}_i$ is the unit vector with $e_i = 1$.

47 Note that the conserved-scalar transformation matrix does not affect the reacting scalars.

where $\mathbf{U}_{\rm sv}$ and $\mathbf{U}^{(0)}$ are found from the SVD of $\boldsymbol{\Upsilon}$ and $\boldsymbol{\Phi}_{\rm c}^{(0)}$, respectively. Likewise, the original composition vector can be found using the inverse transformation:

$$\mathbf{c} = \mathbf{c}^{(0)} + \mathbf{M}^{(0)\rm T} \boldsymbol{\varphi}^{**}. \tag{5.78}$$

Thus, the non-stationary turbulent reacting flow will be completely described by the first $(N_{\Upsilon} + N_{\Phi_{\rm c}^{(0)}})$ components of $\boldsymbol{\varphi}^{**}$. For a *stationary* turbulent reacting flow, $\mathbf{c}^{(0)}$ can be replaced by any of the inlet composition vectors: $\mathbf{c}^{(i)}$ for any $i \in 1, \ldots, N_{\rm in}$. For this case, if $N_{\Phi_{\rm c}^{(0)}} = N_{\rm in}$, then $N_{\Phi_{\rm c}^{(i)}} = N_{\rm in} - 1$ and thus one less conserved-variable scalar will be required to describe completely the stationary turbulent reacting flow.

5.3.2 Definition of mixture fraction

The values of the conserved-variable scalars will depend only on turbulent mixing between the initial contents in the flow domain and the inflowing streams. Thus, under certain conditions described below, it is possible to replace $\boldsymbol{\varphi}_{\rm cv}$ by a *mixture-fraction vector* $\boldsymbol{\xi}$ whose components are non-negative and whose sum is less than or equal to unity. The minimum number of mixture-fraction components that will be required to describe the flow will be denoted by $N_{\rm mf}$ and is bounded by $N_{\Phi_{\rm c}^{(0)}} \leq N_{\rm mf} \leq N_{\rm in}$. The case where $N_{\rm mf} = 0$ corresponds to a *premixed* turbulent reacting flow.[48] The case where $N_{\rm mf} = 1$ corresponds to *binary mixing* and can be described by a single mixture fraction.[49] Likewise, the case where $N_{\rm mf} = 2$ corresponds to *ternary mixing* and can be described by two mixture-fraction components.[50] Higher-order cases can be treated using the same formulation, but rarely occur in practical applications.

The interest in reformulating the conserved-variable scalars in terms of the mixture-fraction vector lies in the fact that relatively simple forms for the mixture-fraction PDF can be employed to describe the reacting scalars. However, if $N_{\Phi_{\rm c}^{(0)}} < N_{\rm mf}$, then the incentive is greatly diminished since more mixture-fraction-component transport equations ($N_{\rm mf}$) would have to be solved than conserved-variable-scalar transport equations ($N_{\Phi_{\rm c}^{(0)}}$). We will thus assume that $N_{\Phi_{\rm c}^{(0)}} = N_{\rm mf}$ and seek to define the mixture-fraction vector only for this case. Nonetheless, in order for the mixture-fraction PDF method to be applicable to the reacting scalars, they must form a *linear mixture* defined in terms of the components of the mixture-fraction vector. In some cases, the existence of linear mixtures is evident from the initial/inlet conditions; however, this need not always be the case. Thus, in this section, a general method for defining the mixture-fraction vector in terms of a *linear-mixture basis* for arbitrary initial/inlet conditions is developed.

[48] In order for $N_{\rm mf} = N_{\Phi_{\rm c}^{(0)}} = 0$, $\boldsymbol{\Phi}_{\rm c}^{(0)}$ must be a zero matrix, i.e., all inlet composition vectors are equal to the initial composition vector.

[49] $N_{\rm mf} = 1$ is the case that is usually associated with non-premixed turbulent reacting flows. Most commercial CFD codes are designed to treat only this case.

[50] The names *binary* and *ternary* come from the number of feed streams that are required for a stationary flow, i.e., two and three, respectively. Some authors also use ternary for the rank-deficient case where $N_{\rm in} = 3$ but $N_{\rm mf} = 1$.

From the definition of $\boldsymbol{\alpha}$ in (5.65), it is clear that when $N_{\Phi_c^{(0)}} = N_{\text{in}}$ we can define the mixture-fraction vector by $\boldsymbol{\xi}(\mathbf{x}, t) = \boldsymbol{\alpha}(\mathbf{x}, t)$.[51] For this case, (5.73) defines an invertible, constant-coefficient linear transformation $\boldsymbol{\Xi}^{(0)}$:

$$\boldsymbol{\xi} = \boldsymbol{\Xi}^{(0)} \left(\mathbf{c}_{\text{cv}} - \mathbf{c}_{\text{cv}}^{(0)} \right) \quad \Longleftrightarrow \quad \mathbf{c}_{\text{cv}} = \mathbf{c}_{\text{cv}}^{(0)} + \boldsymbol{\Xi}^{(0)-1} \boldsymbol{\xi}, \tag{5.79}$$

where $\mathbf{c}_{\text{cv}}$ and $\mathbf{c}_{\text{cv}}^{(0)}$ are the conserved-variable molar concentration vectors found from the partition of $\mathbf{M}^{(0)}\mathbf{c}$ and $\mathbf{M}^{(0)}\mathbf{c}^{(0)}$, respectively, and

$$\boldsymbol{\Xi}^{(0)} = \boldsymbol{\Phi}_{\text{cv}}^{(0)-1}. \tag{5.80}$$

Furthermore, the transport equation for $\boldsymbol{\xi}(\mathbf{x}, t)$ is then exactly the same as (5.65).

The case where $N_{\Phi_c^{(0)}} = N_{\text{mf}} = N_{\text{in}} - 1$ corresponds to flows where the initial/inlet conditions are coupled through a linear relationship of the form

$$\sum_{i=0}^{N_{\text{in}}} \beta_i \mathbf{c}_{\text{cv}}^{(i)} = 0, \quad \text{where} \quad \sum_{i=0}^{N_{\text{in}}} \beta_i = 0, \tag{5.81}$$

and $\mathbf{c}_{\text{cv}}^{(i)}$ is the partition of $\mathbf{M}^{(0)}\mathbf{c}^{(i)}$ ($i \in 0, \ldots, N_{\text{in}}$) corresponding to the conserved-variable scalars. In order to see that this is the case, assume that (5.81) holds with $\beta_0 \neq 0$.[52] We can then write

$$\mathbf{c}_{\text{cv}}^{(0)} = \sum_{i=1}^{N_{\text{in}}} \beta_i^* \mathbf{c}_{\text{cv}}^{(i)}, \quad \text{where} \quad \sum_{i=1}^{N_{\text{in}}} \beta_i^* = 1. \tag{5.82}$$

The initial condition is then a *linear mixture* of the inlet conditions.[53] From (5.64), it follows that the conserved-variable vector $\mathbf{c}_{\text{cv}}$ can be written in terms of the coefficients α_i as

$$\mathbf{c}_{\text{cv}} = \alpha_0 \mathbf{c}_{\text{cv}}^{(0)} + \sum_{i=1}^{N_{\text{in}}} \alpha_i \mathbf{c}_{\text{cv}}^{(i)}. \tag{5.83}$$

Using (5.82), this expression can be rewritten as

$$\begin{aligned} \mathbf{c}_{\text{cv}} &= \alpha_0 \sum_{i=1}^{N_{\text{in}}} \beta_i^* \mathbf{c}_{\text{cv}}^{(i)} + \sum_{i=1}^{N_{\text{in}}} \alpha_i \mathbf{c}_{\text{cv}}^{(i)} \\ &= \sum_{i=1}^{N_{\text{in}}} (\alpha_i + \alpha_0 \beta_i^*) \mathbf{c}_{\text{cv}}^{(i)} \\ &= \sum_{i=1}^{N_{\text{in}}} \gamma_i \mathbf{c}_{\text{cv}}^{(i)}, \end{aligned} \tag{5.84}$$

[51] The vector $\boldsymbol{\alpha}$ is composed of α_i ($i \in 1, \ldots, N_{\text{in}}$).

[52] The same argument can be applied with any non-zero β_i.

[53] Note that *linear mixture* is more restrictive than a *linear combination* due to the constraint on the sum of the coefficients. Below we add one additional constraint: β^* must be non-negative.

where $\gamma_i(\mathbf{x}, t) \equiv \alpha_i(\mathbf{x}, t) + \alpha_0(\mathbf{x}, t)\beta_i^*$ $(i \in 1, \dots, N_{\text{in}})$. Note that at time zero, the initial conditions on the components α_i $(i \in 0, \dots, N_{\text{in}})$ imply

$$\boldsymbol{\gamma}(\mathbf{x} \subset \Omega, 0) = \boldsymbol{\alpha}(\mathbf{x} \in \Omega, 0) + \alpha_0(\mathbf{x} \in \Omega, 0)\boldsymbol{\beta}^* = \boldsymbol{\beta}^*. \tag{5.85}$$

Thus, all components of $\boldsymbol{\beta}^*$ must be non-negative in order for $\boldsymbol{\gamma}$ to be initially non-negative in the flow domain.

By definition of $\boldsymbol{\alpha}$ and $\boldsymbol{\beta}^*$,[54] the sum of the N_{in} components of $\boldsymbol{\gamma}$ is unity. Thus, by extending the definition of *linear mixture* to include the condition that $\boldsymbol{\beta}^*$ must be non-negative,[55] the last $N_{\text{mf}} = N_{\text{in}} - 1$ components of $\boldsymbol{\gamma}$ can be used to define the mixture-fraction vector.[56] The transformation matrix that links $\boldsymbol{\xi}$ to $\mathbf{c}_{\text{cv}}$ can be found by rewriting (5.84) using the fact that the components of $\boldsymbol{\gamma}$ sum to unity:

$$\begin{aligned}
\mathbf{c}_{\text{cv}} &= \gamma_1 \mathbf{c}_{\text{cv}}^{(1)} + \sum_{i=2}^{N_{\text{in}}} \gamma_i \mathbf{c}_{\text{cv}}^{(i)} \\
&= \left(1 - \sum_{i=2}^{N_{\text{in}}} \gamma_i\right) \mathbf{c}_{\text{cv}}^{(1)} + \sum_{i=2}^{N_{\text{in}}} \gamma_i \mathbf{c}_{\text{cv}}^{(i)} \\
&= \mathbf{c}_{\text{cv}}^{(1)} + \sum_{i=2}^{N_{\text{in}}} \gamma_i \left(\mathbf{c}_{\text{cv}}^{(i)} - \mathbf{c}_{\text{cv}}^{(1)}\right) \\
&= \mathbf{c}_{\text{cv}}^{(1)} + \left[(\mathbf{c}_{\text{cv}}^{(2)} - \mathbf{c}_{\text{cv}}^{(1)}) \quad \dots \quad (\mathbf{c}_{\text{cv}}^{(N_{\text{in}})} - \mathbf{c}_{\text{cv}}^{(1)})\right]\boldsymbol{\xi} \\
&\equiv \mathbf{c}_{\text{cv}}^{(1)} + \boldsymbol{\Phi}_{\text{LI}}^{(1)}\boldsymbol{\xi}.
\end{aligned} \tag{5.86}$$

By construction, the $(N_{\text{mf}} \times N_{\text{mf}})$ matrix $\boldsymbol{\Phi}_{\text{LI}}^{(1)}$ is full rank. Thus, the desired linear transformation is

$$\boldsymbol{\xi} = \boldsymbol{\Xi}^{(1)}\left(\mathbf{c}_{\text{cv}} - \mathbf{c}_{\text{cv}}^{(1)}\right) \quad \Longleftrightarrow \quad \mathbf{c}_{\text{cv}} = \mathbf{c}_{\text{cv}}^{(1)} + \boldsymbol{\Xi}^{(1)-1}\boldsymbol{\xi}, \tag{5.87}$$

where

$$\boldsymbol{\Xi}^{(1)} = \boldsymbol{\Phi}_{\text{LI}}^{(1)-1}. \tag{5.88}$$

In addition, the transport equation for $\boldsymbol{\xi}(\mathbf{x}, t)$ has exactly the same form as (5.65), but with appropriately defined initial/inlet conditions. We will illustrate this process for example flows in the next section.

The next case, where $N_{\Phi_c^{(0)}} = N_{\text{mf}} = N_{\text{in}} - 2$, corresponds to flows with a *second* linear mixture of the initial/inlet conditions. Following the same procedure as used above, it is again possible to eliminate a second component of the coefficient vector and thereby

[54] That is, due to the fact that the coefficients sum to unity.

[55] In most practical applications, all but one component of $\boldsymbol{\beta}$ will be non-negative. One can then choose the vector with the negative component to be the left-hand side of (5.82). However, the fact that $\boldsymbol{\beta}^*$ must be non-negative greatly restricts the types of linear dependencies that can be expressed as a mixture-fraction vector of length $N_{\Phi_c^{(0)}}$.

[56] For this problem, the choice of which component of $\boldsymbol{\gamma}$ to eliminate is arbitrary. However, if more than one linear mixture exists between the initial/inlet conditions, it may be necessary to choose more carefully. We illustrate this problem in an example flow below.

to define a mixture-fraction vector with $N_{\rm in} - 2$ components. It should also be apparent to the reader that similar arguments can be applied to define the mixture-fraction vector in subsequent cases (i.e., $N_{\rm mf} = N_{\rm in} - 3$, $N_{\rm mf} = N_{\rm in} - 4$, etc.). Indeed, a general formulation can be developed in terms of the *reciprocal basis vectors* found for $N_{\rm mf}$ linearly independent columns of the $(N_{\rm mf} \times N_{\rm in})$ matrix,

$$\boldsymbol{\Phi}_{\rm cv}^{(k)} \equiv \left[\varphi_{\rm cv}^{(0)} \;\; \cdots \;\; \varphi_{\rm cv}^{(k-1)} \;\; \varphi_{\rm cv}^{(k+1)} \;\; \cdots \;\; \varphi_{\rm cv}^{(N_{\rm in})}\right] \quad \text{for some} \quad k \in 0, \ldots, N_{\rm in}, \tag{5.89}$$

where $\varphi_{\rm cv}^{(i)} \equiv \mathbf{c}_{\rm cv}^{(i)} - \mathbf{c}_{\rm cv}^{(k)}$ for $i \in 0, \ldots, N_{\rm in}$. In (5.89), the kth composition vector serves as a reference point in composition space for which all $N_{\rm mf}$ components of the mixture-fraction vector are null. In general, the reference point must be found by trial and error such that the mixture-fraction vector is well defined. In order to simplify the discussion, here we will take $k = 0$.[57]

Without loss of generality, we can assume that the first $N_{\rm mf}$ columns of $\boldsymbol{\Phi}_{\rm cv}^{(0)}$ are linearly independent.[58] The remaining $(N_{\rm LD} = N_{\rm in} - N_{\rm mf})$ columns are thus linearly dependent on the first $N_{\rm mf}$ columns. We can then write

$$\boldsymbol{\Phi}_{\rm LD}^{(0)} = \boldsymbol{\Phi}_{\rm LI}^{(0)} \mathbf{B}^{(0)}, \tag{5.90}$$

where

$$\boldsymbol{\Phi}_{\rm LD}^{(0)} \equiv \left[\varphi_{\rm cv}^{(N_{\rm mf}+1)} \;\; \cdots \;\; \varphi_{\rm cv}^{(N_{\rm in})}\right], \tag{5.91}$$

$$\boldsymbol{\Phi}_{\rm LI}^{(0)} \equiv \left[\varphi_{\rm cv}^{(1)} \;\; \cdots \;\; \varphi_{\rm cv}^{(N_{\rm mf})}\right], \tag{5.92}$$

and the $(N_{\rm mf} \times N_{\rm LD})$ matrix $\mathbf{B}^{(0)} = [\beta_{ij}^{(0)}]$ is defined by[59]

$$\mathbf{B}^{(0)} \equiv \boldsymbol{\Phi}_{\rm LI}^{(0)-1} \boldsymbol{\Phi}_{\rm LD}^{(0)}. \tag{5.93}$$

Using (5.90), we can rewrite (5.83) as

$$\begin{aligned} \varphi_{\rm cv} &= \sum_{i=1}^{N_{\rm mf}} \alpha_i \varphi_{\rm cv}^{(i)} + \sum_{j=N_{\rm mf}+1}^{N_{\rm in}} \alpha_j \sum_{i=1}^{N_{\rm mf}} \beta_{ij}^{(0)} \varphi_{\rm cv}^{(i)} \\ &= \sum_{i=1}^{N_{\rm mf}} \left(\alpha_i + \sum_{j=N_{\rm mf}+1}^{N_{\rm in}} \alpha_j \beta_{ij}^{(0)} \right) \varphi_{\rm cv}^{(i)} \\ &\equiv \sum_{i=1}^{N_{\rm mf}} \gamma_i \varphi_{\rm cv}^{(i)} \\ &= \boldsymbol{\Phi}_{\rm LI}^{(0)} \boldsymbol{\gamma}, \end{aligned} \tag{5.94}$$

[57] Since the numbering of the initial/inlet conditions is arbitrary, the reference vector can always be renumbered such that $k = 0$.

[58] As noted above, the inlet streams can always be renumbered so that the linearly independent inlet streams are the lowest numbered. In any case, in order to satisfy the necessary conditions for a *mixture-fraction basis*, it will usually be necessary to renumber the columns of $\boldsymbol{\Phi}_{\rm cv}^{(k)}$ to facilitate the search.

[59] The rows of $\boldsymbol{\Phi}_{\rm LI}^{(0)-1}$ are the reciprocal basis vectors.

where

$$\gamma_i(\mathbf{x}, t) \equiv \alpha_i(\mathbf{x}, t) + \sum_{j=N_{\mathrm{mf}}+1}^{N_{\mathrm{in}}} \alpha_j(\mathbf{x}, t)\beta_{ij}^{(0)} \quad \text{for} \quad i \in 1, \ldots, N_{\mathrm{mf}}. \tag{5.95}$$

Note that γ depends on the choice of both the reference vector and the linearly independent columns.

In order for γ to be a mixture-fraction vector, all its components must be non-negative and their sum must be less than or equal to one.[60] From the initial/inlet conditions on $\alpha(\mathbf{x}, t)$, it is easily shown using (5.95) that in order for γ to be a mixture-fraction vector, the components of $\mathbf{B}^{(k)}$ (for some $k \in 0, \ldots, N_{\mathrm{in}}$) must satisfy[61]

$$0 \leq \beta_{ij}^{(k)} \leq 1 \quad \text{for all} \quad i \in 1, \ldots, N_{\mathrm{mf}} \quad \text{and} \quad j \in 1, \ldots, N_{\mathrm{LD}}; \tag{5.96}$$

and[62]

$$0 \leq \sum_{i=1}^{N_{\mathrm{mf}}} \beta_{ij}^{(k)} \leq 1 \quad \text{for all} \quad j \in 1, \ldots, N_{\mathrm{LD}}. \tag{5.97}$$

Since $\mathbf{B}^{(k)}$ depends on the choice of the linearly independent vectors used to form $\mathbf{\Phi}_{\mathrm{LI}}^{(k)}$, all possible combinations must be explored in order to determine if one of them satisfies (5.96) and (5.97). Any set of linearly independent columns of $\mathbf{\Phi}_{\mathrm{cv}}^{(k)}$ that yields a matrix $\mathbf{B}^{(k)}$ satisfying (5.96) and (5.97) will be referred to hereinafter as a *mixture-fraction basis.*

Once a mixture-fraction basis has been found, the linear transformation that yields the mixture-fraction vector is

$$\boldsymbol{\xi} = \boldsymbol{\Xi}^{(k)} \left(\mathbf{c}_{\mathrm{cv}} - \mathbf{c}_{\mathrm{cv}}^{(k)}\right) \quad \Longleftrightarrow \quad \mathbf{c}_{\mathrm{cv}} = \mathbf{c}_{\mathrm{cv}}^{(k)} + \boldsymbol{\Xi}^{(k)-1}\boldsymbol{\xi}, \tag{5.98}$$

where

$$\boldsymbol{\Xi}^{(k)} = \mathbf{\Phi}_{\mathrm{LI}}^{(k)-1}. \tag{5.99}$$

In addition, the transport equation for $\boldsymbol{\xi}(\mathbf{x}, t)$ has exactly the same form as (5.65):

$$\frac{\partial \boldsymbol{\xi}}{\partial t} + U_j \frac{\partial \boldsymbol{\xi}}{\partial x_j} = \Gamma \frac{\partial^2 \boldsymbol{\xi}}{\partial x_j \partial x_j} \tag{5.100}$$

with initial/inlet conditions given by

$$\left[\boldsymbol{\xi}^{(0)} \;\; \cdots \;\; \boldsymbol{\xi}^{(k-1)} \;\; \boldsymbol{\xi}^{(k+1)} \;\; \cdots \;\; \boldsymbol{\xi}^{(N_{\mathrm{in}})}\right] = \left[\mathbf{I} \;\; \mathbf{B}^{(k)}\right] \quad \text{and} \quad \boldsymbol{\xi}^{(k)} = \mathbf{0}. \tag{5.101}$$

Note that (5.101) holds under the assumption that the initial/inlet conditions have been renumbered so that the first N_{mf} correspond to the mixture-fraction basis. By definition, the mixture-fraction vector is always null in the reference stream.

60 The $N_{\mathrm{mf}} + 1$ component is then defined such that the sum is unity.

61 Note that the column indices for $\mathbf{B}^{(k)}$ have been renumbered as compared with (5.95).

62 Note that this condition admits the case where one or more columns of $\mathbf{B}^{(k)}$ is null. This would occur if the reference vector corresponded to more than one initial/inlet condition.

The determination of a mixture-fraction basis is a necessary but not a sufficient condition for using the mixture-fraction PDF method to treat a turbulent reacting flow in the fast-chemistry limit. In order to understand why this is so, note that the mixture-fraction basis is defined in terms of the conserved-variable scalars $\varphi_{\rm cv}$ without regard to the reacting scalars $\varphi_{\rm r}$. Thus, it is possible that a mixture-fraction basis can be found for the $N_{\Phi_{\rm c}^{(k)}}$ conserved-variable scalars that does not apply to the N_Υ reacting scalars. In order to ensure that this is not the case, the linear transformation $\mathbf{M}_{\rm r}$ defined by (5.30) on p. 149 must be applied to the $(K \times N_{\rm in})$ matrix

$$\boldsymbol{\Phi}^{(k)} \equiv \begin{bmatrix} \varphi^{(0)} & \dots & \varphi^{(k-1)} & \varphi^{(k+1)} & \dots & \varphi^{(N_{\rm in})} \end{bmatrix}, \tag{5.102}$$

where $\varphi^{(i)} \equiv \mathbf{c}^{(i)} - \mathbf{c}^{(k)}$ for $i \in 0, \dots, N_{\rm in}$. This operation yields the transformed reacting scalars for the initial/inlet conditions:

$$\boldsymbol{\Phi}_{\rm r}^{(k)} \equiv \begin{bmatrix} \varphi_{\rm r}^{(0)} & \dots & \varphi_{\rm r}^{(k-1)} & \varphi_{\rm r}^{(k+1)} & \dots & \varphi_{\rm r}^{(N_{\rm in})} \end{bmatrix} = \mathbf{M}_{\rm r}\boldsymbol{\Phi}^{(k)}. \tag{5.103}$$

Assuming that the initial and inlet conditions have been renumbered as in (5.101), the sufficient condition can be expressed as

$$\varphi_{\rm r}^{(j)} = \sum_{i=1}^{N_{\rm mf}} \beta_{i\, j-N_{\rm mf}}^{(k)} \varphi_{\rm r}^{(i)} \quad \text{for all} \quad j \in N_{\rm mf}+1, \dots, N_{\rm in}, \quad \text{except} \quad j = k. \tag{5.104}$$

In words, this condition states that $\varphi_{\rm r}^{(j)}$ for $j \in N_{\rm mf}+1, \dots, N_{\rm in}$ (except $j = k$) must be a linear mixture of $\varphi_{\rm r}^{(i)}$ for $i \in 1, \dots, N_{\rm mf}$ with the *same coefficient matrix* $\mathbf{B}^{(k)}$ needed for the mixture-fraction vector. Hereinafter, a mixture-fraction basis that satisfies (5.104) will be referred to as a *linear-mixture basis*.

Note that thus far the reacting-scalar vector $\varphi_{\rm r}$ has not been altered by the mixture-fraction transformation. However, if a linear-mixture basis exists, it is possible to transform the reacting-scalar vector into a new vector $\varphi_{\rm rp}$ whose initial and inlet conditions are null: $\varphi_{\rm rp}^{(i)} = \mathbf{0}$ for all $i \in 0, \dots, N_{\rm in}$. In terms of the mixture-fraction vector, the linear transformation can be expressed as

$$\varphi_{\rm r} = \varphi_{\rm rp} + \boldsymbol{\Phi}_{\rm rp}^{(k)}\boldsymbol{\xi}. \tag{5.105}$$

The $(N_\Upsilon \times N_{\rm mf})$ linear transformation matrix is defined by

$$\boldsymbol{\Phi}_{\rm rp}^{(k)} \equiv \begin{bmatrix} \varphi_{\rm r}^{(1)} & \dots & \varphi_{\rm r}^{(N_{\rm mf})} \end{bmatrix}, \tag{5.106}$$

where we have again assumed that the initial and inlet conditions have been renumbered as in (5.104). The vector $\varphi_{\rm rp}$ has the property that it will be null in the limit where all reaction rate constants are zero (i.e., $k_i^{\rm f} = 0$ for all $i \in 1, \dots, I$). Thus, its magnitude will be a measure of the relative importance of the chemical reactions. For this reason, we will refer to $\varphi_{\rm rp}$ as the *reaction-progress vector*.[63]

[63] In Section 5.5, we also introduce reaction-progress variables for 'simple' chemistry that are defined differently than $\varphi_{\rm rp}$. Although their properties are otherwise quite similar, the reaction-progress variables are always non-negative, which need not be the case for the reaction-progress vector.

$$\boldsymbol{\varphi} = \left(\mathbf{c} - \mathbf{c}^{(k)}\right) \underset{\boldsymbol{\Xi}_{\mathrm{mf}}^{(k)-1}}{\overset{\boldsymbol{\Xi}_{\mathrm{mf}}^{(k)}}{\Longleftrightarrow}} \begin{bmatrix} \boldsymbol{\varphi}_{\mathrm{rp}} \\ \boldsymbol{\xi} \\ \mathbf{0} \end{bmatrix} = \boldsymbol{\varphi}_{\mathrm{mf}}$$

Figure 5.7. When the initial and inlet conditions admit a linear-mixture basis, the molar concentration vector **c** of length K can be partitioned by a linear transformation into three parts: φ_{rp}, a reaction-progress vector of length N_{Υ}; $\boldsymbol{\xi}$, a mixture-fraction vector of length N_{mf}; and **0**, a null vector of length $K - N_{\Upsilon} - N_{\mathrm{mf}}$. The linear transformation matrix $\boldsymbol{\Xi}_{\mathrm{mf}}^{(k)}$ depends on the reference concentration vector $\mathbf{c}^{(k)}$ and the reaction coefficient matrix $\boldsymbol{\Upsilon}$.

If $N_{\mathrm{mf}} \ll N_{\mathrm{in}}$, the process of finding a linear-mixture basis can be tedious. Fortunately, however, in practical applications N_{in} is usually not greater than 2 or 3, and thus it is rarely necessary to search for more than one or two combinations of linearly independent columns for each reference vector. In the rare cases where $N_{\mathrm{in}} > 3$, the linear mixtures are often easy to identify. For example, in a tubular reactor with multiple side-injection streams, the side streams might all have the same inlet concentrations so that $\mathbf{c}^{(2)} = \cdots = \mathbf{c}^{(N_{\mathrm{in}})}$. The stationary flow calculation would then require only $N_{\mathrm{mf}} = 1$ mixture-fraction components to describe mixing between inlet 1 and the $N_{\mathrm{in}} - 1$ side streams. In summary, as illustrated in Fig. 5.7, a turbulent reacting flow for which a linear-mixture basis exists can be completely described in terms of a transformed composition vector φ_{mf} defined by

$$\varphi_{\mathrm{mf}} \equiv \begin{bmatrix} \varphi_{\mathrm{rp}} \\ \boldsymbol{\xi} \\ \mathbf{0} \end{bmatrix} = \boldsymbol{\Xi}_{\mathrm{mf}}^{(k)} \left(\mathbf{c} - \mathbf{c}^{(k)}\right), \tag{5.107}$$

where $k \in 0, \ldots, N_{\mathrm{in}}$ depends on the linear-mixture basis, and

$$\boldsymbol{\Xi}_{\mathrm{mf}}^{(k)} \equiv \begin{bmatrix} \mathbf{I} & -\boldsymbol{\Phi}_{\mathrm{rp}}^{(k)} & \mathbf{0} \\ \mathbf{0} & \mathbf{I} & \mathbf{0} \\ \mathbf{0} & \mathbf{0} & \mathbf{I} \end{bmatrix} \begin{bmatrix} \mathbf{I} & \mathbf{0} & \mathbf{0} \\ \mathbf{0} & \boldsymbol{\Xi}^{(k)} & \mathbf{0} \\ \mathbf{0} & \mathbf{0} & \mathbf{I} \end{bmatrix} \begin{bmatrix} \mathbf{I} & \mathbf{0} \\ \mathbf{0} & \mathbf{U}^{(k)\mathrm{T}} \end{bmatrix} \mathbf{U}_{\mathrm{sv}}^{\mathrm{T}}, \tag{5.108}$$

where $\mathbf{U}^{(k)}$ is found as in (5.70) but using $\mathbf{c}_{\mathrm{c}}^{(k)}$ as the reference vector in (5.69). Likewise, the original composition vector can be found using the inverse transformation:

$$\mathbf{c} = \mathbf{c}^{(k)} + \boldsymbol{\Xi}_{\mathrm{mf}}^{(k)-1} \varphi_{\mathrm{mf}}, \tag{5.109}$$

where

$$\boldsymbol{\Xi}_{\mathrm{mf}}^{(k)-1} \equiv \mathbf{U}_{\mathrm{sv}} \begin{bmatrix} \mathbf{I} & \mathbf{0} \\ \mathbf{0} & \mathbf{U}^{(k)} \end{bmatrix} \begin{bmatrix} \mathbf{I} & \mathbf{0} & \mathbf{0} \\ \mathbf{0} & \boldsymbol{\Xi}^{(k)-1} & \mathbf{0} \\ \mathbf{0} & \mathbf{0} & \mathbf{I} \end{bmatrix} \begin{bmatrix} \mathbf{I} & \boldsymbol{\Phi}_{\mathrm{rp}}^{(k)} & \mathbf{0} \\ \mathbf{0} & \mathbf{I} & \mathbf{0} \\ \mathbf{0} & \mathbf{0} & \mathbf{I} \end{bmatrix}. \tag{5.110}$$

When applying the mixture-fraction transformation in an actual turbulent-reacting-flow calculation, it is important to be aware of the fact that $\boldsymbol{\Xi}_{\mathrm{mf}}^{(k)}$ depends on both the chemical kinetic scheme (through $\boldsymbol{\Upsilon}$) and the inlet and initial conditions (through $\boldsymbol{\Phi}^{(k)}$). Thus, $\boldsymbol{\Xi}_{\mathrm{mf}}^{(k)}$ and $\boldsymbol{\Xi}_{\mathrm{mf}}^{(k)-1}$ can be initialized at the beginning of each calculation, but will otherwise remain constant throughout the rest of the calculation. The mixture-fraction transformation can also be extended to non-isothermal flows by making the appropriate changes in the definition of the transformation matrix $\boldsymbol{\Xi}_{\mathrm{mf}}^{(k)}$ that are needed to accommodate the temperature.

5.3.3 Example flows

In order to illustrate the process of finding a mixture-fraction vector for an actual chemically reacting flow, we will consider the two-step reaction scheme given in (5.21) for five different sets of initial and inlet conditions. For this reaction, $K = 4$, $I = 2$, and $N_\Upsilon = 2$. Thus, there will be two reacting and two conserved scalars after the linear transformation given in (5.107) has been carried out. The number of linear dependencies between the initial and inlet conditions will then determine how many of the conserved scalars are variable and how many of them are constant. Using the conserved-variable vectors, we will attempt to find a mixture-fraction basis, and then check if it is also a linear-mixture basis. For the examples considered here, the mixture-fraction vector has at most two components, and the linear mixtures are easy to recognize from the initial and inlet conditions. Nevertheless, the algorithm described in the previous section is completely general and can be used successfully in systems with multiple linear mixtures that are difficult to recognize from the initial and inlet conditions.

The first set of initial and inlet conditions is given by

$$\boldsymbol{\Phi} = \left[\mathbf{c}^{(0)} \quad \mathbf{c}^{(1)} \quad \mathbf{c}^{(2)}\right] = \begin{bmatrix} 0 & 1 & 0 \\ 0 & 0 & 1 \\ 0 & 0 & 0 \\ 0 & 0 & 0 \end{bmatrix}, \tag{5.111}$$

and corresponds to a non-premixed feed reactor that initially contains no reactants or products and that has two inlet streams. Intuition tells us that, since the initial conditions are not a linear mixture of the inlet conditions,[64] the non-stationary flow cannot be described by a single mixture fraction. We should thus expect to find a mixture-fraction vector with two components.

Any of the columns of $\boldsymbol{\Phi}$ can be chosen as the reference vector. However, we shall see that there is an advantage in choosing one of the two inlet streams. Thus, letting $\mathbf{c}^{(1)}$ be the reference vector, the matrix $\boldsymbol{\Phi}_{\mathrm{c}}^{(1)}$ is easily found to be[65]

$$\boldsymbol{\Phi}_{\mathrm{c}}^{(1)} = \begin{bmatrix} -0.7887 & -1.0774 \\ 0.2113 & 0.7113 \end{bmatrix}. \tag{5.112}$$

The rank of $\boldsymbol{\Phi}_{\mathrm{c}}^{(1)}$ is $N_{\boldsymbol{\Phi}_{\mathrm{c}}^{(1)}} = 2 = N_{\mathrm{in}}$, and hence both conserved scalars are variable. In addition, for this example, the mixture-fraction basis will automatically be a linear-mixture

64 Although the rank of $\boldsymbol{\Phi}$ is two, the linearly dependent vector is not formed by a linear mixture.

65 We have again used MATLAB to perform the computations.

basis. Singular value decomposition (SVD) of $\mathbf{\Phi}_{\mathrm{c}}^{(1)}$, followed by multiplication by $\mathbf{U}^{(1)\mathrm{T}}$ (i.e., (5.72)), yields

$$\mathbf{\Phi}_{\mathrm{LI}}^{(1)} = \begin{bmatrix} 0.7947 & 1.2858 \\ -0.1876 & 0.1159 \end{bmatrix}. \tag{5.113}$$

The transformation matrix needed in (5.87) is thus given by

$$\mathbf{\Xi}^{(1)} = \begin{bmatrix} 0.3478 & -3.8573 \\ 0.5628 & 2.3840 \end{bmatrix}. \tag{5.114}$$

Owing to the choice of $\mathbf{\Phi}_{\mathrm{LI}}^{(1)}$, $\xi_1 = 1$ for the initial conditions, and $\xi_2 = 1$ in the second inlet stream. Thus, the reaction-progress transformation matrix can be found to be

$$\mathbf{\Phi}_{\mathrm{rp}}^{(1)} = \begin{bmatrix} 0.5774 & 0 \\ 0 & -0.5774 \end{bmatrix}. \tag{5.115}$$

The overall transformation matrix $\mathbf{\Xi}_{\mathrm{mf}}^{(1)}$ is then given by

$$\mathbf{\Xi}_{\mathrm{mf}}^{(1)} = \begin{bmatrix} 0 & 0 & 1.7321 & 1.7321 \\ 0 & 0 & 0 & 1.7321 \\ -1 & -1 & -2 & -3 \\ 0 & 1 & 1 & 2 \end{bmatrix}. \tag{5.116}$$

Transforming the initial and inlet conditions, (5.111), according to (5.107) yields

$$\begin{bmatrix} \varphi_{\mathrm{mf}}^{(0)} & \varphi_{\mathrm{mf}}^{(1)} & \varphi_{\mathrm{mf}}^{(2)} \end{bmatrix} = \begin{bmatrix} 0 & 0 & 0 \\ 0 & 0 & 0 \\ 1 & 0 & 0 \\ 0 & 0 & 1 \end{bmatrix}. \tag{5.117}$$

Note that, due to the choice of $\mathbf{c}^{(1)}$ as the reference vector, the mixture-fraction vector $\boldsymbol{\xi}^{(1)}$ (third and fourth components of $\varphi_{\mathrm{mf}}^{(1)}$) is null. The first component of the mixture-fraction vector thus describes mixing between the initial contents of the reactor and the two inlet streams, and the second component describes mixing with the second inlet stream. For a stationary flow $\boldsymbol{\xi}^{(0)} \rightarrow \mathbf{0}$, and only one mixture-fraction component (ξ_2) will be required to describe the flow. Note, however, that if $\mathbf{c}^{(0)}$ had been chosen as the reference vector, a similar reduction would not have occurred. As expected, the inlet and initial values of the two reaction-progress variables are null.

The second set of initial and inlet conditions is given by

$$\mathbf{\Phi} = \begin{bmatrix} \mathbf{c}^{(0)} & \mathbf{c}^{(1)} & \mathbf{c}^{(2)} \end{bmatrix} = \begin{bmatrix} 0.5 & 1 & 0 \\ 0.5 & 0 & 1 \\ 0 & 0 & 0 \\ 0 & 0 & 0 \end{bmatrix}, \tag{5.118}$$

and corresponds to a non-premixed feed reactor that is initially filled with a 50:50 mixture of the two inlet streams.[66] Intuition tells us that, since the initial conditions are a *linear mixture* of the inlet conditions, the flow can be described by a single mixture fraction.

Letting $\mathbf{c}^{(1)}$ again be the reference vector, the matrix $\mathbf{\Phi}_{\text{c}}^{(1)}$ is

$$\mathbf{\Phi}_{\text{c}}^{(1)} = \begin{bmatrix} -0.5387 & -1.0774 \\ 0.3557 & 0.7113 \end{bmatrix}. \tag{5.119}$$

The rank of $\mathbf{\Phi}_{\text{c}}^{(1)}$ is $N_{\Phi_{\text{c}}^{(1)}} = 1 = N_{\text{in}} - 1$, and hence only one conserved scalar is variable. SVD of $\mathbf{\Phi}_{\text{c}}^{(1)}$, followed by multiplication by $\mathbf{U}^{(1)\text{T}}$, yields

$$\mathbf{\Phi}_{\text{cv}}^{(1)} = [0.6455 \quad 1.2910]. \tag{5.120}$$

We can now try either column of $\mathbf{\Phi}_{\text{cv}}^{(1)}$ as a mixture-fraction basis. However, since the matrix $\mathbf{B}^{(1)}$ must have all components less than unity, the second column is the clear choice.[67] Thus, $\mathbf{\Phi}_{\text{LI}}^{(1)} = [1.2910]$, $\mathbf{\Phi}_{\text{LD}}^{(1)} = [0.6455]$ and $\mathbf{B}^{(1)} = [0.5]$.[68] The transformation matrix is then given by $\mathbf{\Xi}^{(1)} = [0.7764]$.

Since we have chosen the second column of $\mathbf{\Phi}_{\text{cv}}^{(1)}$, $\xi_1 = 1$ in the second inlet stream. Thus, the reaction-progress transformation matrix can be found to be

$$\mathbf{\Phi}_{\text{rp}}^{(1)} = \begin{bmatrix} 0 \\ -0.5774 \end{bmatrix}. \tag{5.121}$$

The overall transformation matrix $\mathbf{\Xi}_{\text{mf}}^{(1)}$ is then given by

$$\mathbf{\Xi}_{\text{mf}}^{(1)} = \begin{bmatrix} -0.5774 & -0.5774 & 0.5774 & 0 \\ -0.3464 & -0.3464 & -0.6929 & 0.6929 \\ -0.6 & 0.4 & -0.2 & 0.2 \\ -0.2582 & -0.2582 & -0.5164 & -0.7746 \end{bmatrix}. \tag{5.122}$$

Transforming the initial and inlet conditions yields

$$\left[\boldsymbol{\varphi}_{\text{mf}}^{(0)} \quad \boldsymbol{\varphi}_{\text{mf}}^{(1)} \quad \boldsymbol{\varphi}_{\text{mf}}^{(2)}\right] = \begin{bmatrix} 0 & 0 & 0 \\ 0 & 0 & 0 \\ 0.5 & 0 & 1 \\ 0 & 0 & 0 \end{bmatrix}. \tag{5.123}$$

Again, note that $\boldsymbol{\varphi}_{\text{mf}}^{(1)}$ is null due to the choice of the reference vector. Moreover, note that

$$\boldsymbol{\varphi}_{\text{mf}}^{(0)} = 0.5\boldsymbol{\varphi}_{\text{mf}}^{(2)},$$

i.e., the mixture-fraction basis is also a linear-mixture basis. The first component of the mixture-fraction vector thus describes mixing between the two inlet streams, and its initial

66 The rank of $\mathbf{\Phi}$ is two.

67 If we were to choose the first column instead, then $\mathbf{B}^{(1)} = [2]$. As a general rule, one should always try the vector with the largest magnitude first.

68 Although it is not necessary for this example, we will continue to use matrix notation to remind the reader that matrices will appear in the general case.

value in the reactor is 0.5. Thus (as can be easily observed from the initial and inlet conditions), $\mathbf{c}^{(0)}$ is a linear mixture of $\mathbf{c}^{(1)}$ and $\mathbf{c}^{(2)}$. The final row of (5.123) is null and corresponds to the conserved-constant scalar. The inlet and initial values of the two reaction-progress variables are null, as expected.

The third set of initial and inlet conditions is given by

$$\begin{bmatrix}\mathbf{c}^{(0)} & \mathbf{c}^{(1)} & \mathbf{c}^{(2)}\end{bmatrix} = \begin{bmatrix} 0.3 & 1 & 0 \\ 0.3 & 0 & 1 \\ 0.2 & 0 & 0 \\ 0 & 0 & 0 \end{bmatrix}, \tag{5.124}$$

and is similar to the first set, except that the reactor now initially contains a mixture of the first three reactants. Intuition tells us that, since the initial and inlet conditions are *not* linearly dependent, the flow cannot be described by a single mixture fraction.[69] However, we shall see that, although one-component mixture-fraction bases exist for this set of initial and inlet conditions, they are not linear-mixture bases.[70]

Letting $\mathbf{c}^{(1)}$ again be the reference vector, the matrix $\mathbf{\Phi}_{\mathrm{c}}^{(1)}$ is

$$\mathbf{\Phi}_{\mathrm{c}}^{(1)} = \begin{bmatrix} -0.5387 & -1.0774 \\ 0.3557 & 0.7113 \end{bmatrix}, \tag{5.125}$$

which is exactly the same as (5.119).[71] We can thus use exactly the same mixture-fraction basis as was used for the second set of initial and inlet conditions. Hence, the overall transformation matrix $\mathbf{\Xi}_{\mathrm{mf}}^{(1)}$ is given by (5.122). However, the fact that the mixture-fraction basis has only one component while the initial and inlet conditions exhibit no linear dependency suggests that it may not be a linear-mixture basis. Thus, it is imperative that we check the sufficient condition given in (5.104), and we do this next.

Transforming the initial and inlet conditions yields

$$\begin{bmatrix}\varphi_{\mathrm{mf}}^{(0)} & \varphi_{\mathrm{mf}}^{(1)} & \varphi_{\mathrm{mf}}^{(2)}\end{bmatrix} = \begin{bmatrix} 0.3464 & 0 & 0 \\ 0 & 0 & 0 \\ 0.5 & 0 & 1 \\ 0 & 0 & 0 \end{bmatrix}. \tag{5.126}$$

Note that the reaction-progress vector in the first column is non-zero. Thus, as we suspected, the mixture-fraction basis is *not* a linear-mixture basis. The same conclusion will be drawn for all other mixture-fraction bases found starting from (5.118). For these initial and inlet conditions, a two-component mixture-fraction vector can be found; however, it is of no practical interest since the number of conserved-variable scalars is equal to $N_{\Phi_{\mathrm{c}}^{(k)}} = 1$ $(k \in 0, 1, 2)$. In conclusion, although the mixture fraction can be defined for the

[69] The rank of $\mathbf{\Phi}$ is three.

[70] An arbitrary number of initial and inlet conditions can be found with this property by selecting them from the set of all vectors that generate the same conserved-scalar matrix $\mathbf{\Phi}_{\mathrm{c}}^{(k)} = \mathbf{M}_{\mathrm{c}}\mathbf{\Phi}^{(k)}$, where $\mathbf{M}_{\mathrm{c}}$ is defined in (5.31). For the present example, we see from (5.27) that $\mathbf{c}^{(0)} = [x \quad x \quad (0.5 - x) \quad 0]^{\mathrm{T}}$ with $0 \leq x < 0.5$ is one possibility.

[71] Note, in particular, that $N_{\Phi_{\mathrm{c}}^{(1)}} = 1$.

conserved-variable scalar, the mixture-fraction PDF method cannot be applied to treat the reacting scalars for these initial and inlet conditions.

The fourth set of initial and inlet conditions is given by

$$\boldsymbol{\Phi} = \begin{bmatrix} \mathbf{c}^{(0)} & \mathbf{c}^{(1)} & \mathbf{c}^{(2)} & \mathbf{c}^{(3)} \end{bmatrix} = \begin{bmatrix} 0.25 & 1 & 0 & 0 \\ 0.25 & 0 & 1 & 0 \\ 0.25 & 0 & 0 & 0.5 \\ 0.25 & 0 & 0 & 0.5 \end{bmatrix}, \tag{5.127}$$

and corresponds to a non-premixed feed reactor with three inlet streams. By inspection, we can see that the initial condition is a linear mixture of the three feed streams:[72]

$$\mathbf{c}^{(0)} = 0.25\mathbf{c}^{(1)} + 0.25\mathbf{c}^{(2)} + 0.5\mathbf{c}^{(3)}.$$

No other linear dependency is apparent, and thus we should expect to find a mixture-fraction vector with two components.

Letting $\mathbf{c}^{(1)}$ be the reference vector, the matrix $\boldsymbol{\Phi}_{\mathrm{c}}^{(1)}$ is found to be

$$\boldsymbol{\Phi}_{\mathrm{c}}^{(1)} = \begin{bmatrix} -0.4858 & -1.0774 & -0.4330 \\ 0.5528 & 0.7113 & 0.7500 \end{bmatrix}. \tag{5.128}$$

The rank of $\boldsymbol{\Phi}_{\mathrm{c}}^{(1)}$ is $N_{\Phi_{\mathrm{c}}^{(1)}} = 2 = N_{\mathrm{in}} - 1$, and hence both conserved scalars are variable. SVD of $\boldsymbol{\Phi}_{\mathrm{c}}^{(1)}$, followed by multiplication by $\mathbf{U}^{(1)\mathrm{T}}$, yields

$$\boldsymbol{\Phi}_{\mathrm{cv}}^{(1)} = \begin{bmatrix} 0.7321 & 1.2738 & 0.8273 \\ -0.0756 & 0.2099 & -0.2562 \end{bmatrix}. \tag{5.129}$$

We then choose the second and third columns as the linearly independent basis, which yields

$$\mathbf{B}^{(1)} = \begin{bmatrix} 0.25 \\ 0.5 \end{bmatrix}. \tag{5.130}$$

Thus, since all components of $\mathbf{B}^{(1)}$ are non-negative and their sum is less than unity, the second and third columns of $\boldsymbol{\Phi}_{\mathrm{cv}}^{(1)}$ are a mixture-fraction basis. The mixture-fraction transformation matrix is then given by

$$\boldsymbol{\Xi}^{(1)} = \begin{bmatrix} 0.5124 & 1.6545 \\ 0.4197 & -2.5476 \end{bmatrix}. \tag{5.131}$$

Owing to the choice of second and third columns of $\boldsymbol{\Phi}_{\mathrm{cv}}^{(1)}$, $\xi_1 = 1$ for the second inlet stream, and $\xi_2 = 1$ for the third inlet stream. Thus, the reaction-progress transformation matrix can be found to be

$$\boldsymbol{\Phi}_{\mathrm{rp}}^{(1)} = \begin{bmatrix} 0 & 0.8660 \\ -0.5774 & 0 \end{bmatrix}. \tag{5.132}$$

[72] A necessary condition for a linear mixture in this example is that the rank of $\boldsymbol{\Phi}$ be less than four. The rank is three, indicating one linear dependency.

The overall transformation matrix is then given by

$$\Xi_{\text{mf}}^{(1)} = \begin{bmatrix} -1.1548 & -1.1548 & -0.5774 & -1.7320 \\ 0 & 0.5774 & -1.1548 & 0 \\ -1 & 0 & -1 & -1 \\ 0.6667 & 0.6667 & 1.3333 & 2 \end{bmatrix}. \tag{5.133}$$

Transforming the initial and inlet conditions according to (5.107) yields

$$\left[\varphi_{\text{mf}}^{(0)} \quad \varphi_{\text{mf}}^{(1)} \quad \varphi_{\text{mf}}^{(2)} \quad \varphi_{\text{mf}}^{(3)}\right] = \begin{bmatrix} 0 & 0 & 0 & 0 \\ 0 & 0 & 0 & 0 \\ 0.25 & 0 & 1 & 0 \\ 0.5 & 0 & 0 & 1 \end{bmatrix}. \tag{5.134}$$

Note that, due to the choice of $\mathbf{c}^{(1)}$ as the reference vector, $\varphi_{\text{mf}}^{(1)}$ is again null. We also see that

$$\varphi_{\text{mf}}^{(0)} = 0.25\varphi_{\text{mf}}^{(1)} + 0.25\varphi_{\text{mf}}^{(2)} + 0.5\varphi_{\text{mf}}^{(3)},$$

i.e., the mixture-fraction basis is a linear-mixture basis. The first component of the mixture-fraction vector describes mixing with the second inlet stream, and the second component describes mixing with the third inlet stream. The inlet and initial values of the reaction-progress variables are again null.

The fifth set of initial and inlet conditions is given by

$$\boldsymbol{\Phi} = \left[\mathbf{c}^{(0)} \quad \mathbf{c}^{(1)} \quad \mathbf{c}^{(2)} \quad \mathbf{c}^{(3)}\right] = \begin{bmatrix} 0 & 1 & 0 & 0 \\ 0 & 0 & 1 & 0 \\ 0 & 0 & 0 & 0.5 \\ 0 & 0 & 0 & 0.5 \end{bmatrix}, \tag{5.135}$$

and is similar to the fourth set except that now the reactor is initially empty. By inspection, no linear mixture is apparent,[73] and thus we should expect that it may not be possible to find a mixture-fraction basis. If, indeed, no mixture-fraction basis can be identified, then a two-component mixture-fraction vector cannot be defined for this flow.[74]

In order to show that no mixture-fraction basis exists, it is necessary to check all possible reference vectors. For each choice of the reference vector, there are three possible sets of linearly independent vectors that can be used to compute $\mathbf{B}^{(k)}$. Thus, we must check a total of 12 possible mixture-fraction bases. Starting with $\mathbf{c}^{(0)}$ as the reference vector, the three possible values of $\mathbf{B}^{(0)}$ are

$$\begin{bmatrix} 1 \\ 1.5 \end{bmatrix}, \quad \begin{bmatrix} -0.6667 \\ 0.6667 \end{bmatrix}, \quad \text{and} \quad \begin{bmatrix} -1.5 \\ 1 \end{bmatrix}. \tag{5.136}$$

73 Nevertheless, the rank of $\boldsymbol{\Phi}$ is three.

74 The fact that no two-component mixture-fraction vector exists does not, however, change the fact that the flow can be described by two conserved scalars.

None of these matrices satisfies (5.96). Thus, we next take $\mathbf{c}^{(1)}$ as the reference vector and find the three possible values of $\mathbf{B}^{(1)}$:

$$\begin{bmatrix} -1.5 \\ 1.5 \end{bmatrix}, \quad \begin{bmatrix} 1 \\ 0.6667 \end{bmatrix}, \quad \text{and} \quad \begin{bmatrix} 1 \\ -0.6667 \end{bmatrix}. \tag{5.137}$$

Only the second matrix satisfies (5.96); however, it does not satisfy (5.97). Thus, we next take $\mathbf{c}^{(2)}$ as the reference vector and find the three possible values of $\mathbf{B}^{(2)}$:

$$\begin{bmatrix} -1.5 \\ 1 \end{bmatrix}, \quad \begin{bmatrix} 1.5 \\ 1 \end{bmatrix}, \quad \text{and} \quad \begin{bmatrix} 0.6667 \\ -0.6667 \end{bmatrix}. \tag{5.138}$$

None of these matrices satisfies (5.96). Finally, taking $\mathbf{c}^{(3)}$ as the reference vector, we find the three possible values of $\mathbf{B}^{(3)}$:

$$\begin{bmatrix} 1 \\ -0.6667 \end{bmatrix}, \quad \begin{bmatrix} 1.5 \\ 1.5 \end{bmatrix}, \quad \text{and} \quad \begin{bmatrix} 0.6667 \\ 1 \end{bmatrix}. \tag{5.139}$$

None of these matrices satisfies both (5.96) and (5.97). We can thus conclude that no mixture-fraction basis exists for this set of initial and inlet conditions. Since $N_{\text{in}} = 3$, a three-component mixture-fraction vector exists,[75] but is of no practical interest.

5.3.4 Mixture-fraction PDF

In a turbulent flow for which it is possible to define the mixture-fraction vector, turbulent mixing can be described by the joint one-point mixture-fraction PDF $f_{\boldsymbol{\xi}}(\boldsymbol{\zeta}; \mathbf{x}, t)$. The mean mixture-fraction vector and covariance matrix are defined, respectively, by[76]

$$\langle \boldsymbol{\xi} \rangle \equiv \int_0^1 \cdots \int \boldsymbol{\zeta} f_{\boldsymbol{\xi}}(\boldsymbol{\zeta}; \mathbf{x}, t) \, \mathrm{d}\zeta_1 \cdots \mathrm{d}\zeta_{N_{\text{mf}}}, \tag{5.140}$$

and

$$\langle \xi'_\alpha \xi'_\beta \rangle \equiv \int_0^1 \cdots \int (\zeta_\alpha - \langle \xi_\alpha \rangle)(\zeta_\beta - \langle \xi_\beta \rangle) f_{\boldsymbol{\xi}}(\boldsymbol{\zeta}; \mathbf{x}, t) \, \mathrm{d}\zeta_1 \cdots \mathrm{d}\zeta_{N_{\text{mf}}}. \tag{5.141}$$

Since the mixture-fraction PDF is not known *a priori*, it must be modeled either by solving an appropriate transport equation[77] or by assuming a functional form.

The most widely used approach for approximating $f_{\boldsymbol{\xi}}(\boldsymbol{\zeta}; \mathbf{x}, t)$ is the *presumed PDF method*, in which a known distribution function is chosen to represent the mixture-fraction PDF. We will look at the various possible forms in Section 5.9, where presumed PDF

[75] For example, $\alpha_i(\mathbf{x}, t)$ $(i \in 0, 1, 2)$.

[76] By convention, the mixture-fraction PDF is defined to be null whenever the sum of the components ζ_i is greater than unity. Thus, the upper limit of the integrals in these definitions is set to unity.

[77] We will return to this subject in Chapter 6.

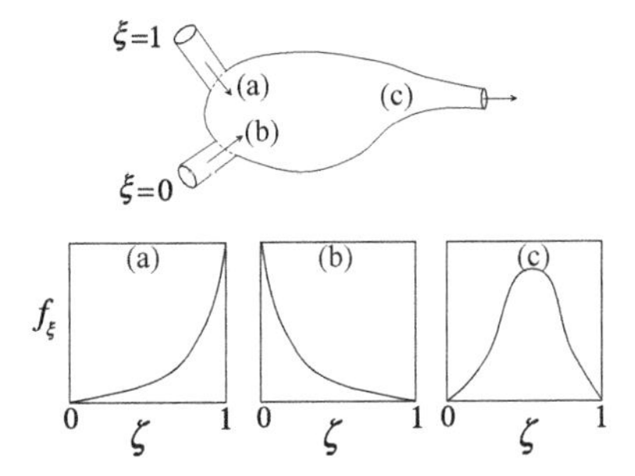

Figure 5.8. The mixture-fraction PDF in turbulent flows with two feed streams (binary mixing) can be approximated by a beta PDF.

methods for the composition vector are discussed. However, for a one-component mixture-fraction vector, a beta PDF is often employed:[78]

$$f_\xi(\zeta;\mathbf{x},t) = \frac{1}{B(a,b)}\zeta^{a-1}(1-\zeta)^{b-1}. \tag{5.142}$$

The beta PDF contains two parameters that are functions of space and time: $a(\mathbf{x},t)$ and $b(\mathbf{x},t)$. The normalization factor $B(a,b)$ is the beta function, and can be expressed in terms of factorials:

$$B(a,b) \equiv \frac{(a-1)!(b-1)!}{(a+b-2)!}. \tag{5.143}$$

The two parameters in (5.142) are related to the mixture-fraction mean and variance by

$$a = \langle\xi\rangle\left[\frac{\langle\xi\rangle(1-\langle\xi\rangle)}{\langle\xi'^2\rangle} - 1\right] \tag{5.144}$$

and

$$b = \frac{1-\langle\xi\rangle}{\langle\xi\rangle}a. \tag{5.145}$$

Thus, if transport equations are solved for $\langle\xi\rangle$ and $\langle\xi'^2\rangle$, then the presumed PDF will be known at every point in the flow domain. Note that the mixture-fraction variance is bounded above by $\langle\xi\rangle(1-\langle\xi\rangle)$, so that a and b are always positive. In the limit where $a = b$ and $a \to \infty$ (i.e., $\langle\xi\rangle = 1/2$ and $\langle\xi'^2\rangle \to 0$), the beta PDF approaches a Gaussian PDF with the same mean and variance.

The beta PDF is widely used in commercial CFD codes to approximate the mixture-fraction PDF for binary mixing. This choice is motivated by the fact that in many of the 'canonical' turbulent mixing configurations (Fig. 5.8) the experimentally observed mixture-fraction PDF is well approximated by a beta PDF. However, it is important to note that all of these flows are stationary with $N_{\text{mf}} = N_{\text{in}} - 1 = 1$, i.e., no linear mixture exists between the inlet conditions. The 'unmixed' PDF is thus well represented by two peaks: one located at $\xi = 0$ and the other at $\xi = 1$, which is exactly the type of behavior exhibited

[78] In order to simplify the notation, on the right-hand side the explicit dependence on $\mathbf{x}$ and t has been omitted.

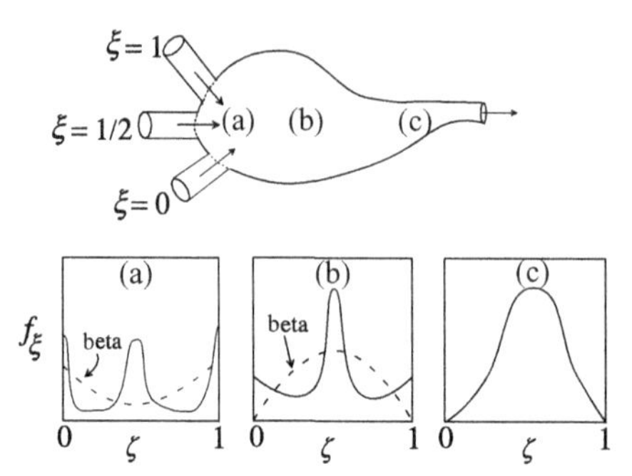

Figure 5.9. The 'unmixed' mixture-fraction PDF in turbulent flows with two feed streams has two peaks that can be approximated by a beta PDF. However, with three feed streams, the 'unmixed' PDF has three peaks, and is therefore poorly approximated by a beta PDF.

by the beta PDF. In more general flows, the 'unmixed' PDF can have multiple peaks. For example, with three inlet streams and $N_{\rm mf} = N_{\rm in} - 2 = 1$, the 'unmixed' PDF would have three peaks (Fig. 5.9): one at $\xi = 0$, one at $\xi = 1$, and one at an intermediate value determined by the linear-mixture basis. The beta PDF would thus be a poor approximation for any flow that has initial and inlet conditions containing one or more linear mixtures. Other types of presumed PDF should therefore be used to approximate the mixture-fraction PDF in flows with initial and inlet conditions containing linear mixtures.[79]

When the mixture-fraction vector has more than one component, the presumed form for the mixture-fraction PDF must be defined such that it will be non-zero only when

$$0 \leq \sum_{i=1}^{N_{\rm mf}} \zeta_i \leq 1. \tag{5.146}$$

For example, a *bi-variate beta PDF* of the form[80]

$$f_{\xi}(\zeta; \mathbf{x}, t) = \mathcal{N} \zeta_1^{a_1 - 1} \zeta_2^{a_2 - 1} (1 - \zeta_1)^{b_1 - 1} (1 - \zeta_2)^{b_2 - 1} (1 - \zeta_1 - \zeta_2)^{b_3 - 1} \tag{5.147}$$

can be employed for a two-component mixture-fraction vector. The five constants (a_1, a_2, b_1, b_2, and b_3) are all positive, and can be related to the two mixture-fraction means and three covariances.[81] In general, a *multi-variate beta PDF* will contain $N_{\rm mf}(N_{\rm mf} + 3)/2$ constants that can be equated to an equal number of mixture-fraction means and covariances. Thus, by solving the transport equations for $\langle \xi_i \rangle$ ($i \in 1, \ldots, N_{\rm mf}$) and $\langle \xi'_j \xi'_k \rangle$ ($j \leq k \in 1, \ldots, N_{\rm mf}$), the mixture-fraction PDF will be defined at every point in the flow domain.[82] However, as noted with the uni-variate beta PDF above, the multi-variate beta PDF cannot exhibit multiple peaks at intermediate values, and thus offers a poor

[79] In commercial CFD codes, the option of using a Gaussian PDF is often available. However, this choice has just one initial peak, and thus is unable to represent the initial conditions accurately.

[80] $\mathcal{N}$ is a normalization constant defined such that the integral over the area defined by $0 \leq \zeta_1 \leq 1$ and $0 \leq \zeta_2 \leq 1 - \zeta_1$ is unity.

[81] $\langle \xi_1 \rangle$, $\langle \xi_2 \rangle$, $\langle \xi'_1 \xi'_1 \rangle$, $\langle \xi'_1 \xi'_2 \rangle$ and $\langle \xi'_2 \xi'_2 \rangle$. Formulae for the means and covariances can be found starting from (5.147). These formulae can be evaluated numerically for fixed values of the parameters.

[82] The problem of determining the parameters given the moments is non-trivial when $2 \leq N_{\rm mf}$. For this case, simple analytical expressions such as (5.144) are not available, and thus the inversion procedure must be carried out numerically.

approximation of flows with initial and inlet conditions containing one or more linear mixtures. For these flows, other more appropriate types of multi-variate presumed PDF should be used.

5.4 Equilibrium-chemistry limit

For elementary chemical reactions, it is sometimes possible to assume that all chemical species reach their chemical-equilibrium values much faster than the characteristic time scales of the flow. Thus, in this section, we discuss how the description of a turbulent reacting flow can be greatly simplified in the *equilibrium-chemistry limit* by reformulating the problem in terms of the mixture-fraction vector.

5.4.1 Treatment of reacting scalars

Having demonstrated the existence of a mixture-fraction vector for certain turbulent reacting flows, we can now turn to the question of how to treat the reacting scalars in the equilibrium-chemistry limit for such flows. Applying the linear transformation given in (5.107), the reaction-progress-vector transport equation becomes

$$\frac{\partial \varphi_{\mathrm{rp}}}{\partial t} + U_j \frac{\partial \varphi_{\mathrm{rp}}}{\partial x_j} = \Gamma \frac{\partial^2 \varphi_{\mathrm{rp}}}{\partial x_j \partial x_j} + \mathbf{S}_{\mathrm{rp}}(\varphi_{\mathrm{rp}}; \boldsymbol{\xi}), \tag{5.148}$$

where the vector function $\mathbf{S}_{\mathrm{rp}}(\varphi_{\mathrm{rp}}; \boldsymbol{\xi})$ contains the first N_{Υ} components of the transformed chemical source term:[83]

$$\begin{bmatrix} \mathbf{S}_{\mathrm{rp}}(\varphi_{\mathrm{rp}}; \boldsymbol{\xi}) \\ \mathbf{0} \end{bmatrix} \equiv \boldsymbol{\Xi}_{\mathrm{mf}}^{(k)} \mathbf{S}_{\mathrm{c}} \left(\mathbf{c}^{(k)} + \boldsymbol{\Xi}_{\mathrm{mf}}^{(k)-1} \begin{bmatrix} \varphi_{\mathrm{rp}} \\ \boldsymbol{\xi} \\ \mathbf{0} \end{bmatrix} \right). \tag{5.149}$$

Note that, given φ_{rp} and $\boldsymbol{\xi}$, the inverse transformation, (5.109), can be employed to find the original composition vector $\mathbf{c}$. In order to simplify the notation, we will develop the theory in terms of φ_{rp}. However, it could just as easily be rewritten in terms of $\mathbf{c}$ using the inverse transformation.

The N_{Υ} eigenvalues of the Jacobian of $\mathbf{S}_{\mathrm{rp}}$ will be equal to the N_{Υ} non-zero eigenvalues of the Jacobian of $\mathbf{S}_{\mathrm{c}}$. Thus, in the equilibrium-chemistry limit, the chemical time scales will obey

$$\tau_1, \ldots, \tau_{N_{\Upsilon}} \ll \tau_{\phi}, \tag{5.150}$$

where τ_{ϕ} is the characteristic scalar-mixing time for the turbulent flow. In this limit, the chemical source term in (5.148) will force the reacting scalars to attain a local chemical equilibrium that depends on $\boldsymbol{\xi}$, but is otherwise unaffected by the flow.[84] Mathematically,

[83] In order to simplify the notation, we drop the superscript (k). Nevertheless, the reader should keep in mind that $\mathbf{S}_{\mathrm{rp}}$ will depend on the linear-mixture basis chosen to define the mixture-fraction vector.

[84] Chemical equilibrium implies that all reactions are *reversible*. The method discussed in this section is thus usually applied to elementary reactions of the form given in (5.1).

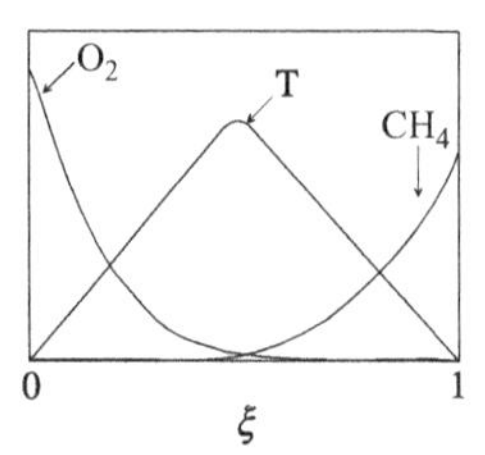

Figure 5.10. Sketch of one-dimensional mixture-fraction chemical lookup table. For any value of the mixture fraction, the reacting scalars can be found from the pre-computed table in a post-processing stage of the flow calculation.

(5.148) can then be replaced by an ODE with only the chemical source term on the right-hand side:

$$\frac{\mathrm{d}\boldsymbol{\varphi}_{\mathrm{fc}}}{\mathrm{d}t} = \mathbf{S}_{\mathrm{rp}}(\boldsymbol{\varphi}_{\mathrm{fc}}; \boldsymbol{\xi}) \tag{5.151}$$

with initial conditions $\boldsymbol{\varphi}_{\mathrm{fc}}(0; \boldsymbol{\xi}) = \mathbf{0}$.

The solution to (5.151) as $t \to \infty$ is the local equilibrium reaction-progress vector:[85]

$$\boldsymbol{\varphi}_{\mathrm{eq}}(\boldsymbol{\xi}) \equiv \boldsymbol{\varphi}_{\mathrm{fc}}(\infty; \boldsymbol{\xi}). \tag{5.152}$$

Thus, in the equilibrium-chemistry limit, the reacting scalars depend on space and time only through the mixture-fraction vector:

$$\boldsymbol{\varphi}_{\mathrm{rp}}(\mathbf{x}, t) = \boldsymbol{\varphi}_{\mathrm{eq}}\left(\boldsymbol{\xi}(\mathbf{x}, t)\right). \tag{5.153}$$

Note that the numerical simulation of the turbulent reacting flow is now greatly simplified. Indeed, the only partial-differential equation (PDE) that must be solved is (5.100) for the mixture-fraction vector, which involves no chemical source term! Moreover, (5.151) is an initial-value problem that depends only on the inlet and initial conditions and is parameterized by the mixture-fraction vector; it can thus be solved independently of (5.100), e.g., in a pre(post)-processing stage of the flow calculation. For a given value of $\boldsymbol{\xi}$, the reacting scalars can then be stored in a *chemical lookup table*, as illustrated in Fig. 5.10.

5.4.2 Application to turbulent reacting flows

The reduction of the turbulent-reacting-flow problem to a *turbulent-scalar-mixing* problem represents a significant computational simplification. However, at high Reynolds numbers, the direct numerical simulation (DNS) of (5.100) is still intractable.[86] Instead, for most practical applications, the Reynolds-averaged transport equation developed in

[85] Typically, solving (5.151) to find $\boldsymbol{\varphi}_{\mathrm{fc}}(\infty; \boldsymbol{\xi})$ is not the best approach. For example, in combusting systems $|\mathbf{S}_{\mathrm{rp}}(\mathbf{0}; \boldsymbol{\xi})| \ll 1$ so that convergence to the equilibrium state will be very slow. Thus, equilibrium thermodynamic methods based on Gibbs free-energy minimization are preferable for most applications.

[86] For a laminar flow problem, (5.100) is tractable; however, the assumption that all scalars have the same molecular diffusivity needed to derive it will usually not be adequate for laminar flows.

Chapter 3 will be employed. Thus, in lieu of $\boldsymbol{\xi}(\mathbf{x}, t)$, only the mixture-fraction means $\langle \boldsymbol{\xi} \rangle$ and covariances $\langle \xi'_i \xi'_j \rangle$ $(i, j \in 1, \ldots, N_{\text{mf}})$ will be available. Given this information, we would then like to compute the reacting-scalar means $\langle \boldsymbol{\varphi}_{\text{rp}} \rangle$ and covariances $\langle \varphi'_{\text{rp}\alpha} \varphi'_{\text{rp}\beta} \rangle$ $(\alpha, \beta \in 1, \ldots, N_\Upsilon)$. However, this computation will require additional information about the *mixture-fraction PDF*. A similar problem arises when a large-eddy simulation (LES) of the mixture-fraction vector is employed. In this case, the resolved-scale mixture-fraction vector $\overline{\boldsymbol{\xi}}(\mathbf{x}, t)$ is known, but the sub-grid-scale (SGS) fluctuations are not resolved. Instead, a transport equation for the SGS mixture-fraction covariance can be solved, but information about the SGS mixture-fraction PDF is still required to compute the resolved-scale reacting-scalar fields.

From (5.152), it follows that the Reynolds-averaged reacting scalars can be found from the mixture-fraction PDF $f_{\boldsymbol{\xi}}(\boldsymbol{\zeta}; \mathbf{x}, t)$ by integration:[87]

$$\langle \boldsymbol{\varphi}_{\text{rp}} \rangle(\mathbf{x}, t) \equiv \int_0^1 \cdots \int \boldsymbol{\varphi}_{\text{eq}}(\boldsymbol{\zeta}) \, f_{\boldsymbol{\xi}}(\boldsymbol{\zeta}; \mathbf{x}, t) \, \mathrm{d}\zeta_1 \cdots \mathrm{d}\zeta_{N_{\text{mf}}}. \tag{5.154}$$

Likewise, the reacting-scalar covariances can be found from

$$\begin{aligned} &\langle \varphi'_{\text{rp}\alpha} \varphi'_{\text{rp}\beta} \rangle(\mathbf{x}, t) \\ &\equiv \int_0^1 \cdots \int (\varphi_{\text{eq}\alpha}(\boldsymbol{\zeta}) - \langle \varphi_{\text{rp}\alpha} \rangle)(\varphi_{\text{eq}\beta}(\boldsymbol{\zeta}) - \langle \varphi_{\text{rp}\beta} \rangle) f_{\boldsymbol{\xi}}(\boldsymbol{\zeta}; \mathbf{x}, t) \, \mathrm{d}\zeta_1 \cdots \mathrm{d}\zeta_{N_{\text{mf}}}. \end{aligned} \tag{5.155}$$

Thus, for a turbulent reacting flow in the equilibrium-chemistry limit, the difficulty of closing the chemical source term is shifted to the problem of predicting $f_{\boldsymbol{\xi}}(\boldsymbol{\zeta}; \mathbf{x}, t)$.

In a CFD calculation, one is usually interested in computing only the reacting-scalar means and (sometimes) the covariances. For binary mixing in the equilibrium-chemistry limit, these quantities are computed from (5.154) and (5.155), which contain the mixture-fraction PDF. However, since the presumed PDF is uniquely determined from the mixture-fraction mean and variance, (5.154) and (5.155) define 'mappings' (or functions) from $\langle \xi \rangle$–$\langle \xi'^2 \rangle$ space:

$$\langle \boldsymbol{\varphi}_{\text{rp}} \rangle = \mathbf{h}_{\text{rp}}(\langle \xi \rangle, \langle \xi'^2 \rangle), \tag{5.156}$$

$$\langle \varphi'_{\text{rp}\alpha} \varphi'_{\text{rp}\alpha} \rangle = h_{\alpha\beta}(\langle \xi \rangle, \langle \xi'^2 \rangle). \tag{5.157}$$

Moreover, the range of the two arguments in the mappings is known in advance to be $0 \leq \langle \xi \rangle \leq 1$ and $0 \leq \langle \xi'^2 \rangle \leq \langle \xi \rangle (1 - \langle \xi \rangle)$. Thus, the reacting-scalar means and covariances can be stored in pre-computed chemical lookup tables parameterized by $x_1 \equiv \langle \xi \rangle$ and $x_2 \equiv \langle \xi'^2 \rangle$. An example of such a table is shown in Fig. 5.11. At any point in the flow domain, the values of x_1 and x_2 will be known from the numerical solution of the turbulent-scalar-mixing problem. Given these values, the reacting-scalar means and covariances can then be efficiently extracted from the chemical lookup tables in a post-processing step.

[87] Note that the expected value of any arbitrary function of the reacting scalars can be computed in a similar manner.

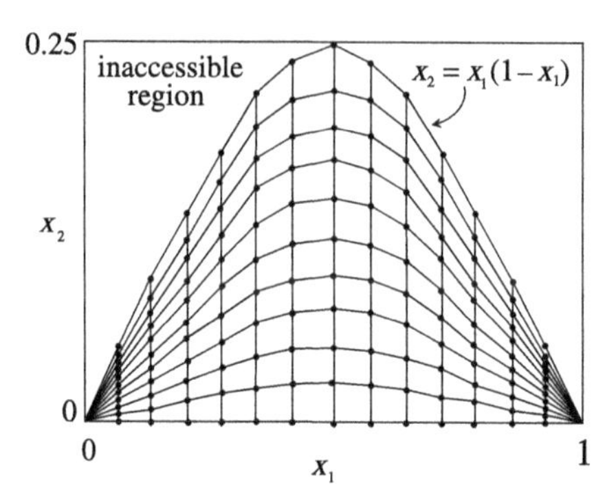

Figure 5.11. Chemical lookup table parameterized in terms of mixture-fraction mean $x_1 \equiv \langle \xi \rangle$ and variance $x_2 \equiv \langle \xi'^2 \rangle$.

In the equilibrium-chemistry limit, the turbulent-reacting-flow problem thus reduces to solving the Reynolds-averaged transport equations for the mixture-fraction mean and variance. Furthermore, if the mixture-fraction field is found from LES, the same chemical lookup tables can be employed to find the SGS reacting-scalar means and covariances simply by setting x_1 equal to the resolved-scale mixture fraction and x_2 equal to the SGS mixture-fraction variance.[88]

In summary, in the equilibrium-chemistry limit, the computational problem associated with turbulent reacting flows is greatly simplified by employing the presumed mixture-fraction PDF method. Indeed, because the chemical source term usually leads to a stiff system of ODEs (see (5.151)) that are solved 'off-line,' the equilibrium-chemistry limit significantly reduces the computational load needed to solve a turbulent-reacting-flow problem. In a CFD code, a second-order transport model for inert scalars such as those discussed in Chapter 3 is utilized to find $\langle \boldsymbol{\xi} \rangle$ and $\langle \xi'_i \xi'_j \rangle$, and the equilibrium compositions $\varphi_{\mathrm{eq}}(\boldsymbol{\xi})$ can be computed in a pre-processing step. As a result, the presumed mixture-fraction PDF method is widely employed in computational combustion codes used for industrial design calculations. However, the equilibrium-chemistry closure for the chemical source term is known to be inaccurate in cases where *finite-rate chemistry* effects are important, e.g., for the prediction of minor species in rich flames near extinction. For these flows, other types of chemical-source-term closures are required that take into account the possibility that some of the chemical time scales may be larger than the turbulent-scalar-mixing time scale (i.e., *finite-rate reactions*).

5.5 Simple chemistry

The methods developed thus far in this chapter apply to arbitrary chemical kinetic expressions. However, in the case of 'simple' chemistry (defined below), it is possible to

[88] The SGS mixture-fraction mean will be equal to the resolved-scale (or filtered) mixture fraction: $\langle \xi \rangle_{\mathrm{sgs}} = \overline{\xi}(\mathbf{x}, t)$. For LES, the SGS reacting-scalar means and covariances will be functions of space and time, e.g., $\langle \varphi_{\mathrm{rp}} \rangle_{\mathrm{sgs}}(\mathbf{x}, t)$.

work out the relationship between the original and transformed scalars analytically without using (numerical) SVD. Thus, in this section, we develop analytical expressions for three common examples of simple chemistry: one-step reaction, competitive-consecutive reactions, and parallel reactions. Before doing so, however, we will introduce the general formulation for treating simple chemistry in terms of *reaction-progress variables*.

5.5.1 General formulation: reaction-progress variables

Chemical reactions for which the rank of the reaction coefficient matrix $\boldsymbol{\Upsilon}$ is equal to the number of reaction rate functions R_i $(i \in 1, \ldots, I)$ (i.e., $N_\Upsilon = I$), can be expressed in terms of I reaction-progress variables Y_i $(i \in 1, \ldots, I)$, in addition to the mixture-fraction vector $\boldsymbol{\xi}$. For these reactions, the chemical source terms for the reaction-progress variables can be found without resorting to SVD of $\boldsymbol{\Upsilon}$. Thus, in this sense, such chemical reactions are 'simple' compared with the general case presented in Section 5.1.

In order to simplify the discussion further, we will only consider the case where the molecular diffusivities of all chemical species are identical. We can then write the linear accumulation and transport terms as a linear operator:

$$\mathcal{L}\{\phi\} \equiv \frac{\partial \phi}{\partial t} + U_j \frac{\partial \phi}{\partial x_j} - \Gamma \frac{\partial^2 \phi}{\partial x_j \partial x_j}, \tag{5.158}$$

so that the scalar transport equation becomes

$$\mathcal{L}\{\mathbf{c}\} \equiv \boldsymbol{\Upsilon}\mathbf{R}. \tag{5.159}$$

The dimensionless vector of reaction-progress variables $\mathbf{Y}$ is then defined to be null in the initial and inlet conditions, and obeys

$$\mathcal{L}\{\mathbf{Y}\} = \mathbf{diag}\,(1/\gamma_1, \ldots, 1/\gamma_I)\,\mathbf{R}, \tag{5.160}$$

where the (positive) constants γ_i are chosen to achieve a 'desirable' scaling (e.g., $0 \leq Y_i \leq 1$), as shown in the examples below.[89] Note that these constants have the same units as $\mathbf{c}$.

By definition, the mixture-fraction vector obeys

$$\mathcal{L}\{\boldsymbol{\xi}\} = \mathbf{0}. \tag{5.161}$$

Thus, the species concentrations can be expressed as a linear combination of $\mathbf{Y}$, $\boldsymbol{\xi}$, and a constant vector $\mathbf{c}_0$:[90]

$$\mathbf{c} = \mathbf{c}_0 + \mathbf{M}_\xi \boldsymbol{\xi} + \mathbf{M}_Y \mathbf{Y}. \tag{5.162}$$

Applying the linear operator to this expression yields

$$\mathcal{L}\{\mathbf{c}\} = \mathbf{M}_Y \mathcal{L}\{\mathbf{Y}\} \quad \Rightarrow \quad \boldsymbol{\Upsilon} = \mathbf{M}_Y \mathbf{diag}\,(1/\gamma_1, \ldots, 1/\gamma_I)\,. \tag{5.163}$$

[89] Alternatively, they can be set equal to a reference inlet concentration.

[90] From the definition of the linear operator, $\mathcal{L}\{\mathbf{c}_0\} = \mathbf{0}$.

From the final expression, it can be seen that $\mathbf{M}_Y$ must be defined by

$$\mathbf{M}_Y \equiv \boldsymbol{\Upsilon}\mathbf{diag}\,(\gamma_1, \ldots, \gamma_I)\,. \tag{5.164}$$

Likewise, the matrix $\mathbf{M}_\xi$ can be found using the method presented in Section 5.3 applied in the limiting case of a *non-reacting system* (i.e., $\mathbf{S} = 0$). In the simplest case (binary mixing), only one mixture-fraction component is required, and $\mathbf{M}_\xi$ is easily found from the species concentrations in the inlet streams.

In addition to 'uncoupling' the reaction rate functions (i.e., (5.160)), the principal advantage of expressing a reacting-flow problem in terms of the reaction-progress variables is the ease with which limiting cases for the reaction rate constants k_i^{f} can be treated. Indeed, in the limit where $k_i^{\mathrm{f}} = 0$, the corresponding reaction-progress variable Y_i will be null. In the opposite limit where $k_i^{\mathrm{f}} \to \infty$, the transport equation for Y_i can be replaced by an equivalent algebraic expression. For example, when $k_1^{\mathrm{f}} \to \infty$, the transport equation yields

$$\lim_{k_1^{\mathrm{f}} \to \infty} \left(\mathcal{L}\{Y_1\} = \gamma_1 R_1 \right) \quad \Rightarrow \quad R_1 = 0, \tag{5.165}$$

which can be used to eliminate Y_1 in the chemical source terms for the other reaction-progress variables. The use of reaction-progress variables is thus especially useful for the treatment of fast *irreversible* reactions (i.e., $k^{\mathrm{r}} = 0$) for which the equilibrium-chemistry limit discussed in Section 5.4 does not apply. We illustrate this procedure below for three examples of binary mixing with simple chemistry.

5.5.2 One-step reaction

Consider first the isothermal[91] one-step reaction

$$\mathrm{A} + r\mathrm{B} \rightarrow s\mathrm{P} \tag{5.166}$$

with reaction rate function

$$R(\mathbf{c}) = k c_{\mathrm{A}} c_{\mathrm{B}}. \tag{5.167}$$

The transport equations for the three chemical species have the form

$$\mathcal{L}\{c_{\mathrm{A}}\} = -R, \tag{5.168}$$

$$\mathcal{L}\{c_{\mathrm{B}}\} = -rR, \tag{5.169}$$

$$\mathcal{L}\{c_{\mathrm{P}}\} = sR. \tag{5.170}$$

The reaction coefficient matrix $\boldsymbol{\Upsilon}$ for this case is rank one. Thus, a linear transformation can be found that generates one reacting scalar and two conserved scalars. Moreover, if the flow system has only two inlet streams, and the initial conditions are a linear mixture of the

[91] As discussed in Section 5.1, the extension to non-isothermal conditions is straightforward under the assumption that the thermodynamic properties are constant.

inlet streams, it is possible to rewrite the transport equations in terms of a one-component mixture-fraction vector (i.e., ξ). For example, with inlet and initial conditions

$$\begin{bmatrix}\mathbf{c}^{(0)} & \mathbf{c}^{(1)} & \mathbf{c}^{(2)}\end{bmatrix} = \begin{bmatrix}(1-\alpha)\mathrm{A}_0 & \mathrm{A}_0 & 0 \\ \alpha\mathrm{B}_0 & 0 & \mathrm{B}_0 \\ 0 & 0 & 0\end{bmatrix}, \tag{5.171}$$

the transport equations can be written in terms of a single reaction-progress variable Y and the mixture fraction ξ. By convention, the reaction-progress variable is null in the inlet streams, and is thus proportional to the amount of product P produced by the reaction.[92]

As shown above, a linear relationship between $\mathbf{c}$ and $(\mathbf{c}_0, Y, \xi)$ can easily be derived starting from (5.162) by letting $\gamma_1 = A_0 B_0/(B_0 + rA_0)$:[93]

$$c_\mathrm{A} = \mathrm{A}_0 \left[1 - \xi - (1 - \xi_\mathrm{st})Y\right], \tag{5.172}$$

$$c_\mathrm{B} = \mathrm{B}_0 \left(\xi - \xi_\mathrm{st} Y\right), \tag{5.173}$$

$$c_\mathrm{P} = s\mathrm{B}_0 \xi_\mathrm{st} Y, \tag{5.174}$$

where the stoichiometric mixture fraction ξ_st is defined by

$$\xi_\mathrm{st} \equiv \frac{r\mathrm{A}_0}{\mathrm{B}_0 + r\mathrm{A}_0}. \tag{5.175}$$

Note that the reaction-progress variable is defined such that $0 \leq Y \leq 1$. However, unlike the mixture fraction, its value will depend on the reaction rate $k^* \equiv k\mathrm{B}_0$. The solution to the reacting-flow problem then reduces to solving two transport equations:

$$\mathcal{L}\{Y\} = \xi_\mathrm{st} k^* \left(\frac{1-\xi}{1-\xi_\mathrm{st}} - Y\right)\left(\frac{\xi}{\xi_\mathrm{st}} - Y\right), \tag{5.176}$$

$$\mathcal{L}\{\xi\} = 0, \tag{5.177}$$

with the following inlet and initial conditions:

$$\left[\begin{bmatrix}Y \\ \xi\end{bmatrix}^{(0)} \begin{bmatrix}Y \\ \xi\end{bmatrix}^{(1)} \begin{bmatrix}Y \\ \xi\end{bmatrix}^{(2)}\right] = \begin{bmatrix}0 & 0 & 0 \\ \alpha & 0 & 1\end{bmatrix}. \tag{5.178}$$

Starting from (5.176), the limiting cases of $k^* = 0$ and $k^* = \infty$ are easily derived (Burke and Schumann 1928). For the first case, the reaction-progress variable is always null. For the second case, the reaction-progress variable can be written in terms of the mixture fraction as

$$Y_\infty(\xi) = \min\left(\frac{\xi}{\xi_\mathrm{st}}, \frac{1-\xi}{1-\xi_\mathrm{st}}\right). \tag{5.179}$$

Thus, just as was the case for equilibrium chemistry, the statistics of the reaction-progress variable depend only on the mixture fraction in this limit. In the infinite-rate chemistry

[92] For the non-isothermal case, the reaction-progress variable is proportional to the temperature.

[93] One has an arbitrary choice of which inlet stream has $\xi = 0$. Here we have chosen the first inlet stream. Thus, $\mathbf{c}_0 = [A_0\, 0\, 0]^\mathrm{T}$.

limit, the one-step reaction problem thus reduces to a turbulent mixing problem that can be described in terms of the mixture-fraction PDF.

One-step chemistry is often employed as an idealized model for combustion chemistry. The primary difference with the results presented above is the strong temperature dependence of the reaction rate constant $k(T)$. For constant-property flows, the temperature can be related to the mixture fraction and reaction-progress variable by a linear expression of the form

$$T = (T_1 - T_0)\xi + T_0 + \beta Y. \tag{5.180}$$

Thus, the turbulent reacting flow can still be described by just two scalars. However, in a combusting flow, the adiabatic temperature rise (β) will be much larger than unity. Hence, the reaction rate constant can be very small at the ambient temperature (T_0), but many orders of magnitude larger at the flame temperature ($T_0 + \beta$). Indeed, unlike the isothermal case (applicable to liquid-phase reacting flows) for which the reaction rate can be large for any physically relevant value of ξ and Y, in combusting flows the reaction rate may differ substantially from zero only near the point $(Y, \xi) = (1, \xi_{st})$. This additional non-linearity is largely responsible for the greater complexity of behaviors exhibited by combusting flows. For example, an adiabatic combusting flow may be indefinitely stable at the point $(0, \xi_{st})$. However, a small 'spark' that raises one point in the flow domain to $(1, \xi_{st})$ will be enough to cause the entire system to explode. In contrast, in a liquid-phase system, the point $(0, \xi_{st})$ is usually highly unstable, and thus no spark is required to achieve complete conversion.

5.5.3 Competitive-consecutive reactions

Consider next the competitive-consecutive reactions

$$\begin{aligned} \mathrm{A} + \mathrm{B} &\rightarrow \mathrm{R} \\ \mathrm{B} + \mathrm{R} &\rightarrow \mathrm{S} \end{aligned} \tag{5.181}$$

with reaction rate functions

$$R_1(\mathbf{c}) = k_1 c_\mathrm{A} c_\mathrm{B}, \tag{5.182}$$

$$R_2(\mathbf{c}) = k_2 c_\mathrm{B} c_\mathrm{R}. \tag{5.183}$$

The transport equations for the four chemical species have the form

$$\mathcal{L}\{c_\mathrm{A}\} = -R_1, \tag{5.184}$$

$$\mathcal{L}\{c_\mathrm{B}\} = -R_1 - R_2, \tag{5.185}$$

$$\mathcal{L}\{c_\mathrm{R}\} = R_1 - R_2, \tag{5.186}$$

$$\mathcal{L}\{c_\mathrm{S}\} = R_2. \tag{5.187}$$

Note that the rank of the reaction coefficient matrix in this case is two.

Using the same inlet/initial conditions as were employed for the one-step reaction, this reaction system can be written in terms of two reaction-progress variables (Y_1, Y_2) and the mixture fraction ξ. A linear relationship between $\mathbf{c}$ and ($\mathbf{c}_0$, $\mathbf{Y}$, ξ) can be derived starting from (5.162) with $\gamma_1 = \gamma_2 = A_0 B_0/(A_0 + B_0)$:

$$c_A = A_0[1 - \xi - (1 - \xi_{st})Y_1], \tag{5.188}$$

$$c_B = B_0\left[\xi - \xi_{st}(Y_1 + Y_2)\right], \tag{5.189}$$

$$c_R = B_0\xi_{st}(Y_1 - Y_2), \tag{5.190}$$

$$c_S = B_0\xi_{st}Y_2, \tag{5.191}$$

where

$$\xi_{st} \equiv \frac{A_0}{B_0 + A_0}. \tag{5.192}$$

The solution to the reacting-flow problem then reduces to solving three transport equations:

$$\mathcal{L}\{Y_1\} = \xi_{st}k_1^*\left(\frac{1-\xi}{1-\xi_{st}} - Y_1\right)\left(\frac{\xi}{\xi_{st}} - Y_1 - Y_2\right), \tag{5.193}$$

$$\mathcal{L}\{Y_2\} = \xi_{st}k_2^*\,(Y_1 - Y_2)\left(\frac{\xi}{\xi_{st}} - Y_1 - Y_2\right), \tag{5.194}$$

$$\mathcal{L}\{\xi\} = 0, \tag{5.195}$$

where $k_1^* \equiv k_1 B_0$ and $k_2^* \equiv k_2 B_0$, with the following inlet and initial conditions:

$$\left[\begin{bmatrix} Y_1 \\ Y_2 \\ \xi \end{bmatrix}^{(0)} \begin{bmatrix} Y_1 \\ Y_2 \\ \xi \end{bmatrix}^{(1)} \begin{bmatrix} Y_1 \\ Y_2 \\ \xi \end{bmatrix}^{(2)}\right] = \begin{bmatrix} 0 & 0 & 0 \\ 0 & 0 & 0 \\ \alpha & 0 & 1 \end{bmatrix}. \tag{5.196}$$

In chemical-engineering applications, the competitive-consecutive reactions are often studied in the limit where $k_1^* \to \infty$. In this limit, the first reaction-progress variable Y_1 can be written in terms of Y_2 and ξ. From the reaction rate expression for Y_1, it can be seen that the limiting value is given by

$$Y_{1\infty}(Y_{2\infty}, \xi) = \min\left(\frac{\xi}{\xi_{st}} - Y_{2\infty}, \frac{1-\xi}{1-\xi_{st}}\right) \tag{5.197}$$

under the condition that $0 \leq Y_{2\infty} \leq \xi/\xi_{st}$. Note that when $c_B = 0$,

$$Y_{1\infty}(Y_{2\infty}, \xi) = \frac{\xi}{\xi_{st}} - Y_{2\infty}, \tag{5.198}$$

i.e., at points in the flow where A is in excess. On the other hand, when $c_A = 0$,

$$Y_{1\infty}(Y_{2\infty}, \xi) = \frac{1-\xi}{1-\xi_{st}}, \tag{5.199}$$

i.e., at points in the flow where B is in excess. If $k_2^* = 0$, $Y_{2\infty} = 0$, and the boundary in mixture-fraction space between these two regions occurs at $\xi = \xi_{st}$.

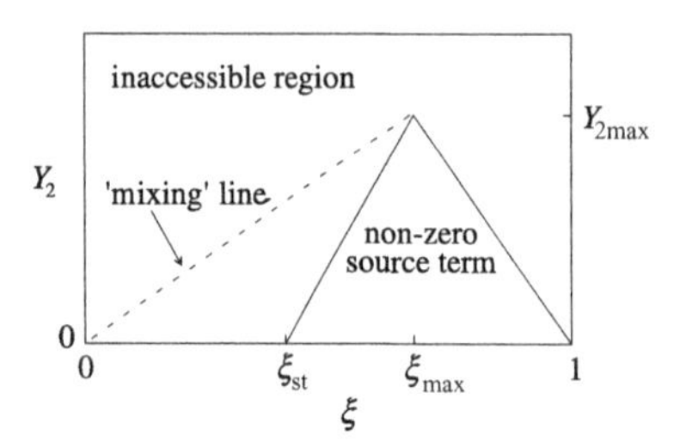

Figure 5.12. The chemical source term for $Y_{2\infty}$ will be non-zero in the triangular region bordered by the line $Y_{2\infty} = 0$ and the two lines found from setting $h_1 = 0$ and $h_2 = 0$. The 'mixing' line corresponds to the upper limit for $Y_{2\infty}$ in the range $0 \le \xi \le \xi_{max}$, and results from micromixing between fluid elements at (0, 0) and $(\xi_{max}, Y_{2\,max})$.

The expression for $Y_{1\infty}$, (5.197), can be used in (5.194) to find the transport equation for $Y_{2\infty}$:

$$\mathcal{L}\{Y_{2\infty}\} = \xi_{st} k_2^* h_1 (Y_{2\infty}, \xi)\, h_2 (Y_{2\infty}, \xi) \tag{5.200}$$

when $c_A(Y_{2\infty}, \xi) = 0$, where

$$h_1 (Y_{2\infty}, \xi) = \frac{1-\xi}{1-\xi_{st}} - Y_{2\infty} \tag{5.201}$$

and

$$h_2 (Y_{2\infty}, \xi) = \frac{\xi - \xi_{st}}{\xi_{st}(1-\xi_{st})} - Y_{2\infty}; \tag{5.202}$$

and $\mathcal{L}\{Y_{2\infty}\} = 0$ when $c_B(Y_{2\infty}, \xi) = 0$. By definition, $Y_{2\infty}$ is non-negative. Thus, the right-hand side of (5.200) must also be non-negative. This condition will occur when both h_1 and h_2 are non-negative,[94] and defines a triangular region in $(\xi, Y_{2\infty})$-space (see Fig. 5.12).

The triangular region can be expressed analytically in terms of two parameters:

$$\xi_{max} \equiv \frac{2\xi_{st}}{1+\xi_{st}} \tag{5.203}$$

and

$$Y_{2\,max} \equiv \frac{1}{1+\xi_{st}} = \frac{1}{2}\frac{\xi_{max}}{\xi_{st}}, \tag{5.204}$$

which correspond to the apex of the triangle. The triangular region is then given by

$$\begin{aligned} 0 \le Y_{2\infty} &\le Y_{2\,max}, \\ \xi_2(Y_{2\infty}) \le \xi &\le \xi_1(Y_{2\infty}), \end{aligned} \tag{5.205}$$

where the limits on ξ are defined from $h_1(Y_{2\infty}, \xi_1) = 0$ and $h_2(Y_{2\infty}, \xi_2) = 0$, respectively. Note also that, due to the final equality in (5.204), the condition appearing after (5.197) required for its validity holds for all values of ξ. Indeed, for $0 \le \xi \le \xi_{max}$ the

[94] It will also occur when both are negative. However, this will yield unphysical values for $Y_{2\infty}$.

maximum value of $Y_{2\infty}$ is equal to $Y_{2\max}\xi$ (see Fig. 5.12), and results from molecular mixing in $(\xi, Y_{2\infty})$-space between fluid elements with concentrations $(\xi, Y_{2\infty}) = (0, 0)$ and $(\xi_{\max}, Y_{2\max})$.

Having established the form of the reaction rate function for $Y_{2\infty}$, we can now look at its limiting behavior when $k_2^* \to \infty$. In this limit, all points in the triangular region where the reaction rate is non-zero (Fig. 5.12) will be forced to the upper boundary. Thus, in the *absence of molecular mixing*, the infinite-rate chemistry forces $Y_{2\infty}$ towards

$$Y_{2\infty}^{\infty}(\xi) = \begin{cases} 0 & \text{if } 0 \leq \xi \leq \xi_{\text{st}} \\ \frac{\xi - \xi_{\text{st}}}{\xi_{\max} - \xi_{\text{st}}} Y_{2\max} & \text{if } \xi_{\text{st}} \leq \xi \leq \xi_{\max} \\ \frac{1-\xi}{1-\xi_{\max}} Y_{2\max} & \text{if } \xi_{\max} \leq \xi \leq 1. \end{cases} \tag{5.206}$$

Physically, the dependence of (5.206) on the mixture fraction can be understood as follows.

- When $0 \leq \xi \leq \xi_{\text{st}}$, species A is in excess so that the first reaction converts all available B to R. Thus, no S is produced.
- When $\xi_{\text{st}} \leq \xi \leq \xi_{\max}$, the first reaction cannot convert all B to R due to the lack of A. The remaining B is completely converted to S.
- When $\xi_{\max} \leq \xi \leq 1$, species B is in excess so that all R is converted to S.

In the presence of molecular mixing, when the mixture fraction is less than $\xi_{\max}$, $Y_{2\infty}$ is no longer bounded above by $Y_{2\infty}^{\infty}(\xi)$. Instead, as noted earlier, the upper bound moves to the 'mixing' line shown in Fig. 5.12. Thus, in a turbulent reacting flow with reaction and mixing, the theoretical upper bound on $Y_{2\infty}$ is

$$Y_{2\infty}^{\max}(\xi) = \begin{cases} \frac{\xi}{\xi_{\max}} Y_{2\max} & \text{if } 0 \leq \xi \leq \xi_{\max} \\ \frac{1-\xi}{1-\xi_{\max}} Y_{2\max} & \text{if } \xi_{\max} \leq \xi \leq 1. \end{cases} \tag{5.207}$$

The actual value of $Y_{2\infty}$ found in an experiment will depend on the relative values of the micromixing rate $\omega_\phi \equiv 1/\tau_\phi$ and k_2^*. When micromixing is much faster than the second reaction ($k_2^* \ll \omega_\phi$), fluctuations in the mixture fraction will be quickly dissipated ($\xi \approx \langle\xi\rangle$) so that the limiting value is

$$Y_{2\infty} = Y_{2\infty}^{\infty}(\langle\xi\rangle). \tag{5.208}$$

In the opposite limit where the second reaction is much faster than micromixing ($\omega_\phi \ll k_2^*$), the limiting value is

$$Y_{2\infty} \leq Y_{2\infty}^{\max}(\langle\xi\rangle). \tag{5.209}$$

Hence, competitive-consecutive reactions will be sensitive to mixing for values of the mean mixture fraction in the range $0 < \langle\xi\rangle < \xi_{\max}$ for which $0 < Y_{2\infty}^{\infty}(\langle\xi\rangle) < Y_{2\infty}^{\max}(\langle\xi\rangle)$.

In most experimental investigations involving competitive-consecutive reactions, species A is in stoichiometric excess, so that $0 < \langle\xi\rangle < \xi_{\text{st}}$. In this range, only $Y_{2\infty}^{\max}$ is non-zero, and thus its value serves as a measure of the rate of micromixing. Ideally, in order to maximize the value of $c_S = B_0\xi_{\text{st}}Y_{2\infty}$ (and thus facilitate its measurement), the

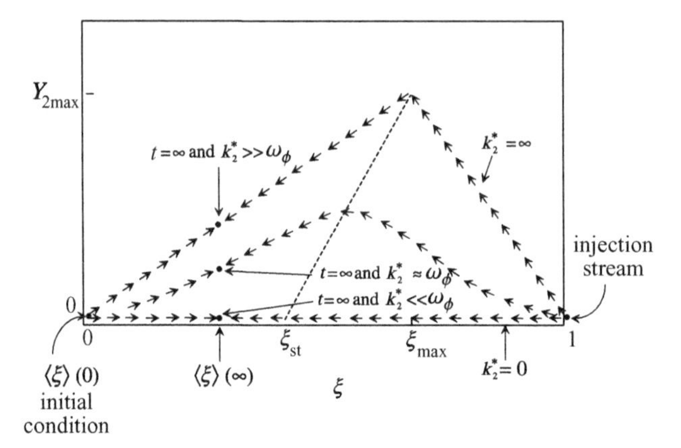

Figure 5.13. In a semi-batch reactor, the mean mixture fraction evolves along a curve starting at (0, 0). Fluid particles in the injection stream start at (1, 0) and move towards $[\langle\xi\rangle(t), Y_{2\infty}(\langle\xi\rangle(t))]$ along a curve that depends on the relative magnitudes of ω_ϕ and k_2^*. If $\omega_\phi \gg k_2^*$, then the fluid particles will immediately move to $[\langle\xi\rangle(t), 0]$ without having time to react. However, if $\omega_\phi \ll k_2^*$, then the fluid particles will react completely so that they first move to $(\xi_{\max}, Y_{2\max})$ before moving along the mixing line to $[\langle\xi\rangle(t), Y_{2\max}\langle\xi\rangle(t)]$. Intermediate values of ω_ϕ will yield trajectories in $(\xi, Y_{2\infty})$-space that lie between these two extremes. The final value of $Y_{2\infty}$ will therefore depend on the rate of micromixing relative to reaction.

flow rates and inlet concentrations should be chosen so that $\langle\xi\rangle$ is less than but nearly equal to ξ_{st} and the product $B_0\xi_{\max}$ is maximized.[95]

The triangular region in Fig. 5.12 can also be used to analyze the effect of micromixing in semi-batch reactors wherein the stream containing B is slowly added to a reactor initially containing only A. In this reactor, the mean mixture fraction is an increasing function of time:

$$\langle\xi\rangle(t) = \frac{q_B t}{V_A + q_B t} \quad \text{for} \quad 0 \leq t \leq t_{\max} \equiv \frac{V_B}{q_B}, \tag{5.210}$$

where V_A is the original volume of reactant A, V_B is the volume of reactant B to be added, and q_B is the flow rate of reactant B. At time $t = 0$, the system starts at the point (0, 0) in $(\xi, Y_{2\infty})$-space. Similarly, the injection stream containing B starts at (1, 0). If micromixing near the injection point is infinitely rapid, fluid particles from the injection stream mix instantaneously with the fluid particles already in the reactor to form a well mixed solution at $[\langle\xi\rangle(t), 0]$. However, if the micromixing is less than instantaneous, then the injected fluid particles will require a finite time to move through the triangular region in Fig. 5.12, and hence the second reaction will have time to increase $Y_{2\infty}$ to a non-zero value. (Sample trajectories are shown in Fig. 5.13.) On the other hand, if the second reaction is infinitely rapid, then the fluid particles will first move along the line from (1, 0) to $(\xi_{\max}, Y_{2\max})$, then along the mixing line to $[\langle\xi\rangle(t), Y_{2\max}\langle\xi\rangle(t)]$. The final value of the second product in the reactor (i.e., $Y_{2\infty}(t_{\max})$) will thus be sensitive to the rate of micromixing near the injection point.

[95] The second condition comes from the fact that $\xi_{\max} = 2\xi_{st}Y_{2\max}$.

5.5.4 Parallel reactions

Consider next the parallel reactions

$$\begin{aligned} \mathrm{A} + \mathrm{B} &\rightarrow \mathrm{R} \\ \mathrm{A} + \mathrm{C} &\rightarrow \mathrm{S} \end{aligned} \tag{5.211}$$

with reaction rate functions

$$R_1(\mathbf{c}) = k_1 c_\mathrm{A} c_\mathrm{B}, \tag{5.212}$$

$$R_2(\mathbf{c}) = k_2 c_\mathrm{A} c_\mathrm{C}. \tag{5.213}$$

The transport equations for the five chemical species have the form

$$\mathcal{L}\{c_\mathrm{A}\} = -R_1 - R_2, \tag{5.214}$$

$$\mathcal{L}\{c_\mathrm{B}\} = -R_1, \tag{5.215}$$

$$\mathcal{L}\{c_\mathrm{C}\} = -R_2, \tag{5.216}$$

$$\mathcal{L}\{c_\mathrm{R}\} = R_1, \tag{5.217}$$

$$\mathcal{L}\{c_\mathrm{S}\} = R_2. \tag{5.218}$$

Note that the rank of the reaction coefficient matrix in this case is again two.

The usual initial and inlet conditions for binary mixing with parallel reactions are

$$\begin{bmatrix} \mathbf{c}^{(0)} & \mathbf{c}^{(1)} & \mathbf{c}^{(2)} \end{bmatrix} = \begin{bmatrix} (1-\alpha)\mathrm{A}_0 & \mathrm{A}_0 & 0 \\ \alpha \mathrm{B}_0 & 0 & \mathrm{B}_0 \\ \alpha \mathrm{C}_0 & 0 & \mathrm{C}_0 \\ 0 & 0 & 0 \\ 0 & 0 & 0 \end{bmatrix}. \tag{5.219}$$

Thus, the relative amounts of the products R and S produced by the system will depend on the relative magnitudes of k_1, k_2, and the rate of micromixing between the two inlet streams.

The parallel reaction system can be written in terms of two reaction-progress variables (Y_1, Y_2) and the mixture fraction ξ. A linear relationship between $\mathbf{c}$ and $(\mathbf{c}_0, \mathbf{Y}, \xi)$ can be derived starting from (5.162) with $\gamma_1 = A_0 B_0/(A_0 + B_0)$ and $\gamma_2 = A_0 C_0/(A_0 + C_0)$:

$$c_\mathrm{A} = \mathrm{A}_0[1 - \xi - (1 - \xi_{s1})Y_1 - (1 - \xi_{s2})Y_2], \tag{5.220}$$

$$c_\mathrm{B} = \mathrm{B}_0\,(\xi - \xi_{s1} Y_1)\,, \tag{5.221}$$

$$c_\mathrm{C} = \mathrm{C}_0\,(\xi - \xi_{s2} Y_2)\,, \tag{5.222}$$

$$c_\mathrm{R} = \mathrm{B}_0 \xi_{s1} Y_1, \tag{5.223}$$

$$c_\mathrm{S} = \mathrm{C}_0 \xi_{s2} Y_2, \tag{5.224}$$

where

$$\xi_{s1} \equiv \frac{A_0}{B_0 + A_0} \tag{5.225}$$

and

$$\xi_{s2} \equiv \frac{A_0}{C_0 + A_0}. \tag{5.226}$$

The solution to the reacting-flow problem thus again reduces to solving three transport equations:

$$\mathcal{L}\{Y_1\} = \xi_{s1} k_1^* \left(\frac{1-\xi}{1-\xi_{s1}} - \frac{1-\xi_{s2}}{1-\xi_{s1}} Y_2 - Y_1 \right) \left(\frac{\xi}{\xi_{s1}} - Y_1 \right), \tag{5.227}$$

$$\mathcal{L}\{Y_2\} = \xi_{s2} k_2^* \left(\frac{1-\xi}{1-\xi_{s2}} - \frac{1-\xi_{s1}}{1-\xi_{s2}} Y_1 - Y_2 \right) \left(\frac{\xi}{\xi_{s2}} - Y_2 \right), \tag{5.228}$$

$$\mathcal{L}\{\xi\} = 0, \tag{5.229}$$

with $k_1^* \equiv k_1 \mathrm{B}_0$ and $k_2^* \equiv k_2 \mathrm{C}_0$, and the following initial and inlet conditions:

$$\left[\begin{bmatrix} Y_1 \\ Y_2 \\ \xi \end{bmatrix}^{(0)} \begin{bmatrix} Y_1 \\ Y_2 \\ \xi \end{bmatrix}^{(1)} \begin{bmatrix} Y_1 \\ Y_2 \\ \xi \end{bmatrix}^{(2)} \right] = \begin{bmatrix} 0 & 0 & 0 \\ 0 & 0 & 0 \\ \alpha & 0 & 1 \end{bmatrix}. \tag{5.230}$$

In chemical-engineering applications, parallel reactions are often studied in the limit where $k_1^* \to \infty$. In this limit, the first reaction-progress variable Y_1 can be written in terms of Y_2 and ξ. From the reaction rate expression for Y_1, it can be seen that the limiting value is given by

$$Y_{1\infty}(Y_{2\infty}, \xi) = \min \left(\frac{\xi}{\xi_{s1}}, \frac{1-\xi}{1-\xi_{s1}} - \frac{1-\xi_{s2}}{1-\xi_{s1}} Y_{2\infty} \right) \tag{5.231}$$

under the condition that $0 \leq Y_{2\infty} \leq (1-\xi)/(1-\xi_{s2})$. Note that when $c_\mathrm{B} = 0$,

$$Y_{1\infty}(Y_{2\infty}, \xi) = \frac{\xi}{\xi_{s1}}, \tag{5.232}$$

i.e., at points in the flow where A is in excess. On the other hand, when $c_\mathrm{A} = 0$,

$$Y_{1\infty}(Y_{2\infty}, \xi) = \frac{1-\xi}{1-\xi_{s1}} - \frac{1-\xi_{s2}}{1-\xi_{s1}} Y_{2\infty}, \tag{5.233}$$

i.e., at points in the flow where B is in excess. If $k_2^* = 0$, $Y_{2\infty} = 0$, and the boundary in mixture-fraction space between these two regions occurs at $\xi = \xi_{s1}$.

The expression for $Y_{1\infty}$, (5.231), can be used in (5.228) to find the transport equation for $Y_{2\infty}$ when $c_\mathrm{B}(Y_{2\infty}, \xi) = 0$:

$$\mathcal{L}\{Y_{2\infty}\} = \xi_{s2} k_2^* h_1(Y_{2\infty}, \xi) h_2(Y_{2\infty}, \xi), \tag{5.234}$$

where

$$h_1(Y_{2\infty}, \xi) = \frac{\xi}{\xi_{s2}} - Y_{2\infty} \tag{5.235}$$

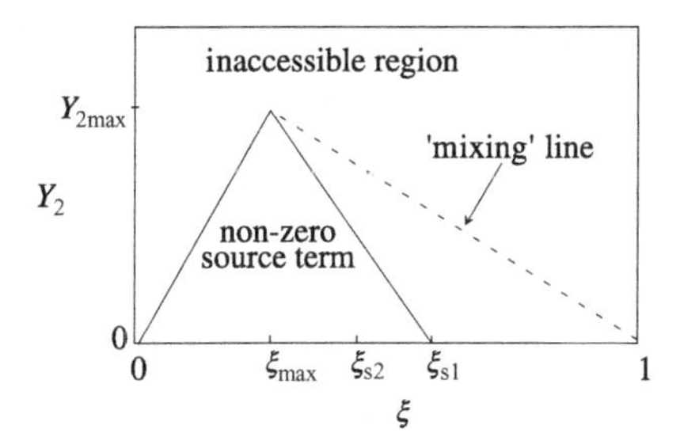

Figure 5.14. The chemical source term for $Y_{2\infty}$ will be non-zero in the triangular region bordered by the line $Y_{2\infty} = 0$ and the two lines found from setting $h_1 = 0$ and $h_2 = 0$. The 'mixing' line corresponds to the upper limit for $Y_{2\infty}$ in the range $\xi_{max} \leq \xi \leq 1$, and results from micromixing between fluid elements at $(1, 0)$ and $(\xi_{max}, Y_{2\,max})$.

and

$$h_2(Y_{2\infty}, \xi) = \frac{\xi_{s1} - \xi}{\xi_{s1}(1 - \xi_{s2})} - Y_{2\infty}; \tag{5.236}$$

and when $c_A(Y_{2\infty}, \xi) = 0$, $\mathcal{L}\{Y_{2\infty}\} = 0$. By definition, $Y_{2\infty}$ is non-negative. Thus, the right-hand side of (5.234) must also be non-negative. This condition will occur when both h_1 and h_2 are non-negative,[96] and defines a triangular region in $(\xi, Y_{2\infty})$-space (see Fig. 5.14).

The triangular region can again be expressed analytically in terms of two parameters:

$$\xi_{max} \equiv \frac{\xi_{s1}\xi_{s2}}{\xi_{s2} + \xi_{s1}(1 - \xi_{s2})} = \frac{A_0}{A_0 + B_0 + C_0} \tag{5.237}$$

and

$$Y_{2\,max} \equiv \frac{\xi_{s1}}{\xi_{s2} + \xi_{s1}(1 - \xi_{s2})} = \frac{\xi_{max}}{\xi_{s2}}, \tag{5.238}$$

which correspond to the apex of the triangle. The triangular region is then given by

$$\begin{aligned} 0 \leq Y_{2\infty} \leq Y_{2\,max}, \\ \xi_{s2} Y_{2\infty} \leq \xi \leq \xi_2(Y_{2\infty}), \end{aligned} \tag{5.239}$$

where the upper limit on ξ is defined from $h_2(Y_{2\infty}, \xi_2) = 0$. Note also that, due to the final equality in (5.238), the condition appearing after (5.231) required for its validity is equivalent to the condition that $0 \leq Y_{2\,max} \leq 1$, which will always hold.[97] Indeed, for $\xi_{max} \leq \xi \leq 1$, the maximum value of $Y_{2\infty}$ is equal to $Y_{2\,max}(\xi - \xi_{max})/(1 - \xi_{max})$ (see Fig. 5.12), and results from molecular mixing between fluid elements with concentrations $(\xi, Y_{2\infty}) = (1, 0)$ and $(\xi_{max}, Y_{2\,max})$.

Having established the form of the reaction rate function for $Y_{2\infty}$, we can now look at its limiting behavior when $k_2^* \to \infty$. In this limit, all points in the triangular region where

[96] It will also occur when both are negative. However, this will yield unphysical values for $Y_{2\infty}$.

[97] By definition, Y_2 can equal unity only if $B_0 = 0$. Likewise, Y_1 can equal unity only if $C_0 = 0$.

the reaction rate is non-zero (Fig. 5.14) will be forced to the upper boundary. Thus, in the *absence of molecular mixing*, the infinite-rate chemistry forces $Y_{2\infty}$ towards

$$Y_{2\infty}^{\infty}(\xi) = \begin{cases} \frac{\xi}{\xi_{\max}} Y_{2\max} & \text{if } 0 \leq \xi \leq \xi_{\max} \\ \frac{\xi_{s1}-\xi}{\xi_{s1}-\xi_{\max}} Y_{2\max} & \text{if } \xi_{\max} \leq \xi \leq \xi_{s1} \\ 0 & \text{if } \xi_{s1} \leq \xi \leq 1. \end{cases} \tag{5.240}$$

Physically, the dependence of (5.240) on the mixture fraction can be understood as follows.

- When $0 \leq \xi \leq \xi_{\max}$, species A is in excess so that the first reaction converts all available B to R. The remaining A is still in excess so that the second reaction then converts all available C to S. The final reaction mixture thus contains only R and S.
- When $\xi_{\max} \leq \xi \leq \xi_{s1}$, the first reaction converts all available B to R. However, there is not enough A remaining to convert all available C to S. The final reaction mixture thus contains C, R, and S.
- When $\xi_{s1} \leq \xi \leq 1$, species B is in excess so that the first reaction converts all available A to R. The final reaction mixture thus contains B, C and R, but no S.

In the presence of molecular mixing, when the mixture fraction is greater than $\xi_{\max}$, $Y_{2\infty}$ is no longer bounded above by $Y_{2\infty}^{\infty}(\xi)$. Instead, as seen in the previous example, the upper bound moves to the 'mixing' line shown in Fig. 5.14. Thus, in a turbulent reacting flow with reaction and mixing, the theoretical upper bound on $Y_{2\infty}$ is

$$Y_{2\infty}^{\max}(\xi) = \begin{cases} \frac{\xi}{\xi_{\max}} Y_{2\max} & \text{if } 0 \leq \xi \leq \xi_{\max} \\ \frac{1-\xi}{1-\xi_{\max}} Y_{2\max} & \text{if } \xi_{\max} \leq \xi \leq 1. \end{cases} \tag{5.241}$$

The actual value of $Y_{2\infty}$ found in an experiment will depend on the relative values of the micromixing rate ω_ϕ and k_2^*. When micromixing is much faster than the second reaction ($k_2^* \ll \omega_\phi$), fluctuations in the mixture fraction will quickly be dissipated ($\xi \approx \langle\xi\rangle$) so that the limiting value is

$$Y_{2\infty} = Y_{2\infty}^{\infty}(\langle\xi\rangle). \tag{5.242}$$

In the opposite limit, where the second reaction is much faster than micromixing ($\omega_\phi \ll k_2^*$), the limiting value is

$$Y_{2\infty} \leq Y_{2\infty}^{\max}(\langle\xi\rangle). \tag{5.243}$$

Hence, parallel reactions will be sensitive to mixing for values of the mean mixture fraction in the range $\xi_{\max} < \langle\xi\rangle < 1$ for which $0 \leq Y_{2\infty}^{\infty}(\langle\xi\rangle) < Y_{2\infty}^{\max}(\langle\xi\rangle)$.

In most experimental investigations involving parallel reactions, species B is in stoichiometric excess, so that $\xi_{s1} < \langle\xi\rangle < 1$. In this range, only $Y_{2\infty}^{\max}$ is non-zero, and thus its value serves as a measure of the rate of micromixing. Ideally, in order to maximize the range of possible values of $c_S = C_0\xi_{s2}Y_{2\infty}$, the flow rates and inlet concentrations should be chosen such that $\xi_{s1} \leq \langle \xi \rangle$ and the product $C_0 \xi_{max}$ is maximized.[98] However, care should be taken to avoid conditions for which ξ_{max} is too close to unity. Indeed, when $\xi_{max} \leq 1$ the maximum value for $Y_{2\infty}$ varies rapidly as a function of ξ, and thus small differences in $\langle \xi \rangle$ will result in large changes in c_S. Since $\langle \xi \rangle$ is controlled by the relative flow rates of the two inlet feed streams, its value will be subject to experimental error due to fluctuations in the flow rates. When $\xi_{max} \leq 1$, small flow-rate fluctuations will be greatly magnified, making it extremely difficult to generate reproducible experimental data. Furthermore, numerical simulations of parallel reactions with $\xi_{max} \leq 1$ will be very sensitive to numerical errors that would make any comparisons between different closures for the chemical source term unreliable.

In conclusion, we have shown in this section that for simple chemistry it is possible to rewrite the transport equations for a reacting flow in terms of the mixture-fraction vector and a vector of reaction-progress variables. Moreover, we have shown that competitive-consecutive and parallel reactions are sensitive to the rate of micromixing when the mean mixture fraction is near the stoichiometric mixture fraction for the first reaction (i.e., ξ_{st} or ξ_{s1}). The dependence of these reactions on the value of the mixture fraction has important ramifications on the modeling of their chemical source terms in turbulent reacting flows. For example, from (5.234) it can be seen that the chemical source term for $Y_{2\infty}$ depends on the inlet concentrations and chemistry (through ξ_{s1}, ξ_{s2}, and k_2^*), all of which are independent of the fluid dynamics. On the other hand, the mixture fraction can be described by a presumed PDF that depends on the local values of $\langle \xi \rangle$ and $\langle \xi'^2 \rangle$, both of which are strongly flow-dependent. Thus, *at a minimum*, a closure for the chemical source term for these reactions must contain information about the reaction rate *as a function of* ξ and a model for the mixture-fraction PDF. For this reason, moment closures (Section 5.2) that only contain information concerning the mean concentrations or the mixture-fraction variance are insufficient for describing the full range of parameter dependencies observed with competitive-consecutive and parallel reactions.[99]

5.6 Lagrangian micromixing models

As briefly discussed in Section 1.2, chemical-reaction engineers recognized early on the need to predict the influence of reactant segregation on the yield of complex reactions. Indeed, the competitive-consecutive and parallel reaction systems analyzed in the previous section have been studied experimentally by numerous research groups (Baldyga and Bourne 1999). However, unlike the mechanical-engineering community, who mainly focused on the fluid-dynamics approach to combustion problems, chemical-reaction

[98] The second condition comes from the fact that $\xi_{max} = \xi_{s2} Y_{2\,max}$.

[99] For example, in an experiment, by varying $\langle \xi \rangle$ one can change the inlet concentrations while holding the mean concentrations constant. For this case, although the reaction rate for $Y_{2\infty}$ would change significantly, a first-order moment closure would predict the same Reynolds-averaged reaction rate for all values of $\langle \xi \rangle$.

engineers began with a Lagrangian perspective based on the residence time distribution (RTD). In this section, we will review some of the simpler Lagrangian micromixing models with the perspective of relating them to chemical-source-term closures based on the joint composition PDF.

5.6.1 IEM model for a stirred reactor

Perhaps the simplest Lagrangian micromixing model is the interaction by exchange with the mean (IEM) model for a CSTR. In addition to the residence time τ, the IEM model introduces a second parameter $t_{\rm m}$ to describe the micromixing time. Mathematically, the IEM model can be written in Lagrangian form by introducing the age α of a fluid particle, i.e., the amount of time the fluid particle has spent in the CSTR since it entered through a feed stream. For a non-premixed CSTR with two feed streams,[100] the species concentrations in a fluid particle can be written as a function of its age as

$$\frac{\mathrm{d}\mathbf{c}^{(n)}}{\mathrm{d}\alpha} = \mathbf{S}(\mathbf{c}^{(n)}) + \frac{1}{t_{\rm m}}\left(\langle\mathbf{c}\rangle - \mathbf{c}^{(n)}\right) \quad \text{for } n = 1, 2, \tag{5.244}$$

where $\mathbf{c}^{(n)}(0)$ is the feed concentration vector for feed stream n, and $\langle\mathbf{c}\rangle$ is the mean concentration vector. The first term on the right-hand side of (5.244) is just the chemical source term evaluated at the fluid particle concentrations. Note that due to the Lagrangian description, the chemical source term is closed. The second term in (5.244) is the micromixing model, and describes the rate of change of the fluid particle concentrations by a simple linear function with a characteristic mixing rate $1/t_{\rm m}$.

The mean concentration vector is found by assuming that the CSTR is homogeneous on large scales (i.e., well macromixed).[101] Using the fact that the age distribution in a well macromixed CSTR is exponential,

$$E(\alpha) = \frac{1}{\tau}\,\mathrm{e}^{-\alpha/\tau}, \tag{5.245}$$

the mean concentration can be found by averaging over all ages and over both inlet feed streams:[102]

$$\langle\mathbf{c}\rangle = \frac{1}{\tau}\int_0^\infty \left[\gamma_1\mathbf{c}^{(1)}(\alpha) + \gamma_2\mathbf{c}^{(2)}(\alpha)\right]\mathrm{e}^{-\alpha/\tau}\,\mathrm{d}\alpha, \tag{5.246}$$

where

$$\gamma_n = \frac{q_n}{q_1 + q_2} \quad \text{for } n = 1, 2 \tag{5.247}$$

is the volume fraction of feed stream n with volumetric flow rate q_n.

[100] The IEM model is easily extended to treat multiple feed streams and time-dependent feed rates (Fox and Villermaux 1990b).

[101] This is usually a poor assumption: because the characteristic mixing time of large scales is longer than that of small scales, both macromixing and micromixing should be accounted for in a CSTR. The only reactor type for which macromixing may be less important than micromixing is the plug-flow reactor (PFR).

[102] Although this definition can be employed to find the mean concentrations, it is numerically unstable for small values of $t_{\rm m}$. In this case, an equivalent integro-differential equation is preferable (Fox 1989; Fox and Villermaux 1990b).

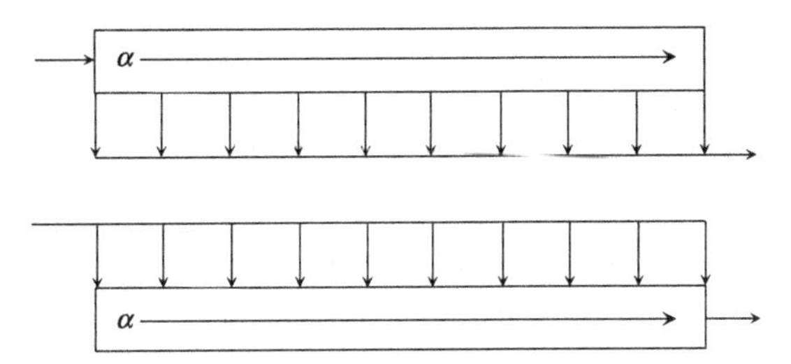

Figure 5.15. Two examples of age-based micromixing models. In the top example, it is assumed that fluid particles remain segregated until the latest possible age. In the bottom example, the fluid particles mix at the earliest possible age. Numerous intermediate mixing schemes are possible, which would result in different predictions for micromixing-sensitive reactions.

Note that the IEM model contains no spatial structure due to the assumption that the vessel is well macromixed. This assumption avoids the need for computing the local mean velocity field and other turbulence quantities. On the other hand, the micromixing parameter $t_{\rm m}$ cannot easily be related to flow conditions. In essence, $t_{\rm m}$ hides all of the flow field information, and thus its value cannot be specified *a priori* for a given CSTR. Nevertheless, due to its simple mathematical form, the IEM model can be applied to complex chemical kinetics to check for sensitivity to micromixing effects. Moreover, bifurcation and stability analysis techniques have been extended to treat the IEM model (Fox 1989; Fox and Villermaux 1990b; Fox *et al.* 1990; Curtis *et al.* 1992). The sensitivity of complex dynamical behavior (e.g., chemical oscillations and chemical chaos) to micromixing can be effectively investigated using the IEM model (Fox and Villermaux 1990a; Fox 1991; Fox *et al.* 1994).

5.6.2 Age-based models

The IEM model is a simple example of an age-based model. Other more complicated models that use the residence time distribution have also been developed by chemical-reaction engineers. For example, two models based on the mixing of fluid particles with different ages are shown in Fig. 5.15. Nevertheless, because it is impossible to map the age of a fluid particle onto a physical location in a general flow, age-based models cannot be used to predict the spatial distribution of the concentration fields inside a chemical reactor. Model validation is thus performed by comparing the predicted outlet concentrations with experimental data.

The relationship between age-based models and models based on the composition PDF can be understood in terms of the joint PDF of composition (ϕ) and age (A):[103]

$$f_{\phi,A}(\psi,\alpha)\,\mathrm{d}\psi\mathrm{d}\alpha \equiv \mathrm{P}[\{\psi \leq \phi < \psi + \mathrm{d}\psi\} \cap \{\alpha \leq A < \alpha + \mathrm{d}\alpha\}]. \tag{5.248}$$

The marginal PDF of the age, $f_A(\alpha)$, will depend on the fluid dynamics and flow geometry.[104] On the other hand, the *conditional* PDF of ϕ given A is provided by the age-based

[103] The age of the fluid particle located at a specific point in the reactor is now treated as a random variable.
[104] In the CRE literature, $f_A(\alpha)$ is known as the internal-age distribution function.

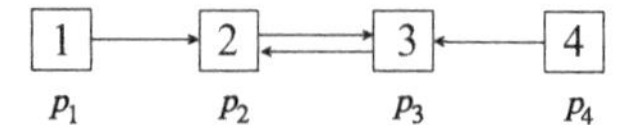

Figure 5.16. The four-environment model divides a *homogeneous* flow into four 'environments,' each with its own local concentration vector. The joint concentration PDF is thus represented by four delta functions. The area under each delta function, as well as its location in concentration space, depends on the rates of exchange between environments. Since the same model can be employed for the mixture fraction, the exchange rates can be partially determined by forcing the four-environment model to predict the correct decay rate for the mixture-fraction variance. The extension to *inhomogeneous* flows is discussed in Section 5.10.

micromixing model. For example, for the IEM model, $f_{\phi|A}(\psi|\alpha)$ is given by

$$f_{\phi|A}(\psi|\alpha) = \sum_{n=1}^{2} \gamma_n \delta(\psi - \phi^n(\alpha)), \tag{5.249}$$

where $\phi^n(\alpha)$ is found by solving an equation like (5.244). Nearly all other Lagrangian micromixing models generate a conditional PDF of this form. The models thus differ only by the method used to compute $\phi^n(\alpha)$. Reynolds-averaged statistics can be computed starting from (5.248). For example, the mean concentration found using (5.249) is given by

$$\begin{aligned} \langle\phi\rangle &= \int_0^\infty \int_0^\infty \psi f_{\phi,A}(\psi,\alpha)\,\mathrm{d}\psi\,\mathrm{d}\alpha \\ &= \int_0^\infty \left(\int_0^\infty \psi f_{\phi|A}(\psi|\alpha)\,\mathrm{d}\psi \right) f_A(\alpha)\,\mathrm{d}\alpha \\ &= \sum_{n=1}^{2} \gamma_n \int_0^\infty \phi^n(\alpha) f_A(\alpha)\,\mathrm{d}\alpha. \end{aligned} \tag{5.250}$$

Note that even if $f_A(\alpha)$ were known for a particular reactor, the functional form of $f_{\phi|A}(\psi|\alpha)$ will be highly dependent on the fluid dynamics and reactor geometry. Indeed, even if one were only interested in $\langle\phi\rangle$ at the reactor outlet where $f_A(\alpha) = E(\alpha)$, the conditional PDF will be difficult to model since it will be highly dependent on the entire flow structure inside the reactor.

Another Lagrangian-based description of micromixing is provided by multi-environment models. In these models, the well macromixed reactor is broken up into sub-grid-scale environments with uniform concentrations. A four-environment model is shown in Fig. 5.16. In this model, environment 1 contains unmixed fluid from feed stream 1; environments 2 and 3 contain partially mixed fluid; and environment 4 contains unmixed fluid from feed stream 2. The user must specify the relative volume of each environment (possibly as a function of age), and the exchange rates between environments. While some qualitative arguments have been put forward to fit these parameters based on fluid dynamics and/or flow visualization, one has little confidence in the general applicability of these rules when applied to scale up or scale down, or to complex reactor geometries.

As shown in Section 5.10, it is possible to reformulate multi-environment models in terms of a multi-peak presumed joint PDF. In this case, it is possible to specify the model parameters to be consistent with closures for scalar mixing based, for example, on the mixture-fraction PDF. However, when applying a Lagrangian multi-environment model, it is still necessary to know how the fluid particle interacts with the surrounding fluid. For example, if one environment represents the bulk fluid in the reactor, a balance equation is required to determine how its concentrations change with time.[105] In general, the terms appearing in this balance equation are unclosed and depend on the flow field and turbulence statistics in a non-trivial manner. Unlike with the fluid-mechanical approach, *ad hoc* closures must then be introduced that cannot be independently validated.

5.6.3 Lagrangian models for the micromixing rate

One of the principal difficulties faced when employing Lagrangian micromixing models is the determination of t_{m} based on properties of the turbulent flow fields. Researchers have thus attempted to use the universal nature of high-Reynolds-number isotropic turbulence to link t_{m} to the turbulence time scales. For example, in the E-model (Baldyga and Bourne 1989) the engulfment rate essentially controls the rate of micromixing and is defined by

$$\frac{1}{t_{\mathrm{m}}} = E \equiv 0.05776 \left(\frac{\varepsilon}{\nu} \right)^{1/2}. \tag{5.251}$$

The E-model is a two-environment model, with environment 1 representing the inlet feed, and environment 2 representing the 'reactor environment.' The E-model differs from the IEM model for a PFR in one important point: the volumes of the environments change with time due to 'engulfment.' In deriving the E-model, it is assumed that micromixing is controlled by the turbulent eddies located at the peak of the turbulence dissipation spectrum. Because the latter scales with the Kolmogorov length scale, the engulfment rate scales as the inverse of the Kolmogorov time scale.

When applying the E-model, it is important to understand the physical interpretation of micromixing employed in its derivation. Unlike the definition in terms of the decay rate of the scalar variance in homogeneous turbulence discussed in Chapter 3, the E-model is based on a physical picture wherein the initial integral scale of the scalar field (L_ϕ) is on the order of the Kolmogorov scale. This can be achieved experimentally by employing a very small feed tube and/or a very low injection velocity. However, such conditions are rarely met in plant-scale reactors, where much higher feed rates are required to maximize throughput. In order to distinguish this situation from the more usual case where the integral scale of the scalar field is on the order of the integral scale of the velocity field ($L_\phi \approx L_u$), the name mesomixing was coined to describe the cascade of scalar eddies from L_u to η. One should thus proceed cautiously when applying (5.251) in CFD calculations of plant-scale reactors.

One common difficulty when applying the E-model is the need to know the turbulent dissipation rate ε for the flow. Moreover, because ε will have an inhomogeneous distribution in most chemical reactors, the problem of finding ε *a priori* is non-trivial. In most

[105] In the IEM model, this information is provided by (5.246).

reported applications of the E-model, a spatially averaged value of ε has been estimated based on an idealized turbulent flow (e.g., a turbulent jet). For a CSTR, this leads to uncertainty in the exact value of ε, and usually a proportionality constant must be fitted to the experimental data. The E-model has also been employed to estimate the value of ε from an experimental value for the reaction yield. However, given the number of assumptions involved, it cannot be expected to be as reliable as a direct estimation based on velocity field measurements. Moreover, in low-Reynolds-number flows the basic assumptions needed to derive (5.251) are no longer valid.[106] For example, its validity at Reynolds numbers for which $E \leq \varepsilon/k$ is dubious.[107]

The importance of including mesomixing in the definition of t_m has been demonstrated using jet-mixer scale-up experiments by Baldyga *et al.* (1995). In these experiments, the micromixing rate was found to scale as

$$\frac{1}{t_\mathrm{m}} \sim \frac{\varepsilon^{1/3}}{L_\phi^{2/3}} \sim \left(\frac{L_u}{L_\phi}\right)^{2/3} \frac{\varepsilon}{k}, \tag{5.252}$$

i.e., t_m has inertial-range scaling as discussed in Chapter 3. On the other hand, for a laboratory-scale reactor, it is possible to decrease L_ϕ to the point where

$$\eta \leq L_\phi \ll L_u, \tag{5.253}$$

in which case Kolmogorov scaling is observed (Baldyga *et al.* 1995). The results of the jet-mixer experiments are thus entirely consistent with classical scalar-mixing theory, and reiterate the fact that no universal micromixing rate is available when $L_\phi \neq L_u$.

5.6.4 Mechanistic models

The lack of a universal model for the micromixing rate has led to mechanistic models that attempt to account for the initial length scale of the scalar field and its interaction with the turbulent flow (Ottino 1980; Ottino 1982; Baldyga and Bourne 1984a; Baldyga and Bourne 1984b; Baldyga and Bourne 1984c; Borghi 1986; Kerstein 1988; Bakker and van den Akker 1996). As discussed in Section 3.1, most mechanistic micromixing models follow a Lagrangian fluid particle that mixes with its environment by the following (serial) steps.

(1) Reduction in size down to the Kolmogorov scale with no change in concentration at a rate that depends on the initial scalar length scale relative to the Kolmogorov scale.
(2) Further reduction in size down to the Batchelor scale with negligible change in concentration at a rate that is proportional to $(\varepsilon/\nu)^{1/2}$.
(3) Molecular diffusion and reaction in Batchelor-scale lamella.

[106] For example, a well defined inertial range exists only for $R_\lambda \geq 240$. Thus the proportionality constant in (5.251) should depend on the Reynolds number in most laboratory-scale experiments.

[107] In other words, for $\mathrm{Re}_L \leq 300$ or $R_\lambda \leq 45$.

The PDF of an inert scalar is unchanged by the first two steps, but approaches the well mixed condition during step (3).[108] The overall rate of mixing will be determined by the slowest step in the process. In general, this will be step (1). Note also that, except in the linear-eddy model (Kerstein 1988), interactions between Lagrangian fluid particles are not accounted for in step (1). This limits the applicability of most mechanistic models to cases where a small volume of fluid is mixed into a much larger volume (i.e., where interactions between fluid particles will be minimal).

When applying a mechanistic model, nearly all of the computational effort resides in step (3).[109] In most mechanistic models, step (3) is modeled by one-dimensional reaction-diffusion equations of the form

$$\frac{\partial \phi_\alpha}{\partial t} = \Gamma_\alpha \frac{\partial^2 \phi_\alpha}{\partial x^2} + S_\alpha(\boldsymbol{\phi}) \quad \text{for } \alpha = 1, \ldots, N_s, \tag{5.254}$$

with *periodic* boundary conditions on an interval $0 \leq x \leq L(t)$:

$$\boldsymbol{\phi}(0, t) = \boldsymbol{\phi}(L(t), t) \quad \text{and} \quad \frac{\partial \boldsymbol{\phi}}{\partial x}(0, t) = \frac{\partial \boldsymbol{\phi}}{\partial x}(L(t), t). \tag{5.255}$$

Since the molecular diffusivities are used in (5.254), the interval length $L(t)$ and the initial conditions will control the rate of molecular diffusion and, subsequently, the rate of chemical reaction. In order to simulate scalar-gradient amplification due to Kolmogorov-scale mixing (i.e., for $1 \ll \text{Sc}$), the interval length is assumed to decrease at a constant rate:

$$L(t) = L(0) \exp\left[-c_B \left(\frac{\varepsilon}{\nu}\right)^{1/2} t\right]. \tag{5.256}$$

The initial length $L(0)$ is usually set equal to the Batchelor length scale λ_B, in which case the initial conditions are

$$\boldsymbol{\phi}(x, 0) = \begin{cases} \langle \boldsymbol{\phi} \rangle_1 & \text{for } 0 \leq x < p_1 \lambda_B \\ \langle \boldsymbol{\phi} \rangle_2 & \text{for } p_1 \lambda_B \leq x < \lambda_B, \end{cases} \tag{5.257}$$

where $0 < p_1 < 1$ is the volume fraction of the feed stream with inlet composition $\langle \boldsymbol{\phi} \rangle_1$. Alternatively, in order to account for interactions between sub-Kolmogorov-scale lamellae, it can be set proportional to the Kolmogorov length scale. In this case, $L(0)$ must be divided into approximately $N = \eta/\lambda_B = \text{Sc}^{1/2}$ lamellae, each of which has the initial condition given by (5.257). The length of each of these lamellae can either be fixed (i.e., equal to λ_B) or selected at random from a distribution function with mean λ_B (see Fox (1994)). However, it is important to note that if a fixed length is employed, then the periodicity of the fields will remain unchanged. Thus, the results will be exactly the same as would be found using a single lamella with $L(0) = \lambda_B$. On the other hand, when the initial lamella thickness is chosen at random (i.e., the initial fields are *aperiodic*), the time evolution

[108] As discussed in Section 4.3, the linear-eddy model solves a one-dimensional reaction-diffusion equation for all length scales. Inertial-range fluid-particle interactions are accounted for by a random rearrangement process.

[109] This leads to significant computational inefficiency since step (3) is not the rate-controlling step. Simplifications have thus been introduced to avoid this problem (Baldyga and Bourne 1989).

of the system will be markedly different. In particular, the lamella size distribution will evolve to a self-similar form, and, more importantly, the micromixing rate for the aperiodic system will scale quite differently than that found for the periodic system (Fox 1994)!

As noted above, the numerical solution of (5.254) is relatively expensive, and in many cases unnecessary. Indeed, at high Reynolds numbers, the time required for $\phi(x, t)$ to become independent of x will be much shorter than the characteristic time for step (1). The reaction-diffusion equation can then be replaced by a much simpler 'mean-field' micromixing model (Baldyga and Bourne 1989) whose computational requirements are orders of magnitude smaller. The only parameter that must be supplied to the mean-field model is a length-scale-dependent micromixing rate. In the context of Reynolds-averaged transport equations, we have seen in Chapter 3 that the micromixing rate is controlled by

$$\omega_\phi = \frac{\varepsilon_\phi}{\langle \phi'^2 \rangle}, \tag{5.258}$$

and thus a length-scale-dependent micromixing rate will be equivalent to a spectral model for the scalar dissipation rate ε_ϕ (see Section 4.5). In particular, since they take into account all of the important processes in turbulent scalar mixing,[110] models for the scalar dissipation rate should enjoy wider applicability than mechanistic models that ignore interactions between Lagrangian fluid particles.

5.6.5 Extension to inhomogeneous flows

Some authors have attempted to extend Lagrangian micromixing models to inhomogeneous flows by coupling them to CFD codes. However, it is not clear what advantages are to be gained by this approach over existing PDF-based methods for turbulent reacting flows. For example, Lagrangian micromixing models follow individual fluid particles as they mix with a homogeneous environment. Because the fluid particles are assumed to be uncoupled, it is possible to follow a series of uncoupled particles instead of a large ensemble of interacting particles. This single-particle representation may be realistic for describing the addition of a very small amount of one reactant into a large volume of a second reactant. However, in cases where the reactant feed rate is large, the 'reactor environment' will be far from homogeneous, and one must keep track of large-scale inhomogeneity in order to model the reactor correctly. In general, large-scale inhomogeneity can be measured by variations in the mixture-fraction mean and/or the scalar variance. In a PDF-based approach, one would have a mixture-fraction PDF that is different at every point in the reactor. In the transported PDF approach discussed in Chapter 6, Lagrangian fluid particles are also employed. However, a large ensemble of *coupled* particles that occupy the entire reactor volume must be tracked in a Lagrangian frame.

In order to overcome the difficulties of following Lagrangian particles in a realistic manner, most Lagrangian micromixing studies have had to eliminate back-mixing and

[110] In other words, mean convection, turbulent diffusion, production, and dissipation.

have assumed that each fluid particle follows an 'average' streamline from the entrance to the exit of the reactor that can be parameterized by the 'age' of the particle. This picture is far from reality in a turbulent flow where turbulent diffusivity and recirculation zones bring fluid particles of many different ages into contact. A potentially more serious difficulty with employing Lagrangian micromixing models in place of PDF-based models is the problem of model validation. Because Lagrangian micromixing models contain many unknown parameters that cannot be independently verified (e.g., using an idealized turbulent-reacting-flow simulation based on DNS), one cannot be confident that the models will work when applied to the more complex flow geometries characteristic of industrial-scale chemical reactors.

5.7 Laminar diffusion flamelets

For fast equilibrium chemistry (Section 5.4), an equilibrium assumption allowed us to write the concentration of all chemical species in terms of the mixture-fraction vector $\mathbf{c}(\mathbf{x}, t) = \mathbf{c}_{\mathrm{eq}}(\boldsymbol{\xi}(\mathbf{x}, t))$. For a turbulent flow, it is important to note that the *local* micromixing rate (i.e., the instantaneous scalar dissipation rate) is a random variable. Thus, while the chemistry may be fast relative to the *mean* micromixing rate, at some points in a turbulent flow the *instantaneous* micromixing rate may be fast compared with the chemistry. This is made all the more important by the fact that fast reactions often take place in thin reaction-diffusion zones whose size may be smaller than the Kolmogorov scale. Hence, the local strain rate (micromixing rate) seen by the reaction surface may be as high as the local Kolmogorov-scale strain rate.

In combustion systems, locally high strain rates can lead to micromixing-induced extinction of the flame – an important source of pollutants in turbulent combustion. The non-equilibrium effects caused by fluctuations in the micromixing rate can be modeled using the concept of *laminar diffusion flamelets* (Peters 1984; Peters 2000). In this model, the instantaneous scalar dissipation rate of the mixture fraction appears as a random variable in a reaction-diffusion equation, thereby directly coupling the local reaction rate to the local micromixing rate. Although the method has been extended to complex chemistry, it is most easily understood in terms of the one-step reaction in a turbulent flow that can be described by a one-component mixture fraction (see Section 5.5). Thus, in the following, we will consider only the one-step reaction case. A complete treatment of the derivation and principal applications of laminar flamelet models can be found in Peters (2000).

5.7.1 Definition of a flamelet

The basic physical model is that of fast reactions occurring in thin quasi-one-dimensional reaction sheets with thickness less than the size of the smallest eddies in the flow. As discussed in Section 5.5, this picture will be valid for *non-isothermal* one-step reactions when the reaction rate constant $k^{\mathrm{f}}(T)$ is near zero at the ambient temperature, but very

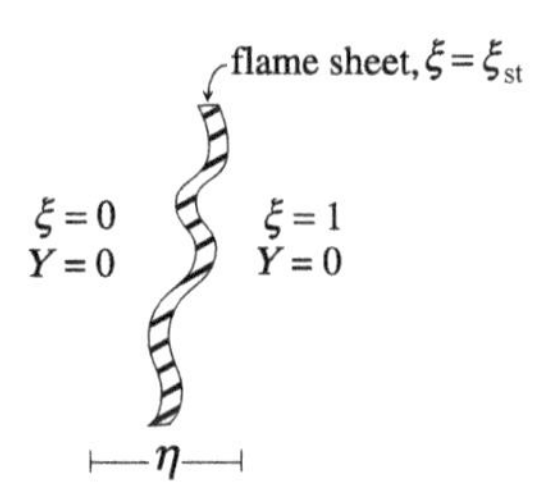

Figure 5.17. A laminar diffusion flamelet occurs between two regions of unmixed fluid. On one side, the mixture fraction is unity, and on the other side it is null. If the reaction rate is localized near the stoichiometric value of the mixture fraction ξ_{st}, then the reaction will be confined to a thin reaction zone that is small compared with the Kolmogorov length scale.

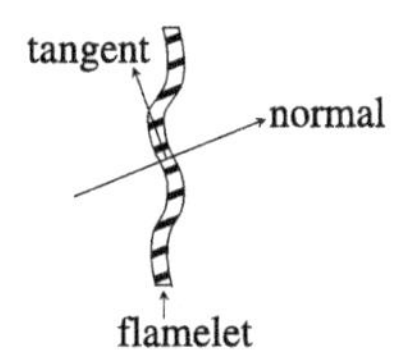

Figure 5.18. The diffusion flamelet can be approximated by a one-dimensional transport equation that describes the change in the direction normal to the stoichiometric surface. The rate of change in the tangent direction is assumed to be negligible since the flamelet thickness is small compared with the Kolmogorov length scale. The flamelet approximation is valid when the reaction separates regions of unmixed fluid. Thus, the boundary conditions on each side are known, and can be uniquely expressed in terms of ξ.

large at the flame temperature. For this case, the reaction rate is significant only near the stoichiometric surface defined by the mixture fraction: $\xi(\mathbf{x}, t) = \xi_{st}$. On the other hand, the flamelet model will not apply for the *isothermal*[111] one-step reaction when $\text{Sc} \approx 1$ since the reaction zone will be large compared with the smallest scales in the flow. However, for $\text{Sc} \gg 1$, the flamelet model should also be applicable to the isothermal case. The assumed structure of the laminar diffusion flamelet is shown in Fig. 5.17.

Given the thinness of a diffusion flamelet, it is possible to neglect as a first approximation curvature effects, and to establish a local coordinate system centered at the reaction interface. By definition, x_1 is chosen to be normal to the reaction surface. Furthermore, because the reaction zone is thin compared with the Kolmogorov scale, gradients with respect to x_2 and x_3 will be much smaller than gradients in the x_1 direction (i.e., the curvature is small).[112] Thus, as shown in Fig. 5.18, the scalar fields will be locally one-dimensional.

For the one-step reaction

$$\text{A} + \text{B} \rightarrow \text{P}, \tag{5.259}$$

[111] Or, more generally, to one-step reactions with a relatively weak dependency on T. This may occur in combusting flows where the adiabatic temperature rise is small; for example, in a 'rich' flame near extinction.

[112] Flamelet 'wrinkling' will be caused by vorticity, which is negligible below the Kolmogorov scale.

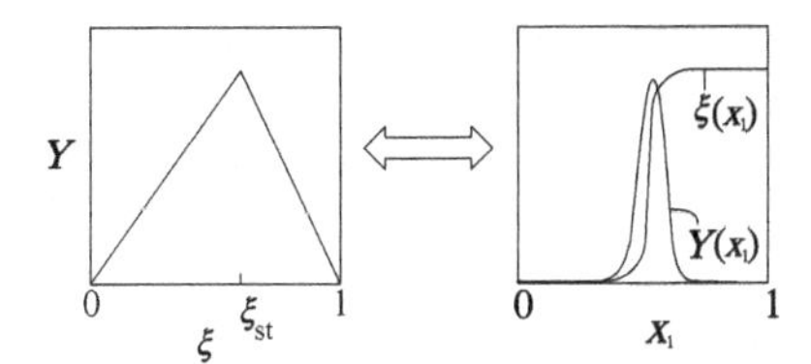

Figure 5.19. The flamelet can be parameterized in terms of the mixture fraction, which changes monotonically from 0 to 1 across the flamelet.

the mixture fraction is defined by

$$\xi = \frac{c_B - c_A + A_0}{A_0 + B_0}, \tag{5.260}$$

where A_0 and B_0 are the inlet concentrations. The transport equation for the mixture fraction is given by

$$\frac{\partial \xi}{\partial t} + U_i \frac{\partial \xi}{\partial x_i} = \Gamma \frac{\partial^2 \xi}{\partial x_i \partial x_i}. \tag{5.261}$$

By definition, on the 'fuel' side of the reaction zone $\xi = 0$, while on the 'oxidizer' side $\xi = 1$. As illustrated in Fig. 5.19, when moving across the reaction zone from 'fuel' to 'oxidizer' the mixture fraction increases monotonically.

The non-isothermal one-step reaction can then be described by a reaction-progress variable Y that is null when $\xi = 0$ or $\xi = 1$. As shown in Section 5.5, the transport equation for the reaction-progress variable is given by

$$\frac{\partial Y}{\partial t} + U_i \frac{\partial Y}{\partial x_i} = \Gamma \frac{\partial^2 Y}{\partial x_i \partial x_i} + S_Y(Y, \xi), \tag{5.262}$$

where $S_Y(Y, \xi)$ is the transformed chemical source term. By definition, the flamelet model will be applicable when

$$S_Y(Y, \xi) \ll 1 \quad \text{for } \xi \neq \xi_{st}, \quad \text{but} \quad S_Y(Y_{max}, \xi_{st}) \gg 1. \tag{5.263}$$

Thus, (5.262) will differ significantly from (5.261) only near the stoichiometric surface where $Y = Y_{max}$ and $\xi = \xi_{st} = A_0/(A_0 + B_0)$.

The governing equation for the flamelet model is then found starting from a change of variables:[113]

$$t \rightarrow \tau \tag{5.264}$$

and

$$x_1 \rightarrow \xi, \tag{5.265}$$

which re-parameterizes the direction normal to the stoichiometric surface in terms of the mixture fraction. Note that by replacing the spatial variable x_1 with the mixture fraction,

[113] As shown in Peters (2000), a rigorous derivation of the flamelet equations can be carried out using a two-scale asymptotic analysis.

the flamelet model becomes independent of the spatial location in the flow.[114] The space and time derivatives can then be expressed in terms of the new variables as follows:

$$\frac{\partial}{\partial t} = \frac{\partial}{\partial \tau} + \frac{\partial \xi}{\partial t} \frac{\partial}{\partial \xi} \tag{5.266}$$

and

$$\frac{\partial}{\partial x_1} = \frac{\partial \xi}{\partial x_1} \frac{\partial}{\partial \xi}, \tag{5.267}$$

where the x_2 and x_3 spatial derivatives are neglected.

By conditioning on the mixture fraction (i.e., on the event where $\xi(\mathbf{x}, t) = \zeta$), the reaction-progress-variable transport equation can be rewritten in terms of $Y(\zeta, \tau)$:[115]

$$\frac{\partial Y}{\partial \tau} = \frac{1}{2} \chi(\zeta, \tau) \frac{\partial^2 Y}{\partial \zeta^2} + S_Y(Y, \zeta), \tag{5.268}$$

where the instantaneous *conditional* scalar dissipation rate of the mixture fraction is defined by

$$\chi(\zeta, \tau) = 2 \left\langle \Gamma \frac{\partial \xi}{\partial x_i} \frac{\partial \xi}{\partial x_i} \middle| \xi(\mathbf{x}, t) = \zeta \right\rangle \tag{5.269}$$

and quantifies the local micromixing rate. The terms on the right-hand side of (5.268) represent instantaneous micromixing and chemical reaction. The boundary conditions for the reaction-progress variable in mixture-fraction space are $Y(0, \tau) = Y(1, \tau) = 0$.

5.7.2 Stationary laminar flamelet model

Generally, near the stoichiometric surface, both terms in (5.268) are large in magnitude and opposite in sign (Peters 2000). A quasi-stationary state is thus quickly established wherein the accumulation term on the left-hand side is negligible. The stationary laminar flamelet (SLF) model is found by simply neglecting the accumulation term, and ignoring the time dependency of $\chi(\zeta, \tau)$:

$$0 = \frac{1}{2} \chi(\zeta) \frac{\partial^2 Y}{\partial \zeta^2} + S_Y(Y, \zeta). \tag{5.270}$$

Thus, given an appropriate form for $\chi(\zeta)$, the SLF model can be solved to find $Y(\zeta)$ with boundary conditions $Y(0) = Y(1) = 0$.[116] Note that if $S_Y(0, \zeta) = 0$, the SLF model admits the solution $Y(\zeta) = 0$, which corresponds to a non-reacting system (i.e., 'flame extinction'). Likewise, when $\chi(\zeta) \gg S_Y(Y, \zeta)$, the solution will be dominated by the diffusion term so that $Y(\zeta) \approx 0$. In the opposite limit, the solution will be dominated by the

[114] This procedure is similar to the one discussed in Section 5.4 for equilibrium chemistry where the spatial information is contained only in the mixture fraction.

[115] In this equation, Y should be replaced by the conditional expected value of Y given $\xi = \zeta$. However, based on the structure of the flamelet, it can be assumed that the conditional PDF is a delta function centered at $Y(\zeta, \tau)$.

[116] The SLF model generates a two-point boundary-value problem for which standard numerical techniques exist.

reaction term so that $Y(\zeta)$ will approach the infinite-rate chemistry limit ((5.179), p. 183). The SLF model can thus be employed to study flame extinction as a function of $\chi(\xi_{st})$.

The form for $\chi(\zeta)$ is usually taken from the one-dimensional laminar counterflow diffusion layer model (Peters 2000):

$$\chi(\zeta) = \chi^* \exp(-2(\mathrm{erf}^{-1}(2\zeta - 1))^2), \tag{5.271}$$

where χ^* is the (random) scalar dissipation rate at $\xi = 0.5$. In most applications, χ^* is assumed to have a log-normal distribution. The SLF solution can thus be parameterized in terms of two random variables: $Y(\xi, \chi^*)$. If the joint PDF of ξ and χ^* is known, then any statistics involving Y can be readily computed. For example, the mean reaction-progress variable is found from

$$\langle Y \rangle(\mathbf{x}, t) = \int_0^1 \int_0^\infty Y(\zeta, \eta) f_{\xi,\chi^*}(\zeta, \eta; \mathbf{x}, t) \, \mathrm{d}\eta \, \mathrm{d}\zeta. \tag{5.272}$$

Application of the SLF model thus reduces to predicting the joint PDF of the mixture fraction and the scalar dissipation rate. As noted above, in combusting flows flame extinction will depend on the value of χ^*. Thus, unlike the equilibrium-chemistry method (Section 5.4), the SLF model can account for flame extinction due to local fluctuations in the scalar dissipation rate.

5.7.3 Joint mixture fraction, dissipation rate PDF

In almost all reported applications of the flamelet model, ξ and χ^* have been assumed to be independent so that the joint PDF can be written in terms of the marginal PDFs:

$$f_{\xi,\chi^*}(\zeta, \eta; \mathbf{x}, t) = f_\xi(\zeta; \mathbf{x}, t) f_{\chi^*}(\eta; \mathbf{x}, t), \tag{5.273}$$

where $f_\xi(\zeta; \mathbf{x}, t)$ and $f_{\chi^*}(\eta; \mathbf{x}, t)$ are the one-point, one-time marginal PDFs of mixture fraction and scalar dissipation rate, respectively. The mixture-fraction PDF can be approximated using presumed PDF methods.

The scalar-dissipation PDF is usually approximated by a log-normal form:

$$f_{\chi^*}(\eta; \mathbf{x}, t) = \frac{1}{\sigma(2\pi)^{1/2}} \frac{1}{\eta} \exp\left[-\frac{1}{2\sigma^2} (\ln(\eta) - \mu)^2 \right], \tag{5.274}$$

where the mean and variance of the scalar dissipation rate are related to the parameters $\mu(\mathbf{x}, t)$ and $\sigma(\mathbf{x}, t)$ by

$$\langle \chi \rangle = \exp(\mu + 0.5\sigma^2), \tag{5.275}$$

and

$$\langle \chi'^2 \rangle = \langle \chi \rangle^2 [\exp(\sigma^2) - 1]. \tag{5.276}$$

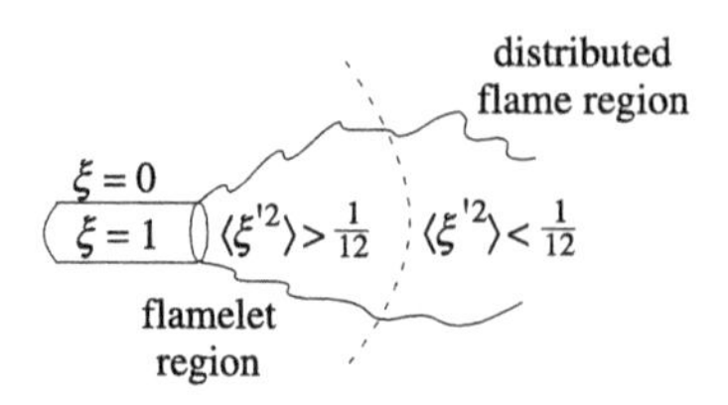

Figure 5.20. The flamelet model requires the existence of unmixed regions in the flow. This will occur only when the mixture-fraction PDF is non-zero at $\xi = 0$ and $\xi = 1$. Normally, this condition is only satisfied near inlet zones where micromixing is poor. Beyond these zones, the flamelets begin to interact through the boundary conditions, and the assumptions on which the flamelet model is based no longer apply.

As discussed in Chapter 3, classical scalar-mixing theory yields a mean scalar dissipation rate of

$$\langle \chi \rangle = 2 \frac{\varepsilon}{k} \langle \xi'^2 \rangle, \tag{5.277}$$

and experimental data suggest that $2 \leq \sigma \leq 4$. The two parameters appearing in (5.274) can thus be directly linked to the local turbulence statistics and mixture-fraction variance.

5.7.4 Extension to inhomogeneous flows

Because χ^* appears as a parameter in the flamelet model, in numerical implementations a *flamelet library* (Pitsch and Peters 1998; Peters 2000) is constructed that stores $Y(\xi, \chi^*)$ for $0 \leq \xi \leq 1$ in a lookup table parameterized by $\langle \xi \rangle$, $\langle \xi'^2 \rangle$, and χ^*. Based on the definition of a flamelet, at any point in the flow the reaction zone is assumed to be *isolated* so that no interaction occurs between individual flamelets. In order for this to be true, the probabilities of finding $\xi = 0$ and $\xi = 1$ must both be non-zero.

For example, for a beta PDF with $\langle \xi \rangle = 1/2$ ($a = b$ in (5.142)), zones of unmixed fluid will exist only if

$$a = \frac{1}{2} \left[\frac{1}{4\langle \xi'^2 \rangle} - 1 \right] \leq 1. \tag{5.278}$$

Thus, the flamelet model is applicable only in regions of the flow where fluctuations are large, i.e, the variance must satisfy

$$\frac{1}{12} \leq \langle \xi'^2 \rangle. \tag{5.279}$$

As shown in Fig. 5.20, such regions normally occur only near the inlet zones where micromixing is poor. Further downstream, interaction between flamelets will become significant, and the assumptions on which the flamelet model is based will no longer apply.[117] Reactors with recirculation zones are also problematic for flamelet models. For these reactors, partially reacted fluid is brought back to mix with the feed streams so that the simple non-premixed flow model no longer applies.

[117] Although the boundary conditions are different, the same comments apply to the lamellar mixing models proposed by Ottino (1981). Indeed, as soon as the boundary values are no longer equal to the concentrations found in unmixed regions, lamellae interactions should not be ignored.

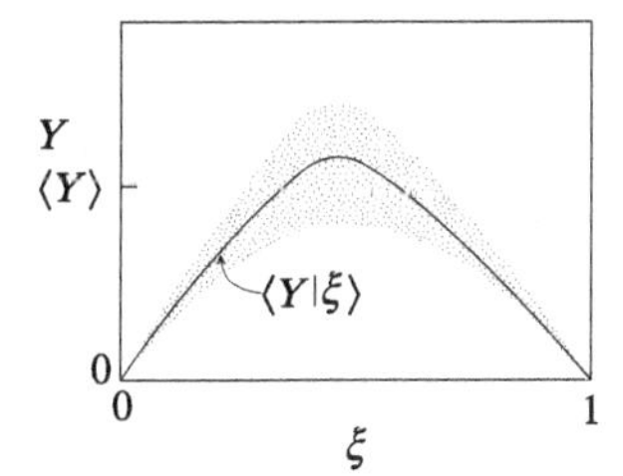

Figure 5.21. Scatter plot of concentration in a turbulent reacting flow conditioned on the value of the mixture fraction. Although large fluctuations in the unconditional concentration are present, the conditional fluctuations are considerably smaller. In the limit where the conditional fluctuations are negligible, the chemical source term can be closed using the conditional scalar means.

Many liquid-phase chemical reactions that are sensitive to micromixing will occur in reactors that have significant non-flamelet regions. The flamelet model is, however, widely applicable to combustion systems (Bray and Peters 1994; Peters 2000), in general, and premixed combustion systems, in particular. In the latter, chemical reactions take place between thin zones separating fuel and burnt gases. For premixed flames, a single reaction-progress variable is used in place of the mixture fraction, and has a strongly bi-modal PDF due to chemical reactions. For further details on applying flamelet models to combusting systems, the reader is referred to the excellent monograph on turbulent combustion by Peters (2000).

5.8 Conditional-moment closures

For the equilibrium-chemistry limit, we have seen in Section 5.4 that the reaction-progress vector can be re-parameterized in terms of the mixture fraction, i.e., $\varphi_{\text{rp}}(\mathbf{x}, t) = \varphi_{\text{eq}}(\boldsymbol{\xi}(\mathbf{x}, t))$. It has been observed experimentally and from DNS that even for many *finite-rate* reactions the scatter plot of $\phi_\alpha(\mathbf{x}, t)$ versus $\xi(\mathbf{x}, t) = \zeta$ for all α often exhibits relatively small fluctuations around the mean *conditioned* on a given value of the mixture fraction, i.e., around $\langle \phi(\mathbf{x}, t) | \xi(\mathbf{x}, t) = \zeta \rangle$. An example of such a scatter plot is shown in Fig. 5.21. In the limit where fluctuations about the conditional mean are negligible, the Reynolds-averaged chemical source term can be written as $\langle \mathbf{S}(\phi) | \boldsymbol{\zeta} \rangle = \mathbf{S}(\langle \phi | \boldsymbol{\zeta} \rangle)$. Thus, a model for the conditional means, combined with a presumed mixture-fraction PDF, would suffice to close the Reynolds-averaged turbulent-reacting-flow equations. In this section, we will thus look at two methods for modeling the conditional means. First, however, we will review the general formulation of conditional random variables.

5.8.1 General formulation: conditional moments

The scalar mean conditioned on the mixture-fraction vector can be denoted by

$$\mathbf{Q}(\boldsymbol{\zeta}; \mathbf{x}, t) \equiv \langle \phi(\mathbf{x}, t) | \boldsymbol{\zeta} \rangle, \tag{5.280}$$

and the conditional fluctuations may be denoted by

$$\phi^*(\mathbf{x}, t) = \phi(\mathbf{x}, t) - \mathbf{Q}(\zeta; \mathbf{x}, t), \tag{5.281}$$

where (see Section 3.2) the conditional means are defined in terms of the conditional composition PDF:

$$\langle \phi(\mathbf{x}, t) | \zeta \rangle \equiv \int_{-\infty}^{+\infty} \cdots \int \psi f_{\phi|\xi}(\psi|\zeta; \mathbf{x}, t) \, \mathrm{d}\psi. \tag{5.282}$$

By definition, the unconditional scalar means can be found from $\mathbf{Q}(\zeta; \mathbf{x}, t)$ and the mixture-fraction-vector PDF:

$$\begin{aligned} \langle \phi \rangle(\mathbf{x}, t) &\equiv \int_0^1 \cdots \int \int_{-\infty}^{+\infty} \cdots \int \psi f_{\phi,\xi}(\psi, \zeta; \mathbf{x}, t) \, \mathrm{d}\psi \, \mathrm{d}\zeta \\ &= \int_0^1 \cdots \int \left(\int_{-\infty}^{+\infty} \cdots \int \psi f_{\phi|\xi}(\psi|\zeta; \mathbf{x}, t) \, \mathrm{d}\psi \right) f_\xi(\zeta; \mathbf{x}, t) \, \mathrm{d}\zeta \\ &= \int_0^1 \cdots \int \mathbf{Q}(\zeta; \mathbf{x}, t) f_\xi(\zeta; \mathbf{x}, t) \, \mathrm{d}\zeta. \end{aligned} \tag{5.283}$$

Thus, knowledge of $\mathbf{Q}(\zeta; \mathbf{x}, t)$ and $f_\xi(\zeta; \mathbf{x}, t)$ suffice for computing the scalar means.

The higher-order conditional scalar moments are defined in terms of the conditional fluctuations. For example, the conditional scalar covariances are defined by

$$\begin{aligned} \langle \phi_\alpha^* \phi_\beta^* | \zeta \rangle(\mathbf{x}, t) &\equiv \int_{-\infty}^{+\infty} \cdots \int \psi_\alpha^* \psi_\beta^* f_{\phi|\xi}(\psi|\zeta; \mathbf{x}, t) \, \mathrm{d}\psi \\ &= \int_{-\infty}^{+\infty} \cdots \int \psi_\alpha \psi_\beta f_{\phi|\xi}(\psi|\zeta; \mathbf{x}, t) \, \mathrm{d}\psi \\ &\quad - Q_\alpha(\zeta; \mathbf{x}, t) Q_\beta(\zeta; \mathbf{x}, t) \\ &= \langle \phi_\alpha \phi_\beta | \zeta \rangle(\mathbf{x}, t) - Q_\alpha(\zeta; \mathbf{x}, t) Q_\beta(\zeta; \mathbf{x}, t), \end{aligned} \tag{5.284}$$

and are a measure of the degree of scatter about the conditional scalar means. If the conditional fluctuations are negligible, then

$$|\langle \phi_\alpha^* \phi_\beta^* | \zeta \rangle| \ll 1 \quad \text{for all } \alpha \text{ and } \beta, \tag{5.285}$$

and the conditional second-order moments can be approximated by

$$\langle \phi_\alpha \phi_\beta | \zeta \rangle(\mathbf{x}, t) \approx Q_\alpha(\zeta; \mathbf{x}, t) Q_\beta(\zeta; \mathbf{x}, t). \tag{5.286}$$

Extending (5.286) to conditional scalar moments of arbitrary order, the conditional chemical source term can be approximated by

$$\langle \mathbf{S}(\phi(\mathbf{x}, t))|\zeta\rangle \approx \mathbf{S}(\mathbf{Q}(\zeta; \mathbf{x}, t)). \tag{5.287}$$

Given a closure for $\mathbf{Q}(\zeta; \mathbf{x}, t)$ and the mixture-fraction PDF, the Reynolds-averaged chemical source term is readily computed:

$$\langle \mathbf{S}(\phi)\rangle = \int_0^1 \cdots \int \mathbf{S}(\mathbf{Q}(\zeta; \mathbf{x}, t)) f_{\xi}(\zeta; \mathbf{x}, t)\, \mathrm{d}\zeta. \tag{5.288}$$

This observation suggests that a moment-closure approach based on the conditional scalar moments may be more successful than one based on unconditional moments. Because adequate models are available for the mixture-fraction PDF, conditional-moment closures focus on the development of methods for finding a general expression for $\mathbf{Q}(\zeta; \mathbf{x}, t)$.

Finally, note that all of the expressions developed above for the composition vector also apply to the reaction-progress vector φ_{rp} or the reaction-progress variables $\mathbf{Y}$. Thus, in the following, we will develop closures using the form that is most appropriate for the chemistry under consideration.

5.8.2 Closures based on presumed conditional moments

For simple chemistry, a form for $\mathbf{Q}(\zeta; \mathbf{x}, t)$ can sometimes be found based on linear interpolation between two limiting cases. For example, for the one-step reaction discussed in Section 5.5, we have seen that the chemical source term can be rewritten in terms of a reaction-progress variable Y and the mixture fraction ξ. By taking the conditional expectation of (5.176) and applying (5.287), the chemical source term for the conditional reaction-progress variable can be found to be

$$\langle S_Y(Y, \xi)|\zeta\rangle \approx \xi_{\mathrm{st}} k^* \left(\frac{\zeta}{\xi_{\mathrm{st}}} - \langle Y|\zeta\rangle\right)\left(\frac{1-\zeta}{1-\xi_{\mathrm{st}}} - \langle Y|\zeta\rangle\right). \tag{5.289}$$

In Section 5.5, we have seen that in the limit where $k^* = 0$, $Y = 0$, and thus $\langle Y|\zeta\rangle = 0$. In the opposite limit where $k^* = \infty$, $Y = Y_\infty(\xi)$, and thus $\langle Y|\zeta\rangle = Y_\infty(\zeta)$, where the function $Y_\infty(\xi)$ is given by (5.179) on p. 183.

The actual value of $\langle Y|\zeta\rangle$ for finite k^* must lie somewhere between the two limits. Mathematically, this constraint can be expressed as

$$\langle Y|\zeta\rangle = \gamma_Y Y_\infty(\zeta), \tag{5.290}$$

where $0 \leq \gamma_Y \leq 1$ is a linear-interpolation parameter. *A priori*, the value of the interpolation parameter is unknown. However, it can be determined in terms of computable quantities by forcing (5.290) to yield the correct unconditional mean:

$$\begin{aligned}\langle Y\rangle &= \int_0^1 \langle Y|\zeta\rangle f_\xi(\zeta;\mathbf{x},t)\,\mathrm{d}\zeta \\ &= \gamma_Y\langle Y_\infty(\xi)\rangle,\end{aligned} \tag{5.291}$$

i.e., $\gamma_Y \equiv \langle Y\rangle/\langle Y_\infty(\xi)\rangle$. Thus, the right-hand side of (5.289) can be written in terms of ζ and $\langle Y\rangle$:

$$\langle S_Y(Y,\xi)|\zeta\rangle \approx \xi_{\mathrm{st}}k^* h(\zeta,\langle Y\rangle), \tag{5.292}$$

where

$$h(\zeta,\langle Y\rangle) = \left(\frac{\zeta}{\xi_{\mathrm{st}}} - \frac{\langle Y\rangle}{\langle Y_\infty(\xi)\rangle}Y_\infty(\zeta)\right)\left(\frac{1-\zeta}{1-\xi_{\mathrm{st}}} - \frac{\langle Y\rangle}{\langle Y_\infty(\xi)\rangle}Y_\infty(\zeta)\right). \tag{5.293}$$

Given (5.292) and the mixture-fraction PDF, the chemical source term in the Reynolds-averaged transport equation for $\langle Y\rangle$ is closed:

$$\langle S_Y(Y,\xi)\rangle = \xi_{\mathrm{st}}k^*\int_0^1 h(\zeta,\langle Y\rangle)f_\xi(\zeta;\mathbf{x},t)\,\mathrm{d}\zeta. \tag{5.294}$$

Thus, if a presumed beta PDF is employed to represent $f_\xi(\zeta;\mathbf{x},t)$, the *finite-rate* one-step reaction can be modeled in terms of transport equations for only three scalar moments: $\langle\xi\rangle$, $\langle\xi'^2\rangle$, and $\langle Y\rangle$. The simple closure based on linear interpolation thus offers a highly efficient computational model as compared with other closures.

Looking back over the steps required to derive (5.290), it is immediately apparent that the same method can be applied to treat any reaction scheme for which *only one* reaction rate function is finite. The method has thus been extended by Baldyga (1994) to treat competitive-consecutive (see (5.181)) and parallel (see (5.211)) reactions in the limiting case where $k_1 \to \infty$.[118] For both reaction systems, the conditional moments are formulated in terms of $Y_{2\infty}$ and can be written as

$$\langle Y_{2\infty}|\zeta\rangle = \frac{\langle Y_{2\infty}\rangle}{\langle Y_{2\infty}^{\max}(\xi)\rangle}Y_{2\infty}^{\max}(\zeta), \tag{5.295}$$

where $Y_{2\infty}^{\max}(\zeta)$ is given by (5.241), p. 192.[119] Thus, the chemical source term in the Reynolds-averaged transport equation for $\langle Y_{2\infty}\rangle$ is closed:[120]

$$\langle S_{Y_{2\infty}}(Y_{2\infty},\xi)\rangle = \int_0^1 S_{Y_{2\infty}}(\langle Y_{2\infty}|\zeta\rangle,\zeta)f_\xi(\zeta;\mathbf{x},t)\,\mathrm{d}\zeta. \tag{5.296}$$

[118] Baldyga (1994) does not formulate the method in terms of conditional moments or reaction-progress variables. The resulting expressions thus appear to be more complex. However, they are identical to the expressions presented here when reformulated in terms of reaction-progress variables. In addition to simplicity, formulating the problem in terms of reaction-progress variables has the advantages that the two limiting cases are easily found and the linear-interpolation formula is trivial.

[119] Note that the definition of $\xi_{\max}$ is different for the competitive-consecutive and parallel reactions.

[120] The functional forms for $S_{Y_{2\infty}}$ are given by the right-hand sides of (5.200) (competitive-consecutive) and (5.234) (parallel).

As with (5.294), the right-hand side of this expression will depend only on $\langle\xi\rangle$, $\langle\xi'^2\rangle$, and $\langle Y_{2\infty}\rangle$.

An *ad hoc* extension of the method presented above can be formulated for complex chemistry written in terms of $\boldsymbol{\varphi}_{\rm rp}$ and $\boldsymbol{\xi}$. In the absence of chemical reactions, $\boldsymbol{\varphi}_{\rm rp} = \mathbf{0}$. Thus, if a second 'limiting' case can be identified, interpolation parameters can be defined to be consistent with the unconditional means. In combusting flows, the obvious second limiting case is the equilibrium-chemistry limit where $\boldsymbol{\varphi}_{\rm rp} = \boldsymbol{\varphi}_{\rm eq}(\boldsymbol{\xi})$ (see Section 5.4). The components of the conditional reacting-progress vector can then be approximated by (no summation is implied on α)

$$\langle\varphi_{{\rm rp}\alpha}|\boldsymbol{\zeta}\rangle = \frac{\langle\varphi_{{\rm rp}\alpha}\rangle}{\langle\varphi_{{\rm eq}\alpha}(\boldsymbol{\xi})\rangle}\varphi_{{\rm eq}\alpha}(\boldsymbol{\zeta}), \tag{5.297}$$

and the conditional chemical source term can be approximated by

$$\langle\mathbf{S}_{\rm rp}(\boldsymbol{\varphi}_{\rm rp};\boldsymbol{\xi})|\boldsymbol{\zeta}\rangle \approx \mathbf{S}_{\rm rp}(\langle\boldsymbol{\varphi}_{\rm rp}|\boldsymbol{\zeta}\rangle;\boldsymbol{\zeta}). \tag{5.298}$$

Given $\boldsymbol{\varphi}_{\rm eq}(\boldsymbol{\zeta})$ and a presumed mixture-fraction vector PDF, the Reynolds-averaged chemical source term then depends only on $\langle\boldsymbol{\varphi}_{\rm rp}\rangle$, $\langle\boldsymbol{\xi}\rangle$, and $\langle\xi'_\alpha\xi'_\beta\rangle$.

In (5.297), the interpolation parameter is defined separately for each component. Note, however, that unlike the earlier examples, there is no guarantee that the interpolation parameters will be bounded between zero and one. For example, the equilibrium concentration of intermediate species may be negligible despite the fact that these species can be abundant in flows dominated by finite-rate chemistry. Thus, although (5.297) provides a convenient closure for the chemical source term, it is by no means guaranteed to produce accurate predictions! A more reliable method for determining the conditional moments is the formulation of a transport equation that depends explicitly on turbulent transport and chemical reactions. We will look at this method for both homogeneous and inhomogeneous flows below.

5.8.3 Conditional scalar mean: homogeneous flow

If the conditional fluctuations $\boldsymbol{\varphi}^*_{\rm rp}$ are neglected, the homogeneous conditional scalar mean $\mathbf{Q}(\boldsymbol{\zeta};t) = \langle\boldsymbol{\varphi}_{\rm rp}|\boldsymbol{\zeta}\rangle$ is governed by (Klimenko 1990; Bilger 1993) (summation is implied with respect to j and k)

$$\frac{\partial\mathbf{Q}}{\partial t} = \left\langle\Gamma\frac{\partial\xi_j}{\partial x_i}\frac{\partial\xi_k}{\partial x_i}\right|\boldsymbol{\zeta}\left.\right\rangle\frac{\partial^2\mathbf{Q}}{\partial\zeta_j\partial\zeta_k} + \mathbf{S}_{\rm rp}(\mathbf{Q},\boldsymbol{\zeta}), \tag{5.299}$$

where we have employed (5.287) to close the chemical source term. The first term on the right-hand side of (5.299) contains the conditional joint scalar dissipation rate of the mixture-fraction vector. As discussed below, this term must be chosen to be consistent with the form of the mixture-fraction PDF in order to ensure that (5.299) yields the correct expression for the unconditional means.

Since (5.299) is solved in 'mixture-fraction' space, the independent variables ζ_i are bounded by hyperplanes defined by pairs of axes and the hyperplane defined by $\sum_{i=1}^{N_{\rm mf}} \zeta_i = 1$. At the vertices (i.e., $\mathcal{V} \equiv \{\boldsymbol{\zeta} = (\mathbf{0}, \text{and } e_i \ (i \in 1, \dots, N_{\rm mf}))\}$, where e_i is the Cartesian unit vector for the ith axis), the conditional mean reaction-progress vector is null:[121]

$$\mathbf{Q}(\boldsymbol{\zeta} \in \mathcal{V}; t) = \mathbf{0}. \tag{5.300}$$

On the other hand, on the bounding hypersurfaces the normal 'diffusive' flux must be null. However, this condition will result 'naturally' from the fact that the conditional joint scalar dissipation rate must be zero-flux in the normal direction on the bounding hypersurfaces in order to satisfy the transport equation for the mixture-fraction PDF.[122]

For a non-premixed homogeneous flow, the initial conditions for (5.299) will usually be trivial: $\mathbf{Q}(\boldsymbol{\zeta}; t) = \mathbf{0}$. Given the chemical kinetics and the conditional scalar dissipation rate, (5.299) can thus be solved to find $\langle \varphi_{\rm rp} | \boldsymbol{\zeta} \rangle$. The unconditional means $\langle \varphi_{\rm rp} \rangle$ are then found by averaging with respect to the mixture-fraction PDF. All applications reported to date have dealt with the simplest case where the mixture-fraction vector has only one component. For this case, (5.299) reduces to a simple boundary-value problem that can be easily solved using standard numerical routines. However, as discussed next, even for this simple case care must be taken in choosing the conditional scalar dissipation rate.

5.8.4 Conditional scalar dissipation rate

As shown in Chapter 6, the mixture-fraction PDF in a homogeneous flow $f_\xi(\zeta; t)$ obeys a simple transport equation:

$$\frac{\partial f_\xi}{\partial t} = -\frac{\partial^2}{\partial \zeta^2}[Z(\zeta, t) f_\xi], \tag{5.301}$$

where

$$Z(\zeta, t) \equiv \left\langle \Gamma \frac{\partial \xi}{\partial x_i} \frac{\partial \xi}{\partial x_i} \middle| \zeta \right\rangle \tag{5.302}$$

is proportional to the conditional scalar dissipation rate. Note that, given $f_\xi(\zeta; t)$, (5.301) can be solved for $Z(\zeta, t)$:

$$Z(\zeta, t) = -\frac{1}{f_\xi(\zeta; t)} \frac{\partial}{\partial t} \int_0^\zeta F_\xi(s; t) \, \mathrm{d}s, \tag{5.303}$$

[121] This boundary condition does not ensure that the unconditional means will be conserved if the chemical source term is set to zero (or if the flow is non-reacting with non-zero initial conditions $\mathbf{Q}(\boldsymbol{\zeta}; 0) \neq \mathbf{0}$). Indeed, as shown in the next section, the mean values will only be conserved if the conditional scalar dissipation rate is chosen to be exactly consistent with the mixture-fraction PDF. An alternative boundary condition can be formulated by requiring that the first term on the right-hand side of (5.299) (i.e., the 'diffusive' term) has zero expected value with respect to the mixture-fraction PDF. However, it is not clear how this 'global' condition can be easily implemented in the solution procedure for (5.299).

[122] More specifically, the condition that the probability flux at the boundaries is zero and the condition that the mean mixture-fraction vector is constant in a homogeneous flow lead to 'natural' boundary conditions (Gardiner 1990) for the mixture-fraction PDF governing equation.

where the cumulative distribution function is defined by

$$F_\xi(\zeta;t) \equiv \mathrm{P}[0 \leq \xi(\mathbf{x},t) < \zeta] = \int_0^\zeta f_\xi(s;t)\,\mathrm{d}s. \tag{5.304}$$

Thus, $Z(\zeta, t)$ is completely determined by $f_\xi(\zeta, t)$. As a consequence, as was first pointed out by Tsai and Fox (1995a), *the conditional scalar dissipation rate cannot be chosen independently of the mixture-fraction PDF.*

If (5.303) is disregarded and the functional form for the conditional scalar dissipation rate is chosen based on other considerations, an error in the unconditional scalar means will result. Defining the product of the conditional scalar means and the mixture-fraction PDF by

$$\mathbf{X}(\zeta, t) \equiv \mathbf{Q}(\zeta;t) f_\xi(\zeta;t), \tag{5.305}$$

the unconditional scalar means can be found by integration:

$$\langle \varphi_{\mathrm{rp}} \rangle = \int_0^1 \mathbf{X}(\zeta, t)\,\mathrm{d}\zeta. \tag{5.306}$$

The governing equation for $\mathbf{X}(\zeta, t)$ follows from (5.299) and (5.301):

$$\frac{\partial \mathbf{X}}{\partial t} = Z^* \frac{\partial^2 \mathbf{Q}}{\partial \zeta^2} f_\xi - \mathbf{Q} \frac{\partial^2}{\partial \zeta^2}[Z f_\xi] + \langle \mathbf{S}_{\mathrm{rp}} | \zeta \rangle f_\xi, \tag{5.307}$$

where Z^* is the chosen form for the conditional scalar dissipation rate.[123]

Integrating (5.307) results in the governing equation for the scalar means in a homogeneous flow:[124]

$$\frac{\mathrm{d}\langle \varphi_{\mathrm{rp}} \rangle}{\mathrm{d}t} = \int_0^1 (Z^* - Z) \frac{\partial^2 \mathbf{Q}}{\partial \zeta^2} f_\xi \,\mathrm{d}\zeta + \langle \mathbf{S}_{\mathrm{rp}} \rangle. \tag{5.308}$$

Note that if the conditional scalar dissipation rate is chosen correctly (i.e., $Z^* = Z$), then the first term on the right-hand side of this expression is null. However, if Z^* is inconsistent with f_ξ, then the scalar means will be erroneous due to the term

$$\mathbf{E}(t) \equiv \int_0^1 (Z^* - Z) \frac{\partial^2 \mathbf{Q}}{\partial \zeta^2} f_\xi \,\mathrm{d}\zeta. \tag{5.309}$$

Note that the magnitude of $\mathbf{E}(t)$ will depend on the chemistry through $\mathbf{Q}$, and will be zero in a non-reacting flow.

In order to avoid this error, (5.307) can be rewritten (with $Z^* = Z$) as

$$\frac{\partial \mathbf{X}}{\partial t} = \frac{\partial}{\partial \zeta}\left[Z f_\xi \frac{\partial \mathbf{Q}}{\partial \zeta} - \mathbf{Q} \frac{\partial (Z f_\xi)}{\partial \zeta} \right] + \langle \mathbf{S}_{\mathrm{rp}} | \zeta \rangle f_\xi. \tag{5.310}$$

[123] Note, however, that in order to conserve probability and to keep the mixture-fraction mean constant, the product $Z^* f_\xi$ and its derivative with respect to ζ must be null at both $\zeta = 0$ and $\zeta = 1$.

[124] Integration by parts yields this result under the condition that $\mathbf{Q}(0;t) = \mathbf{Q}(1;t) = 0$ and $Z(0,t) f_\xi(0;t) = Z(1,t) f_\xi(1;t) = 0$.

The error term in the unconditional means will then be zero if the following two conditions hold:

$$Z(1, t) f_\xi(1; t) = Z(0, t) f_\xi(0; t) = 0 \tag{5.311}$$

and

$$\mathbf{Q}(1; t) \frac{\partial (Z f_\xi)}{\partial \zeta}(1; t) = \mathbf{Q}(0; t) \frac{\partial (Z f_\xi)}{\partial \zeta}(0; t) = 0. \tag{5.312}$$

Any choice for $Z(\zeta, t)$ that satisfies these conditions will yield the correct expression for the unconditional scalar means. If, for example, f_ξ is modeled by a beta PDF ((5.142), p. 175) and $Z(\zeta, t) \propto \zeta(1 - \zeta)$ ((6.149), p. 286), then

$$Z f_\xi \propto \zeta^a (1 - \zeta)^b, \quad \text{where } a, b > 0, \tag{5.313}$$

and

$$\frac{\partial Z f_\xi}{\partial \zeta} \propto a \zeta^{a-1} (1 - \zeta)^b - b \zeta^a (1 - \zeta)^{b-1}. \tag{5.314}$$

Condition (5.311) is satisfied due to (5.313). On the other hand, condition (5.312) requires $\mathbf{Q}(0; t) = \mathbf{0}$ when $0 < a < 1$, and $\mathbf{Q}(1; t) = \mathbf{0}$ when $0 < b < 1$. These requirements are satisfied when $\mathbf{Q}$ is defined in terms of the reaction-progress vector, (5.300). Note also that condition (5.312) will be satisfied for all values of a and b if $\mathbf{X}(0, t) = \mathbf{X}(1, t) = \mathbf{0}$.[125] Thus, for applications, the terms on the right-hand side of (5.310) involving $\mathbf{Q}$ can be rewritten in terms of $\mathbf{X}$.

In most reported applications, $Z(\zeta, t)$ is usually assumed to have one of the following three forms:

$$2Z(\zeta, t) = \begin{cases} \varepsilon_\xi(t) & \text{for a Gaussian PDF} \\ N \varepsilon_\xi(t) \exp[-2(\text{erf}^{-1}(2\zeta - 1))^2] & \text{for an AMC PDF} \\ \varepsilon_\xi(t) \left(1 + \frac{\sqrt{2}|\zeta - \langle \zeta \rangle|}{(\langle \xi'^2 \rangle(t))^{1/2}}\right) & \text{for a Laplace PDF,} \end{cases} \tag{5.315}$$

where ε_ξ is the scalar dissipation rate for the mixture fraction, and the corresponding mixture-fraction PDF is noted for each form. Strictly speaking, the (unbounded) Gaussian PDF cannot be used for the mixture fraction. The amplitude mapping closure (AMC) is the most widely used form, and closely resembles the beta PDF. The exact conditional scalar dissipation rate for the beta PDF is unwieldy, but can be found from (5.303) using a numerical procedure described by Tsai and Fox (1995a) (see also Tsai and Fox (1998)).

5.8.5 Extension to inhomogeneous flows

The transport equation for $\mathbf{Q}$ can be extended to inhomogeneous flows (Klimenko 1990; Bilger 1993; Klimenko 1995; Klimenko and Bilger 1999) and to LES (Bushe and Steiner

[125] In other words, either the mixture-fraction PDF or the conditional reaction-progress vector (but not necessarily both) must be zero on the boundaries of mixture-fraction space.

1999). Ignoring the conditional fluctuations and differential-diffusion effects, the transport equation for $\mathbf{Q}(\zeta;\mathbf{x},t)$ becomes

$$\frac{\partial \mathbf{Q}}{\partial t} + \frac{\partial}{\partial x_i}[\langle U_i|\zeta\rangle \mathbf{Q}] = Z(\zeta,t)\frac{\partial^2 \mathbf{Q}}{\partial \zeta^2} + \Gamma \frac{\partial^2 \mathbf{Q}}{\partial x_i \partial x_i} + \Gamma \frac{\partial^2 \mathbf{Q}}{\partial x_i \partial \zeta}\frac{\partial \langle \xi \rangle}{\partial x_i} + \mathbf{S}(\mathbf{Q}). \tag{5.316}$$

For high-Reynolds-number flows, the two terms on the right-hand side involving Γ will be negligible.[126] The inhomogeneous transport equation will thus have only one additional term due to the conditional velocity $\langle \mathbf{U}|\zeta\rangle$.

The conditional velocity also appears in the inhomogeneous transport equation for $f_\xi(\zeta;\mathbf{x},t)$, and is usually closed by a simple 'gradient-diffusion' model. Given the mixture-fraction PDF, (5.316) can be closed in this manner by first decomposing the velocity into its mean and fluctuating components:

$$\langle \mathbf{U}|\zeta\rangle = \langle \mathbf{U}\rangle + \langle \mathbf{u}|\zeta\rangle, \tag{5.317}$$

and then employing the gradient-diffusion model for the conditional fluctuations:

$$\langle u_i|\zeta\rangle = -\frac{\Gamma_T}{f_\xi}\frac{\partial f_\xi}{\partial x_i}. \tag{5.318}$$

Note, however, that the right-hand side of (5.318) will be a complicated function of ζ, $\mathbf{x}$, and t, which will be difficult to employ for actual calculations. In most applications, a simpler linear form has been used:[127]

$$\langle u_i|\zeta\rangle = -\frac{\Gamma_T}{\langle \xi'^2\rangle}(\zeta - \langle \xi\rangle)\frac{\partial \langle \xi\rangle}{\partial x_i}. \tag{5.319}$$

While inconsistent with the closure for the mixture-fraction PDF, (5.319) does yield the usual gradient-diffusion model for the scalar flux, i.e., for $\langle u_i \xi\rangle$. However, it will not predict the correct behavior in certain limiting cases, e.g., when the mixture-fraction mean is constant, but the mixture-fraction variance depends on $\mathbf{x}$.[128]

Despite the availability of closures for the conditional velocity, the extension of conditional-moment closures to inhomogeneous reacting flows is still problematic. For example, in combusting flows with finite-rate chemistry the conditional fluctuations can be very large (e.g., due to ignition and extinction phenomena). Conditional-moment closures that ignore such fluctuations cannot be expected to yield satisfactory predictions under these conditions (Cha *et al.* 2001).[129] Moreover, for inhomogeneous flows with significant back-mixing, the concentrations of fluid particles in back-mixed zones will be strongly

126 Although $Z(\zeta,t)$ involves Γ, it will not be negligible!

127 Non-linear forms can also be developed in terms of power series in $(\zeta - \langle\xi\rangle)$ by starting from (5.318).

128 A two-term power series expression can be derived to handle this case, but it will again fail in cases where the skewness depends on $\mathbf{x}$, but the mean and variance are constant. However, note that the beta PDF can be successfully handled with the two-term form since all higher-order moments depend on the mean and variance. By accounting for the entire shape of the mixture-fraction PDF, (5.318) will be applicable to all forms of the mixture-fraction PDF.

129 Obviously, conditional moments of higher order could also be modeled. However, as with moment closures, the unclosed terms in the higher-order transport equation are more and more difficult to close.

dependent on their flow history and the chemical kinetics.[130] In the context of conditional-moment closures, back-mixing will also be a significant source of conditional fluctuations. For this reason, the neglect of conditional fluctuations in the chemical-source-term closure may yield unsatisfactory predictions in complex flows.

5.9 Presumed PDF methods

Various authors have attempted to extend mixture-fraction PDF methods to handle finite-rate reactions. The principal difficulty lies in the fact that the joint PDF of the reaction-progress variables must be presumed to close the model. The transport equations for the moments of the reaction-progress variables contain unclosed chemical-reaction source terms. Thus, strong assumptions must be made concerning the statistical dependence between the mixture fraction and the reaction-progress variables in order to close the problem. In most cases, it is extremely difficult to justify both the assumed form of the joint reaction-progress-variable PDF and the assumed statistical dependence of the reaction-progress variables on the mixture-fraction vector. For example, in many of the proposed closures, the reaction-progress variables are assumed to be independent of the mixture fraction, which is unlikely to be the case. Nevertheless, in this section we will first review presumed PDF methods for kinetic schemes that can be described by a single reaction-progress variable (e.g., see Section 5.5), and then look at possible extensions to more complex schemes requiring multiple reaction-progress variables.

5.9.1 Single reaction-progress variable

For simple chemistry, we have seen in Section 5.5 that limiting cases of general interest exist that can be described by a single reaction-progress variable, in addition to the mixture fraction.[131] For these flows, the chemical source term can be closed by assuming a form for the joint PDF of the reaction-progress variable Y and the mixture fraction ξ. In general, it is easiest to decompose the joint PDF into the product of the conditional PDF of Y and the mixture-fraction PDF:[132]

$$f_{Y,\xi}(y,\zeta;\mathbf{x},t) = f_{Y|\xi}(y|\zeta;\mathbf{x},t)f_{\xi}(\zeta;\mathbf{x},t). \tag{5.320}$$

Since acceptable forms are available for the mixture-fraction PDF, the chemical source term $\langle S_Y(Y,\xi)\rangle$ can be closed by modeling $f_{Y|\xi}(y|\zeta;\mathbf{x},t)$.

By definition, the reaction-progress variable is bounded below by zero. Likewise, we have seen in Section 5.5 that a single reaction-progress variable is bounded above by a

[130] A fluid particle that has spent a significant time in the flow will have reacted further than a particle that has only recently entered. Thus, for the same value of the mixture fraction, their concentrations can be very different!

[131] For premixed flows, the mixture fraction is not applicable. Nonetheless, the methods in this section can still be employed to model the PDF of the reaction-progress variable.

[132] One could also do the same with the mixture-fraction-vector PDF. However, it is much more difficult to find acceptable forms for the conditional PDF.

(piece-wise linear) function $Y_{\max}(\xi)$. Thus, if we redefine the reaction-progress variable as

$$Y^* \equiv \frac{Y}{Y_{\max}(\xi)}, \tag{5.321}$$

then the conditional PDF of Y^* will be non-zero on the unit interval for all values of ξ. We can then approximate the conditional PDF by a beta PDF:

$$f_{Y^*|\xi}(y|\zeta;\mathbf{x},t) \approx \frac{1}{B(a_1,b_1)} y^{a_1-1}(1-y)^{b_1-1}. \tag{5.322}$$

In theory, the parameters a_1 and b_1 can be functions of ζ, i.e., by relating them to the conditional moments $\langle Y^*|\zeta\rangle$ and $\langle Y^{*2}|\zeta\rangle$. However, this would require us to solve transport equations for the conditional moments, which is rarely justified. Instead, the assumption is made that Y^* and ξ are independent so that

$$f_{Y^*|\xi}(y|\zeta;\mathbf{x},t) = f_{Y^*}(y;\mathbf{x},t). \tag{5.323}$$

This assumption leads to relationships for a_1 and b_1 of the form of (5.144) and (5.145), respectively, but with the mixture-fraction mean and variance replaced by the mean and variance of Y^*. The joint PDF can thus be closed by solving four transport equations for the means and variances of Y^* and ξ.

A transport equation for Y^* could be derived starting from its definition in (5.321). However, the resulting expression would be unduly complicated and not necessarily agree with our assumption of independence between Y^* and ξ.[133] Instead, Y^* can be treated as any other scalar so that the transport equation for $\langle Y^*\rangle$ has the form of (3.88) on p. 81 with $\phi_\alpha = Y^*$, and a chemical source term given by

$$\langle S^*(Y^*,\xi)\rangle = \int_0^1\int_0^1 S_Y(Y_{\max}(\zeta)y,\zeta) f_{Y^*}(y;\mathbf{x},t) f_\xi(\zeta;\mathbf{x},t)\,\mathrm{d}y\,\mathrm{d}\zeta, \tag{5.324}$$

where S_Y is the chemical source term for Y. Note that (5.324) is closed since f_{Y^*} and f_ξ are known functions. Likewise, the transport equation for the variance of Y^* has the form of (3.136) on p. 90 with $\phi_\alpha = \phi_\beta = Y^*$, and a chemical source term given by

$$\langle Y^{*'} S^*(Y^*,\xi)\rangle = \int_0^1\int_0^1 (y - \langle Y^*\rangle) S_Y(Y_{\max}(\zeta)y,\zeta) f_{Y^*}(y;\mathbf{x},t) f_\xi(\zeta;\mathbf{x},t)\,\mathrm{d}y\,\mathrm{d}\zeta. \tag{5.325}$$

Note that due to the assumption that ξ and Y^* are independent, a covariance transport equation is not required to close the chemical source term. However, it would be of the form of (3.136) and have a non-zero chemical source term $\langle \xi' S^*(Y^*,\xi)\rangle$. Thus, since the covariance should be null due to independence, the covariance equation could in theory be solved to check the validity of the independence assumption.

[133] In other words, if Y^* is really independent of ξ, then the turbulent transport terms in its transport equation should not depend on ξ.

Unlike for the mixture fraction, the initial and inlet conditions for the mean and variance of Y^* will be zero. Thus, the chemical source term will be responsible for the generation of non-zero values of the mean and variance of the reaction-progress variable inside the reactor. Appealing again to the assumption of independence, the mean reaction-progress variable is given by

$$\langle Y \rangle = \langle Y^* \rangle \langle Y_{\max}(\xi) \rangle, \tag{5.326}$$

and its variance is given by

$$\langle Y'^2 \rangle = \langle Y^{*2} \rangle \langle Y_{\max}^2(\xi) \rangle - \langle Y \rangle^2. \tag{5.327}$$

Thus, the turbulent-reacting-flow problem can be completely closed by assuming independence between Y^* and ξ, and assuming simple forms for their marginal PDFs. In contrast to the conditional-moment closures discussed in Section 5.8, the presumed PDF method does account for the effect of fluctuations in the reaction-progress variable. However, the independence assumption results in conditional fluctuations that depend on ζ only through $Y_{\max}(\zeta)$. The conditional fluctuations thus contain no information about 'local' events in mixture-fraction space (such as ignition or extinction) that are caused by the mixture-fraction dependence of the chemical source term.

5.9.2 Multiple reaction-progress variables

The extension of the ideas presented above to multiple reaction-progress variables is complicated by the fact that upper bounds must be found for each variable as a function of ζ. However, assuming that this can be done, it then suffices to define new reaction-progress variables as done in (5.321). The new reaction-progress variables are then assumed to be mutually independent, and independent of the mixture fraction. Using this assumption, the joint PDF of the reaction-progress variables and the mixture fraction is expressed as the product of the marginal PDFs. The marginal PDF for each variable can be approximated by a beta PDF. As in (5.324) and (5.325), the chemical source terms in the transport equations for the means and variances of the reaction-progress variables will be closed. Thus, the overall turbulent-reacting-flow problem is closed and can be solved for the means and variances of the reaction-progress variables and mixture fraction. Note that this implementation requires a number of strong assumptions of dubious validity. In particular, due to the chemical source term, the reaction-progress variables are usually strongly coupled and, hence, far from independent. Thus, although the chemical source term will be closed, the quality of the resulting predictions is by no means guaranteed!

The method just described for treating multiple reacting-progress variables has the distinct disadvantage that the upper bounds must be found *a priori*. For a complex reaction scheme, this may be unduly difficult, if not impossible. This fact, combined with the desire to include the correlations between the reacting scalars, has led to the development of even simpler methods based on a presumed joint PDF for the composition vector

ϕ. In Section 3.3, the general transport equations for the means, (3.88), and covariances, (3.136), of ϕ are derived. These equations contain a number of unclosed terms that must be modeled. For high-Reynolds-number flows, we have seen that simple models are available for the turbulent transport terms (e.g., the gradient-diffusion model for the scalar fluxes). Invoking these models,[134] the transport equations become

$$\frac{\partial \langle \phi \rangle}{\partial t} + \langle U_i \rangle \frac{\partial \langle \phi \rangle}{\partial x_i} = \frac{\partial}{\partial x_i} \left(\Gamma_{\mathrm{T}} \frac{\partial \langle \phi \rangle}{\partial x_i} \right) + \langle \mathbf{S}(\phi) \rangle \tag{5.328}$$

and[135]

$$\begin{aligned} &\frac{\partial \langle \phi'_\alpha \phi'_\beta \rangle}{\partial t} + \langle U_i \rangle \frac{\partial \langle \phi'_\alpha \phi'_\beta \rangle}{\partial x_i} \\ &= \frac{\partial}{\partial x_i} \left(\Gamma_{\mathrm{T}} \frac{\partial \langle \phi'_\alpha \phi'_\beta \rangle}{\partial x_i} \right) + 2\Gamma_{\mathrm{T}} \frac{\partial \langle \phi_\alpha \rangle}{\partial x_i} \frac{\partial \langle \phi_\beta \rangle}{\partial x_i} - C_\phi \frac{\varepsilon}{k} \langle \phi'_\alpha \phi'_\beta \rangle \\ &\quad + \langle \phi'_\beta S_\alpha(\phi) \rangle + \langle \phi'_\alpha S_\beta(\phi) \rangle. \end{aligned} \tag{5.329}$$

Thus, the only unclosed terms are those involving the chemical source term.

As discussed in Section 5.1, the chemical source term can be written in terms of reaction rate functions $R_i(\phi)$. These functions, in turn, can be expressed in terms of two *non-negative* functions, (5.6), corresponding to the forward and reverse reactions:

$$R_i(\phi) = \max\left[0, R_i^{\mathrm{f}}(\phi)\right] - \max\left[0, R_i^{\mathrm{r}}(\phi)\right], \tag{5.330}$$

where we have explicitly set the reaction rates on the right-hand side to zero whenever they are evaluated with (non-physical) negative components of ϕ. By definition, the composition vector should always be non-negative, i.e., the probability of finding a negative component should be null. Nevertheless, since it is much simpler to assume a continuous function (such as a joint Gaussian PDF) for the joint composition PDF, we can use (5.330) to filter out the effect of negative components when closing the chemical source term. Indeed, using (5.330), the Reynolds-averaged chemical source term can be written in terms of

$$\begin{aligned} \langle R_i(\phi) \rangle &= \int_0^{+\infty} \cdots \int R_i(\psi) f_\phi(\psi; \mathbf{x}, t)\, \mathrm{d}\psi \\ &= \int_{-\infty}^{+\infty} \cdots \int \left(\max\left[0, R_i^{\mathrm{f}}(\psi)\right] - \max\left[0, R_i^{\mathrm{r}}(\psi)\right] \right) f_\phi(\psi; \mathbf{x}, t)\, \mathrm{d}\psi, \end{aligned} \tag{5.331}$$

which will be closed if $f_\phi(\psi; \mathbf{x}, t)$ can be approximated by a simple functional form that depends only on $\langle \phi \rangle$ and $\langle \phi'_\alpha \phi'_\beta \rangle$.

[134] Given the strong assumptions needed to model the joint PDF, the use of more refined models would not be justified.

[135] Since we have assumed that all scalars have the same molecular diffusivity, $\gamma_{\alpha\beta} = 1$ in (3.136).

A simple functional form that can be used to approximate the joint PDF of the N_s composition variables is the joint Gaussian PDF:

$$f_{\phi}(\psi; \mathbf{x}, t) \approx [(2\pi)^{N_s}|\mathbf{C}|]^{-1/2} \exp\left[-\frac{1}{2}(\psi - \langle\phi\rangle)^{\mathrm{T}}\mathbf{C}^{-1}(\psi - \langle\phi\rangle)\right], \tag{5.332}$$

where the covariance matrix $\mathbf{C}$ is defined by

$$\mathbf{C} \equiv [\langle\phi'_{\alpha}\phi'_{\beta}\rangle] = \langle(\phi - \langle\phi\rangle)(\phi - \langle\phi\rangle)^{\mathrm{T}}\rangle. \tag{5.333}$$

However, care must be taken to avoid the singularity that occurs when $\mathbf{C}$ is not full rank. In general, the rank of $\mathbf{C}$ will be equal to the number of random variables needed to define the joint PDF. Likewise, its rank deficiency will be equal to the number of random variables that can be expressed as linear functions of other random variables. Thus, the covariance matrix can be used to decompose the composition vector into its linearly independent and linearly dependent components. The joint PDF of the linearly independent components can then be approximated by (5.332).

The covariance matrix is positive semi-definite and symmetric. Thus, it can be written in terms of eigenvalues and eigenvectors as

$$\mathbf{C} = \mathbf{U}\boldsymbol{\Lambda}\mathbf{U}^{\mathrm{T}}, \tag{5.334}$$

where $\mathbf{U}$ is an ortho-normal matrix and

$$\boldsymbol{\Lambda} = \mathbf{diag}\left(\lambda_1, \ldots, \lambda_{N_{\mathrm{C}}}, 0, \ldots, 0\right) \tag{5.335}$$

contains the $N_{\mathrm{C}} = \mathrm{rank}(\mathbf{C})$ positive (real) eigenvalues.[136] The desired linear transformation of the composition vector is defined by

$$\phi^* \equiv \begin{bmatrix} \phi_{\mathrm{in}} \\ \phi_{\mathrm{dep}} \end{bmatrix} \equiv \boldsymbol{\Sigma}^{-1}\mathbf{U}^{\mathrm{T}}\left(\phi - \langle\phi\rangle\right), \tag{5.336}$$

where

$$\boldsymbol{\Sigma} \equiv \mathbf{diag}\left(\lambda_1^{1/2}, \ldots, \lambda_{N_{\mathrm{C}}}^{1/2}, 1, \ldots, 1\right). \tag{5.337}$$

Note that, by definition, $\langle\phi^*\rangle = 0$. Moreover, starting from (5.336), it is easily shown that $\langle\phi_{\mathrm{in}}\phi_{\mathrm{in}}^{\mathrm{T}}\rangle = \mathbf{I}$ and $\langle\phi_{\mathrm{dep}}\phi_{\mathrm{dep}}^{\mathrm{T}}\rangle = \mathbf{0}$. It then follows from the assumed joint Gaussian PDF for ϕ that $\phi_{\mathrm{dep}} = \mathbf{0}$, and that the joint PDF of ϕ_{in} is given by

$$\begin{aligned} f_{\phi_{\mathrm{in}}}(\psi; \mathbf{x}, t) &\approx (2\pi)^{-N_{\mathrm{C}}/2} \exp\left(-\frac{1}{2}\psi^{\mathrm{T}}\psi\right) \\ &= \prod_{i=1}^{N_{\mathrm{C}}} \frac{1}{2\pi^{1/2}} \mathrm{e}^{-\psi_i^2/2}. \end{aligned} \tag{5.338}$$

The eigenvalue/eigenvector decomposition of the covariance matrix thus allows us to redefine the problem in terms of N_{C} independent, standard normal random variables ϕ_{in}.

[136] If $N_{\mathrm{C}} = 0$, then $\phi = \langle\phi\rangle$ so that the chemical source term is closed.

In summary, given the covariance matrix $\mathbf{C}$, an eigenvalue/eigenvector decomposition can be carried out to find $\mathbf{U}$ and $\mathbf{\Lambda}$. These matrices define a linear transformation

$$\phi = \mathbf{M}\phi_{\text{in}} + \langle\phi\rangle. \tag{5.339}$$

The Reynolds-averaged chemical source term in (5.328) is then given by

$$\begin{aligned}\langle \mathbf{S}(\phi)\rangle &= \int_{-\infty}^{+\infty}\cdots\int \mathbf{S}(\mathbf{M}\psi + \langle\phi\rangle) f_{\phi_{\text{in}}}(\psi;\mathbf{x},t)\,\mathrm{d}\psi \\ &= \int_{-\infty}^{+\infty}\cdots\int \mathbf{S}(\mathbf{M}\psi + \langle\phi\rangle)\prod_{i=1}^{N_C}\frac{1}{2\pi^{1/2}}\mathrm{e}^{-\psi_i^2/2}\,\mathrm{d}\psi,\end{aligned} \tag{5.340}$$

with the reaction rate functions given by (5.330). In a similar manner, the chemical source term in (5.329) will also be closed.

The joint Gaussian presumed PDF method described above would appear to be a relatively simple method to account for scalar correlations in the chemical-source-term closure. However, it should be noted that the Gaussian form is often far from reality during the initial stages of mixing. Indeed, fast chemical reactions often lead to strongly bi-modal PDF shapes that are poorly approximated by a joint Gaussian PDF. In some cases, it may be possible to improve the closure by first decomposing the composition vector into a reaction-progress vector φ_{rp} and the mixture-fraction vector $\boldsymbol{\xi}$, and then employing the joint Gaussian presumed PDF method only for φ_{rp}. However, unless one is willing to formulate and solve transport equations for the means and covariances of φ_{rp} *conditioned on the mixture-fraction vector*, an assumption of independence between the reaction-progress vector and the mixture-fraction vector must be invoked. As seen in Section 5.8, the conditional means of the reaction-progress variables are usually strongly dependent on the mixture fraction. Thus, the independence assumption will largely negate any improvements that might be gained by modeling the joint PDF. For these reasons, presumed joint PDF methods that explicitly account for the chemical source term by employing even a crude approximation of the joint PDF are desirable. In the next section, we will look at such an approximation based on the extension of the Lagrangian micromixing models introduced in Section 5.6.

5.10 Multi-environment presumed PDF models

In Section 5.6, Lagrangian micromixing models based on 'mixing environments' were introduced. In terms of the joint composition PDF, nearly all such models can be expressed mathematically as a multi-peak delta function. The principal advantage of this type of model is the fact that the chemical source term is closed, and thus it is not necessary to integrate with respect to the joint composition PDF in order to evaluate the

Reynolds-averaged chemical source term.[137] However, the multi-peak form of the presumed PDF requires particular attention to the definition of the micromixing terms when the model is extended to inhomogeneous flow (Fox 1998), or to homogeneous flows with uniform mean scalar gradients.[138] In this section, we first develop the general formulation for multi-environment presumed PDF models in homogeneous flows, and then extend the model to inhomogeneous flows for the particular cases with two and four mixing environments. Tables summarizing all such models that have appeared in the literature are given at the end of this section.

5.10.1 General formulation

In a multi-environment micromixing model, the presumed composition PDF has the following form:

$$f_{\phi}(\psi;\mathbf{x},t) = \sum_{n=1}^{N_e} p_n(\mathbf{x},t) \prod_{\alpha=1}^{N_s} \delta\left[\psi_\alpha - \langle\phi_\alpha\rangle_n(\mathbf{x},t)\right], \tag{5.341}$$

where N_e is the number of environments, $p_n(\mathbf{x},t)$ is the probability[139] of environment n, and $\langle\phi\rangle_n(\mathbf{x},t)$ is the mean[140] composition vector in environment n. For a homogeneous flow in the absence of mean scalar gradients, the model equations for $\mathbf{p}(t)$ and $\langle\phi\rangle_n(t)$ are given by

$$\frac{\mathrm{d}\mathbf{p}}{\mathrm{d}t} = \gamma\mathbf{G}(\mathbf{p}) \tag{5.342}$$

and[141]

$$\frac{\mathrm{d}\langle\mathbf{s}\rangle_n}{\mathrm{d}t} = \gamma\mathbf{M}^{(n)}(\mathbf{p}, \langle\mathbf{s}\rangle_1, \dots, \langle\mathbf{s}\rangle_{N_e}) + p_n\mathbf{S}(\langle\phi\rangle_n), \tag{5.343}$$

where $\langle\mathbf{s}\rangle_n \equiv p_n\langle\phi\rangle_n$ (no summation is implied) is the probability-weighted mean composition vector in environment n, $\gamma\mathbf{G}$ is the rate of change of $\mathbf{p}$ due to micromixing, and $\gamma\mathbf{M}^{(n)}$ is the rate of change of $\langle\mathbf{s}\rangle_n$ due to micromixing.

The specification of $\mathbf{G}$ and $\mathbf{M}^{(n)}$ depends on the definition of the micromixing model. However, the conservation of probability requires that the sum of the probabilities be unity. Thus, from (5.342), we find

$$\sum_{n=1}^{N_e} G_n(\mathbf{p}) = 0. \tag{5.344}$$

[137] More precisely, due to the form of the presumed PDF, the integral reduces to a sum over all environments.
[138] Multi-environment presumed PDF models are generally *not recommended* for homogeneous flows with uniform mean gradients. Indeed, their proper formulation will require the existence of a mixture-fraction vector that, by definition, cannot generate a uniform mean scalar gradient in a homogeneous flow.
[139] For constant-density flows, the probabilities are equal to the volume fractions.
[140] More precisely, $\langle\phi\rangle_n(\mathbf{x},t)$ is the *conditional* expected value of $\phi(\mathbf{x},t)$ given that it is sampled from environment n.
[141] The model can be written in terms of $\langle\phi\rangle_n$. However, we shall see that the extension to inhomogeneous flows is trivial when $\langle\mathbf{s}\rangle_n$ is used.

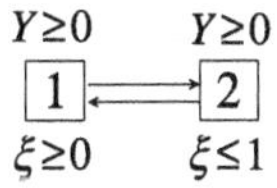

$Y \geq 0$ $Y \geq 0$
1 2
$\xi \geq 0$ $\xi \leq 1$

Figure 5.22. In the two-environment E-model, environment 1 initially contains unmixed fluid with $\xi = 1$, and environment 2 contains unmixed fluid with $\xi = 0$. Chemical reactions occur in environment 1 when $0 < \xi < 1$.

Likewise, the mean composition vector can be found from (5.341), and is given by

$$\langle \phi \rangle = \sum_{n=1}^{N_e} p_n \langle \phi \rangle_n = \sum_{n=1}^{N_e} \langle \mathbf{s} \rangle_n. \tag{5.345}$$

The condition that micromixing must leave the means unchanged in (5.343) then requires

$$\sum_{n=1}^{N_e} \mathbf{M}^{(n)}(\mathbf{p}, \langle \mathbf{s} \rangle_1, \ldots, \langle \mathbf{s} \rangle_{N_e}) = 0, \tag{5.346}$$

so that (5.343) reduces to

$$\frac{\mathrm{d}\langle \phi \rangle}{\mathrm{d}t} = \langle \mathbf{S} \rangle. \tag{5.347}$$

As done below for two examples, expressions can also be derived for the scalar variance starting from the model equations. For the homogeneous flow under consideration, micromixing controls the variance decay rate, and thus γ can be chosen to agree with a particular model for the scalar dissipation rate. For inhomogeneous flows, the definitions of $\mathbf{G}$ and $\mathbf{M}^{(n)}$ must be modified to avoid 'spurious' dissipation (Fox 1998). We will discuss the extension of the model to inhomogeneous flows after looking at two simple examples.

As an example of a two-environment micromixing model, consider the E-model shown in Fig. 5.22, which can be expressed as[142]

$$G_1 = -G_2 = p_1 p_2, \tag{5.348}$$

$$\mathbf{M}^{(1)} = -\mathbf{M}^{(2)} = p_1 \langle \mathbf{s} \rangle_2, \tag{5.349}$$

where G_1 controls the rate of growth of environment 1. In the E-model,[143] the relation $p_2 = 1 - p_1$ is used to eliminate the equation for p_2. The E-model equations can then be

[142] Note that the E-model is 'asymmetric' in the sense that environment 1 always grows in probability, while environment 2 always decreases. In general, this is not a desirable feature for a CFD-based micromixing model, and can be avoided by adding a probability flux from environment 1 to environment 2, or by using three environments and letting environment 2 represent pure fluid that mixes with environment 1 to form environment 3. Examples of these models are given in Tables 5.1–5.5 at the end of this section.

[143] Since, in most applications, $\mathbf{S}(\langle \phi \rangle_2) = \mathbf{0}$, environment 1 is referred to as the *reaction zone*, and environment 2 as the *environment*.

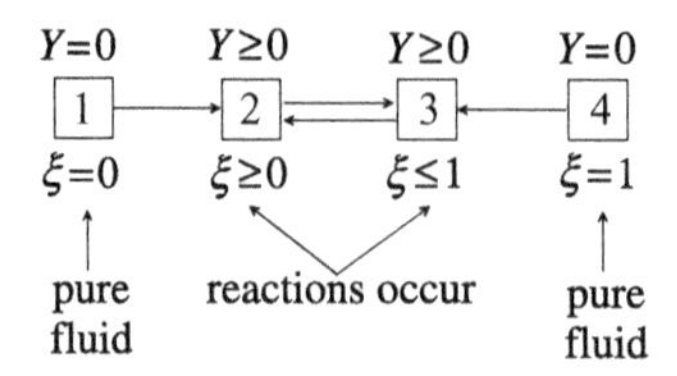

Figure 5.23. In the four-environment generalized mixing model, environment 1 contains unmixed fluid with $\xi = 0$, and environment 4 contains unmixed fluid with $\xi = 1$. Chemical reactions occur in environments 2 and 3 where $0 \leq \xi \leq 1$.

manipulated to yield

$$\frac{\mathrm{d}p_1}{\mathrm{d}t} = \gamma(1 - p_1)p_1, \tag{5.350}$$

$$\frac{\mathrm{d}\langle\phi\rangle_1}{\mathrm{d}t} = \gamma(1 - p_1)(\langle\phi\rangle_2 - \langle\phi\rangle_1) + \mathbf{S}(\langle\phi\rangle_1), \tag{5.351}$$

$$\frac{\mathrm{d}\langle\phi\rangle_2}{\mathrm{d}t} = \mathbf{S}(\langle\phi\rangle_2). \tag{5.352}$$

Moreover, using the definition of the second moment of a *non-reacting* scalar:

$$\langle\phi^2\rangle = p_1\langle\phi\rangle_1^2 + p_2\langle\phi\rangle_2^2 = \frac{\langle s\rangle_1^2}{p_1} + \frac{\langle s\rangle_2^2}{p_2}, \tag{5.353}$$

the E-model yields the following expression for the variance:

$$\frac{\mathrm{d}\langle\phi'^2\rangle}{\mathrm{d}t} = -\gamma\langle\phi'^2\rangle. \tag{5.354}$$

Thus, γ is related to the scalar dissipation rate by

$$\gamma = \frac{\varepsilon_\phi}{\langle\phi'^2\rangle}. \tag{5.355}$$

The homogeneous E-model will thus be completely defined once a model has been chosen for ε_ϕ.

As an example of a four-environment model,[144] consider the generalized mixing model proposed by Villermaux and Falk (1994) shown in Fig. 5.23. The probability exchange rates $\mathbf{r}$ control the probability fluxes between environments.[145] The micromixing terms for the probabilities can be expressed as

$$G_1 = -r_1, \tag{5.356}$$

$$G_2 = r_1 - r_2 + r_3, \tag{5.357}$$

$$G_3 = r_2 - r_3 + r_4, \tag{5.358}$$

[144] A three-environment model can be generated from this model by taking $p_2 = p_3$ and $\langle\mathbf{s}\rangle_2 = \langle\mathbf{s}\rangle_3$. An example is given in Table 5.3.

[145] See Villermaux and Falk (1994) for a discussion on how to choose the exchange rates.

and

$$G_4 = -r_4. \tag{5.359}$$

Likewise, the micromixing terms in the equation for the weighted scalars are given by[146]

$$\mathbf{M}^{(1)} = -r_1 \langle \boldsymbol{\phi} \rangle_1, \tag{5.360}$$

$$\mathbf{M}^{(2)} = r_1 \langle \boldsymbol{\phi} \rangle_1 - r_2 \langle \boldsymbol{\phi} \rangle_2 + r_3 \langle \boldsymbol{\phi} \rangle_3, \tag{5.361}$$

$$\mathbf{M}^{(3)} = r_2 \langle \boldsymbol{\phi} \rangle_2 - r_3 \langle \boldsymbol{\phi} \rangle_3 + r_4 \langle \boldsymbol{\phi} \rangle_4, \tag{5.362}$$

and

$$\mathbf{M}^{(4)} = -r_4 \langle \boldsymbol{\phi} \rangle_4. \tag{5.363}$$

Model specification thus reduces to fitting functional forms for $\mathbf{r}$ expressed in terms of the probabilities $\mathbf{p}$. For example, the probability exchange rates can be chosen to be linear functions:

$$r_n = p_n \quad \text{for all } n, \tag{5.364}$$

or non-linear functions (i.e., as in the E-model):

$$r_1 = p_1(1 - p_1), \tag{5.365}$$

$$r_2 = p_2, \tag{5.366}$$

$$r_3 = p_3, \tag{5.367}$$

and

$$r_4 = p_4(1 - p_4). \tag{5.368}$$

As for the E-model above, γ in the four-environment model can be fixed by forcing the variance decay rate of a non-reacting scalar to agree with a particular model for the scalar dissipation rate. For the four-environment model, the inert-scalar variance obeys

$$\frac{\mathrm{d}\langle \phi'^2 \rangle}{\mathrm{d}t} = -\gamma h(\mathbf{p}, \langle \phi \rangle_1, \ldots, \langle \phi \rangle_4), \tag{5.369}$$

where

$$h(\mathbf{p}, \langle \phi \rangle_1, \ldots, \langle \phi \rangle_4) \equiv r_1 \left(\langle \phi \rangle_1 - \langle \phi \rangle_2 \right)^2 + (r_2 + r_3) \left(\langle \phi \rangle_2 - \langle \phi \rangle_3 \right)^2 + r_4 \left(\langle \phi \rangle_3 - \langle \phi \rangle_4 \right)^2. \tag{5.370}$$

Thus, the micromixing rate is given by

$$\gamma = \frac{\varepsilon_\phi}{h(\mathbf{p}, \langle \phi \rangle_1, \ldots, \langle \phi \rangle_4)}. \tag{5.371}$$

[146] The implied rule for finding $\mathbf{M}^{(n)}$ given G_n expressed in terms of $\mathbf{r}$ is completely general: replace r_m with $r_m \langle \boldsymbol{\phi} \rangle_m$. If multiple arrows leave an environment, the concentration vector should represent the environment from which the arrow originates.

Alternatively, one can use $\gamma = 0.5\varepsilon/k$ so that in the limit of large t where $r_2 = r_3 = 0.5$, the scalar variance obeys

$$\frac{\mathrm{d}\langle\phi'^2\rangle}{\mathrm{d}t} = -2\frac{\varepsilon}{k}\langle\phi'^2\rangle. \tag{5.372}$$

Then, for shorter times, the scalar dissipation rate is modeled by[147]

$$\varepsilon_\phi = \frac{1}{2}h(\mathbf{p}, \langle\phi\rangle_1, \ldots, \langle\phi\rangle_4)\frac{\varepsilon}{k}. \tag{5.373}$$

For longer times, the scalar dissipation rate reverts to the asymptotic form given in (5.372).

5.10.2 Extension to inhomogeneous flows

The principal advantage of employing presumed PDF methods over Lagrangian micromixing models is the ability to couple them with Eulerian turbulence models. From the discussion above, the relationship between micromixing models and presumed PDF methods is transparent when written in terms of multi-scalar presumed PDF closures for homogeneous scalar fields. The extension to *inhomogeneous* scalar mixing requires the addition of the convective and turbulent-diffusion terms:[148]

$$\frac{\partial \mathbf{p}}{\partial t} + \langle U_i\rangle\frac{\partial \mathbf{p}}{\partial x_i} = \frac{\partial}{\partial x_i}\left(\Gamma_{\mathrm{T}}\frac{\partial \mathbf{p}}{\partial x_i}\right) + \gamma\mathbf{G}(\mathbf{p}) + \mathbf{G}_{\mathrm{s}}(\mathbf{p}) \tag{5.374}$$

and

$$\begin{aligned}\frac{\partial\langle\mathbf{s}\rangle_n}{\partial t} + \langle U_i\rangle\frac{\partial\langle\mathbf{s}\rangle_n}{\partial x_i} = {}& \frac{\partial}{\partial x_i}\left(\Gamma_{\mathrm{T}}\frac{\partial\langle\mathbf{s}\rangle_n}{\partial x_i}\right) + \gamma\mathbf{M}^{(n)}(\mathbf{p}, \langle\mathbf{s}\rangle_1, \ldots, \langle\mathbf{s}\rangle_{N_{\mathrm{e}}}) \\ & + \mathbf{M}_{\mathrm{s}}^{(n)}(\mathbf{p}, \langle\mathbf{s}\rangle_1, \ldots, \langle\mathbf{s}\rangle_{N_{\mathrm{e}}}) + p_n\mathbf{S}(\langle\phi\rangle_n),\end{aligned} \tag{5.375}$$

where two additional 'micromixing' terms, $\mathbf{G}_{\mathrm{s}}$ and $\mathbf{M}_{\mathrm{s}}^{(n)}$, have been added to eliminate the 'spurious' dissipation rate in the mixture-fraction-variance transport equation. A completely general treatment of the spurious dissipation terms is given in Appendix B, where it is shown that additional terms are needed in order to predict the higher-order moments (e.g, the mixture-fraction skewness and kurtosis) correctly.

Note that summing (5.375) over all environments yields the correct transport equation for the mean compositions:

$$\frac{\partial\langle\phi\rangle}{\partial t} + \langle U_i\rangle\frac{\partial\langle\phi\rangle}{\partial x_i} = \frac{\partial}{\partial x_i}\left(\Gamma_{\mathrm{T}}\frac{\partial\langle\phi\rangle}{\partial x_i}\right) + \langle\mathbf{S}(\phi)\rangle. \tag{5.376}$$

On the other hand, the form of the scalar-variance transport equation will depend on $\mathbf{G}_{\mathrm{s}}$ and $\mathbf{M}_{\mathrm{s}}^{(n)}$. For example, applying the model to the mixture fraction ξ in the *absence of*

[147] In particular, at $t = 0$ where $r_2 = r_3 = 0$ and $\langle\phi\rangle_1 = \langle\phi\rangle_2$ and $\langle\phi\rangle_3 = \langle\phi\rangle_4$, the scalar dissipation rate will be null as it should be for completely unmixed fluid.

[148] A more sophisticated scalar-flux model could be employed in place of the gradient-diffusion model. However, given the degree of approximation inherent in the multi-environment model, it is probably unwarranted.

micromixing (and with $\mathbf{G}_s$ and $\mathbf{M}_s^{(n)}$ null), yields a mixture-fraction-variance transport equation of the form

$$\frac{\partial \langle \xi'^2 \rangle}{\partial t} + \langle U_i \rangle \frac{\partial \langle \xi'^2 \rangle}{\partial x_i} = \frac{\partial}{\partial x_i} \left(\Gamma_T \frac{\partial \langle \xi'^2 \rangle}{\partial x_i} \right) + 2\Gamma_T \frac{\partial \langle \xi \rangle}{\partial x_i} \frac{\partial \langle \xi \rangle}{\partial x_i} - 2\Gamma_T \left\langle \frac{\partial \xi}{\partial x_i} \frac{\partial \xi}{\partial x_i} \right\rangle_{N_e}, \tag{5.377}$$

where

$$\left\langle \frac{\partial \xi}{\partial x_i} \frac{\partial \xi}{\partial x_i} \right\rangle_{N_e} = \sum_{n=1}^{N_e} p_n \frac{\partial \langle \xi \rangle_n}{\partial x_i} \frac{\partial \langle \xi \rangle_n}{\partial x_i}. \tag{5.378}$$

Comparing (5.377) with (3.105) on p. 85 in the high-Reynolds-number limit (and with $\varepsilon_\phi = 0$), it can be seen that (5.378) is a *spurious* dissipation term.[149] This model artifact results from the presumed form of the joint composition PDF. Indeed, in a transported PDF description of inhomogeneous scalar mixing, the scalar PDF relaxes to a continuous (Gaussian) form. Although this relaxation process cannot be represented exactly by a finite number of delta functions, $\mathbf{G}_s$ and $\mathbf{M}_s^{(n)}$ can be chosen to eliminate the spurious dissipation term in the mixture-fraction-variance transport equation.[150]

The choice of $\mathbf{G}_s$ and $\mathbf{M}_s^{(n)}$ needed to eliminate the spurious dissipation term will be model-dependent. However, in every case,

$$\sum_{n=1}^{N_e} G_{sn}(\mathbf{p}) = 0 \tag{5.379}$$

and

$$\sum_{n=1}^{N_e} \mathbf{M}_s^{(n)}(\mathbf{p}, \langle \mathbf{s} \rangle_1, \dots, \langle \boldsymbol{\phi} \rangle_{N_e}) = \mathbf{0}. \tag{5.380}$$

Furthermore, as a general rule, the correction term should not affect the composition vector in any environment where the mixture fraction $\langle \xi \rangle_n$ is constant. Application of this rule yields:

$$\text{if } \langle \xi \rangle_n \text{ is constant, then } \mathbf{M}_s^{(n)} = \langle \boldsymbol{\phi} \rangle_n G_{sn}. \tag{5.381}$$

As an example, consider again the E-model for which the mixture fraction is constant in environment 2 (see (5.350)). We then have

$$\mathbf{M}_s^{(1)} = -\mathbf{M}_s^{(2)} = -\langle \boldsymbol{\phi} \rangle_2 G_{s2} = \langle \boldsymbol{\phi} \rangle_2 G_{s1}, \tag{5.382}$$

[149] Note, however, that in the absence of micromixing, $\langle \xi \rangle_n$ is constant so that this term will be null. Nevertheless, when micromixing is present, the spurious scalar dissipation term will be non-zero, and thus decrease the scalar variance for inhomogeneous flows.

[150] The agreement between higher-order moments will not be assured. However, this is usually not as important as agreement with the mixture-fraction variance, which controls the rate of micromixing relative to chemical reactions. A complete treatment of this problem is given in Appendix B.

so that all of the mixing terms in (5.374) and (5.375) are closed, and involve two parameters: γ and G_{s1}. Letting $\langle \xi \rangle_2 = 0$, the transport equations for p_1 and $\langle s \rangle_1 = p_1 \langle \xi \rangle_1$ become

$$\frac{\partial p_1}{\partial t} + \langle U_i \rangle \frac{\partial p_1}{\partial x_i} = \frac{\partial}{\partial x_i} \left(\Gamma_T \frac{\partial p_1}{\partial x_i} \right) + \gamma p_1 (1 - p_1) + G_{s1}, \tag{5.383}$$

$$\frac{\partial \langle s \rangle_1}{\partial t} + \langle U_i \rangle \frac{\partial \langle s \rangle_1}{\partial x_i} = \frac{\partial}{\partial x_i} \left(\Gamma_T \frac{\partial \langle s \rangle_1}{\partial x_i} \right), \tag{5.384}$$

and G_{s1} must be chosen to eliminate the spurious dissipation term in the mixture-fraction-variance transport equation. The boundary conditions follow from the inlet flow configuration. For example, in the inflow stream containing only environment 2, $p_1 = 0$ (and thus $\langle s \rangle_1 = 0$). Likewise, in the inflow stream containing only environment 1, $p_1 = 1$ and $\langle s \rangle_1 = \langle \xi \rangle_1 = 1$.

The mixture-fraction-variance transport equation can be found starting from (5.383) and (5.384):[151]

$$\begin{aligned} \frac{\partial \langle \xi'^2 \rangle}{\partial t} + \langle U_i \rangle \frac{\partial \langle \xi'^2 \rangle}{\partial x_i} = & \frac{\partial}{\partial x_i} \left(\Gamma_T \frac{\partial \langle \xi'^2 \rangle}{\partial x_i} \right) + 2\Gamma_T \frac{\partial \langle \xi \rangle}{\partial x_i} \frac{\partial \langle \xi \rangle}{\partial x_i} \\ & - 2\Gamma_T p_1 \frac{\partial \langle \xi \rangle_1}{\partial x_i} \frac{\partial \langle \xi \rangle_1}{\partial x_i} - G_{s1} \langle \xi \rangle_1^2 - \gamma \langle \xi'^2 \rangle. \end{aligned} \tag{5.385}$$

The spurious dissipation term (third term on the right-hand side) will be eliminated if $G_{s1} = -\gamma_s p_1$, where γ_s is defined by

$$\gamma_s = \frac{2\Gamma_T}{\langle \xi \rangle_1^2} \frac{\partial \langle \xi \rangle_1}{\partial x_i} \frac{\partial \langle \xi \rangle_1}{\partial x_i}. \tag{5.386}$$

Note that (5.383) then becomes

$$\frac{\partial p_1}{\partial t} + \langle U_i \rangle \frac{\partial p_1}{\partial x_i} = \frac{\partial}{\partial x_i} \left(\Gamma_T \frac{\partial p_1}{\partial x_i} \right) + \gamma p_1 (1 - p_1) - \gamma_s p_1, \tag{5.387}$$

so that the spurious dissipation correction term slows down the rate of micromixing for inhomogeneous flows. Likewise, the scalar dissipation rate will be correct if

$$\gamma = \frac{\varepsilon_\xi}{\langle \xi'^2 \rangle}, \tag{5.388}$$

where ε_ξ is provided by a separate model for the scalar dissipation rate. All the parameters in the inhomogeneous E-model are now fixed, and the transport equations for p_1, $\langle \mathbf{s} \rangle_1$, and $\langle \mathbf{s} \rangle_2$ can be solved with appropriate boundary conditions. The terms in the transport equations are summarized in Table 5.1.[152]

[151] In fact, since the E-model involves two functions p_1 and $\langle s \rangle_1$, it is equivalent to solving transport equations for $\langle \xi \rangle$ and $\langle \xi^2 \rangle$, i.e., $\langle s \rangle_1 = \langle \xi \rangle$ and $p_1 = \langle \xi \rangle^2 / \langle \xi^2 \rangle$.

[152] Note that the effect of the spurious dissipation term can be non-trivial. For example, consider the case where at $t = 0$ the system is initialized with $p_1 = 1$, but with $\langle \xi \rangle_1$ varying as a function of $\mathbf{x}$. The micromixing term in (5.387) will initially be null, but γ_s will be non-zero. Thus, p_2 will be formed in order to generate the correct distribution for the mixture-fraction variance. By construction, the composition vector in environment 2 will be constant and equal to $\langle \boldsymbol{\phi} \rangle_2$.

Table 5.1. *The terms in the transport equations for the E-model.* Note that, by construction, $\langle \xi \rangle_2 = 0$. For cases where no reactions occur in environment 2, $\langle \phi \rangle_2$ is constant, and the transport equation for $\langle \mathbf{s} \rangle_2$ is not needed. A separate model must be provided for the scalar dissipation rate ε_ξ.

Model variables	$\gamma \mathbf{G}, \gamma \mathbf{M}^{(n)}$	$\mathbf{G}_\mathrm{s}, \mathbf{M}_\mathrm{s}^{(n)}$
p_1	$\gamma p_1 p_2$	$-\gamma_\mathrm{s} p_1$
$\langle \mathbf{s} \rangle_1$	$\gamma p_1 \langle \mathbf{s} \rangle_2$	$-\gamma_\mathrm{s} p_1 \langle \phi \rangle_2$
$\langle \mathbf{s} \rangle_2$	$-\gamma p_1 \langle \mathbf{s} \rangle_2$	$\gamma_\mathrm{s} p_1 \langle \phi \rangle_2$
$p_2 = 1 - p_1,\ \langle \xi'^2 \rangle = p_1 p_2 \langle \xi \rangle_1^2,\ \langle \phi \rangle_2 = \langle \mathbf{s} \rangle_2 / p_2$		
$\gamma = \frac{\varepsilon_\xi}{\langle \xi'^2 \rangle},\ \gamma_\mathrm{s} = \frac{2 p_1 p_2 \Gamma_\mathrm{T}}{\langle \xi'^2 \rangle} \frac{\partial \langle \xi \rangle_1}{\partial x_i} \frac{\partial \langle \xi \rangle_1}{\partial x_i}$		

Applying the same reasoning, the micromixing terms for the inhomogeneous four-environment model can also be specified. From (5.381), we find

$$\mathbf{M}_\mathrm{s}^{(1)} = \langle \phi \rangle_1 G_{\mathrm{s}1}, \tag{5.389}$$

and

$$\mathbf{M}_\mathrm{s}^{(4)} = \langle \phi \rangle_4 G_{\mathrm{s}4}. \tag{5.390}$$

We can then let $G_{\mathrm{s}2} = -\gamma_{\mathrm{s}2} p_2$ and $G_{\mathrm{s}3} = -\gamma_{\mathrm{s}3} p_3$, and[153]

$$G_{\mathrm{s}1} = -G_{\mathrm{s}2}, \qquad G_{\mathrm{s}4} = -G_{\mathrm{s}3}, \tag{5.391}$$

$$\mathbf{M}_\mathrm{s}^{(2)} = -\mathbf{M}_\mathrm{s}^{(1)}, \qquad \mathbf{M}_\mathrm{s}^{(3)} = -\mathbf{M}_\mathrm{s}^{(4)}. \tag{5.392}$$

Elimination of the spurious dissipation term then requires

$$\gamma_{\mathrm{s}2} = \frac{2\Gamma_\mathrm{T}}{(\langle \xi \rangle_1 - \langle \xi \rangle_2)^2} \frac{\partial \langle \xi \rangle_2}{\partial x_i} \frac{\partial \langle \xi \rangle_2}{\partial x_i} \tag{5.393}$$

and

$$\gamma_{\mathrm{s}3} = \frac{2\Gamma_\mathrm{T}}{(\langle \xi \rangle_3 - \langle \xi \rangle_4)^2} \frac{\partial \langle \xi \rangle_3}{\partial x_i} \frac{\partial \langle \xi \rangle_3}{\partial x_i}, \tag{5.394}$$

where $\langle \xi \rangle_1$ and $\langle \xi \rangle_4$ are constant (e.g., $\langle \xi \rangle_1 = 1$ and $\langle \xi \rangle_4 = 0$). Finally, the micromixing rate is fixed by letting

$$\gamma = \frac{\varepsilon_\xi}{h(\mathbf{p}, \langle \boldsymbol{\xi} \rangle_1, \dots, \langle \boldsymbol{\xi} \rangle_4)} \tag{5.395}$$

(see (5.370)). A separate model must be provided for the scalar dissipation rate ε_ξ.

Five examples of multi-environment presumed PDF models are summarized in Tables 5.1–5.5. For each case, the flow is assumed to be non-premixed with a mixture

[153] Unlike for the E-model, other choices are possible that satisfy the constraints. For example, correction terms between environments 2 and 3 similar to those in Table 5.2 could be included (see Table 5.5).

Table 5.2. *The terms in the transport equations for a 'symmetric' two-environment model.*
A separate model must be provided for the scalar dissipation rate ε_{ξ}.

Model variables	$\gamma \mathbf{G}$, $\gamma \mathbf{M}^{(n)}$	$\mathbf{G}_{\mathrm{s}}$, $\mathbf{M}_{\mathrm{s}}^{(n)}$
p_1	0	$\gamma_{\mathrm{s}2} p_2 - \gamma_{\mathrm{s}1} p_1$
$\langle \mathbf{s} \rangle_1$	$\gamma (p_1 \langle \mathbf{s} \rangle_2 - p_2 \langle \mathbf{s} \rangle_1)$	$\gamma_{\mathrm{s}2} p_2 \langle \boldsymbol{\phi} \rangle_1 - \gamma_{\mathrm{s}1} p_1 \langle \boldsymbol{\phi} \rangle_2$
$\langle \mathbf{s} \rangle_2$	$\gamma (p_2 \langle \mathbf{s} \rangle_1 - p_1 \langle \mathbf{s} \rangle_2)$	$\gamma_{\mathrm{s}1} p_1 \langle \boldsymbol{\phi} \rangle_2 - \gamma_{\mathrm{s}2} p_2 \langle \boldsymbol{\phi} \rangle_1$

$$p_2 = 1 - p_1, \ \langle \xi'^2 \rangle = p_1 p_2 \left(\langle \xi \rangle_1 - \langle \xi \rangle_2 \right)^2, \ \langle \boldsymbol{\phi} \rangle_n = \langle \mathbf{s} \rangle_n / p_n \text{ with } n = 1, 2$$
$$\gamma = \frac{\varepsilon_{\xi}}{2 \langle \xi'^2 \rangle}, \ \gamma_{\mathrm{s}n} = \frac{p_1 p_2 \Gamma_{\mathrm{T}}}{\langle \xi'^2 \rangle} \frac{\partial \langle \xi \rangle_n}{\partial x_i} \frac{\partial \langle \xi \rangle_n}{\partial x_i} \text{ with } n = 1, 2$$

Table 5.3. *The terms in the transport equations for a three-environment model.*
Environment 3 is formed by mixing between environments 1 and 2. Thus, by construction, $\langle \xi \rangle_1 = 1$ and $\langle \xi \rangle_2 = 0$. For cases where no reactions occur in environment 1 (2), $\langle \boldsymbol{\phi} \rangle_1$ ($\langle \boldsymbol{\phi} \rangle_2$) is constant, and the transport equation for $\langle \mathbf{s} \rangle_1$ ($\langle \mathbf{s} \rangle_2$) is not needed. A separate model must be provided for the scalar dissipation rate ε_{ξ}.

Model variables	$\gamma \mathbf{G}$, $\gamma \mathbf{M}^{(n)}$	$\mathbf{G}_{\mathrm{s}}$, $\mathbf{M}_{\mathrm{s}}^{(n)}$
p_1	$-\gamma p_1 (1 - p_1)$	$\gamma_{\mathrm{s}} p_3$
p_2	$-\gamma p_2 (1 - p_2)$	$\gamma_{\mathrm{s}} p_3$
$\langle \mathbf{s} \rangle_1$	$-\gamma (1 - p_1) \langle \mathbf{s} \rangle_1$	$\gamma_{\mathrm{s}} p_3 \langle \boldsymbol{\phi} \rangle_1$
$\langle \mathbf{s} \rangle_2$	$-\gamma (1 - p_2) \langle \mathbf{s} \rangle_2$	$\gamma_{\mathrm{s}} p_3 \langle \boldsymbol{\phi} \rangle_2$
$\langle \mathbf{s} \rangle_3$	$\gamma [(1 - p_1) \langle \mathbf{s} \rangle_1 + (1 - p_2) \langle \mathbf{s} \rangle_2]$	$-\gamma_{\mathrm{s}} p_3 (\langle \boldsymbol{\phi} \rangle_1 + \langle \boldsymbol{\phi} \rangle_2)$

$$p_3 = 1 - p_1 - p_2, \ \langle \boldsymbol{\phi} \rangle_n = \langle \mathbf{s} \rangle_n / p_n \text{ with } n = 1, 2$$
$$\gamma = \frac{\varepsilon_{\xi}}{p_1 (1 - p_1)(1 - \langle \xi \rangle_3)^2 + p_2 (1 - p_2) \langle \xi \rangle_3^2}, \ \gamma_{\mathrm{s}} = \frac{2 \Gamma_{\mathrm{T}}}{(1 - \langle \xi \rangle_3)^2 + \langle \xi \rangle_3^2} \frac{\partial \langle \xi \rangle_3}{\partial x_i} \frac{\partial \langle \xi \rangle_3}{\partial x_i}$$
$$\langle \xi'^2 \rangle = p_1 (1 - p_1) - 2 p_1 p_3 \langle \xi \rangle_3 + p_3 (1 - p_3) \langle \xi \rangle_3^2$$

fraction $\langle \xi \rangle_n$ for each environment. As is evident from these examples, multi-environment presumed PDF models offer significant flexibility in the choice of the number of environments N_{e} and the micromixing functions $\mathbf{G}$ and $\mathbf{M}^{(n)}$. However, due to the fact that a separate weighted composition vector must be solved for each environment, they are most attractive for use with chemical kinetic schemes that require only a small number of environments (e.g., isothermal reactions such as the ones discussed in Section 5.5). In general, N_{e} will determine the highest-order moment of the mixture fraction for which the model can be forced to agree with the corresponding transport equation. Thus, for example, with $N_{\mathrm{e}} = 1$ the mean mixture fraction is predicted correctly, but not the variance. With $N_{\mathrm{e}} = 2$, it is also possible to predict the correct mixture-fraction variance, but not the skewness, which requires $N_{\mathrm{e}} = 3$.

A rational (but not necessarily easy to implement) rule for choosing the micromixing functions given N_{e} is to require that the corresponding mixture-fraction moment equation

Table 5.4. *The terms in the transport equations for a four-environment model.*

Environments 2 and 3 are formed by mixing between environments 1 and 4. Thus, by construction, $\langle \xi \rangle_1 = 1$ and $\langle \xi \rangle_4 = 0$. For cases where no reactions occur in environment 1 (4), $\langle \phi \rangle_1$ ($\langle \phi \rangle_4$) is constant, and the transport equation for $\langle \mathbf{s} \rangle_1$ ($\langle \mathbf{s} \rangle_4$) is not needed. A separate model must be provided for the scalar dissipation rate ε_ξ.

Model variables	$\gamma \mathbf{G}$, $\gamma \mathbf{M}^{(n)}$	$\mathbf{G}_s$, $\mathbf{M}_s^{(n)}$
p_1	$-\gamma p_1(1 - p_1)$	$\gamma_{s2} p_2$
p_2	$\gamma[p_1(1 - p_1) - p_2 + p_3]$	$-\gamma_{s2} p_2$
p_3	$\gamma[p_2 - p_3 + p_4(1 - p_4)]$	$-\gamma_{s3} p_3$
$\langle \mathbf{s} \rangle_1$	$-\gamma(1 - p_1)\langle \mathbf{s} \rangle_1$	$\gamma_{s2} p_2 \langle \phi \rangle_1$
$\langle \mathbf{s} \rangle_2$	$\gamma[(1 - p_1)\langle \mathbf{s} \rangle_1 - \langle \mathbf{s} \rangle_2 + \langle \mathbf{s} \rangle_3]$	$-\gamma_{s2} p_2 \langle \phi \rangle_1$
$\langle \mathbf{s} \rangle_3$	$\gamma[\langle \mathbf{s} \rangle_2 - \langle \mathbf{s} \rangle_3 + (1 - p_4)\langle \mathbf{s} \rangle_4]$	$-\gamma_{s3} p_3 \langle \phi \rangle_4$
$\langle \mathbf{s} \rangle_4$	$-\gamma(1 - p_4)\langle \mathbf{s} \rangle_4$	$\gamma_{s3} p_3 \langle \phi \rangle_4$

$$p_4 = 1 - p_1 - p_2 - p_3,\ \langle \phi \rangle_n = \langle \mathbf{s} \rangle_n / p_n \text{ with } n = 1, 4$$

$$\gamma = \frac{\varepsilon_\xi}{p_1(1-p_1)(1-\langle \xi \rangle_2)^2 + (p_2+p_3)((\langle \xi \rangle_2 - \langle \xi \rangle_3)^2 + p_4(1-p_4)\langle \xi \rangle_3^2}$$

$$\gamma_{s2} = \frac{2\Gamma_T}{(1-\langle \xi \rangle_2)^2} \frac{\partial \langle \xi \rangle_2}{\partial x_i} \frac{\partial \langle \xi \rangle_2}{\partial x_i},\ \gamma_{s3} = \frac{2\Gamma_T}{\langle \xi \rangle_3^2} \frac{\partial \langle \xi \rangle_3}{\partial x_i} \frac{\partial \langle \xi \rangle_3}{\partial x_i}$$

$$\langle \xi'^2 \rangle = p_1(1 - p_1) + p_2(1 - p_2)\langle \xi \rangle_2^2 + p_3(1 - p_3)\langle \xi \rangle_3^2 - 2p_1 p_2 \langle \xi \rangle_2 - 2p_1 p_3 \langle \xi \rangle_3 - 2p_2 p_3 \langle \xi \rangle_2 \langle \xi \rangle_3$$

Table 5.5. *The terms in the transport equations for a 'symmetric' four-environment model that reduces to the 'symmetric' two-environment model in the limit where $p_1 = p_4 = 0$.*

By construction, $\langle \xi \rangle_1 = 1$ and $\langle \xi \rangle_4 = 0$. For cases where no reactions occur in environment 1 (4), $\langle \phi \rangle_1$ ($\langle \phi \rangle_4$) is constant and the transport equation for $\langle \mathbf{s} \rangle_1$ ($\langle \mathbf{s} \rangle_4$) is not needed. A separate model must be provided for the scalar dissipation rate ε_ξ.

Model variables	$\gamma \mathbf{G}$, $\gamma \mathbf{M}^{(n)}$	$\mathbf{G}_s$, $\mathbf{M}_s^{(n)}$
p_1	$-\gamma p_1(1 - p_1)$	0
p_2	$\gamma p_1(1 - p_1)$	$\gamma_{s3} p_3 - \gamma_{s2} p_2$
p_3	$\gamma p_4(1 - p_4)$	$\gamma_{s2} p_2 - \gamma_{s3} p_3$
$\langle \mathbf{s} \rangle_1$	$-\gamma(1 - p_1)\langle \mathbf{s} \rangle_1$	0
$\langle \mathbf{s} \rangle_2$	$\gamma[(1 - p_1)\langle \mathbf{s} \rangle_1 - p_3\langle \mathbf{s} \rangle_2 + p_2\langle \mathbf{s} \rangle_3]$	$\gamma_{s3} p_3 \langle \phi \rangle_2 - \gamma_{s2} p_2 \langle \phi \rangle_3$
$\langle \mathbf{s} \rangle_3$	$\gamma[p_3\langle \mathbf{s} \rangle_2 - p_2\langle \mathbf{s} \rangle_3 + (1 - p_4)\langle \mathbf{s} \rangle_4]$	$\gamma_{s2} p_2 \langle \phi \rangle_3 - \gamma_{s3} p_3 \langle \phi \rangle_2$
$\langle \mathbf{s} \rangle_4$	$-\gamma(1 - p_4)\langle \mathbf{s} \rangle_4$	0

$$p_4 = 1 - p_1 - p_2 - p_3,\ \langle \phi \rangle_n = \langle \mathbf{s} \rangle_n / p_n \text{ with } n = 2, 3$$

$$\gamma = \frac{\varepsilon_\xi}{p_1(1-p_1)(1-\langle \xi \rangle_2)^2 + 2p_2 p_3(\langle \xi \rangle_2 - \langle \xi \rangle_3)^2 + p_4(1-p_4)\langle \xi \rangle_3^2}$$

$$\gamma_{s2} = \frac{2\Gamma_T}{(\langle \xi \rangle_2 - \langle \xi \rangle_3)^2} \frac{\partial \langle \xi \rangle_2}{\partial x_i} \frac{\partial \langle \xi \rangle_2}{\partial x_i},\ \gamma_{s3} = \frac{2\Gamma_T}{(\langle \xi \rangle_2 - \langle \xi \rangle_3)^2} \frac{\partial \langle \xi \rangle_3}{\partial x_i} \frac{\partial \langle \xi \rangle_3}{\partial x_i}$$

$$\langle \xi'^2 \rangle = p_1(1 - p_1) + p_2(1 - p_2)\langle \xi \rangle_2^2 + p_3(1 - p_3)\langle \xi \rangle_3^2 - 2p_1 p_2 \langle \xi \rangle_2 - 2p_1 p_3 \langle \xi \rangle_3 - 2p_2 p_3 \langle \xi \rangle_2 \langle \xi \rangle_3$$

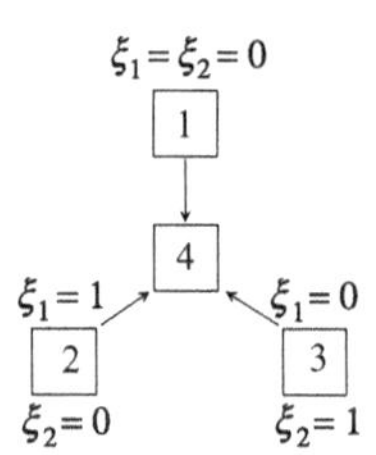

Figure 5.24. A four-environment mixing model can be developed for reactors with three feed streams. In environment 1, the two components of the mixture-fraction vector are null: $\boldsymbol{\xi} = \mathbf{0}$. In environment 2, $\xi_1 = 1$ and $\xi_2 = 0$, while, in environment 3, $\xi_1 = 0$ and $\xi_2 = 1$. Chemical reactions will then occur in environment 4 where $0 \leq \xi_1, \xi_2 \leq 1$.

is satisfied. Nevertheless, even when applying this rule, the micromixing functions will not be unique. Thus, within a family of acceptable functions, one should select functions that provide the best agreement with the next higher-order moment (Fox 1998). For example, with $N_e = 2$ one can choose between functions that keep the mixture-fraction skewness constant, cause it to increase, or cause it to decrease. However, based on the behavior of the beta PDF, only functions that cause the mixture-fraction skewness to decrease should be retained. A completely general method for choosing the spurious dissipation terms is discussed in Appendix B.

Multi-environment presumed PDF models can also be easily extended to treat cases with more than two feed streams. For example, a four-environment model for a flow with three feed streams is shown in Fig. 5.24. For this flow, the mixture-fraction vector will have two components, ξ_1 and ξ_2. The micromixing functions should thus be selected to agree with the variance transport equations for both components. However, in comparison with multi-variable presumed PDF methods for the mixture-fraction vector (see Section 5.3), the implementation of multi-environment presumed PDF models in CFD calculations of chemical reactors with multiple feed streams is much simpler.

In summary, multi-environment presumed PDF models offer a simple description of the joint composition PDF that generates a closed form for the chemical source term. The number of environments and the micromixing functions can be chosen to yield good agreement with other presumed PDF models for the mixture fraction (e.g., the beta PDF). When extended to inhomogeneous flows, care must be taken to eliminate the spurious dissipation terms that result from the discrete form of the presumed PDF. (In general, it is possible to eliminate completely the spurious dissipation term for the mixture-fraction variance when $2 \leq N_e$.) However, due to the ease with which multi-environment presumed PDF models can be added to an existing CFD code, the extension to inhomogeneous flows offers a powerful alternative to other closures.

The ability of multi-environment presumed PDF models to predict mean compositions in a turbulent reacting flow will depend on a number of factors. For example, their use with chemical kinetic schemes that are highly sensitive to the shape (and not just the low-order moments) of the joint composition PDF will be problematic for small N_e. Such reactions may include *non-isothermal* reactions, such as the one-step reaction discussed in Section 5.5 for which the reaction rate is significant only near the stoichiometric point (ξ_{st}, Y_{st}).[154] On the other hand, multi-environment presumed PDF models should be effective for treating isothermal (e.g., liquid-phase) reactions that are sensitive to micromixing.

Note finally that, for any given value of the mixture fraction (i.e., $\xi = \zeta$), the multi-environment presumed PDF model discussed in this section will predict a unique value of ϕ. In this sense, the multi-environment presumed PDF model provides a simple description of the conditional means $\langle \phi | \zeta \rangle$ at N_e discrete values of ζ. An obvious extension of the method would thus be to develop a multi-environment *conditional* PDF to model the conditional joint composition PDF $f_{\phi|\xi}(\psi|\zeta; \mathbf{x}, t)$. We look at models based on this idea below.

5.10.3 Multi-environment conditional PDF models

In a multi-environment conditional PDF model, it is assumed that the composition vector can be partitioned (as described in Section 5.3) into a reaction-progress vector φ_{rp} and a mixture-fraction vector $\boldsymbol{\xi}$. The presumed conditional PDF for the reaction-progress vector then has the form:[155]

$$f_{\varphi_{rp}|\xi}(\psi|\boldsymbol{\zeta}; \mathbf{x}, t) = \sum_{n=1}^{N_e} p_n(\mathbf{x}, t) \prod_{\alpha=1}^{N_Y} \delta[\psi_\alpha - \langle \varphi_{rp\alpha} | \boldsymbol{\zeta} \rangle_n(\mathbf{x}, t)], \tag{5.396}$$

where N_e is the number of environments, p_n is the probability of environment n, and $\langle \varphi_{rp} | \boldsymbol{\zeta} \rangle_n$ is the conditional mean reaction-progress vector in environment n. Note that it is assumed that the mixture-fraction PDF is known so that only mixing in reaction-progress-variable space need be described by the micromixing model.

In Section 5.8, transport equations for the conditional moments $\langle \varphi_{rp} | \boldsymbol{\zeta} \rangle$ were presented. These equations represent the limit $N_e = 1$, and must be generalized for cases where $N_e > 1$. Defining the weighted conditional moments by[156]

$$\mathbf{Q}^{(n)}(\boldsymbol{\zeta}; \mathbf{x}, t) \equiv p_n(\mathbf{x}, t) \langle \varphi_{rp} | \boldsymbol{\zeta} \rangle_n(\mathbf{x}, t), \tag{5.397}$$

the homogeneous model equations for $\mathbf{p}$ and $\mathbf{X}^{(n)}$ are[157]

$$\frac{d\mathbf{p}}{dt} = \gamma \mathbf{G}(\mathbf{p}) \tag{5.398}$$

[154] *Ad hoc* extensions may be possible for this case by fixing the compositions in one environment at the stoichiometric point, and modeling the probability. On the other hand, by making N_e large, the results will approach those found using transported PDF methods.

[155] In a more general formulation, p_n could depend on the mixture-fraction vector.

[156] Note that the conditional means can be found by summing over all environments: $\mathbf{Q} = \langle \varphi_{rp} | \boldsymbol{\zeta} \rangle = \sum_{n=1}^{N_e} \mathbf{Q}^{(n)}$.

[157] Following the example of (5.307), the model equations are written in terms of the mixture-fraction-PDF-weighted conditional means. The integral of (5.399) over mixture-fraction space will eliminate the first two terms on the right-hand side, leaving only the mean chemical source term as required.

and

$$\frac{\partial \mathbf{X}^{(n)}}{\partial t} = \frac{\partial}{\partial \zeta_j}\left[D_{jk}\frac{\partial \mathbf{Q}^{(n)}}{\partial \zeta_k} - \mathbf{Q}^{(n)}\frac{\partial D_{jk}}{\partial \zeta_k}\right] + \gamma \mathbf{M}^{(n)}(\mathbf{p}, \mathbf{X}^{(1)}, \dots, \mathbf{X}^{(N_e)}) + p_n \mathbf{S}_{rp}(\langle \varphi_{rp}|\zeta\rangle_n, \zeta) f_\xi, \tag{5.399}$$

where the (known) coefficients in the conditional 'diffusion' term are defined by

$$D_{jk}(\zeta; t) \equiv \left\langle \Gamma \frac{\partial \xi_j}{\partial x_i}\frac{\partial \xi_k}{\partial x_i} \middle| \zeta; t \right\rangle f_\xi(\zeta; t), \tag{5.400}$$

and mixture-fraction-PDF-weighted conditional means are defined by

$$\mathbf{X}^{(n)}(\zeta; t) \equiv \mathbf{Q}^{(n)}(\zeta; t) f_\xi(\zeta; t). \tag{5.401}$$

The other terms appearing in (5.398) and (5.399) are as follows: $\gamma\mathbf{G}$ is the rate of change of $\mathbf{p}$ due to micromixing, $\gamma\mathbf{M}^{(n)}$ is the rate of change of $\mathbf{X}^{(n)}$ due to micromixing, and f_ξ is the mixture-fraction PDF.

One advantage of formulating the model in terms of $\mathbf{X}^{(n)}$ instead of $\mathbf{Q}^{(n)}$ is that it is no longer necessary to use the conditional scalar dissipation rate in (5.400) that corresponds to the mixture-fraction PDF. Indeed, the only requirement is that D_{jk} – like $\mathbf{X}^{(n)}$ – must be 'zero-flux' at the boundaries of mixture-fraction space. Note that (except for the chemical source term) by summing over n in (5.399) and applying the multi-variate form of (5.301) on p. 212, the model reduces to (5.310), p. 213, which describes the conditional means in a homogeneous flow. Similarly, for the inhomogeneous case, transport terms can be added to the model so that it agrees with (5.316) on p. 215. The conditional mean chemical source term, on the other hand, is given by

$$\langle \mathbf{S}_{rp}|\zeta\rangle = \sum_{n=1}^{N_e} p_n \mathbf{S}_{rp}(\langle \varphi_{rp}|\zeta\rangle_n, \zeta) \neq \mathbf{S}_{rp}(\mathbf{Q}, \zeta). \tag{5.402}$$

The multi-environment conditional PDF model thus offers a simple description of the effect of fluctuations about the conditional expected values on the chemical source term.

The connection between the multi-environment conditional and unconditional PDF models can be made by noting that

$$\langle \mathbf{s}_{rp}\rangle_n = \int_{\mathcal{V}} \mathbf{X}^{(n)}\, \mathrm{d}\zeta, \tag{5.403}$$

where the integral is over all of mixture-fraction space (denoted by $\mathcal{V}$). The subscript rp is a reminder that, unlike $\langle \mathbf{s}\rangle_n$ in (5.343), $\langle \mathbf{s}_{rp}\rangle_n$ does not include the mixture-fraction vector.[158] Integrating (5.399) over $\mathcal{V}$, and applying integration by parts to the 'diffusion'

[158] For this reason, mixing in the conditional PDF model is 'orthogonal' to mixing in the unconditional model. Thus, γ in the conditional model need not be the same as in the unconditional model, where its value controls the mixture-fraction-variance decay rate.

terms, yields

$$\frac{\mathrm{d}\langle \mathbf{s}_{\mathrm{rp}}\rangle_n}{\mathrm{d}t} = \gamma \int_{\mathcal{V}} \mathbf{M}^{(n)}\left(\mathbf{p}, \mathbf{X}^{(1)}, \ldots, \mathbf{X}^{(N_e)}\right) \mathrm{d}\zeta + p_n \int_{\mathcal{V}} \mathbf{S}_{\mathrm{rp}}(\langle \boldsymbol{\varphi}_{\mathrm{rp}}|\boldsymbol{\zeta}\rangle_n, \boldsymbol{\zeta}) f_{\xi}(\boldsymbol{\zeta}; t)\, \mathrm{d}\boldsymbol{\zeta}. \tag{5.404}$$

If we then assume (as is usually the case) that the micromixing term is linear in $\mathbf{X}^{(n)}$, it follows that

$$\mathbf{M}^{(n)}(\mathbf{p}, \langle \mathbf{s}_{\mathrm{rp}}\rangle_1, \ldots, \langle \mathbf{s}_{\mathrm{rp}}\rangle_{N_e}) = \int_{\mathcal{V}} \mathbf{M}^{(n)}\left(\mathbf{p}, \mathbf{X}^{(1)}, \ldots, \mathbf{X}^{(N_e)}\right) \mathrm{d}\boldsymbol{\zeta}, \tag{5.405}$$

which agrees with the corresponding term in (5.343). The reaction term, on the other hand, does not reduce to the corresponding term in (5.343) unless the chemical source term is linear. However, this lack of agreement is expected since the conditional PDF model contains a more detailed description of the composition PDF.

Because the micromixing terms describe mixing at a fixed value of the mixture fraction, choosing these terms in (5.398) and (5.399) is more problematic than for the unconditional case. For example, since we have assumed that $\mathbf{p}$ does not depend on the mixture fraction, the model for $\gamma\mathbf{G}$ must also be independent of mixture fraction. However, the model for $\gamma\mathbf{M}^{(n)}$ will, in general, depend on $\boldsymbol{\zeta}$. Likewise, as with the unconditional model, the micromixing rate (γ) and 'spurious dissipation' terms must be chosen to ensure that the model yields the correct transport equation for the conditional second-order moments (e.g., $\langle \phi^2_{\mathrm{rp}\alpha}|\boldsymbol{\zeta}\rangle$). Thus, depending on the number of environments employed, it may be difficult to arrive at consistent closures.

Despite these difficulties, the multi-environment conditional PDF model is still useful for describing simple non-isothermal reacting systems (such as the one-step reaction discussed in Section 5.5) that cannot be easily treated with the unconditional model. For the non-isothermal, one-step reaction, the reaction-progress variable Y in the (unreacted) feed stream is null, and the system is essentially non-reactive unless an ignition source is provided. Letting $Y_\infty(\xi)$ (see (5.179), p. 183) denote the fully reacted conditional progress variable, we can define a two-environment model based on the E-model:[159]

$$\frac{\mathrm{d}p_1}{\mathrm{d}t} = \gamma(1 - p_1)p_1 \tag{5.406}$$

and

$$\begin{aligned}\frac{\partial X^{(1)}}{\partial t} &= D\frac{\partial^2}{\partial \zeta^2}\left(\frac{1}{f_\xi}X^{(1)}\right) - \left(\frac{1}{f_\xi}\frac{\partial^2 D}{\partial \zeta^2}\right)X^{(1)} \\ &\quad + \gamma p_1(1 - p_1)Y_\infty(\zeta)f_\xi + p_1 S_Y\left(\langle Y|\zeta,\rangle_1, \zeta\right) f_\xi,\end{aligned} \tag{5.407}$$

where $S_Y(Y, \xi)$ is the chemical source term given by (5.176) on p. 183 with $k^*(Y, \xi)$.[160]

[159] By definition, the reaction rate in environment 2 is null, so that $\langle Y|\zeta\rangle_2 = Y_\infty(\zeta)$ is independent of time. Likewise, $X^{(2)} = p_2 Y_\infty f_\xi$ so that $M^{(1)} = p_1 X^{(2)}$.

[160] In the system under consideration, $k^*(0, \xi) = 0$.

From its definition, the boundary conditions for $X^{(1)}$ are $X^{(1)}(0;t) = X^{(1)}(1;t) = 0$. On the other hand, the functional form for $D(\zeta;t)$ must be supplied, and must have the property that $D(0;t) = D(1;t) = 0$. If a beta PDF is used for the mixture fraction (i.e., (5.142) on p. 175 with parameters $a(t)$ and $b(t)$, which depend on the mixture-fraction mean and variance), a convenient choice[161] is

$$D(\zeta;t) \equiv \left\langle \Gamma \frac{\partial \xi}{\partial x_i} \frac{\partial \xi}{\partial x_i} \middle| \zeta;t \right\rangle f_\xi(\zeta;t) \tag{5.408}$$

$$= \varepsilon_\xi(t) h(\langle \xi \rangle, \langle \xi'^2 \rangle) \zeta^a (1-\zeta)^b, \tag{5.409}$$

where $h(\langle \xi \rangle, \langle \xi'^2 \rangle)$ can be found by using the fact that

$$\frac{\mathrm{d}\langle \xi'^2 \rangle}{\mathrm{d}t} = -2 \int_0^1 D(\zeta;t)\,\mathrm{d}\zeta = -\varepsilon_\xi(t). \tag{5.410}$$

Thus, once a model for ε_ξ has been chosen, (5.407) can be solved to find $X^{(1)}$.

At time zero, $X^{(1)} = 0$ and $0 \ll p_1 < 1$, where $p_2 = 1 - p_1$ represents the 'size of the spark' used to ignite the reaction. As time proceeds, micromixing of the reaction-progress variable (which is controlled by γ) will increase $\langle Y|\zeta\rangle_1 = X^{(1)}/(p_1 f_\xi)$ until, eventually, it becomes large enough to ignite the reaction in environment 1. However, if the reaction rate is very localized in mixture-fraction space and $\langle \xi \rangle$ lies outside this range (i.e., $S_Y(Y, \langle \xi \rangle) \approx 0$), micromixing of the mixture fraction (as measured by $\langle \xi'^2 \rangle$) may prevent ignition or cause extinction before all the reactants are consumed.

For large times, $p_1 \to 1$, so that the mean reaction-progress variable predicted by (5.407) will be given by

$$\langle Y \rangle_\infty = \lim_{t\to\infty} \langle Y | \langle \xi \rangle \rangle_1(t) \leq Y_\infty(\langle \xi \rangle). \tag{5.411}$$

Thus, the final product mixture will depend on the relative importance of mixing and reaction in determining $\langle Y|\zeta\rangle_1(t)$. Finally, note that since the second environment was necessary to describe the ignition source, this simple description of ignition and extinction would not be possible with a one-environment model (e.g., the conditional moment closure).

As compared with the other closures discussed in this chapter, computation studies based on the presumed conditional PDF are relatively rare in the literature. This is most likely because of the difficulties of deriving and solving conditional moment equations such as (5.399). Nevertheless, for chemical systems that can exhibit multiple 'reaction branches' for the same value of the mixture fraction,[162] these methods may offer an attractive alternative to more complex models (such as transported PDF methods). Further research to extend multi-environment conditional PDF models to inhomogeneous flows should thus be pursued.

[161] If the conditional Laplacian has a linear form: $2\langle \Gamma \nabla^2 \xi | \zeta \rangle = \varepsilon_\xi(\langle \xi \rangle - \zeta)/\langle \xi'^2 \rangle$, this choice yields a beta PDF for the mixture fraction. However, a better choice would be to use the true conditional scalar dissipation (Tsai and Fox 1995a; Tsai and Fox 1998).

[162] An example of such a system is the so-called 'interacting-flamelet' regime for partially premixed combustion.

5.10.4 Extension to LES

In Section 4.2, the LES composition PDF was introduced to describe the effect of residual composition fluctuations on the chemical source term. As noted there, the LES composition PDF is a conditional PDF for the composition vector given that the filtered velocity and filtered compositions are equal to $\overline{\mathbf{U}}^*$ and $\overline{\phi}^*$, respectively. The LES composition PDF is denoted by $f_{\phi|\overline{\mathbf{U}},\overline{\phi}}(\psi|\overline{\mathbf{U}}^*, \overline{\phi}^*; \mathbf{x}, t)$, and a closure model is required to describe it.

Multi-environment presumed PDF models can be developed for the LES composition PDF using either the unconditional, (5.341), or conditional, (5.396), form. However, in order to simplify the discussion, here we will use the unconditional form to illustrate the steps needed to develop the model. The LES composition PDF can be modeled by[163]

$$f_{\phi|\overline{\mathbf{U}},\overline{\phi}}(\psi|\overline{\mathbf{U}}^*, \overline{\phi}^*) = \sum_{n=1}^{N_e} p_n \prod_{\alpha=1}^{N_s} \delta[\psi_\alpha - \langle \phi_\alpha | \overline{\mathbf{U}}^*, \overline{\phi}^* \rangle_n], \tag{5.412}$$

where for simplicity we will assume that[164]

$$\begin{aligned} \overline{\phi}^* &= \langle \phi | \overline{\mathbf{U}}^*, \overline{\phi}^* \rangle \\ &= \sum_{n=1}^{N_e} p_n \langle \phi | \overline{\mathbf{U}}^*, \overline{\phi}^* \rangle_n \\ &= \sum_{n=1}^{N_e} \langle \mathbf{s} | \overline{\mathbf{U}}^*, \overline{\phi}^* \rangle_n . \end{aligned} \tag{5.413}$$

Thus, in analogy to (5.341), $\overline{\phi}^*$ will replace $\langle \phi \rangle$ in the multi-environment LES model. Hereinafter, in order to simplify the notation, we will denote all conditional variables with a superscript * as follows: $\langle \mathbf{s} \rangle_n^* \equiv \langle \mathbf{s} | \overline{\mathbf{U}}^*, \overline{\phi}^* \rangle_n$.

It is now necessary to formulate transport equations for $\mathbf{p}$ and $\langle \mathbf{s} \rangle_n^*$. However, by making the following substitutions, the *same transport equations as are used in the multi-environment PDF model* can be used for the multi-environment LES model:

$$\langle \mathbf{U} \rangle \longrightarrow \overline{\mathbf{U}}^*, \tag{5.414}$$

$$\langle \xi \rangle \longrightarrow \overline{\xi}^*, \tag{5.415}$$

$$\langle \xi'^2 \rangle \longrightarrow \langle \xi'^2 \rangle^*, \tag{5.416}$$

$$\langle \phi \rangle \longrightarrow \overline{\phi}^*, \tag{5.417}$$

$$\langle \phi \rangle_n \longrightarrow \langle \phi \rangle_n^* = \frac{\langle \mathbf{s} \rangle_n^*}{p_n}, \tag{5.418}$$

$$\Gamma_T \longrightarrow \Gamma_{sgs}, \tag{5.419}$$

[163] By definition, LES is inhomogeneous at the scale of the filter. Thus, to simplify the notation, hereinafter we will omit the explicit dependence on $\mathbf{x}$ and t.

[164] As discussed in Section 4.2, the conditional mean compositions will, in general, depend on the filter so that $\langle \phi | \overline{\mathbf{U}}^*, \overline{\phi}^* \rangle = \overline{\phi}^*$ need not be true. However, if the equality does not hold, it is then necessary to model the difference. Given the simplicity of the multi-environment presumed PDF, such a complication does not seem warranted.

and

$$\varepsilon_\xi \longrightarrow \varepsilon_\xi^*, \tag{5.420}$$

where $\langle \xi'^2 \rangle^*$ is the residual mixture-fraction variance, and ε_ξ^* is the residual mixture-fraction-variance dissipation rate. The SGS turbulent diffusion coefficient Γ_{sgs} is defined by (4.34) on p. 109. The only other significant modification that must be introduced to the original model is to let the rate parameters γ and γ_{s} be functions of the filtered velocity and/or filtered compositions. We will denote the modified rate parameters by γ^* and γ_{s}^*.

The modified transport equations for the multi-environment LES model are thus

$$\frac{\partial \mathbf{p}}{\partial t} + \overline{U}_i^* \frac{\partial \mathbf{p}}{\partial x_i} = \frac{\partial}{\partial x_i}\left(\Gamma_{\text{sgs}} \frac{\partial \mathbf{p}}{\partial x_i}\right) + \gamma^* \mathbf{G}(\mathbf{p}) + \mathbf{G}_{\text{s}}^*(\mathbf{p}) \tag{5.421}$$

and

$$\begin{aligned} \frac{\partial \langle \mathbf{s} \rangle_n^*}{\partial t} + \overline{U}_i^* \frac{\partial \langle \mathbf{s} \rangle_n^*}{\partial x_i} = {} & \frac{\partial}{\partial x_i}\left(\Gamma_{\text{sgs}} \frac{\partial \langle \mathbf{s} \rangle_n^*}{\partial x_i}\right) + \gamma^* \mathbf{M}^{(n)}(\mathbf{p}, \langle \mathbf{s} \rangle_1^*, \ldots, \langle \mathbf{s} \rangle_{N_{\text{e}}}^*) \\ & + \mathbf{M}_{\text{s}}^{*(n)}(\mathbf{p}, \langle \mathbf{s} \rangle_1^*, \ldots, \langle \mathbf{s} \rangle_{N_{\text{e}}}^*) + p_n \mathbf{S}(\langle \boldsymbol{\phi} \rangle_n^*), \end{aligned} \tag{5.422}$$

where the functions $\mathbf{G}_{\text{s}}^*$ and $\mathbf{M}_{\text{s}}^{*(n)}$ are the same as before, but with γ_{s}^* used in place of γ_{s}. In summary, by making the appropriate substitutions, any of the models in Tables 5.1–5.5 can be used in (5.421) and (5.422).

Note that summing (5.422) over all environments yields the transport equation for the filtered compositions:

$$\frac{\partial \overline{\boldsymbol{\phi}}^*}{\partial t} + \overline{U}_i^* \frac{\partial \overline{\boldsymbol{\phi}}^*}{\partial x_i} = \frac{\partial}{\partial x_i}\left(\Gamma_{\text{sgs}} \frac{\partial \overline{\boldsymbol{\phi}}^*}{\partial x_i}\right) + \overline{\mathbf{S}(\boldsymbol{\phi})}, \tag{5.423}$$

wherein the filtered chemical source term is modeled by

$$\overline{\mathbf{S}(\boldsymbol{\phi})} = \sum_{n=1}^{N_{\text{e}}} p_n \mathbf{S}(\langle \boldsymbol{\phi} \rangle_n^*). \tag{5.424}$$

Hence, it is not necessary to solve a separate LES transport equation for the filtered compositions. Indeed, in the limit of one environment, (5.422) reduces to the LES model for the filtered compositions with the simplest possible closure for the chemical source term (i.e., one that neglects all SGS fluctuations).

The model for the residual mixture-fraction variance found from (5.421) and (5.422) has the form

$$\frac{\partial \langle \xi'^2 \rangle^*}{\partial t} + \overline{U}_i^* \frac{\partial \langle \xi'^2 \rangle^*}{\partial x_i} = \frac{\partial}{\partial x_i}\left(\Gamma_{\text{sgs}} \frac{\partial \langle \xi'^2 \rangle^*}{\partial x_i}\right) + 2\Gamma_{\text{sgs}} \frac{\partial \overline{\xi}^*}{\partial x_i} \frac{\partial \overline{\xi}^*}{\partial x_i} - \varepsilon_\xi^*. \tag{5.425}$$

From the terms on the right-hand side of (5.425), it can be seen that residual mixture-fraction fluctuations are produced by spatial gradients in the filtered mixture fraction, and are destroyed by molecular dissipation at sub-grid scales. The multi-environment LES model will thus be completely defined once a model for ε_ξ^* has been chosen. The simplest possible model can be developed using the characteristic filtered strain rate defined in (4.26) on p. 106:

$$\varepsilon_\xi^* = C_\phi |\bar{S}| \langle \xi'^2 \rangle^*. \tag{5.426}$$

The mixing parameter C_ϕ must be chosen to yield the correct mixture-fraction-variance dissipation rate. However, inertial-range scaling arguments suggest that its value should be near unity.[165]

The procedure followed above can be used to develop a multi-environment conditional LES model starting from (5.396). In this case, all terms in (5.399) will be conditioned on the filtered velocity and filtered compositions,[166] in addition to the residual mixture-fraction vector $\boldsymbol{\xi}' \equiv \boldsymbol{\xi} - \bar{\boldsymbol{\xi}}$. In the case of a one-component mixture fraction, the latter can be modeled by a presumed beta PDF with mean $\bar{\xi}^*$ and variance $\langle \xi'^2 \rangle^*$. LES transport equations must then be added to solve for the mixture-fraction mean and variance. Despite this added complication, all model terms carry over from the original model. The only remaining difficulty is to extend (5.399) to cover inhomogeneous flows.[167] As with the conditional-moment closure discussed in Section 5.8 (see (5.316) on p. 215), this extension will be non-trivial, and thus is not attempted here.

5.11 Transported PDF methods

Of all of the methods reviewed thus far in this book, only DNS and the linear-eddy model require no closure for the molecular-diffusion term or the chemical source term in the scalar transport equation. However, we have seen that both methods are computationally expensive for three-dimensional inhomogeneous flows of practical interest. For all of the other methods, closures are needed for either scalar mixing or the chemical source term. For example, classical micromixing models treat chemical reactions exactly, but the fluid dynamics are overly simplified. The extension to multi-scalar presumed PDFs comes the closest to providing a flexible model for inhomogeneous turbulent reacting flows. Nevertheless, the presumed form of the joint scalar PDF in terms of a finite collection of delta functions may be inadequate for complex chemistry. The next step – computing the shape of the joint scalar PDF from its transport equation – comprises transported PDF methods and is discussed in detail in the next chapter. Some of the properties of transported PDF methods are listed here.

165 The similarity model for the residual mixture-fraction variance results by ignoring the transport terms in (5.425) and using (5.426): $\langle \xi'^2 \rangle^* = [2(C_{s\phi}\Delta)^2/C_\phi](\nabla\bar{\xi}^*)\cdot(\nabla\bar{\xi}^*)$.

166 In this case, the filtered compositions are partitioned into the filtered mixture-fraction vector and the filtered reaction-progress variables.

167 Alternatively, one can attempt to formulate an 'algebraic' model by assuming that the spatial/temporal transport terms are null for the conditional reaction-progress variables. However, care must be taken to ensure that the correct filtered reaction-progress variables are predicted by the resulting model.

- Transported PDF methods combine an exact treatment of chemical reactions with a closure for the turbulence field. (Transported PDF methods can also be combined with LES.) They do so by solving a balance equation for the joint one-point, velocity, composition PDF wherein the chemical-reaction terms are in closed form. In this respect, transported PDF methods are similar to micromixing models.
- Unlike presumed PDF methods, transported PDF methods do not require *a priori* knowledge of the joint PDF. The effect of chemical reactions on the joint PDF is treated exactly. The key modeled term in transported PDF methods is the molecular mixing term (i.e., the micromixing term), which describes how molecular diffusion modifies the shape of the joint PDF.
- The one-point PDF contains no length-scale information so that scalar dissipation must be modeled. In particular, the mixing time scale must be related to turbulence time scales through a model for the scalar dissipation rate.

The numerical methods employed to solve the transported PDF transport equation are very different from standard CFD codes. In essence, the joint PDF is represented by a large collection of 'notional particles.' The idea is similar to the presumed multi-scalar PDF method discussed in the previous section. The principal difference is that the notional particles move in real and composition space by well defined stochastic models. Some of the salient features of transported PDF codes are listed below.

- Transported PDF codes are more CPU intensive than moment and presumed PDF closures, but are still tractable for engineering calculations.
- The Monte-Carlo solvers developed for transported PDF methods are highly parallelizable and scale nearly linearly on multiprocessor computers.
- While both Eulerian and Lagrangian Monte-Carlo codes have been implemented, Lagrangian code offers many advantages, especially when velocity is a random variable.
- In theory, an arbitrary number of scalars could be used in transported PDF calculations. In practice, applications are limited by computer memory. In most applications, a reaction lookup table is used to store pre-computed changes due to chemical reactions, and models are limited to five to six chemical species with arbitrary chemical kinetics. Current research efforts are focused on 'smart' tabulation schemes capable of handling larger numbers of chemical species.

Transported PDF methods are continuing to develop in various directions (i.e, compressible flows, LES turbulence models, etc.). A detailed overview of transported PDF methods is presented in the following two chapters.

6

PDF methods for turbulent reacting flows

The methods presented in Chapter 5 attempt to close the chemical source term by making *a priori* assumptions concerning the form of the joint composition PDF. In contrast, the methods discussed in this chapter involve solving a transport equation for the joint PDF in which the chemical source term appears in closed form. In the literature, this type of approach is referred to as *transported PDF* or *full PDF* methods. In this chapter, we begin by deriving the fundamental transport equation for the one-point joint velocity, composition PDF. We then look at modeling issues that arise from this equation, and introduce the Lagrangian PDF formulation as a natural starting point for developing transported PDF models. The simulation methods that are used to 'solve' for the joint PDF are presented in Chapter 7.

6.1 Introduction

As we saw in Chapter 1, the one-point joint velocity, composition PDF contains random variables representing the three velocity components and all chemical species at a particular spatial location. The restriction to a one-point description implies the following.

- The joint PDF contains no information concerning local velocity and/or scalar gradients. A two-point description would be required to describe the gradients.[1]
- All non-linear terms involving spatial gradients require transported PDF *closures*. Examples of such terms are viscous dissipation, pressure fluctuations, and scalar dissipation.

[1] Alternatively, a joint velocity, scalar, scalar-gradient PDF could be employed. However, this only moves the closure problem to a higher multi-point level.

The one-point joint *composition* PDF contains random variables representing all chemical species at a particular spatial location. It can be found from the joint velocity, composition PDF by integrating over the entire phase space of the velocity components. The loss of instantaneous velocity information implies the following.

- There is no direct information on the velocity field, and thus a turbulence model is required to provide this information.
- There is no direct information on scalar transport due to velocity fluctuations. A PDF scalar-flux model is required to describe turbulent scalar transport.[2]
- There is no information on the instantaneous scalar dissipation rate and its coupling to the turbulence field. A transported PDF micromixing model is required to determine the effect of molecular diffusion on both the shape of the PDF and the rate of scalar-variance decay.

One of the principal attractions of employing transported PDF methods for turbulent reacting flows is the fact that all one-point terms are treated exactly. The most notable one-point term is the chemical source term $\mathbf{S}(\boldsymbol{\phi})$. The absence of a chemical-source-term closure problem implies that, in principle, arbitrarily complex kinetic schemes can be treated without the closure difficulties discussed in Chapter 5. In this respect, transported PDF methods are a natural extension of the micromixing models introduced in Section 5.6. One important advantage of transported PDF methods over micromixing models is the ability to validate directly the transported PDF closures term by term against DNS data. This provides greater assurance as to the general validity of the transported PDF closures, as well as pointing to weaknesses that require further model refinement.

6.1.1 Velocity, composition PDF

The joint velocity, composition PDF is defined in terms of the probability of observing the event where the velocity and composition random fields at point $\mathbf{x}$ and time t fall in the differential neighborhood of the fixed values $\mathbf{V}$ and $\boldsymbol{\psi}$:

$$\begin{aligned} & f_{\mathbf{U},\boldsymbol{\phi}}(\mathbf{V}, \boldsymbol{\psi}; \mathbf{x}, t)\,\mathrm{d}\mathbf{V}\,\mathrm{d}\boldsymbol{\psi} \\ & \quad \equiv \text{Prob}\,\{(\mathbf{V} < \mathbf{U}(\mathbf{x}, t) < \mathbf{V} + \mathrm{d}\mathbf{V}) \cap (\boldsymbol{\psi} < \boldsymbol{\phi}(\mathbf{x}, t) < \boldsymbol{\psi} + \mathrm{d}\boldsymbol{\psi})\}\,. \end{aligned} \tag{6.1}$$

The transport equation for $f_{\mathbf{U},\boldsymbol{\phi}}(\mathbf{V}, \boldsymbol{\psi}; \mathbf{x}, t)$ will be derived in the next section. Before doing so, it is important to point out one of the key properties of transported PDF methods:

If $f_{\mathbf{U},\boldsymbol{\phi}}(\mathbf{V}, \boldsymbol{\psi}; \mathbf{x}, t)$ were known, then all one-point statistics of $\mathbf{U}$ and $\boldsymbol{\phi}$ would also be known.

[2] The usual choice is an extension of the classical gradient-diffusion model; hence, it cannot describe counter-gradient or non-gradient diffusion.

For example, given $f_{\mathbf{U},\phi}(\mathbf{V}, \psi; \mathbf{x}, t)$, the expected value of any function $Q(\mathbf{U}, \phi)$ of the random variables[3] $\mathbf{U}$ and ϕ can be found by integrating over velocity–composition phase space:[4]

$$\langle Q(\mathbf{U}, \phi)\rangle = \iint_{-\infty}^{+\infty} Q(\mathbf{V}, \psi) f_{\mathbf{U},\phi}(\mathbf{V}, \psi; \mathbf{x}, t)\, \mathrm{d}\mathbf{V}\, \mathrm{d}\psi, \tag{6.2}$$

where, in order to simplify the notation, we have denoted the integrals over velocity and composition phase space by a single integral sign.[5] This implies that, given $f_{\mathbf{U},\phi}(\mathbf{V}, \psi; \mathbf{x}, t)$, all of the usual velocity and/or scalar statistics associated with second-order RANS models are known:

$\langle \mathbf{U}\rangle$ mean velocity,

$\langle \phi\rangle$ mean composition,

$\langle u_i u_j\rangle$ Reynolds stresses,

$\langle \phi'_\alpha \phi'_\beta\rangle$ composition covariances,

$\langle u_i \phi'_\alpha\rangle$ scalar fluxes,

and

$\langle \mathbf{S}(\phi)\rangle$ mean chemical source term.

Moreover, higher-order statistics are also available, e.g.,

$\langle u_i u_j u_k\rangle$ triple correlations.

Thus, solutions to the transported PDF equation will provide more information than is available from second-order RANS models without the problem of closing the chemical source term.

On the other hand, the expected values of all non-linear functions of velocity and/or scalar gradients require two-point information, and thus cannot be found from $f_{\mathbf{U},\phi}(\mathbf{V}, \psi; \mathbf{x}, t)$. For example, the scalar dissipation rate involves the square of the scalar gradients:

$$\left\langle \frac{\partial \phi'}{\partial x_i} \frac{\partial \phi'}{\partial x_i} \right\rangle,$$

and cannot be computed from the one-point composition PDF. Transported PDF closures can be developed for these terms by extending the joint PDF description to include, for example, the scalar dissipation as a random variable. However, this process introduces new multi-point terms that must be modeled.

[3] The function Q cannot be an operator that introduces multi-point terms such as gradients or space/time-convolution integrals.

[4] Note that the expected value on the left-hand side is an implicit function of $\mathbf{x}$ and t. This is generally true of all expected values generated by transported PDF methods. However, for notational simplicity, an explicit dependence is indicated only when needed to avoid confusion. In that case, we would write $\langle Q\rangle(\mathbf{x}, t)$.

[5] Except when needed to avoid confusion, we will use this simplified notation throughout this chapter.

6.1.2 Composition PDF

The joint composition PDF $f_{\phi}(\psi; \mathbf{x}, t)$ can be found by integrating the joint velocity, composition PDF over velocity phase space:

$$f_{\phi}(\psi; \mathbf{x}, t) = \int_{-\infty}^{+\infty} f_{\mathbf{U},\phi}(\mathbf{V}, \psi; \mathbf{x}, t)\, \mathrm{d}\mathbf{V}. \tag{6.3}$$

Many of the reported applications of transported PDF methods have employed the composition PDF in lieu of the velocity, composition PDF. The principal advantage of using the latter is an improved description of the turbulent velocity field and its coupling to the composition fields, i.e., the scalar flux (Pope 1994b) and the molecular mixing term (Fox 1996b). In the composition PDF approach, an RANS turbulence model must be coupled to the PDF solver. Nevertheless, from the composition PDF all one-point scalar statistics can be computed exactly; in particular, the chemical source term $\langle \mathbf{S}(\phi) \rangle$. For turbulent-reacting-flow calculations for which the turbulence model does not play a crucial role in determining scalar statistics, the composition PDF provides a slightly more economical approach while still treating the chemical source term exactly. Thus, we will discuss both approaches in this chapter.

6.2 Velocity, composition PDF transport equation

The joint velocity, composition PDF transport equation can be derived starting from the transport equations[6] for $\mathbf{U}$ and ϕ given in Chapter 1:[7]

$$\frac{\mathrm{D}U_i}{\mathrm{D}t} \equiv \frac{\partial U_i}{\partial t} + U_j \frac{\partial U_i}{\partial x_j} = A_i, \tag{6.4}$$

where the right-hand side is found from the Navier–Stokes equation:

$$A_i \equiv \nu \frac{\partial^2 U_i}{\partial x_j \partial x_j} - \frac{1}{\rho} \frac{\partial p}{\partial x_i} + g_i. \tag{6.5}$$

Likewise, for the composition fields,

$$\frac{\mathrm{D}\phi_\alpha}{\mathrm{D}t} \equiv \frac{\partial \phi_\alpha}{\partial t} + U_j \frac{\partial \phi_\alpha}{\partial x_j} = \Theta_\alpha, \tag{6.6}$$

where the right-hand side is defined by[8]

$$\Theta_\alpha \equiv \Gamma_\alpha \frac{\partial^2 \phi_\alpha}{\partial x_j \partial x_j} + S_\alpha(\phi). \tag{6.7}$$

[6] The transport equations are written in terms of the convected derivative to simplify the derivation. Likewise, the right-hand sides are denoted by a single operator to simplify the notation. We will again assume that the velocity field is solenoidal.

[7] The derivation method is completely general and can easily be extended to include other random variables, e.g., ϵ or ϵ_ϕ.

[8] We have used Fick's law of diffusion with separate molecular diffusivities for each species. However, most PDF models for molecular mixing do not include differential-diffusion effects.

There are several equivalent approaches[9] that can be followed to derive the transport equation for $f_{\mathbf{U},\phi}(\mathbf{V}, \psi; \mathbf{x}, t)$. We shall use the one first suggested by Pope (1985), which is based on finding two independent expressions for the mean convected derivation of an arbitrary function of the velocity and compositions. The two forms are subsequently equated and, since the function is arbitrary, the remaining terms yield the transported PDF equation. The principal advantage of using this approach is that the required passage from a multi-point to a conditional single-point PDF can be very easily located in the manipulations, and hopefully more clearly understood.

6.2.1 Mean convected derivative: first form

We start by considering an arbitrary measurable[10] one-point[11] scalar function of the *random fields* $\mathbf{U}$ and ϕ: $Q(\mathbf{U}, \phi)$. Note that, based on this definition, Q is also a random field parameterized by $\mathbf{x}$ and t. For each realization of a turbulent flow, Q will be different, and we can define its expected value using the probability distribution for the ensemble of realizations.[12] Nevertheless, the expected value of the convected derivative of Q can be expressed in terms of partial derivatives of the *one-point* joint velocity, composition PDF:[13]

$$\begin{aligned}\left\langle \frac{\mathrm{D}Q}{\mathrm{D}t} \right\rangle &= \frac{\partial \langle Q \rangle}{\partial t} + \frac{\partial \langle U_i Q \rangle}{\partial x_i} \\ &= \frac{\partial}{\partial t} \iint_{-\infty}^{+\infty} Q(\mathbf{V}, \psi) f_{\mathbf{U},\phi}(\mathbf{V}, \psi; \mathbf{x}, t) \, \mathrm{d}\mathbf{V} \, \mathrm{d}\psi \\ &\quad + \frac{\partial}{\partial x_j} \iint_{-\infty}^{+\infty} V_j Q(\mathbf{V}, \psi) f_{\mathbf{U},\phi}(\mathbf{V}, \psi; \mathbf{x}, t) \, \mathrm{d}\mathbf{V} \, \mathrm{d}\psi.\end{aligned} \tag{6.8}$$

The expected value on the left-hand side is taken with respect to the entire ensemble of random fields. However, as shown for the velocity derivative starting from (2.82) on p. 45, only two-point information is required to estimate a derivative.[14] The first equality then follows from the fact that the expected value and derivative operators commute. In the two integrals after the second equality, only $f_{\mathbf{U},\phi}(\mathbf{V}, \psi; \mathbf{x}, t)$ depends on $\mathbf{x}$ and t

9 For other alternatives, see Dopazo (1994) and Pope (2000).

10 The term *measurable* (Billingsley 1979) is meant to exclude 'poorly behaved' functions for which the probability cannot be defined. In addition, we will need to exclude functions which 'blow up' so quickly at infinity that higher-order moments are undefined.

11 Hence, we exclude functions (or, more precisely, operators) which require multi-point information such as $Q(\mathbf{U}, \phi) = \mathbf{U}(t_1, \mathbf{x}) \cdot \mathbf{U}(t_2, \mathbf{x})$.

12 In other words, each member of the ensemble represents the entire time/space history of a single experiment. For a turbulent flow, each experiment will result in a different time/space history, and the infinite collection of all such experiments constitutes an ensemble. Note that one realization contains an enormous amount of information (i.e., at least as much as is produced by a DNS of a turbulent flow which saves the fields at every time step).

13 We again assume constant-density flow so that $\mathbf{U}$ is solenoidal. The first equality requires this assumption.

14 If Q were a multi-point operator, this simplification would not be possible.

(i.e., $\mathbf{V}$ and ψ are integration variables). Thus, the right-hand side simplifies to

$$\left\langle \frac{DQ}{Dt} \right\rangle = \iint_{-\infty}^{+\infty} Q(\mathbf{V}, \psi) \left\{ \frac{\partial f_{\mathbf{U},\phi}}{\partial t} + V_j \frac{\partial f_{\mathbf{U},\phi}}{\partial x_j} \right\} \mathrm{d}\mathbf{V}\, \mathrm{d}\psi. \tag{6.9}$$

In summary, due to the linear nature of the derivative operator, it is possible to express the expected value of a convected derivative of Q in terms of temporal and spatial derivatives of the one-point joint velocity, composition PDF. Two-point information about the random fields $\mathbf{U}$ and ϕ is needed only to prove that the expected value and derivative operators commute, and does not appear in the final expression (i.e., (6.9)).

6.2.2 Mean convected derivative: second form

The expected value of the convected derivative of Q can also be written in a second independent form starting with[15]

$$\begin{aligned} \frac{DQ}{Dt}(\mathbf{U}, \phi) &= \frac{\partial Q}{\partial U_i}\frac{DU_i}{Dt} + \frac{\partial Q}{\partial \phi_i}\frac{D\phi_i}{Dt} \\ &= \frac{\partial Q}{\partial U_i} A_i + \frac{\partial Q}{\partial \phi_i}\Theta_i, \end{aligned} \tag{6.10}$$

where the second equality follows by substituting the transport equations (6.4) and (6.6) for $\mathbf{U}$ and ϕ. Note also that we have changed from a Greek to a Roman index on ϕ_i in order to imply summation over all composition variables. Taking the expected value of this expression yields

$$\left\langle \frac{DQ}{Dt} \right\rangle = \left\langle \frac{\partial Q}{\partial U_i} A_i \right\rangle + \left\langle \frac{\partial Q}{\partial \phi_i}\Theta_i \right\rangle. \tag{6.11}$$

Note that A_i and Θ_i will, in general, depend on multi-point information from the random fields $\mathbf{U}$ and ϕ. For example, they will depend on the velocity/scalar gradients and the velocity/scalar Laplacians. Since these quantities are not contained in the one-point formulation for $\mathbf{U}(\mathbf{x}, t)$ and $\phi(\mathbf{x}, t)$, we will lump them all into an unknown random vector $\mathbf{Z}(\mathbf{x}, t)$.[16] Denoting the one-point joint PDF of $\mathbf{U}$, ϕ, and $\mathbf{Z}$ by $f_{\mathbf{U},\phi,\mathbf{Z}}(\mathbf{V}, \psi, \mathbf{z}; \mathbf{x}, t)$, we can express it in terms of an unknown conditional joint PDF and the known joint velocity, composition PDF:

$$f_{\mathbf{U},\phi,\mathbf{Z}}(\mathbf{V}, \psi, \mathbf{z}) = f_{\mathbf{Z}|\mathbf{U},\phi}(\mathbf{z}|\mathbf{V}, \psi) f_{\mathbf{U},\phi}(\mathbf{V}, \psi). \tag{6.12}$$

[15] Note that the extension to other flow quantities is formally trivial: application of the chain rule would add the corresponding terms on the right-hand side of the final equality.

[16] Conceptually, since a random field can be represented by a Taylor expansion about the point $(\mathbf{x}, t)$, the random vector $\mathbf{Z}(\mathbf{x}, t)$ could have as its components the infinite set of partial derivatives of $\mathbf{U}$ and ϕ of all orders greater than zero with respect to x_1, x_2, x_3, and t evaluated at $(\mathbf{x}, t)$.

The expected value of the term involving A_i can then be written as

$$\left\langle \frac{\partial Q}{\partial U_i} A_i \right\rangle \equiv \iiint_{-\infty}^{+\infty} \frac{\partial Q}{\partial V_i}(\mathbf{V}, \boldsymbol{\psi}) A_i(\mathbf{V}, \boldsymbol{\psi}, \mathbf{z}) f_{\mathbf{U},\boldsymbol{\phi},\mathbf{Z}}(\mathbf{V}, \boldsymbol{\psi}, \mathbf{z}) \, \mathrm{d}\mathbf{z} \, \mathrm{d}\mathbf{V} \, \mathrm{d}\boldsymbol{\psi}$$
$$= \iint_{-\infty}^{+\infty} \frac{\partial Q}{\partial V_i}(\mathbf{V}, \boldsymbol{\psi}) \langle A_i | \mathbf{V}, \boldsymbol{\psi} \rangle f_{\mathbf{U},\boldsymbol{\phi}}(\mathbf{V}, \boldsymbol{\psi}) \, \mathrm{d}\mathbf{V} \, \mathrm{d}\boldsymbol{\psi}, \tag{6.13}$$

where the conditional expected value is defined by

$$\langle A_i | \mathbf{V}, \boldsymbol{\psi} \rangle \equiv \int_{-\infty}^{+\infty} A_i(\mathbf{V}, \boldsymbol{\psi}, \mathbf{z}) f_{\mathbf{Z}|\mathbf{U},\boldsymbol{\phi}}(\mathbf{z} | \mathbf{V}, \boldsymbol{\psi}) \, \mathrm{d}\mathbf{z}. \tag{6.14}$$

Note that, since the dependence on all other variables has been integrated out, $\langle A_i | \mathbf{V}, \boldsymbol{\psi} \rangle$ is a function of only $\mathbf{V}$ and $\boldsymbol{\psi}$ (in addition to an implicit dependence on $\mathbf{x}$ and t). A similar expression can be derived for the term involving Θ_i:

$$\left\langle \frac{\partial Q}{\partial \phi_i} \Theta_i \right\rangle = \iint_{-\infty}^{+\infty} \frac{\partial Q}{\partial \psi_i}(\mathbf{V}, \boldsymbol{\psi}) \langle \Theta_i | \mathbf{V}, \boldsymbol{\psi} \rangle f_{\mathbf{U},\boldsymbol{\phi}}(\mathbf{V}, \boldsymbol{\psi}) \, \mathrm{d}\mathbf{V} \, \mathrm{d}\boldsymbol{\psi}. \tag{6.15}$$

The next step is to use integration by parts to rewrite the right-hand sides of (6.13) and (6.15). Integration by parts is required for each variable used in the partial derivative of Q, i.e., V_i and ψ_i. The first integral on the right-hand side thus has one less integration than the second, the missing integration just being the one for which the integrand is evaluated at infinity:[17]

$$\left\langle \frac{\partial Q}{\partial U_i} A_i \right\rangle = \iint_{-\infty}^{+\infty} Q(\mathbf{V}, \boldsymbol{\psi}) \langle A_i | \mathbf{V}, \boldsymbol{\psi} \rangle f_{\mathbf{U},\boldsymbol{\phi}}(\mathbf{V}, \boldsymbol{\psi}) \, \mathrm{d}\mathbf{V}_{\neq i} \, \mathrm{d}\boldsymbol{\psi} \Bigg|_{V_i=-\infty}^{V_i=+\infty}$$
$$- \iint_{-\infty}^{+\infty} Q(\mathbf{V}, \boldsymbol{\psi}) \frac{\partial}{\partial V_i} [\langle A_i | \mathbf{V}, \boldsymbol{\psi} \rangle f_{\mathbf{U},\boldsymbol{\phi}}] \, \mathrm{d}\mathbf{V} \, \mathrm{d}\boldsymbol{\psi}, \tag{6.16}$$

and

$$\left\langle \frac{\partial Q}{\partial \phi_i} \Theta_i \right\rangle = \iint_{-\infty}^{+\infty} Q(\mathbf{V}, \boldsymbol{\psi}) \langle \Theta_i | \mathbf{V}, \boldsymbol{\psi} \rangle f_{\mathbf{U},\boldsymbol{\phi}}(\mathbf{V}, \boldsymbol{\psi}) \, \mathrm{d}\mathbf{V} \, \mathrm{d}\boldsymbol{\psi}_{\neq i} \Bigg|_{\psi_i=-\infty}^{\psi_i=+\infty}$$
$$- \iint_{-\infty}^{+\infty} Q(\mathbf{V}, \boldsymbol{\psi}) \frac{\partial}{\partial \psi_i} [\langle \Theta_i | \mathbf{V}, \boldsymbol{\psi} \rangle f_{\mathbf{U},\boldsymbol{\phi}}] \, \mathrm{d}\mathbf{V} \, \mathrm{d}\boldsymbol{\psi}. \tag{6.17}$$

The first terms on the right-hand sides of (6.16) and (6.17) are related to the probability flux at infinity. For all 'well behaved' PDF and all 'well behaved' functions[18] $Q(\mathbf{U}, \boldsymbol{\phi})$,

[17] For example, $\mathrm{d}\mathbf{V}_{\neq 2} = \mathrm{d}V_1 \mathrm{d}V_3$.

[18] A well behaved function is one that grows slowly enough at infinity so that the integrand is null. Since the PDF typically falls off exponentially at infinity, any function that can be expressed as a convergent power series will be well behaved.

the flux at infinity must be null so that these terms are null. The second form, (6.11), of the mean convected derivative of Q thus reduces to

$$\left\langle \frac{\mathrm{D}Q}{\mathrm{D}t} \right\rangle = - \iint_{-\infty}^{+\infty} Q(\mathbf{V}, \psi) \left\{ \frac{\partial}{\partial V_i} [\langle A_i | \mathbf{V}, \psi \rangle f_{\mathbf{U},\phi}] + \frac{\partial}{\partial \psi_i} [\langle \Theta_i | \mathbf{V}, \psi \rangle f_{\mathbf{U},\phi}] \right\} \mathrm{d}\mathbf{V} \, \mathrm{d}\psi. \tag{6.18}$$

The implied summations on the right-hand side range over $\{V_i : i \in 1, 2, 3\}$ and $\{\psi_i : i \in 1, \ldots, N_s\}$, where $N_s = K + 1$, and K is the number of chemical species.

6.2.3 Joint PDF transport equation: final form

Combining (6.9) and (6.18), and using the fact that the equality must hold for arbitrary choices of Q, leads to the joint velocity, composition PDF transport equation:[19]

$$\frac{\partial f_{\mathbf{U},\phi}}{\partial t} + V_i \frac{\partial f_{\mathbf{U},\phi}}{\partial x_i} = - \frac{\partial}{\partial V_i} [\langle A_i | \mathbf{V}, \psi \rangle f_{\mathbf{U},\phi}] - \frac{\partial}{\partial \psi_i} [\langle \Theta_i | \mathbf{V}, \psi \rangle f_{\mathbf{U},\phi}]. \tag{6.19}$$

From this expression, it can be seen that the joint PDF evolves by transport in

(i) real space, due to the fluctuating velocity field $\mathbf{V}$;
(ii) velocity phase space, due to the conditional acceleration term $\langle A_i | \mathbf{V}, \psi \rangle$;
(iii) composition phase space, due to the conditional reaction/diffusion term $\langle \Theta_i | \mathbf{V}, \psi \rangle$.

However, in order to use (6.19) to solve for the joint PDF, closures must be supplied for the conditional acceleration and reaction/diffusion terms. For simplicity, we will refer to these terms as the *conditional fluxes*.

6.2.4 Conditional fluxes: the unclosed terms

The conditional fluxes play a key role in the PDF transport equation. Writing them out using the definitions of A_i and Θ_α yields[20]

$$\langle A_i | \mathbf{V}, \psi \rangle = \left\langle \left(\nu \frac{\partial^2 U_i}{\partial x_j \partial x_j} - \frac{1}{\rho} \frac{\partial p'}{\partial x_i} \right) | \mathbf{V}, \psi \right\rangle - \frac{1}{\rho} \frac{\partial \langle p \rangle}{\partial x_i} + g_i \tag{6.20}$$

and

$$\langle \Theta_\alpha | \mathbf{V}, \psi \rangle = \left\langle \Gamma_\alpha \frac{\partial^2 \phi_\alpha}{\partial x_j \partial x_j} | \mathbf{V}, \psi \right\rangle + S_\alpha(\psi). \tag{6.21}$$

The first term on the right-hand side of (6.20) is unclosed, and corresponds to the effects of viscous dissipation and pressure fluctuations. The closure of this term is the principal

[19] It is now obvious that the addition of other random variables simply leads to additional conditional flux terms on the right-hand side of the PDF transport equation. The major challenge is thus to find appropriate closures for the conditional fluxes.

[20] Note that, unlike unconditional expected values, conditional expected values do not usually commute with derivatives. For example, $\langle \nabla^2 \mathbf{U} | \mathbf{V} \rangle \neq \nabla^2 \langle \mathbf{U} | \mathbf{V} \rangle = \mathbf{0}$, whereas $\langle \nabla^2 \mathbf{U} \rangle = \nabla^2 \langle \mathbf{U} \rangle$.

modeling challenge for the transported PDF description of turbulence. The first term on the right-hand side of (6.21) is also unclosed, and corresponds to the effects of molecular diffusion. The closure of the term is the principal modeling challenge for the transported PDF description of turbulent mixing. We shall see later that both of these terms can be extracted from DNS data for model comparison. Finally, note that the chemical source term in (6.21) appears in closed form. Thus, unlike in RANS models for reacting scalars, the chemical source term is treated exactly.

In contrast to moment closures, the models used to close the conditional fluxes typically involve *random processes*. The choice of the models will directly affect the evolution of the shape of the PDF, and thus indirectly affect the moments of the PDF. For example, once closures have been selected, all one-point statistics involving $\mathbf{U}$ and ϕ can be computed by deriving moment transport equations starting from the transported PDF equation. Thus, in Section 6.4, we will look at the relationship between (6.19) and RANS transport equations. However, we will first consider the composition PDF transport equation.

6.3 Composition PDF transport equation

We have seen that the joint velocity, composition PDF treats both the velocity and the compositions as random variables. However, as noted in Section 6.1, it is possible to carry out transported PDF simulations using only the composition PDF. By definition, $f_{\phi}(\psi; \mathbf{x}, t)$ can be found from $f_{\mathbf{U},\phi}(\mathbf{V}, \psi; \mathbf{x}, t)$ using (6.3). The same definition can be used with the transported PDF equation derived in Section 6.2 to find a transport equation for $f_{\phi}(\psi; \mathbf{x}, t)$.

6.3.1 Derivation of the transport equation

The transport equation for the joint composition PDF can be found by integrating the velocity, composition PDF transport equation, (6.19), over velocity phase space:

$$\int_{-\infty}^{+\infty} \left\{ \frac{\partial f_{\mathbf{U},\phi}}{\partial t} + V_i \frac{\partial f_{\mathbf{U},\phi}}{\partial x_i} = -\frac{\partial}{\partial V_i} [\langle A_i | \mathbf{V}, \psi \rangle f_{\mathbf{U},\phi}] - \frac{\partial}{\partial \psi_i} [\langle \Theta_i | \mathbf{V}, \psi \rangle f_{\mathbf{U},\phi}] \right\} \mathrm{d}\mathbf{V}. \tag{6.22}$$

Taking each contribution individually yields the following simplifications.

Accumulation:

$$\begin{aligned} \int_{-\infty}^{+\infty} \frac{\partial f_{\mathbf{U},\phi}}{\partial t} \mathrm{d}\mathbf{V} &= \frac{\partial}{\partial t} \left\{ \int_{-\infty}^{+\infty} f_{\mathbf{U},\phi}(\mathbf{V}, \psi) \, \mathrm{d}\mathbf{V} \right\} \\ &= \frac{\partial f_{\phi}}{\partial t}. \end{aligned} \tag{6.23}$$

Convection:[21]

$$
\begin{aligned}
\int_{-\infty}^{+\infty} V_i \frac{\partial f_{\mathbf{U},\phi}}{\partial x_i} \, d\mathbf{V} &= \frac{\partial}{\partial x_i} \left\{ \int_{-\infty}^{+\infty} V_i f_{\mathbf{U},\phi}(\mathbf{V}, \psi) \, d\mathbf{V} \right\} \\
&= \frac{\partial}{\partial x_i} \left\{ \int_{-\infty}^{+\infty} V_i f_{\mathbf{U}|\phi}(\mathbf{V}|\psi) f_\phi(\psi) \, d\mathbf{V} \right\} \\
&= \frac{\partial}{\partial x_i} [\langle U_i | \psi \rangle f_\phi] \\
&= \frac{\partial}{\partial x_i} [(\langle U_i \rangle + \langle u_i | \psi \rangle) f_\phi] \\
&= \langle U_i \rangle \frac{\partial f_\phi}{\partial x_i} + \frac{\partial}{\partial x_i} [\langle u_i | \psi \rangle f_\phi].
\end{aligned} \tag{6.24}
$$

Acceleration:[22]

$$
\int_{-\infty}^{+\infty} \frac{\partial}{\partial V_i} [\langle A_i | \mathbf{V}, \psi \rangle f_{\mathbf{U},\phi}(\mathbf{V}, \psi)] \, d\mathbf{V} = 0. \tag{6.25}
$$

Reaction/diffusion:[23]

$$
\begin{aligned}
&\int_{-\infty}^{+\infty} \frac{\partial}{\partial \psi_i} [\langle \Theta_i | \mathbf{V}, \psi \rangle f_{\mathbf{U},\phi}(\mathbf{V}, \psi)] \, d\mathbf{V} \\
&= \frac{\partial}{\partial \psi_i} \left\{ \int_{-\infty}^{+\infty} \langle \Theta_i | \mathbf{V}, \psi \rangle f_{\mathbf{U},\phi}(\mathbf{V}, \psi) \, d\mathbf{V} \right\} \\
&= \frac{\partial}{\partial \psi_i} \left\{ \int_{-\infty}^{+\infty} \langle \Theta_i | \mathbf{V}, \psi \rangle f_{\mathbf{U}|\phi}(\mathbf{V}|\psi) f_\phi(\psi) \, d\mathbf{V} \right\} \\
&= \frac{\partial}{\partial \psi_i} [\langle \Theta_i | \psi \rangle f_\phi] \\
&= \frac{\partial}{\partial \psi_i} [\langle \Gamma_i \nabla^2 \phi'_i | \psi \rangle f_\phi] + \frac{\partial}{\partial \psi_i} [(\Gamma_i \nabla^2 \langle \phi_i \rangle + S_i(\psi)) f_\phi].
\end{aligned} \tag{6.26}
$$

Collecting all terms, the composition PDF transport equation reduces to

$$
\begin{aligned}
&\frac{\partial f_\phi}{\partial t} + \langle U_i \rangle \frac{\partial f_\phi}{\partial x_i} + \frac{\partial}{\partial x_i} [\langle u_i | \psi \rangle f_\phi] \\
&= -\frac{\partial}{\partial \psi_i} [\langle \Gamma_i \nabla^2 \phi'_i | \psi \rangle f_\phi] - \frac{\partial}{\partial \psi_i} [(\Gamma_i \nabla^2 \langle \phi_i \rangle + S_i(\psi)) f_\phi].
\end{aligned} \tag{6.27}
$$

The composition PDF thus evolves by convective transport in real space due to the mean velocity (macromixing), by convective transport in real space due to the scalar-conditioned velocity fluctuations (mesomixing), and by transport in composition space due to molecular mixing (micromixing) and chemical reactions. Note that any of the molecular mixing models to be discussed in Section 6.6 can be used to close the micromixing term. The chemical source term is closed; thus, only the mesomixing term requires a new model.

[21] The key step here is to use the conditional PDF to eliminate the velocity dependence. However, this generates a new unclosed term. Note that we assume the mean velocity field to be solenoidal in the last line.

[22] We again use integration by parts to show that this term is null.

[23] Again, the key step here is to use the conditional PDF to eliminate the velocity dependence.

6.3.2 Scalar-conditioned velocity fluctuations

The only new unclosed term that appears in the composition PDF transport equation is $\langle u_i|\psi\rangle$. The exact form of this term will depend on the flow. However, if the velocity and scalar fields are Gaussian, then the scalar-conditioned velocity can be expressed in terms of the scalar flux and the scalar covariance matrix:

$$\langle u_i|\psi\rangle = \langle u_i\phi^{\mathrm{T}}\rangle\langle\phi'\phi'^{\mathrm{T}}\rangle^{-1}(\psi - \langle\phi\rangle). \tag{6.28}$$

In general, the scalar fields will not be Gaussian. Thus, analogous to what is done in second-moment closure methods, a gradient-diffusion model is usually employed to close this term:

$$\langle u_i|\psi\rangle = -\frac{\Gamma_{\mathrm{T}}}{f_\phi}\frac{\partial f_\phi}{\partial x_i}. \tag{6.29}$$

This model generates a turbulent-diffusivity term in (6.27) which transports the composition PDF in real space:[24]

$$\frac{\partial f_\phi}{\partial t} + \langle U_i\rangle\frac{\partial f_\phi}{\partial x_i} = \frac{\partial}{\partial x_i}\left[\Gamma_{\mathrm{T}}\frac{\partial f_\phi}{\partial x_i}\right] - \frac{\partial}{\partial\psi_i}[\langle\Gamma_i\nabla^2\phi'_i|\psi\rangle f_\phi] - \frac{\partial}{\partial\psi_i}[(\Gamma_i\nabla^2\langle\phi_i\rangle + S_i(\psi))f_\phi], \tag{6.30}$$

where the turbulent diffusivity $\Gamma_{\mathrm{T}}(\mathbf{x}, t)$ depends on the local values of k and ε.

6.3.3 Relationship to Lagrangian micromixing models

When applying the composition PDF transport equation to real flows, the mean velocity $\langle\mathbf{U}\rangle$, the turbulent kinetic energy k, and turbulence dissipation rate ε must be supplied by a separate turbulence model (or input from experimental data). Nevertheless, unlike traditional chemical-reaction-engineering models which also treat the chemical source term $\mathbf{S}(\psi)$ exactly (i.e., Lagrangian micromixing models discussed in Section 5.6), the transport equation for f_ϕ contains the simultaneous contributions of mixing processes at all scales (i.e., macro, meso, and micromixing). Note, however, that the difficulty of treating scalar length-scale distribution effects must still be addressed in the micromixing model (e.g., through the model for the joint scalar dissipation rates $\boldsymbol{\varepsilon}$). We will look at the computational advantages/difficulties of employing the composition PDF as opposed to the joint velocity, composition PDF in Chapter 7. In general, the composition PDF approach will suffice for flows that can be adequately described at the level of k–ε, gradient-diffusion models. On the other hand, flows which require a full second-order RANS model are best treated by the joint velocity, composition PDF approach.

[24] The diffusion term in this expression differs from the Fokker–Planck equation. This difference leads to a Γ_{T}-dependent term in the drift term of the corresponding stochastic differential equation (6.177), p. 294.

6.4 Relationship to RANS transport equations

The transported PDF equation contains more information than an RANS turbulence model, and can be used to derive the latter. We give two example derivations ($\langle \mathbf{U} \rangle$ and $\langle \mathbf{u}\mathbf{u}^{\mathrm{T}} \rangle$) below, but the same procedure can be carried out to find any one-point statistic of the velocity and/or composition fields.[25]

6.4.1 RANS mean velocity transport equation

The RANS transport equation for the mean velocity $\langle \mathbf{U} \rangle$ can be derived by multiplying the transported PDF equation (6.19) by V_i and integrating over phase space:

$$\iint_{-\infty}^{+\infty} V_i \left\{ \frac{\partial f_{\mathbf{U},\phi}}{\partial t} + V_j \frac{\partial f_{\mathbf{U},\phi}}{\partial x_j} \right. \\ \left. = -\frac{\partial}{\partial V_j}[\langle A_j | \mathbf{V}, \boldsymbol{\psi} \rangle f_{\mathbf{U},\phi}] - \frac{\partial}{\partial \psi_j}[\langle \Theta_j | \mathbf{V}, \boldsymbol{\psi} \rangle f_{\mathbf{U},\phi}] \right\} \mathrm{d}\mathbf{V}\, \mathrm{d}\boldsymbol{\psi}. \tag{6.31}$$

Integrating each term individually yields the following simplifications. Note that for the first two terms we make use of the fact that only $f_{\mathbf{U},\phi}$ depends on $\mathbf{x}$ and t. Integration by parts is used to simplify the remaining terms.

Accumulation:

$$\begin{aligned} \iint_{-\infty}^{+\infty} V_i \frac{\partial f_{\mathbf{U},\phi}}{\partial t} \mathrm{d}\mathbf{V}\, \mathrm{d}\boldsymbol{\psi} &= \frac{\partial}{\partial t} \iint_{-\infty}^{+\infty} V_i f_{\mathbf{U},\phi}(\mathbf{V}, \boldsymbol{\psi}; \mathbf{x}, t)\, \mathrm{d}\mathbf{V}\, \mathrm{d}\boldsymbol{\psi} \\ &= \frac{\partial \langle U_i \rangle}{\partial t}. \end{aligned} \tag{6.32}$$

Convection:

$$\begin{aligned} \iint_{-\infty}^{+\infty} V_i V_j \frac{\partial f_{\mathbf{U},\phi}}{\partial x_j} \mathrm{d}\mathbf{V}\, \mathrm{d}\boldsymbol{\psi} &= \frac{\partial}{\partial x_j} \iint_{-\infty}^{+\infty} V_i V_j f_{\mathbf{U},\phi}(\mathbf{V}, \boldsymbol{\psi}; \mathbf{x}, t)\, \mathrm{d}\mathbf{V}\, \mathrm{d}\boldsymbol{\psi} \\ &= \frac{\partial \langle U_i U_j \rangle}{\partial x_j}. \end{aligned} \tag{6.33}$$

Acceleration:

$$\iint_{-\infty}^{+\infty} V_i \frac{\partial}{\partial V_j}[\langle A_j | \mathbf{V}, \boldsymbol{\psi} \rangle f_{\mathbf{U},\phi}]\, \mathrm{d}\mathbf{V}\, \mathrm{d}\boldsymbol{\psi}$$

$$= \iint_{-\infty}^{+\infty} V_i \langle A_j | \mathbf{V}, \boldsymbol{\psi} \rangle f_{\mathbf{U},\phi}(\mathbf{V}, \boldsymbol{\psi}; \mathbf{x}, t)\, \mathrm{d}\mathbf{V}_{\neq j}\, \mathrm{d}\boldsymbol{\psi} \Bigg|_{V_j=-\infty}^{V_j=+\infty}$$

[25] A joint statistic of particular interest is the scalar flux: $\langle u_i \phi_\alpha \rangle$. The reader is encouraged to derive its transport equation, and to compare the result to (3.102) on p. 84.

$$
\begin{aligned}
& - \iint_{-\infty}^{+\infty} \langle A_i | \mathbf{V}, \boldsymbol{\psi} \rangle f_{\mathbf{U},\boldsymbol{\phi}}(\mathbf{V}, \boldsymbol{\psi}; \mathbf{x}, t) \, \mathrm{d}\mathbf{V} \, \mathrm{d}\boldsymbol{\psi} \\
&= -\langle A_i \rangle.
\end{aligned} \tag{6.34}
$$

Reaction/diffusion:

$$
\begin{aligned}
& \iint_{-\infty}^{+\infty} V_i \frac{\partial}{\partial \psi_j} [\langle \Theta_j | \mathbf{V}, \boldsymbol{\psi} \rangle f_{\mathbf{U},\boldsymbol{\phi}}] \, \mathrm{d}\mathbf{V} \, \mathrm{d}\boldsymbol{\psi} \\
&= \left. \iint_{-\infty}^{+\infty} V_i \langle \Theta_j | \mathbf{V}, \boldsymbol{\psi} \rangle f_{\mathbf{U},\boldsymbol{\phi}}(\mathbf{V}, \boldsymbol{\psi}) \, \mathrm{d}\mathbf{V} \, \mathrm{d}\boldsymbol{\psi}_{\neq j} \right|_{\psi_j = -\infty}^{\psi_j = +\infty} \\
&= 0.
\end{aligned} \tag{6.35}
$$

Note that for any statistic involving only the velocity, the reaction/diffusion term will always be zero. Likewise, for any statistic involving only the composition, the acceleration term will always be zero.

Collecting all of the terms, the RANS transport equation for the mean velocity simplifies to

$$
\frac{\partial \langle U_i \rangle}{\partial t} + \frac{\partial \langle U_i U_j \rangle}{\partial x_j} = \langle A_i \rangle. \tag{6.36}
$$

The right-hand side of this expression can be written as

$$
\langle A_i \rangle = \nu \frac{\partial^2 \langle U_i \rangle}{\partial x_j \partial x_j} - \frac{1}{\rho} \frac{\partial \langle p \rangle}{\partial x_i} + g_i. \tag{6.37}
$$

The term $\langle U_i U_j \rangle$ can be simplified by introducing the Reynolds stresses:

$$
\langle U_i U_j \rangle = \langle U_i \rangle \langle U_j \rangle + \langle u_i u_j \rangle. \tag{6.38}
$$

Finally, using the fact that the mean velocity field is solenoidal, the RANS mean velocity transport equation reduces to

$$
\frac{\partial \langle U_i \rangle}{\partial t} + \langle U_j \rangle \frac{\partial \langle U_i \rangle}{\partial x_j} + \frac{\partial \langle u_i u_j \rangle}{\partial x_j} = \nu \frac{\partial^2 \langle U_i \rangle}{\partial x_j \partial x_j} - \frac{1}{\rho} \frac{\partial \langle p \rangle}{\partial x_i} + g_i. \tag{6.39}
$$

Note that the same equation was found by Reynolds averaging of the Navier–Stokes equation ((2.93), p. 47).

6.4.2 Reynolds-stress transport equation

In the transported PDF formulation, the Reynolds stresses are found starting from

$$\iint_{-\infty}^{+\infty} V_i V_j \left\{ \frac{\partial f_{\mathbf{U},\phi}}{\partial t} + V_k \frac{\partial f_{\mathbf{U},\phi}}{\partial x_k} \right.$$
$$\left. = -\frac{\partial}{\partial V_k}[\langle A_k|\mathbf{V},\psi\rangle f_{\mathbf{U},\phi}] - \frac{\partial}{\partial \psi_k}[\langle \Theta_k|\mathbf{V},\psi\rangle f_{\mathbf{U},\phi}] \right\} \mathrm{d}\mathbf{V}\,\mathrm{d}\psi. \tag{6.40}$$

Manipulation of this expression (as was done above for the mean velocity) leads to terms for accumulation and convection that are analogous to (6.32) and (6.33), respectively, but with $\langle U_i U_j\rangle$ in place of $\langle U_i\rangle$. On the other hand, the acceleration term yields

$$\iint_{-\infty}^{+\infty} V_i V_j \frac{\partial}{\partial V_k}[\langle A_k|\mathbf{V},\psi\rangle f_{\mathbf{U},\phi}]\,\mathrm{d}\mathbf{V}\,\mathrm{d}\psi = -\langle U_i A_j\rangle - \langle U_j A_i\rangle. \tag{6.41}$$

The resulting transport equation for $\langle U_i U_j\rangle$ has the form

$$\frac{\partial \langle U_i U_j\rangle}{\partial t} + \frac{\partial \langle U_i U_j U_k\rangle}{\partial x_k} = \langle U_i A_j\rangle + \langle U_j A_i\rangle. \tag{6.42}$$

At this point, the next step is to decompose the velocity into its mean and fluctuating components, and to substitute the result into the left-hand side of (6.42). In doing so, the triple-correlation term $\langle u_i u_j u_k\rangle$ will appear. Note that if the joint velocity PDF were known (i.e., by solving (6.19)), then the triple-correlation term could be computed exactly. This is not the case for the RANS turbulence models discussed in Section 4.4 where a model is required to close the triple-correlation term.

In transported PDF methods (Pope 2000), the closure model for $\langle A_i|\mathbf{V},\psi\rangle$ will be a known function[26] of $\mathbf{V}$. Thus, $\langle U_i A_j\rangle$ will be closed and will depend on the moments of $\mathbf{U}$ and their spatial derivatives.[27] Moreover, Reynolds-stress models derived from the PDF transport equation are guaranteed to be realizable (Pope 1994b), and the corresponding consistent scalar flux model can easily be found. We shall return to this subject after looking at typical conditional acceleration and conditional diffusion models.

6.5 Models for conditional acceleration

The conditional acceleration can be decomposed into mean and fluctuating components:

$$\langle A_i|\mathbf{V},\psi\rangle = \langle A_i'|\mathbf{V},\psi\rangle + \nu\frac{\partial^2\langle U_i\rangle}{\partial x_j \partial x_j} - \frac{1}{\rho}\frac{\partial\langle p\rangle}{\partial x_i} + g_i, \tag{6.43}$$

[26] For passive scalars, $\langle A_i|\mathbf{V},\psi\rangle$ will be independent of ψ. For active scalars, e.g., temperature dependence in stratified flows, $\langle A_i|\mathbf{V},\psi\rangle$ would also depend on the value of ψ.

[27] We shall see that transported PDF closures for the velocity field are usually linear in $\mathbf{V}$. Thus $\langle U_i A_j\rangle$ will depend only on the first two moments of $\mathbf{U}$. In general, non-linear velocity models could be formulated, in which case arbitrary moments of $\mathbf{U}$ would appear in the Reynolds-stress transport equation.

where the unclosed fluctuating component is defined by

$$\langle A'_i | \mathbf{V}, \psi \rangle \equiv \left\langle \nu \frac{\partial^2 u_i}{\partial x_j \partial x_j} \middle| \mathbf{V}, \psi \right\rangle - \left\langle \frac{1}{\rho} \frac{\partial p'}{\partial x_i} \middle| \mathbf{V}, \psi \right\rangle. \tag{6.44}$$

The two terms in $\langle A'_i | \mathbf{V}, \psi \rangle$ represent the effects of the fluctuating viscous forces and the fluctuating pressure field, respectively. The molecular viscous-dissipation term in (6.43) is negligible at high Reynolds numbers. However, it becomes important in boundary layers where the Reynolds number is low, and must be included in the boundary conditions as described below.

6.5.1 Velocity PDF: decoupling from the scalar field

For constant-density flows, the scalar fields have no direct effect on the velocity field so that the conditional acceleration is independent of composition:

$$\langle A'_i | \mathbf{V}, \psi \rangle = \langle A'_i | \mathbf{V} \rangle. \tag{6.45}$$

Thus, the conditional acceleration model can be developed by considering only the PDF transport equation for $f_{\mathbf{U}}$:[28]

$$\frac{\partial f_{\mathbf{U}}}{\partial t} + V_i \frac{\partial f_{\mathbf{U}}}{\partial x_i} = -\frac{\partial}{\partial V_i} \left[\langle A_i | \mathbf{V} \rangle f_{\mathbf{U}} \right]. \tag{6.46}$$

Closure of the conditional acceleration requires a model for

$$\langle A'_i | \mathbf{V} \rangle \equiv \left\langle \nu \frac{\partial^2 u_i}{\partial x_j \partial x_j} \middle| \mathbf{V} \right\rangle - \left\langle \frac{1}{\rho} \frac{\partial p'}{\partial x_i} \middle| \mathbf{V} \right\rangle. \tag{6.47}$$

Note that, since the joint velocity PDF will be known from the solution of (6.46), the model can be formulated in terms of $\mathbf{V}$, the moments of $\mathbf{U}$ and their gradients, or any arbitrary function of $\mathbf{V}$. However, as with Reynolds-stress models, in practice (Pope 2000) the usual choice of functional dependencies is limited to

$$\mathbf{V}, \quad \langle \mathbf{U} \rangle, \quad \langle u_i u_j \rangle, \quad \frac{\partial \langle \mathbf{U} \rangle}{\partial x_i}, \quad \text{and} \quad \varepsilon,$$

where ε can be treated as a deterministic variable, or modeled as a random process using a separate stochastic model (Pope 2000).[29]

6.5.2 Velocity PDF closures

A general modeling strategy that has been successfully employed to model the joint velocity PDF in a wide class of turbulent flows[30] is to develop stochastic models which

[28] This expression is found by integrating the PDF transport equation for $f_{\mathbf{U},\phi}$, (6.19), over composition phase space.

[29] For a detailed overview of proposed methods for modeling the conditional acceleration, see Dopazo (1994) and Pope (2000).

[30] For an overview on this modeling strategy, see Pope (2000). For an alternative view based on statistical physics, see Minier and Pozorski (1997).

are linear in $\mathbf{V}$, and which yield viable Reynolds-stress models for $\langle u_i u_j \rangle$ (Haworth and Pope 1986). Mathematically, these models have the form[31]

$$\langle A_i' | \mathbf{V} \rangle = G_{ij}(V_j - \langle U_j \rangle) - \frac{C_0 \varepsilon}{2 f_{\mathbf{U}}} \frac{\partial f_{\mathbf{U}}}{\partial V_i}, \tag{6.48}$$

where $G_{ij}(\mathbf{x}, t)$ is a second-order tensor which depends on (Pope 2000)

$$\langle u_i u_j \rangle, \quad \frac{\partial \langle \mathbf{U} \rangle}{\partial x_i}, \quad \text{and} \quad \varepsilon.$$

Note that the expected value of A_i' is null:[32]

$$\begin{aligned}
\langle A_i' \rangle &= \int_{-\infty}^{+\infty} \langle A_i' | v \rangle f_{\mathbf{U}}(\mathbf{V}; \mathbf{x}, t) \, \mathrm{d}\mathbf{V} \\
&= G_{ij} \int_{-\infty}^{+\infty} \left(V_j - \langle U_j \rangle \right) f_{\mathbf{U}}(\mathbf{V}; \mathbf{x}, t) \, \mathrm{d}\mathbf{V} - \frac{C_0 \varepsilon}{2} \int_{-\infty}^{+\infty} \frac{\partial f_{\mathbf{U}}}{\partial V_i} \, \mathrm{d}\mathbf{V} \\
&= 0.
\end{aligned} \tag{6.49}$$

When the linear model is inserted into (6.46), a linear Fokker–Planck equation (Gardiner 1990) results:

$$\begin{aligned}
\frac{\partial f_{\mathbf{U}}}{\partial t} + V_i \frac{\partial f_{\mathbf{U}}}{\partial x_i} + \left(\nu \frac{\partial^2 \langle U_i \rangle}{\partial x_j \partial x_j} - \frac{1}{\rho} \frac{\partial \langle p \rangle}{\partial x_i} + g_i \right) \frac{\partial f_{\mathbf{U}}}{\partial V_i} \\
= -\frac{\partial}{\partial V_i} [G_{ij}(V_j - \langle U_j \rangle) f_{\mathbf{U}}] + \frac{C_0 \varepsilon}{2} \frac{\partial^2 f_{\mathbf{U}}}{\partial V_i \partial V_i}.
\end{aligned} \tag{6.50}$$

Homogeneous, linear Fokker–Planck equations are known to admit a multi-variate Gaussian PDF as a solution.[33] Thus, this closure scheme ensures that a joint Gaussian velocity PDF will result for statistically stationary, homogeneous turbulent flow.

6.5.3 Corresponding Reynolds-stress models

The closed PDF transport equation given above can be employed to derive a transport equation for the Reynolds stresses. The velocity–pressure gradient and the dissipation terms in the corresponding Reynolds-stress model result from

$$\begin{aligned}
\langle u_j A_i' \rangle + \langle u_i A_j' \rangle &= \langle u_j \nu \nabla^2 u_i \rangle + \langle u_i \nu \nabla^2 u_j \rangle \\
&\quad - \frac{1}{\rho} \left\langle u_j \frac{\partial p'}{\partial x_i} \right\rangle - \frac{1}{\rho} \left\langle u_i \frac{\partial p'}{\partial x_j} \right\rangle \\
&= -\varepsilon_{ij} + \nu \nabla^2 \langle u_i u_j \rangle + \Pi_{ij},
\end{aligned} \tag{6.51}$$

[31] We shall see that the last term leads to a random noise term in the Lagrangian PDF model for the velocity.
[32] Integration by parts is again used to show that the last term is null.
[33] In the literature on stochastic processes, the above Fokker–Planck equation describes a multi-variate Ornstein–Uhlenbeck process. For a discussion on the existence of Gaussian solutions to this process, see Gardiner (1990).

where the Π_{ij} and ε_{ij} are defined after (2.105) on p. 49. The expected values on the left-hand side can be easily computed from the proposed closure. For example,

$$\begin{aligned}\langle u_j A_i' \rangle &= \int_{-\infty}^{+\infty} (V_j - \langle U_j \rangle) \langle A_i' | \mathbf{V} \rangle f_{\mathbf{U}}(\mathbf{V}; \mathbf{x}, t) \, \mathrm{d}\mathbf{V} \\ &= G_{ik} \int_{-\infty}^{+\infty} (V_j - \langle U_j \rangle)(V_k - \langle U_k \rangle) f_{\mathbf{U}}(\mathbf{V}; \mathbf{x}, t) \, \mathrm{d}\mathbf{V} \\ &\quad - \frac{C_0 \varepsilon}{2} \int_{-\infty}^{+\infty} V_j \frac{\partial f_{\mathbf{U}}}{\partial V_i} \, \mathrm{d}\mathbf{V} \\ &= G_{ik} \langle u_j u_k \rangle + \frac{C_0 \varepsilon}{2} \delta_{ij}. \end{aligned} \tag{6.52}$$

The linear model thus yields

$$\langle u_j A_i' \rangle + \langle u_i A_j' \rangle = G_{ik} \langle u_j u_k \rangle + G_{jk} \langle u_i u_k \rangle + C_0 \varepsilon \delta_{ij}. \tag{6.53}$$

Note that, due to the assumed form of (6.48), the resulting dissipation term is isotropic.

The final form of the Reynolds-stress transport equation will then depend on the choice of G_{ij}. For example, in the simplified Langevin model (SLM)

$$G_{ij} = -\left(\frac{1}{2} + \frac{3}{4} C_0 \right) \frac{\varepsilon}{k} \delta_{ij}, \tag{6.54}$$

so that

$$\langle u_j A_i' \rangle + \langle u_i A_j' \rangle = -\left(1 + \frac{3}{2} C_0 \right) \frac{\epsilon}{k} \langle u_i' u_j' \rangle + C_0 \varepsilon \delta_{ij}. \tag{6.55}$$

The corresponding Reynolds-stress model has the form

$$\frac{\partial \langle u_i u_j \rangle}{\partial t} + \langle U_k \rangle \frac{\partial \langle u_i u_j \rangle}{\partial x_k} + \frac{\partial \langle u_i u_j u_k \rangle}{\partial x_k} = \mathcal{P}_{ij} - \left(\frac{1}{2} + \frac{3}{4} C_0 \right) \frac{\varepsilon}{k} \langle u_i u_j \rangle + C_0 \varepsilon \delta_{ij}, \tag{6.56}$$

where $\mathcal{P}_{ij}$ is the production term for the Reynolds stresses derived in Chapter 2. The standard choice of $C_0 = 2.1$ leads to the Reynolds-stress model proposed by Rotta (1951). Given its ease of implementation, the SLM has been widely used in transported PDF calculations. However, if the complexity of the flow requires the predictive properties of a more detailed Reynolds-stress model, the generalized Langevin model (GLM) described below should be employed.

6.5.4 Generalized Langevin model

The GLM was introduced by Haworth and Pope (1986), and has a second-order tensor of the form[34]

$$G_{ij} = \frac{\varepsilon}{k} \left(\alpha_1 \delta_{ij} + \alpha_2 b_{ij} + \alpha_3 b_{ij}^2 \right) + H_{ijkl} \frac{\partial \langle U_k \rangle}{\partial x_l}, \tag{6.57}$$

[34] For further details on the GLM, see Pope (1994b) and Pope (2000).

where the fourth-order tensor $\mathbf{H}$ is a linear function of anisotropy tensor $\mathbf{b}$ and contains nine model parameters:[35]

$$H_{ijkl} = \beta_1 \delta_{ij}\delta_{kl} + \beta_2 \delta_{ik}\delta_{jl} + \beta_3 \delta_{il}\delta_{jk} + \gamma_1 \delta_{ij} b_{kl} + \gamma_2 \delta_{ik} b_{jl} + \gamma_3 \delta_{il} b_{jk} + \gamma_4 \delta_{kl} b_{ij} + \gamma_5 \delta_{jl} b_{ik} + \gamma_6 \delta_{jk} b_{il}. \tag{6.58}$$

The GLM thus contains 12 model parameters which can be chosen to agree with any realizable Reynolds-stress model (Pope 1994b). Pope and co-workers have made detailed comparisons between the GLM and turbulent-flow data. In general, the agreement is good for flows where the corresponding Reynolds-stress model performs adequately.

Finally, it is important to reiterate that while the Reynolds-stress model requires a closure for the triple-correlation term $\langle u_i u_j u_k \rangle$, the PDF transport equation does not:

Transport by velocity fluctuations is treated exactly at the joint velocity PDF level.

We shall see that a conditional acceleration model in the form of (6.48) is equivalent to a stochastic Lagrangian model for the velocity fluctuations whose characteristic correlation time is proportional to ε/k. As discussed below, this implies that the scalar flux $\langle u_i \phi' \rangle$ will be closed at the joint velocity, composition PDF level, and thus that a consistent scalar-flux transport equation can be derived from the PDF transport equation.

Despite the ability of the GLM to reproduce any realizable Reynolds-stress model, Pope (2002b) has shown that it is not consistent with DNS data for homogeneous turbulent shear flow. In order to overcome this problem, and to incorporate the Reynolds-number effects observed in DNS, a stochastic model for the acceleration can be formulated (Pope 2002a; Pope 2003). However, it remains to be seen how well such models will perform for more complex inhomogeneous flows. In particular, further research is needed to determine the functional forms of the coefficient matrices in both homogeneous and inhomogeneous turbulent flows.

6.5.5 Extension to velocity, composition PDF

As noted earlier, the extension of the conditional acceleration model in (6.48) to the joint velocity, composition PDF is trivial:

$$\langle A_i' | \mathbf{V}, \psi \rangle = G_{ij}(V_j - \langle U_j \rangle) - \frac{C_0 \varepsilon}{2 f_{\mathbf{U},\phi}} \frac{\partial f_{\mathbf{U},\phi}}{\partial V_i}. \tag{6.59}$$

The implications on the scalar-flux model, however, are profound. Indeed, one finds[36]

$$\Pi_i^\alpha = \langle \phi_\alpha A_i' \rangle = G_{ij} \langle u_j \phi_\alpha \rangle, \tag{6.60}$$

[35] The definition of H_{ijkl} has been extended to include terms that are quadratic in b_{ij} (Wouters *et al.* 1996).

[36] Strictly speaking, the first equality is only valid at high Reynolds numbers where the scalar- and viscous-dissipation terms are uncorrelated (i.e., locally isotropic).

so that the pressure-scrambling term in the scalar-flux transport equation ((3.102), p. 84) is closed (Pope 1994b). A consistent scalar-flux model will thus use the same expression for G_{ij} in the pressure-scrambling term as is used in the Reynolds-stress model. This fact is not widely appreciated in the second-order RANS modeling community where inconsistent scalar-flux models are often employed with sophisticated Reynolds-stress models. Physically, however, this result is very satisfying: since scalar mixing is controlled by integral-scale velocity fluctuations, the scalar-flux transport model should be completely determined by the conditional-acceleration closure.[37]

6.5.6 Coupling with mean pressure field

The mean pressure field $\langle p \rangle$ appears as a closed term in the conditional acceleration $\langle A_i | \mathbf{V}, \psi \rangle$. Nevertheless, it must be computed from a Poisson equation found by taking the divergence of the mean velocity transport equation:

$$-\frac{1}{\rho}\frac{\partial^2 \langle p \rangle}{\partial x_i \partial x_i} = \frac{\partial^2 \langle u_i u_j \rangle}{\partial x_i \partial x_j} + \frac{\partial \langle U_j \rangle}{\partial x_i}\frac{\partial \langle U_i \rangle}{\partial x_j}. \tag{6.61}$$

Since the mean velocity and Reynolds-stress fields are known given the joint velocity PDF $f_{\mathbf{U}}(\mathbf{V}; \mathbf{x}, t)$, the right-hand side of this expression is closed. Thus, in theory, a standard Poisson solver could be employed to find $\langle p \rangle(\mathbf{x}, t)$. However, in practice, $\langle \mathbf{U} \rangle(\mathbf{x}, t)$ and $\langle u_i u_j \rangle(\mathbf{x}, t)$ must be estimated from a finite-sample Lagrangian particle simulation (Pope 2000), and therefore are subject to considerable statistical noise. The spatial derivatives on the right-hand side of (6.61) are consequently even noisier, and therefore are of no practical use when solving for the mean pressure field. The development of numerical methods to overcome this difficulty has been one of the key areas of research in the development of stand-alone transported PDF codes.[38]

At least three approaches have been proposed to solve for the mean pressure field that avoid the noise problem. The first approach is to extract the mean pressure field from a simultaneous consistent[39] Reynolds-stress model solved using a standard CFD solver.[40] While this approach does alleviate the noise problem, it is intellectually unsatisfying since it leads to a redundancy in the velocity model.[41] The second approach seeks to overcome the noise problem by computing the so-called 'particle-pressure field' in an equivalent, but superior, manner (Delarue and Pope 1997). Moreover, this approach leads to a truly

[37] We shall see in the next section that the conditional-diffusion closure may contribute an additional dissipation term. However, since at high Reynolds numbers the velocity and scalar fields should be locally isotropic, this term will be negligible.

[38] Stand-alone refers to the capability of determining the mean pressure field based on the Monte-Carlo simulation without resorting to an external CFD code.

[39] In many reported applications, a k–ε CFD code has been employed, which gives an inconsistent description of the Reynolds stresses. Wouters (1998) has shown that this leads to other subtle problems, most notably in the scalar-flux model.

[40] The resulting PDF code is thus not a stand-alone code. For details on implementation, see Correa and Pope (1992).

[41] One can argue, however, that the improved description of the triple-correlation term in the Reynolds-stress model and the improved scalar-flux model justify this redundancy.

stand-alone transported PDF code. The key idea is the recognition that the role of the mean pressure field in constant-density flows is to ensure mass continuity. Dampening out the local fluctuations in the density resulting from statistical noise is thus equivalent to finding the correct mean pressure field. The third approach avoids the need to solve (6.60) by invoking a hybrid approach which uses a grid-based CFD code to solve for $\langle \mathbf{U} \rangle$, and a particle-based PDF code to solve for the fluctuating velocity field (Muradoglu *et al.* 1999). Since the mean pressure field appears only in (6.39) and not in the Reynolds-stress transport equation, a hybrid CFD–PDF method is completely consistent, non-redundant, and computationally more efficient than a stand-alone PDF code. We will look at the underlying numerical methods in more detail in Chapter 7.

6.5.7 Wall boundary conditions for velocity PDF

Most of the reported applications of the generalized Langevin model have been to free shear flows where boundary conditions pose little difficulty. The extension to wall-bounded flows poses new challenges, and work on imposing correct boundary conditions on $f_{\mathbf{U}}(\mathbf{V}; \mathbf{x}, t)$ near the wall has been reported by Dreeben and Pope (1997a); see also Dreeben and Pope (1997b) and Dreeben and Pope (1998). As with Reynolds-stress models, the treatment of thin shear flows is now well understood. However, the extension to more complex flows (e.g., impinging jets) has yet to be considered. Nevertheless, it can be expected that as Reynolds-stress models are improved to treat such flows, the improvements can be carried over directly to the PDF transport equation. More details on near-wall modeling can be found in Pope (2000).

6.5.8 Large-eddy PDF methods

As discussed above, the GLM was developed in the spirit of Reynolds-stress modeling. An obvious extension is to devise large-eddy-based closures for the conditional acceleration. For this case, it is natural to decompose the instantaneous velocity into its resolved and unresolved components:[42]

$$\mathbf{U} = \bar{\mathbf{U}} + \mathbf{u}'. \tag{6.62}$$

The resolved velocity $\bar{\mathbf{U}}$ would then be found from an LES simulation, and the LES velocity PDF (defined in Section 4.2) would be written in terms of the unresolved velocity $\mathbf{u}'$. Alternatively, the filtered density function (FDF) approach can be used with a variant

[42] Theoretically, because the resolved scales are known (from LES) but random (depending on initial conditions), the PDF for the unresolved scales will be conditioned on a given realization of the resolved scales. The Reynolds-averaged velocity $\langle \mathbf{U} \rangle$ would thus be found by first computing the conditional mean of the unresolved scales for a given realization $\langle \mathbf{u}' | \bar{\mathbf{V}} \rangle(\mathbf{x}, t)$ (the result will depend on both $\mathbf{x}$ and t, even if the Reynolds averages are homogeneous and stationary), and then by averaging the conditional mean over all possible realizations of the resolved scales, i.e., $\langle \mathbf{U} \rangle = \langle \bar{\mathbf{U}} \rangle_{\mathrm{R}} + \langle \langle \mathbf{u}' | \bar{\mathbf{U}} \rangle \rangle_{\mathrm{R}}$, where $\langle \cdot \rangle_{\mathrm{R}}$ denotes the ensemble average over all possible realizations of $\bar{\mathbf{U}}(\mathbf{x}, t)$. Higher-order statistics would be computed similarly. In practice, this would require multiple large-eddy simulations or time/space averaging in homogeneous directions.

of the generalized Langevin model used to model the unresolved velocity fluctuations (Colucci *et al.* 1998; Gicquel *et al.* 2002). In practice, the principal difficulty when applying large-eddy PDF methods will undoubtedly be the computational cost. Indeed, multiple LES realization will be computationally prohibitive so one will be forced to resort to time averaging based on an assumption that large-eddy simulations are ergodic over a sufficiently long time scale. However, it is important to note that the computational cost of LES–PDF methods is independent of the Reynolds number. Thus, for example, the LES–PDF approach will be more expensive than standard LES by a constant factor, which is approximately the same as the additional cost of RANS–PDF methods over RANS models. Despite the computational challenges, relatively low-Reynolds-number LES–PDF simulations have been reported by Colucci *et al.* (1998), Jaberi *et al.* (1999), and Gicquel *et al.* (2002).

6.5.9 Velocity, wavenumber PDF models

In an effort to improve the description of the Reynolds stresses in the rapid distortion turbulence (RDT) limit, the velocity PDF description has been extended to include directional information in the form of a random wave vector by Van Slooten and Pope (1997). The added directional information results in a transported PDF model that corresponds to the directional spectrum of the velocity field in wavenumber space. The model thus represents a bridge between Reynolds-stress models and more detailed spectral turbulence models. Due to the exact representation of spatial transport terms in the PDF formulation, the extension to inhomogeneous flows is straightforward (Van Slooten *et al.* 1998), and maintains the exact solution in the RDT limit. The model has yet to be extensively tested in complex flows (see Van Slooten and Pope 1999); however, it has the potential to improve greatly the turbulence description for high-shear flows. More details on this modeling approach can be found in Pope (2000).

6.6 Models for conditional diffusion

Because the chemical source term is treated exactly in transported PDF methods, the conditional reaction/diffusion term,

$$\langle \Theta_\alpha | \mathbf{V}, \psi \rangle = \langle \Gamma_\alpha \nabla^2 \phi_\alpha | \mathbf{V}, \psi \rangle + S_\alpha(\psi), \tag{6.63}$$

will require a closure only for the conditional diffusion (or *molecular mixing*) term:

$$\langle \Gamma_\alpha \nabla^2 \phi_\alpha | \mathbf{V}, \psi \rangle = \langle \Gamma_\alpha \nabla^2 \phi'_\alpha | \mathbf{V}, \psi \rangle + \Gamma_\alpha \nabla^2 \langle \phi_\alpha \rangle. \tag{6.64}$$

While the form of this term is the same as the viscous-dissipation term in the conditional acceleration, the modeling approach is very different. Indeed, while the velocity field in a homogeneous turbulent flow is well described by a multi-variate Gaussian process, the scalar fields are very often bounded and, hence, non-Gaussian. Moreover, joint scalar scatter plots usually cover a bounded region of composition space that is completely determined by molecular mixing and chemistry.[43] Thus, when constructing a suitable molecular mixing model, the compositions must remain in the allowable region of composition space. This rules out the use of *linear* Langevin models[44] like (6.48), which would allow the scalar variables to spread arbitrarily in all directions.

6.6.1 Some useful constraints

Before discussing in detail specific molecular mixing models, it is useful to first state a few important constraints that can be derived by computing expected values. The first constraint follows from[45]

$$\langle\langle \Gamma_\alpha \nabla^2 \phi'_\alpha | \mathbf{U}, \phi \rangle\rangle = \langle \Gamma_\alpha \nabla^2 \phi'_\alpha \rangle = 0, \tag{6.65}$$

and can be stated in words as

> (I) *The molecular mixing model must leave the scalar mean unchanged.*

A second constraint can be found by multiplying the molecular mixing term by ϕ'_β and averaging:

$$\begin{aligned} \langle \phi'_\beta \langle \Gamma_\alpha \nabla^2 \phi'_\alpha | \mathbf{U}, \phi \rangle\rangle &= \Gamma_\alpha \langle \phi'_\beta \nabla^2 \phi'_\alpha \rangle \\ &= -\Gamma_\alpha \langle (\nabla \phi'_\alpha) \cdot (\nabla \phi'_\beta) \rangle + \Gamma_\alpha \nabla \cdot \langle \phi'_\beta \nabla \phi'_\alpha \rangle. \end{aligned} \tag{6.66}$$

The first term on the right-hand side of this expression is proportional to the joint scalar dissipation rate $\varepsilon_{\alpha\beta}$ (defined in (3.139), p. 90), while the second term corresponds to molecular transport in real space.[46] Thus,

$$\langle \phi'_\beta \Gamma_\alpha \nabla^2 \phi_\alpha \rangle = -\frac{1}{2} \left(\frac{\Gamma_\alpha}{\Gamma_\beta} \right)^{1/2} \varepsilon_{\alpha\beta} + \Gamma_\alpha \nabla \cdot \langle \phi'_\beta \nabla \phi'_\alpha \rangle, \tag{6.67}$$

and the second constraint can be stated in words as

> (II) *The molecular mixing model must yield the correct joint scalar dissipation rate.*

The third constraint has to do with local isotropy of the scalar and velocity fields (Fox 1996b; Pope 1998). At high Reynolds numbers, the small scales of the velocity and scalar

[43] Turbulent mixing (i.e., the scalar flux) transports fluid elements in real space, but leaves the scalars unchanged in composition space. This implies that in the absence of molecular diffusion and chemistry the one-point composition PDF in homogeneous turbulence will remain unchanged for all time. Contrast this to the velocity field which quickly approaches a multi-variate Gaussian PDF due, mainly, to the fluctuating pressure term in (6.47).

[44] A non-linear, multi-variate Langevin model wherein the diffusion coefficients depend on ψ can be formulated for scalar mixing (Fox 1999).

[45] The outermost $\langle\cdot\rangle$ on the left-hand side denotes the expected value with respect to $f_{\mathbf{U},\phi}$.

[46] At high Reynolds numbers, the second term is negligible.

fields should be nearly uncorrelated. Thus, the expected value of the product of a velocity gradient and a scalar gradient will be null. The third constraint can be found by multiplying the molecular mixing term by u_i and averaging:

$$\begin{aligned}\langle u_i \langle \Gamma_\alpha \nabla^2 \phi'_\alpha | \mathbf{U}, \boldsymbol{\phi} \rangle \rangle &= \Gamma_\alpha \langle u_i \nabla^2 \phi'_\alpha \rangle \\ &= -\Gamma_\alpha \langle (\boldsymbol{\nabla} u_i) \cdot (\boldsymbol{\nabla} \phi'_\alpha) \rangle + \Gamma_\alpha \boldsymbol{\nabla} \cdot \langle u_i \boldsymbol{\nabla} \phi'_\alpha \rangle .\end{aligned} \tag{6.68}$$

Invoking local isotropy, the first term on the right-hand side is null. The second term corresponds to molecular transport in real space, which is negligible at high Reynolds numbers. Thus, local isotropy requires

$$\langle u_i \Gamma_\alpha \nabla^2 \phi'_\alpha \rangle = 0, \tag{6.69}$$

and the third constraint can be stated in words as

(III) *The molecular mixing model must be uncorrelated with the velocity at high Reynolds numbers.*

Note, however, that in the presence of a mean scalar gradient the local isotropy condition is known to be incorrect (see Warhaft (2000) for a review of this topic). Although most molecular mixing models do not account for it, the third constraint can be modified to read

(III′) *The molecular mixing model must yield the correct local scalar isotropy.*

We shall see that constraints (I) and (II) are always taken into consideration when developing molecular mixing models, but that constraint (III) has been largely ignored. This is most likely because almost all of the existing molecular mixing models have been developed in the context of the joint composition PDF, i.e., for

$$\langle \Gamma_\alpha \nabla^2 \phi'_\alpha | \boldsymbol{\psi} \rangle ,$$

where constraint (III) is not applicable. Nevertheless, by introducing velocity-conditioned scalar statistics, constraint (III) can, with some difficulty, be incorporated into numerical simulations of any existing molecular mixing model, and will have a significant effect on the predicted mixing rate and on the scalar flux. We will look at this issue in more detail after considering specific molecular mixing models. In addition to these constraints, there also exists a number of 'desirable' properties of molecular mixing models, most of which have been deduced for experimental or DNS data.[47]

6.6.2 Desirable properties for mixing models

In general, we would like a molecular mixing model to have the following properties.

[47] The division between 'constraints' and 'desirable properties' is somewhat vague. For example, the boundedness property (ii) is in fact an essential property of any mixing model used for reacting scalars. It could thus rightly be considered as a constraint, and not just a desirable property.

(i) The PDF of inert scalars should relax to a multi-variate Gaussian form in homogeneous turbulence for arbitrary initial conditions.[48]
(ii) All scalars must remain in the allowable region as determined by mixing and chemistry.[49]
(iii) Any conserved linear combination of scalars must be maintained.[50]
(iv) Scalar mixing should be local in composition space.[51]
(v) The evolution of the scalar PDF should depend on the length-scale distribution of the scalar field.[52]
(vi) Any known Re, Sc, or Da dependence should be included in the model parameters.[53]

The reasoning behind each of these properties will be illustrated in the next section. We will then look at three simple molecular mixing models (namely, the CD, the IEM, and the FP models) and discuss why each is not completely satisfactory. For convenience, the list of constraints and desirable properties is summarized in Table 6.1.

6.6.3 Physical basis for desirable properties

Although much effort has gone into searching for molecular mixing models that improve upon the existing models,[54] no model completely satisfies all of the desirable properties

[48] The scalars must remain bounded (e.g., the mixture fraction must lie between 0 and 1), but as the variances decrease to zero the shape of the joint PDF about $\langle \phi \rangle$ should be asymptotically Gaussian.

[49] In terms of the joint PDF, it must remain exactly zero outside the allowable region. Note that the allowable region may grow in certain directions when differential diffusion is important. For example, for the extreme case with $\Gamma_1 > \Gamma_2 = 0$ and $\phi_1(\mathbf{x}, 0) = \phi_2(\mathbf{x}, 0)$, the joint PDF $f_{\phi_1,\phi_2}(\psi_1, \psi_2; t)$ will initially be non-zero only for $\psi_1 = \psi_2$. However, for $t > 0$ it will spread over the area bounded by the lines $\psi_1 = \psi_2$ and $\psi_1 = \langle \phi_1 \rangle$. Nevertheless, as pointed out by Pope (2000), the allowable region will generally shrink in the absence of differential-diffusion effects.

[50] For example, element balances are not affected by chemistry and, if the molecular diffusion constants are identical, when combined linearly will remain conserved (Pope 1983). This property eliminates models that let each scalar randomly fluctuate according to independent random processes. However, it does not exclude non-linear Langevin models with correlated noise terms. Based on the linear form of the scalar transport equation, Pope (1983) has proposed a more stringent *strong independence condition* that requires that, for any scalar field, the mixing model be unaffected by the other scalar fields. A related *weak independence condition* requires that this be true only for *uncorrelated* sets of scalar fields. For example, if due to *initial and boundary conditions* some scalar fields are correlated and have the same molecular diffusivity, then the covariance matrix will have non-zero off-diagonal components for the correlated scalars. For this case, the weak independence condition would allow the mixing model for a particular scalar to be expressed in terms of any of the scalars with which it is correlated. Note that the weak independence condition is consistent with the fact that the eigenvectors of the covariance matrix can be used to define a set of uncorrelated scalars (see (5.336), p. 220), which will then satisfy the strong independence condition. Thus, since the strong independence condition does not account for scalar-field correlation that results from the initial and boundary conditions, we will require that mixing models satisfy only the weak independence condition.

[51] Local mixing is best defined in terms of stochastic models. However, this condition is meant to rule out models based on 'jump processes' where the scalar variables jump large distances in composition space for arbitrarily small dt. It also rules out 'interactions' between points in composition space and global statistics such as the mean.

[52] In other words, the model should recognize differences in shape of the scalar spectra and their effect on the scalar dissipation rate.

[53] This is rarely done in practice. For example, all commonly used models ignore possible effects of chemical reactions on the scalar-mixing process. Compare this with the flamelet model, where mixing and reactions are tightly coupled.

[54] For a review, see Dopazo (1994).

Table 6.1. *Constraints and desirable properties of molecular mixing models.*

(I)	Scalar means must remain unchanged
(II)	Joint scalar dissipation rate must be correct
(III)	Velocity and scalar gradients must be uncorrelated
(III′)	Local scalar isotropy must be correct
(i)	Inert-scalar PDF should relax to Gaussian form
(ii)	All scalars must remain in the allowable region
(iii)	Conserved linear combinations must be maintained
(iv)	Mixing should be local in composition space
(v)	Mixing rate should depend on scalar length scales
(vi)	Re, Sc, and Da dependencies should be taken into account

listed in Table 6.1. Thus, in order to understand the relative significance of each property, we will take each one separately and look at its physical basis.

(i) Why Gaussian?

Before the advent of DNS, little was known concerning the evolution of the scalar PDF. One of the first detailed DNS studies (Eswaran and Pope 1988) demonstrated that

- the PDF of an inert scalar evolves through a series of 'universal' shapes that are similar to a beta PDF;[55]
- the scalar dissipation rate strongly depends on the initial scalar length-scale distribution (i.e., the initial scalar spectrum);[56]
- the limiting form of the scalar PDF is nearly Gaussian.[57]

In another DNS study of two-scalar mixing (Juneja and Pope 1996), similar conclusions were drawn for the joint PDF of two inert scalars. These observations suggest that the development of molecular mixing models can proceed in two separate steps.

(1) Generate a mixing model that predicts the correct joint scalar PDF shape for *a given scalar covariance matrix*, including the asymptotic collapse to a Gaussian form.
(2) Couple it with a model for the *joint scalar dissipation rate* that predicts the correct scalar covariance matrix, including the effect of the initial scalar length-scale distribution.

[55] In this context, where the scalar mean is constant, universal implies that the shape of the scalar PDF at the same value of the scalar variance is identical regardless of the initial scalar spectrum.

[56] Since the scalar dissipation rate controls the time dependence of the scalar variance, a good model for the scalar dissipation rate is crucial for predicting the form of the scalar PDF at a particular time t.

[57] Similar behavior is seen in a purely diffusive system when the initial scalar spectrum is sufficiently broad (Fox 1994).

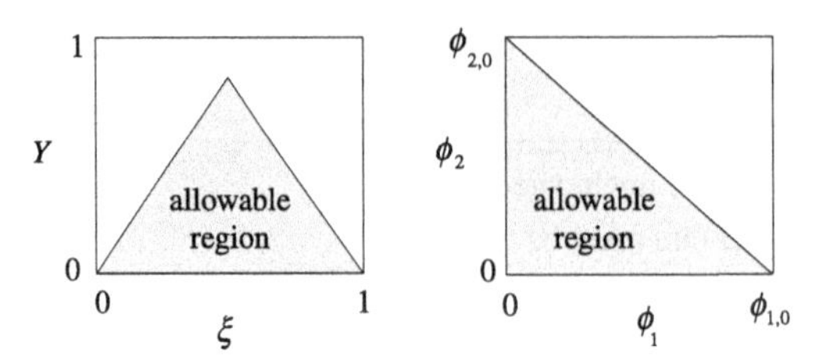

Figure 6.1. Sketch of allowable region for a one-step reaction.

Most molecular mixing models concentrate on step (1). However, for chemical-reactor applications, step (2) can be very important since the integral length scales of the scalar and velocity fields are often unequal ($L_\phi \neq L_u$) due to the feed-stream configuration. In the FP model (discussed below), step (1) is handled by the shape matrix $\mathbf{H}$, while step (2) requires an appropriate model for ε.

(ii) Why worry about the allowable region?

Consideration of the allowable region comes from the very nature of chemical reacting flows.

- All chemical species concentrations are, by definition, positive: $\boldsymbol{\phi} \geq \mathbf{0}$. Moreover, the maximum ($\boldsymbol{\phi}_{\max}$) and minimum ($\boldsymbol{\phi}_{\min}$) concentrations observed in a particular system will depend on the initial conditions, the extents of reaction, and the concentrations of other species.[58]
- All chemical reactions follow stoichiometric 'pathways' in composition space.[59] Along any pathway, the total number of moles of any element is conserved. This results in the system of linear constraints given by (5.10) on p. 144.
- Reaction-stoichiometry plus element-conservation constraints put non-trivial bounds on the composition vector $\boldsymbol{\phi}$. The interior of these bounds forms the allowable region (see Fig. 6.1).

Any molecular mixing model that takes the composition vector out of the allowable region violates stoichiometry and mass balances. Since the boundaries of the allowable region can be complicated and time-dependent, general Langevin-type mixing models are difficult to construct. (The FP model is an example of such a mixing model.) In order to remain in the allowable region, the diffusion coefficient in a Langevin-type model must be zero-flux across the boundary, and the drift vector must be pointed into the interior of the allowable region.

(iii) Why linearity?

Application of the element-conservation constraints, in particular, and linear transformations, in general, to the transport equations forces us to seek mixing models that preserve

[58] For finite-rate chemistry, the concentration bounds will thus be time-dependent. As seen in Section 5.5, the dependence of the upper bounds of the reaction-progress variables on the mixture fraction is usually non-trivial.

[59] These pathways are defined by the chemical source term $\mathbf{S}_c$ in the absence of other transport processes.

linearity. If the molecular-diffusion coefficients are identical, we have seen in Section 5.1 that since

$$\mathbf{\Lambda S}_{\rm c} = \mathbf{0}, \tag{6.70}$$

the transport equations for the chemical species imply that

$$\frac{\mathrm{D}\mathbf{c}_{\rm c}}{\mathrm{D}t} = \Gamma \nabla^2 \mathbf{c}_{\rm c}, \tag{6.71}$$

where $\mathbf{c}_{\rm c}$ is the (conserved) element 'concentration' vector. Thus, a viable mixing model must satisfy

$$\mathbf{\Lambda} \langle \Gamma \nabla^2 \mathbf{c} | \mathbf{c} \rangle = \langle \Gamma \nabla^2 \mathbf{c}_{\rm c} | \mathbf{c}_{\rm c} \rangle, \tag{6.72}$$

where the conditioning on $\mathbf{c}$ (or $\mathbf{c}_{\rm c}$) implies that all statistics appearing in the model are computed using the PDF of $\mathbf{c}$ (or $\mathbf{c}_{\rm c}$). In general, (6.72) will hold if the mixing model is linear in $\mathbf{c}$ and employs the same mixing time for every scalar (e.g., the IEM model). Nevertheless, as we will see for the FP model, stochastic models can also be developed that retain linearity and still allow the user to specify the joint scalar dissipation rate.[60] However, when using stochastic diffusion models, the 'boundary conditions' in composition space must be taken into account by the use of a shape function in the definition of the diffusion matrix.

Deterministic models that are non-linear in $\mathbf{c}$ will be limited to specific applications. For example, in the generalized IEM (GIEM) model (Tsai and Fox 1995a; Tsai and Fox 1998), which is restricted to binary mixing, the mixture fraction appears non-linearly:[61]

$$\langle \Gamma \nabla^2 \mathbf{c} | \mathbf{c}, \zeta \rangle = \alpha(\zeta, t)(\boldsymbol{\beta} - \mathbf{c}), \tag{6.73}$$

where $\boldsymbol{\beta}(t)$ is chosen to conserve the scalar means:[62]

$$\boldsymbol{\beta}(t) \equiv \frac{\langle \alpha(\xi, t)\mathbf{c} \rangle}{\langle \alpha(\xi, t) \rangle}. \tag{6.74}$$

In this manner, the non-relaxing property of the IEM model is avoided, and $\alpha(\zeta, t)$ can be chosen such that the limiting mixture-fraction PDF is Gaussian. Indeed, from DNS it is known that the conditional diffusion for the mixture fraction has a non-linear form that varies with time (see Fig. 6.2). In the GIEM model, this behavior is modeled by

$$\langle \Gamma \nabla^2 \xi | \zeta \rangle = \alpha(\zeta, t)(\beta_\xi - \zeta), \tag{6.75}$$

where $\beta_\xi(t)$ is the time-dependent intersection point with the ζ-axis, and $\alpha(\zeta, t)$ quantifies the deviations from linearity. Note that the GIEM model is in some ways similar to the

60 At first glance, an extension of the IEM model would appear also to predict the correct scalar covariance:

$$\frac{\mathrm{d}\boldsymbol{\phi}}{\mathrm{d}t} = -\frac{1}{2}\mathbf{S}_\Gamma \boldsymbol{\varepsilon} \mathbf{S}_\Gamma^{-1} \mathbf{S}_\phi^{-1} \mathbf{U}_\rho \mathbf{S}_\rho \mathbf{U}_\rho^{\rm T} \mathbf{S}_\phi^{-1} (\boldsymbol{\phi} - \langle \boldsymbol{\phi} \rangle).$$

However, if the correlation matrix $\boldsymbol{\rho}$ is rank-deficient, but the scalar dissipation matrix $\boldsymbol{\varepsilon}$ is full rank, the IEM model cannot predict the increase in rank of $\boldsymbol{\rho}$ due to molecular diffusion. In other words, the last term on the right-hand side of (6.105), p. 278, due to the diffusion term in the FP model will not be present in the IEM model.

61 The GIEM model violates the strong independence condition proposed by Pope (1983). However, since in binary mixing the scalar fields are correlated with the mixture fraction, it does satisfy the weak independence condition.

62 The expected value on the left-hand side is with respect to the joint PDF $f_{\mathbf{c},\xi}(\mathbf{c}, \zeta; \mathbf{x}, t)$.

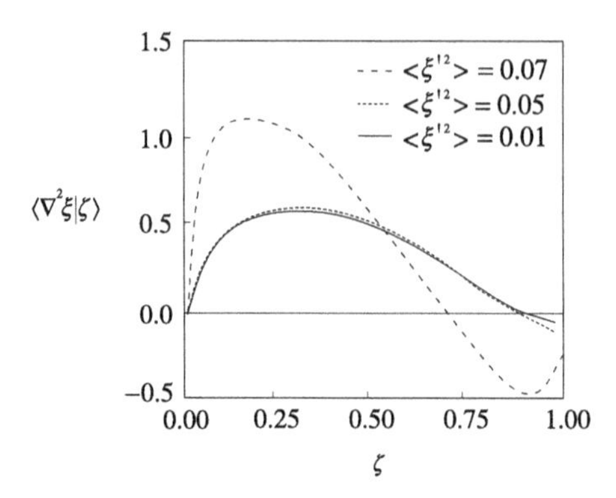

Figure 6.2. Sketch of conditional scalar Laplacian for $\langle\xi\rangle = 0.90$ and three values of the variance.

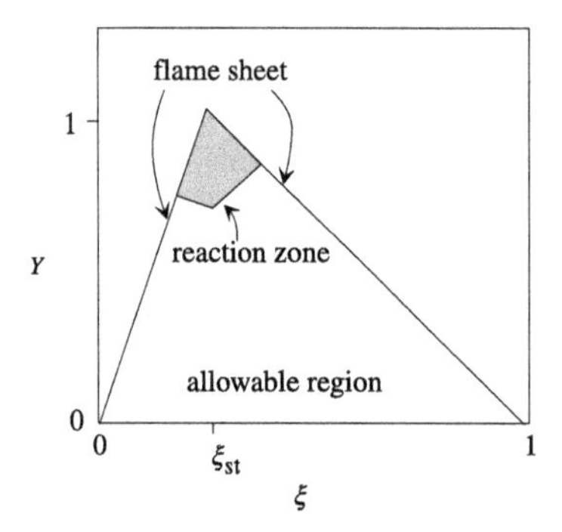

Figure 6.3. Sketch of the flame sheet, reaction zone, and allowable region for a one-step localized reaction.

limiting case of the FP model appearing in (6.145) on p. 285. For both models, the non-linearity needed to relax the PDF shape is expressed solely in terms of the mixture fraction, while all other variables appear linearly.

(iv) Why should mixing be local in composition space?

As will be shown for the CD model, early mixing models used stochastic jump processes to describe turbulent scalar mixing. However, since the mixing model is supposed to mimic molecular diffusion, which is continuous in space and time, jumping in composition space is inherently unphysical. The flame-sheet example (Norris and Pope 1991; Norris and Pope 1995) provides the best illustration of what can go wrong with non-local mixing models. For this example, a one-step reaction is described in terms of a reaction-progress variable Y and the mixture fraction ξ, and the reaction rate is localized near the stoichiometric point. In Fig. 6.3, the reaction zone is the box below the 'flame-sheet' lines in the upper left-hand corner. In physical space, the points with $\xi = 0$ are initially assumed to be separated from the points with $\xi = 1$ by a thin flame sheet centered at

$\xi = \xi_{st}$ and $Y = 1$ in composition space.[63] Since the reaction rate is assumed to be very large, all points that fall in the reaction zone will be immediately moved to the flame sheet defined by

$$Y(\xi) = \min\left(\frac{\xi}{\xi_{st}}, \frac{1-\xi}{1-\xi_{st}}\right), \tag{6.76}$$

where ξ_{st} is the stoichiometric value of the mixture fraction. Outside the reaction zone, the reaction rate is assumed to be null. Thus, molecular mixing and reaction should keep all points on the flame sheet. In essence, points on one side of the flame sheet cannot 'communicate' with points on the other side without passing through the flame. However, non-local mixing models (e.g., the CD model) generate unphysical points that lie in the allowable region, but not on the flame sheet.

In order to overcome the problem of non-local mixing, Subramaniam and Pope (1998) proposed the Euclidean minimum spanning tree (EMST) model which involves only interactions between neighboring fluid particles in composition space.[64] In the flame-sheet example, nearest neighbors lie precisely on the curve $Y(\xi)$ given by (6.76). For the flame-sheet example, the minimum spanning tree is one-dimensional, even though the allowable region in composition space is two-dimensional. The EMST model thus forces all particles to remain on the flame sheet during mixing (Subramaniam and Pope 1999). In contrast, the IEM model will pull particles towards the mean compositions, and hence off the flame sheet. In a homogeneous system with no mean scalar gradients, both the EMST model and the IEM model share the property that if, at time zero, the joint composition PDF has a one-dimensional support,[65] then it will remain nearly one-dimensional for all time. For the flame-sheet example, the support of the joint composition PDF found with the EMST model will be close to (6.76). In contrast, the support found with the IEM model will be given by (6.76) only at time zero, and will evolve with time. Indeed, since in the IEM model the mixing times for all scalars are assumed to be identical (see (6.84) and (6.85)), the homogeneous flame-sheet case reduces to

$$\frac{d\xi}{dt} = \frac{C_\phi \varepsilon}{2k}\left(\langle\xi\rangle - \xi\right) \tag{6.77}$$

$$\frac{dY}{dt} = \frac{C_\phi \varepsilon}{2k}\left(\langle Y\rangle - Y\right) + S_Y(Y, \xi), \tag{6.78}$$

where $S_Y(Y, \xi)$ has the form of the right-hand side of (5.176) on p. 183 but with $k^* = 0$ outside the reaction zone. Since $\langle\xi\rangle$ is constant, the mixture fraction in the IEM model

63 This initial condition is rather idealized. In reality, one would expect to see partially premixed zones with $\xi = \xi_{st}$ and $Y = 0$ which will move towards $Y = 1$ along the stoichiometric line. The movement along lines of constant ξ corresponds to premixed combustion, and occurs at a rate that is controlled by the interaction between molecular diffusion and chemical reactions (i.e., the laminar flame speed).

64 As discussed in Subramaniam and Pope (1998), the EMST model does not satisfy the independence and linearity properties proposed by Pope (1983).

65 The support of a PDF is the sub-set of composition space whereon the PDF is non-zero. Thus, a one-dimensional support is a curve in two- (or higher) dimensional space. The intermittency feature in the EMST model will cause the support to 'thicken' around the stoichiometric point, so that it will be locally two-dimensional.

evolves as

$$\xi(t) = \langle \xi \rangle + (\xi(0) - \langle \xi \rangle) \exp\left(-\frac{C_\phi \varepsilon}{2k} t\right). \tag{6.79}$$

Likewise, if $\langle Y \rangle(0)$ lies well below the reaction zone, the IEM model will collapse all points outside the reaction zone towards the mean values without passing through the reaction zone, i.e., the flame will be quenched even when the local reaction rate is infinite.[66] Such unphysical behavior is avoided with the EMST model (Subramaniam and Pope 1999).

Despite the success of the EMST model for the flame-sheet example, its applicability to other examples is limited by the fact that it does not allow fluid particles to attain points in the allowable region that are not consistent with the initial conditions. For example, consider a non-reacting homogeneous system with initial conditions $(Y, \xi)_0$ lying on the flame sheet given by (6.76).[67] For this system, turbulent mixing will generate initial diffusion layers with different characteristic length scales[68] or, in other words, the joint scalar dissipation rates:

$$\epsilon_\phi = \begin{bmatrix} \epsilon_\xi & \epsilon_{\xi,Y} \\ \epsilon_{\xi,Y} & \epsilon_Y \end{bmatrix} \tag{6.80}$$

will be random variables whose joint PDF cannot be represented by a delta function. Nevertheless, like the one-point joint PDF for ξ and Y, the EMST model contains no information concerning the joint scalar dissipation rates and thus will predict that the support of the joint PDF remains nearly one-dimensional for all time. In reality, initial diffusion layers with high scalar dissipation rates will diffuse much faster than diffusion layers with low rates, and the support of the joint scalar PDF will grow to fill the entire two-dimensional allowable region. Thus, the lack of information concerning the joint scalar dissipation rates in the EMST model[69] can also lead to unphysical predictions. Nonetheless, in defense of the EMST model, it should be noted that homogeneous mixing is a rather idealized situation. Most practical applications will involve inhomogeneous flows for which initial conditions play a minor role. For example, the EMST model has been successfully applied to model diffusion flames near extinction (Xu and Pope 1999).

Based on the above examples, we can conclude that while localness is a desirable property, it is not sufficient for ensuring physically realistic predictions. Indeed, a key ingredient that is missing in all mixing models described thus far (except the FP and EMST[70] models) is a description of the conditional joint scalar dissipation rates $\langle \epsilon | \phi \rangle$ and *their dependence on the chemical source term*. For example, from the theory of premixed turbulent flames, we can expect that $\langle \epsilon_Y | Y, \xi \rangle$ will be strongly dependent on the chemical

[66] This will happen regardless of the magnitude of the mixing time k/ε. Thus, it does not correspond to flame quenching at high scalar dissipation rates (small mixing time) seen in real flames with finite-rate chemistry.

[67] This initial condition is easily accomplished in DNS by first generating the mixture-fraction field $\xi(\mathbf{x}, 0)$, and then using (6.76) to define $Y(\mathbf{x}, 0)$.

[68] In other words, the scalars will have spectra that are continuous over a significant range of wavenumbers.

[69] The same is true for the IEM model.

[70] Although $\langle \epsilon | \phi \rangle$ is not explicitly input into the EMST model, a form is implied by the interaction matrix, which is affected by chemistry through the PDF shape.

source term. On the other hand, $\langle \epsilon_\xi | Y, \xi \rangle$ will be less dependent on the chemistry (Fox *et al.* 2002). Thus, the characteristic mixing times in ξ–Y composition space will be directionally dependent. In the limit of an infinitely fast, one-step reaction, the scalar gradients will be perfectly correlated (see (6.143) on p. 284) so that

$$\langle \epsilon_\phi | Y, \xi \rangle = \langle \epsilon_\xi | \xi \rangle \mathbf{g}_1(\xi) \mathbf{g}_1^{\mathrm{T}}(\xi), \tag{6.81}$$

where

$$\mathbf{g}_1(\xi) \equiv \begin{bmatrix} 1 \\ \frac{\partial \langle Y | \xi \rangle}{\partial \xi} \end{bmatrix}, \tag{6.82}$$

and $\langle Y | \xi \rangle$ is given by (6.76). Since the rank of $\langle \epsilon_\phi | Y, \xi \rangle$ is one in this limit, the FP model (like the EMST model) will conserve a one-dimensional support for the joint composition PDF. However, for finite-rate chemistry the rank of $\langle \epsilon_\phi | Y, \xi \rangle$ will be two, and the FP model will allow fluid-particle compositions to spread throughout the entire allowable region (Fox *et al.* 2002). Note, however, that this does not imply that particle compositions will be equally distributed in the allowable region. Indeed, depending on the local value of $\langle \epsilon_\phi | Y, \xi \rangle$ and the chemical source term $S_Y(Y, \xi)$, the distribution in composition space may be highly non-uniform (Fox *et al.* 2002). In order to improve existing models, a major challenge is to find accurate models for $\langle \epsilon_\phi | \phi \rangle$ which account for the combined effects of chemistry and molecular mixing.

(v) Why account for the scalar length-scale distribution?

We have already considered this question, in part, in our discussion of models for the scalar dissipation rate in Chapter 3. There we saw that the integral scales of the velocity and scalar fields need not be the same, in which case a spectral description that takes into account the evolution of the scalar length-scale distribution is crucial for predicting the decay rate of the scalar variance (Kosály 1989; Mell *et al.* 1991; Cremer *et al.* 1994). Micromixing models used in chemical reaction engineering deal with this problem by looking at multi-scale Lagrangian descriptions of scalar mixing (Baldyga 1989). In any case, when dealing with mixing-sensitive chemical reactions, the predicted product distribution (i.e., yield) can be very sensitive to the choice of model for the mean scalar dissipation rate. Likewise, in models for turbulent combustion (e.g., the flamelet model in Section 5.7), the mixture-fraction length-scale distribution often appears in the form of the fluctuating scalar dissipation rate ϵ_ξ. For high values of ϵ_ξ, a flamelet will be quenched, and complete extinction of the flame can result for certain forms of the scalar-dissipation PDF. More generally, all commonly used models for premixed[71] and non-premixed turbulent flames can be derived starting from the one-point joint PDF of the composition ϕ and the joint scalar dissipation rate ϵ_ϕ.

Despite the obvious advantages of using a transported PDF model for f_{ϕ,ϵ_ϕ}, in practice it is very difficult to find general models for the unclosed terms in the PDF transport equation

[71] In particular, premixed combustion models based on the flame surface density (Veynante and Vervisch 2002).

(for an example, see Meyers and O'Brien (1981), Dopazo (1994), and Fox (1994)). A less ambitious (and more tractable!) route is usually followed wherein multi-scale models are derived for the mean scalar dissipation rate ε_ϕ (Fox 1995), or for the mean scalar dissipation rate conditioned on the turbulent dissipation rate $\langle \epsilon_\phi | \epsilon \rangle$ (Fox 1997; Fox 1999).[72] However, to date, models of this type have not taken into account the effect of the chemical source term, and thus they are limited to describing turbulent mixing of inert scalars (albeit with differential diffusion). As seen for the FP model where a shape matrix is used, i.e.,

$$\langle \epsilon_\phi | \phi, \epsilon \rangle = \mathbf{H}(\phi) \langle \epsilon_\phi | \epsilon \rangle \mathbf{H}^{\mathrm{T}}(\phi), \tag{6.83}$$

an accurate model for $\langle \epsilon_\phi | \epsilon \rangle$ which includes chemistry effects is a crucial ingredient in the advancement of molecular mixing models. Unlike for inert scalars, multi-scale models for reacting scalars (or, more accurately, reaction-progress variables) may be strongly influenced by the balance between Batchelor-scale production due to chemical reactions and spectral transport back to large scales (see, for example, O'Brien (1966), O'Brien (1968a), O'Brien (1968b), O'Brien (1971), Jiang and O'Brien (1991), Tsai and O'Brien (1993)). However, due to the formidable difficulties of formulating spectral models for reacting scalars, progress in this area has been slow, and mainly limited to one-step reactions.

(vi) Why include Re, Sc and Da dependencies?

As discussed in Chapter 3, at very high Reynolds numbers, turbulent mixing theory predicts that the scalar dissipation rate will be independent of Re and Sc. Thus, most molecular models ignore all dependencies on these parameters, even at moderate Reynolds numbers. In general, the inclusion of dependencies on Re, Sc, or Da is difficult and, most likely, will have to be done on a case-by-case basis.

Examples where the Sc dependence may be significant include liquid-phase reacting flows at moderate Reynolds numbers for which the Schmidt number is very large (Dahm and Dimotakis 1990; Dowling and Dimotakis 1990; Miller 1991; Miller and Dimotakis 1991; Miller and Dimotakis 1996), and gas-phase reacting flows for which the molecular-diffusion coefficients of some species differ by an order of magnitude (Smith 1994). At moderate Reynolds numbers, differential-diffusion effects have been observed experimentally, particularly for trace species (Smith *et al.* 1995; Saylor and Sreenivasan 1998).

Examples where the Da dependence is important include the flame sheet discussed at length above, turbulent jet flames (Barlow *et al.* 1990), partially premixed flames that occur due to local extinction in zones of high shear (Peters 2000), and transported PDF modeling of premixed flames (Borghi 1986; Borghi 1988; Borghi 1990). For the latter, Mantel and Borghi (1994) have proposed a model for the dissipation rate of the progress variable (ε_Y) which includes the effect of chemistry. For isothermal reactions, the importance

[72] Models conditioned on the turbulent dissipation rate attempt to describe non-stationary effects due to the fluctuating strain-rate field, and thus should be adequate for flamelet applications which require a model for the mixture-fraction dissipation rate at the stoichiometric surface.

of Da effects is less obvious.[73] For example, using the GIEM model of Tsai and Fox (1998), it can be shown that for the competitive-consecutive reaction, (5.181), the mean concentrations are well predicted if the mixture-fraction scalar dissipation rate is known. However, the generality of this result has not been tested for other reaction schemes.

6.6.4 Three simple mixing models

A rather large number of molecular mixing models have been proposed in the literature (Curl 1963; Villermaux and Devillon 1972; Dopazo and O'Brien 1974; Janicka *et al.* 1979; Pope 1982; Borghi 1988; Valiño and Dopazo 1990; Pope 1991b; Valiño and Dopazo 1991; Fox 1992; Miller *et al.* 1993; Fox 1994; Tsai and Fox 1995a; Subramaniam and Pope 1998; Tsai and Fox 1998; Fox and Yeung 2003). However, instead of looking at the properties of each individually, here we will concentrate on three simple models with distinct properties.[74] The first model is the coalescence-dispersion (CD) model which uses a stochastic 'jump' process to imitate mixing. The second model is the widely used interaction by exchange with the mean (IEM) model, which is a linear deterministic process. The third model is the Fokker–Planck (FP) model, which uses a stochastic diffusion process to imitate mixing. As we shall see, the mathematical form of a model strongly affects its ability to satisfy the constraints and desirable properties given in Table 6.1.

CD model

The CD model was first proposed by Curl (1963) to describe coalescence and breakage of a dispersed two-fluid system. In each *mixing event*, two fluid particles with distinct compositions first 'coalesce' and then 'disperse' with identical compositions.[75] Written in terms of the two compositions ϕ_A and ϕ_B, a CD mixing event can be expressed mathematically as[76]

$$\begin{matrix} (\phi_{A1}, \phi_{B1})_1 \\ (\phi_{A2}, \phi_{B2})_2 \end{matrix} \xrightarrow{\text{coalescence}} \begin{pmatrix} \phi_A^* = (\phi_{A1} + \phi_{A2})/2 \\ \phi_B^* = (\phi_{B1} + \phi_{B2})/2 \end{pmatrix} \xrightarrow{\text{dispersion}} \begin{matrix} (\phi_A^*, \phi_B^*)_1 \\ (\phi_A^*, \phi_B^*)_2 \end{matrix},$$

where $(\cdot, \cdot)_1$ and $(\cdot, \cdot)_2$ are the compositions of fluid particles 1 and 2, respectively. Note that, since the final compositions are a linear combination of the initial compositions, the mean composition $\langle \phi \rangle$ remains unchanged. Constraint (I) is thus satisfied by the CD model. In the CD model, mixing events occur with a frequency characteristic of turbulent mixing. Thus, it is possible to reproduce the correct scalar dissipation rate for an inert scalar so that constraint (II) is partially satisfied.[77] In principle, it is also possible to condition

[73] Recall that the one-step isothermal reaction has a non-zero chemical source term for $Y = 0$. Thus, premixing for the fast-reaction limit yields immediate conversion to the equilibrium limit, regardless of the local scalar dissipation rate.

[74] Most of the other models appearing in the literature are found by modifying one of these three models.

[75] Several variants of the CD model exist wherein only 'partial' coalescence occurs. The final particle compositions are then only partially mixed, and thus will not be identical.

[76] The CD model can also be written in terms of a convolution integral with respect to the joint composition PDF. However, its properties are more easily understood in terms of a mechanistic model for a mixing event.

[77] In other words, the CD model assumes that all scalars have the same micromixing rate.

mixing events on the fluid-particle velocities in order to satisfy constraint (III). However, this is rarely done.

In terms of the desirable properties, the CD model has some serious deficiencies.[78] For example, in the limit of many CD events, the scalar PDF approaches a limiting form that is not Gaussian (the tails are too short). Thus, the CD model does not exhibit property (i). On the other hand, since it only involves linear combinations of the current values of ϕ, and the mixing time is the same for every scalar, the CD model does exhibit properties (ii) and (iii).[79] Property (iv) is the biggest deficiency of the CD model: during a mixing event the fluid-particle compositions 'jump' discontinuously in severe violation of the localness principle. While this is not a serious drawback for some applications, it is particularly problematic in flame-sheet calculations where the reaction rate is large only over a thin region in composition space. For this case, the CD model causes regions of fresh (cold) fuel to jump into regions of fresh (cold) oxidizer without reacting. In reality, this is impossible since all fuel must pass through the high-temperature flame sheet where it is burnt before it can reach the cold oxidizer. Mainly for this reason, the CD model has fallen out of favor for transported PDF simulations of reacting flows. Nevertheless, modified versions of the CD model (Dopazo 1994) are still used by many researchers in the combustion community for diffusion-flame simulations. Finally, properties (v) and (vi) are not explicitly treated in the CD model. However, it would be possible to build property (v) into the mixing frequency model by specifying a spectral dependence for the scalar dissipation rate.

IEM model

The IEM model[80] has been widely employed in both chemical-reaction engineering (Villermaux and Devillon 1972) and computational combustion (Dopazo 1994) due (mainly) to its simple form. The IEM model assumes a linear relaxation of the scalar towards its mean value:[81]

$$\langle \Gamma_\alpha \nabla^2 \phi'_\alpha | \psi \rangle = \frac{\varepsilon_\alpha}{2\langle \phi'^2_\alpha \rangle} (\langle \phi_\alpha \rangle - \psi_\alpha), \tag{6.84}$$

where ε_α is found from a separate model for the scalar dissipation rate. In most reported applications, a simple 'scale-similarity' model has been used:

$$\frac{\varepsilon_\alpha}{\langle \phi'^2_\alpha \rangle} = C_\phi \frac{\varepsilon}{k} \tag{6.85}$$

with $C_\phi \approx 2$. Note that this implies that all scalars mix at the same rate so that differential diffusion is not accounted for in the IEM model. In terms of the three constraints, the IEM model readily satisfies (I) and partially satisfies (II). However, in order to satisfy constraint

78 Modifications have been made to improve the CD model (Janicka *et al.* 1979; Pope 1982; Dopazo 1994), but its fundamental properties remain the same.

79 The CD model also satisfies the strong independence condition proposed by Pope (1983).

80 Also known in the literature as the linear-mean-square-estimate (LMSE) model (Dopazo and O'Brien 1974).

81 Note that since the right-hand side depends only on ϕ_α, the IEM model satisfies the strong independence condition proposed by Pope (1983).

(III), a velocity-conditioned version (VCIEM) must be introduced:

$$\langle \Gamma_\alpha \nabla^2 \phi'_\alpha | \mathbf{V}, \psi \rangle = \frac{\varepsilon_\alpha}{2\langle \phi'^2_\alpha \rangle}(\langle \phi_\alpha | \mathbf{V} \rangle - \psi_\alpha). \tag{6.86}$$

Fox (1996b) has investigated the properties of the VCIEM model, and has shown that correlation between the velocity and scalar fields due to the turbulent scalar flux can have a significant effect on the molecular mixing rate predicted by (6.86).

Although the IEM model is widely employed in transported PDF calculations, its principal shortcoming is the fact that it leaves the shape of the PDF unchanged in homogeneous turbulence. In the absence of mean scalar gradients, the composition PDF retains its original shape and never relaxes (let alone to a Gaussian form!). The IEM model thus does not exhibit property (i). On the other hand, due to its linear form, properties (ii) and (iii) are satisfied by the IEM model. Property (iv), however, is more problematic. Although the IEM model predicts that composition variables vary continuously in composition space, they do so under the influence of the mean composition (i.e., a global property of the composition field). Compared with the discontinuous behavior of the CD model, this violation of property (iv) is rather 'mild,' and does not automatically disqualify it as a viable mixing model. Property (v) can be introduced into the IEM model by using a spectral model for the scalar dissipation rate (Tsai and Fox 1996a). Finally, as with the CD model, the IEM model does not take into account property (vi). In defense of the IEM model, it is important to note that, for inhomogeneous scalar fields, turbulent velocity fluctuations will generate a Gaussian PDF in the absence of the molecular mixing model.[82] Thus, in many transported PDF simulations, the shape of the composition PDF will often be dominated by the scalar-flux term, and the non-Gaussian behavior exhibited by the IEM model will be of lesser importance.

FP model

In order to go beyond the simple description of mixing contained in the IEM model, it is possible to formulate a Fokker–Planck equation for scalar mixing that includes the effects of differential diffusion (Fox 1999).[83] Originally, the FP model was developed as an extension of the IEM model for a single scalar (Fox 1992). At high Reynolds numbers,[84] the conditional scalar Laplacian can be related to the conditional scalar dissipation rate by (Pope 2000)

$$\langle \Gamma \nabla^2 \phi' | \psi \rangle = \frac{1}{2 f_\phi} \frac{\partial}{\partial \psi}(\langle \epsilon_\phi | \psi \rangle f_\phi). \tag{6.87}$$

Since it leads to a diffusion equation with a *negative diffusion coefficient* (Pope 2000), the use of this expression directly in the composition PDF transport equation is of little

[82] For an example of this, see Fox (1996b).

[83] The model proposed by Fox (1999) also accounts for fluctuations in the joint scalar dissipation rate. Here we will look only at the simpler case, where the $\varepsilon_{\alpha\beta}$ is deterministic.

[84] In this limit, molecular-scale terms can be neglected. The relationship (6.87) is exact for statistically homogeneous scalars in the absence of uniform mean scalar gradients. This is the case considered in Fox (1999), and Fox (1994), and can occur even in the absence of turbulence.

interest. However, the diffusion process can be 'regularized' by rewriting (6.87) as[85]

$$\begin{aligned} \langle \Gamma \nabla^2 \phi' | \psi \rangle &= 2\langle \Gamma \nabla^2 \phi' | \psi \rangle - \langle \Gamma \nabla^2 \phi' | \psi \rangle \\ &= 2\langle \Gamma \nabla^2 \phi' | \psi \rangle - \frac{1}{2 f_\phi} \frac{\partial}{\partial \psi} (\langle \epsilon_\phi | \psi \rangle f_\phi) \\ &= \frac{\varepsilon_\phi}{\langle \phi'^2 \rangle} (\langle \phi \rangle - \psi) - \frac{1}{2 f_\phi} \frac{\partial}{\partial \psi} (\langle \epsilon_\phi | \psi \rangle f_\phi), \end{aligned} \tag{6.88}$$

where the final equality follows by substituting the IEM model for the conditional scalar Laplacian.[86] The resulting Fokker–Planck equation for f_ϕ has the form of a non-linear diffusion process (Gardiner 1990):

$$\frac{\partial f_\phi}{\partial t} = \frac{\varepsilon_\phi}{\langle \phi'^2 \rangle} \frac{\partial}{\partial \psi} [(\psi - \langle \phi \rangle) f_\phi] + \frac{1}{2} \frac{\partial^2}{\partial \psi^2} (\langle \epsilon_\phi | \psi \rangle f_\phi), \tag{6.89}$$

with a *non-negative* diffusion coefficient $\langle \epsilon_\phi | \psi \rangle$. The solution to (6.89) is thus well behaved, and the shape of f_ϕ is determined by the choice of $\langle \epsilon_\phi | \psi \rangle$.[87]

Applying the same procedure for multiple (N_s) scalars, leads to an expression of the form

$$\sum_{\alpha=1}^{N_s} \frac{\partial}{\partial \psi_\alpha} [\langle \nabla^2 \phi'_\alpha | \boldsymbol{\psi} \rangle f_\phi] = \sum_{\alpha=1}^{N_s} \sum_{\beta=1}^{N_s} \frac{1}{2(\Gamma_\alpha \Gamma_\beta)^{1/2}} \frac{\partial^2}{\partial \psi_\alpha \partial \psi_\beta} (\langle \epsilon_{\alpha\beta} | \boldsymbol{\psi} \rangle f_\phi). \tag{6.90}$$

This expression does not determine the mixing model uniquely. However, by specifying that the diffusion matrix in the resulting FP equation must equal the conditional joint scalar dissipation rate,[88] the FP model for the molecular mixing term in the form of (6.48) becomes

$$\langle \Gamma_\alpha \nabla^2 \phi'_\alpha | \boldsymbol{\psi} \rangle = -\frac{1}{2} \sum_{\beta=1}^{N_s} \left[M_{\alpha\beta} (\psi_\beta - \langle \phi_\beta \rangle) + \frac{1}{f_\phi} \frac{\partial}{\partial \psi_\beta} (\langle \epsilon_{\alpha\beta} | \boldsymbol{\psi} \rangle f_\phi) \right]. \tag{6.91}$$

The coefficient matrix[89] $\mathbf{M} \equiv [M_{\alpha\beta}]$ can be found by invoking (6.67) and using the expression for the covariance decay rate:[90]

$$\begin{aligned} \frac{\mathrm{d}\langle \phi'_\alpha \phi'_\beta \rangle}{\mathrm{d}t} &= \langle \Gamma_\alpha \phi'_\beta \nabla^2 \phi'_\alpha \rangle + \langle \Gamma_\beta \phi'_\alpha \nabla^2 \phi'_\beta \rangle \\ &= -\gamma_{\alpha\beta} \varepsilon_{\alpha\beta}, \end{aligned} \tag{6.92}$$

[85] More generally, we can write $\langle \Gamma \nabla^2 \phi' | \psi \rangle = (1 + c) \langle \Gamma \nabla^2 \phi' | \psi \rangle - c \langle \Gamma \nabla^2 \phi' | \psi \rangle$, where $c > 0$ controls the 'relaxation' rate of the composition PDF. Comparison with Lagrangian scalar time series (Yeung 2001) suggests that $c \approx 1$ yields good agreement with DNS. Note that $c = 0$ would yield the IEM model. Similar comments also apply to (6.91).

[86] Note that it is not necessary to use the IEM model for $\langle \Gamma \nabla^2 \phi' | \psi \rangle$. For example, a non-linear expression could be employed (Pope and Ching 1993; Ching 1996; Warhaft 2000). However, the linear form of the IEM model greatly simplifies the determination of the model constants.

[87] For example, if $\langle \epsilon_\phi | \psi \rangle = \varepsilon_\phi$, then the stationary PDF will be Gaussian. On the other hand, a beta PDF will result by choosing (6.149).

[88] In terms of the equivalent stochastic differential equation (6.106), p. 279, this choice yields a diffusion matrix of the form $\mathbf{B}(\phi) = \mathbf{S}_\Gamma \mathbf{G}(\phi)$, where the matrix $\mathbf{G}$ does not depend on the molecular-diffusion coefficients.

[89] All matrices used in this section are defined with respect to composition space. Thus, square matrices will be $N_S \times N_S$.

[90] This expression results from (3.136) on p. 90.

where $\gamma_{\alpha\beta}$ is defined in (3.140) on p. 91. This yields the following expression for $\mathbf{M}$:[91]

$$\mathbf{M} = \left(\mathbf{S}_\Gamma \boldsymbol{\varepsilon} \mathbf{S}_\Gamma^{-1} + \boldsymbol{\varepsilon}\right)\mathbf{C}^{-1}, \tag{6.93}$$

where the matrices on the right-hand side are defined by

$$\mathbf{S}_\Gamma \equiv \mathbf{diag}\left(\Gamma_1^{1/2}, \ldots, \Gamma_{N_s}^{1/2}\right), \tag{6.94}$$

$$\boldsymbol{\varepsilon} \equiv [\varepsilon_{\alpha\beta}], \tag{6.95}$$

and $\mathbf{C} \equiv \langle \boldsymbol{\phi}' \boldsymbol{\phi}'^{\mathrm{T}} \rangle$ is the scalar covariance matrix.

The first term on the right-hand side of (6.93) can be interpreted as a linear model for the scalar Laplacian:

$$\langle \boldsymbol{\Gamma} \nabla^2 \boldsymbol{\phi}' | \boldsymbol{\psi} \rangle = -\frac{1}{2} \mathbf{S}_\Gamma \boldsymbol{\varepsilon} \mathbf{S}_\Gamma^{-1} \mathbf{C}^{-1} \boldsymbol{\phi}'. \tag{6.96}$$

This model is consistent with (6.67), and can be seen as a multi-variate version of the IEM model. The role of the second term ($\boldsymbol{\varepsilon}\mathbf{C}^{-1}$) is simply to compensate for the additional 'diffusion' term in (6.91). Note that, like with the flamelet model and the conditional-moment closure discussed in Chapter 5, in the FP model the conditional joint scalar dissipation rates $\langle \epsilon_{\alpha\beta} | \boldsymbol{\psi} \rangle$ must be provided by the user. Since these functions have many independent variables, and can be time-dependent due to the effects of transport and chemistry, specifying appropriate functional forms for general applications will be non-trivial. However, in specific cases where the scalar fields are perfectly correlated, appropriate functional forms can be readily established. We will return to this question with specific examples below.

The scalar covariance matrix in (6.93) can be decomposed as follows:

$$\mathbf{C} = \mathbf{S}_\phi \boldsymbol{\rho} \mathbf{S}_\phi, \tag{6.97}$$

where the diagonal variance matrix is defined by

$$\mathbf{S}_\phi \equiv \mathbf{diag}\left(\langle \phi_1'^2 \rangle^{1/2}, \ldots, \langle \phi_{N_s}'^2 \rangle^{1/2}\right), \tag{6.98}$$

and the scalar correlation matrix is defined by

$$\boldsymbol{\rho} \equiv \left[\frac{\langle \phi'_\alpha \phi'_\beta \rangle}{\left(\langle \phi_\alpha'^2 \rangle \langle \phi_\beta'^2 \rangle\right)^{1/2}} \right]. \tag{6.99}$$

Unlike the velocity field, scalar fields are often perfectly correlated so that the correlation matrix $\boldsymbol{\rho}$ will be rank-deficient.[92] When this occurs, the coefficient matrix $\mathbf{M}$ will not be properly defined by (6.93), and so the FP model must be modified to handle perfectly correlated scalars. As shown in Section 5.9 for the multi-variate Gaussian presumed PDF,

[91] Note that the matrix $2\left[\gamma_{\alpha\beta}\varepsilon_{\alpha\beta}\right] = \mathbf{S}_\Gamma \boldsymbol{\varepsilon} \mathbf{S}_\Gamma^{-1} + \mathbf{S}_\Gamma^{-1} \boldsymbol{\varepsilon} \mathbf{S}_\Gamma$.

[92] In particular applications, cases where one or more variances $\langle \phi_\alpha'^2 \rangle$ are null may also occur. For these cases, no mixing model is needed. It thus suffices to set the corresponding components in $\mathbf{S}_\phi^{-1}$ equal to zero in (6.93).

the singular correlation matrix can be avoided by introducing a linear change of variables:[93]

$$\phi^{\dagger} = \mathbf{U}_{\rho}^{\mathrm{T}} \left(\phi - \langle \phi \rangle \right), \tag{6.100}$$

and

$$\phi = \langle \phi \rangle + \mathbf{U}_{\rho} \phi^{\dagger}, \tag{6.101}$$

where $\mathbf{U}_{\rho}$ is the ortho-normal eigenvector matrix[94] of $\boldsymbol{\rho}$ (compare with (5.334) on p. 220), and $\phi^{\dagger}$ has $N_{\rho} = \mathrm{rank}(\boldsymbol{\rho})$ non-zero components:

$$\phi^{\dagger} = \begin{bmatrix} \phi_{\mathrm{in}}^{\dagger} \\ \mathbf{0} \end{bmatrix}. \tag{6.102}$$

It then follows that

$$\mathbf{M} = \left(\mathbf{S}_{\Gamma} \boldsymbol{\varepsilon} \mathbf{S}_{\Gamma}^{-1} + \boldsymbol{\varepsilon} \right) \mathbf{S}_{\phi}^{-1} \mathbf{U}_{\rho} \mathbf{S}_{\rho} \mathbf{U}_{\rho}^{\mathrm{T}} \mathbf{S}_{\phi}^{-1}, \tag{6.103}$$

where $\mathbf{S}_{\rho}$ is defined in terms of the non-zero eigenvalues of $\boldsymbol{\rho}$ in order to avoid singularity:[95]

$$\mathbf{S}_{\rho} \equiv \mathbf{diag}(1/\lambda_1, \ldots, 1/\lambda_{N_{\rho}}, 0, \ldots, 0), \tag{6.104}$$

where λ_i are the N_{ρ} non-zero eigenvalues of $\boldsymbol{\rho}$.

The definition of $\mathbf{M}$ in (6.103) results in a covariance equation of the form

$$\frac{\mathrm{d}\mathbf{C}}{\mathrm{d}t} = -[\gamma_{\alpha\beta} \varepsilon_{\alpha\beta}] \mathbf{U}_1 \mathbf{U}_1^{\mathrm{T}} + \boldsymbol{\varepsilon} \mathbf{U}_2 \mathbf{U}_2^{\mathrm{T}}, \tag{6.105}$$

where $\mathbf{U}_{\rho} = [\mathbf{U}_1 \ \mathbf{U}_2]$ is the orthogonal decomposition of the eigenvector matrix into non-zero- and zero-eigenvalue sub-spaces, respectively. The first term on the right-hand side of (6.105) changes the components of $\mathbf{C}$, but not its rank. On the other hand, the second term will increase the rank of $\mathbf{C}$ since its effect will be to make some of the zero eigenvalues increase in magnitude.[96] Note that the second term results from the 'noise' term in (6.91), which depends on scalar-gradient correlations at small scales. Thus, an increase in the rank of $\mathbf{C}$ will often be associated with differential-diffusion effects which force the scalar gradients to de-correlate (Yeung 1998a; Fox 1999).

In order to illustrate the properties of the FP model, it is easier to rewrite it in terms of an equivalent stochastic differential equation (SDE) (Arnold 1974; Risken 1984; Gardiner

[93] Note that unlike (5.336) on p. 220, we do not rescale the variables. Thus, $\langle \phi^{\dagger} \phi^{\dagger\mathrm{T}} \rangle = \boldsymbol{\Lambda}$, where $\boldsymbol{\Lambda}$ is the eigenvalue matrix for $\mathbf{C}$.

[94] Note that, since the correlation can change with time, $\mathbf{U}_{\rho}$ will in general depend on t.

[95] In a numerical implementation of the FP model, $\mathbf{S}_{\rho}$ is found by replacing eigenvalues which are smaller than some minimum value with one, and there is no need to put the eigenvalues/eigenvectors in descending order.

[96] This can be proven by deriving a differential equation for $\boldsymbol{\Lambda}$ and using the fact that the matrix $\boldsymbol{\varepsilon}$ is non-negative.

1990):[97]

$$\mathrm{d}\phi = -\frac{1}{2}\mathbf{M}(\phi - \langle\phi\rangle)\,\mathrm{d}t + \mathbf{B}(\phi)\,\mathrm{d}\mathbf{W}(t), \tag{6.106}$$

where $\mathrm{d}\mathbf{W}(t)$ is a multi-variate Wiener process. The diffusion matrix $\mathbf{B}(\phi)$ is related to $\langle\epsilon|\phi\rangle \equiv [\langle\epsilon_{\alpha\beta}|\phi\rangle]$ by

$$\begin{aligned}\mathbf{B}(\phi)\mathbf{B}(\phi)^{\mathrm{T}} &= \langle\epsilon|\phi\rangle \\ &= \mathbf{S}_{\mathrm{g}}(\phi)\mathbf{G}(\phi)\mathbf{S}_{\mathrm{g}}(\phi)^{\mathrm{T}},\end{aligned} \tag{6.107}$$

where $\mathbf{S}_{\mathrm{g}}$ is defined using the conditional scalar dissipation rates:

$$\mathbf{S}_{\mathrm{g}}(\phi) \equiv \mathbf{diag}(\langle\epsilon_{11}|\phi\rangle^{1/2}, \langle\epsilon_{22}|\phi\rangle^{1/2}, \dots), \tag{6.108}$$

and the conditional gradient-correlation matrix is defined by

$$\mathbf{G}(\phi) \equiv \left[\frac{\langle\epsilon_{\alpha\beta}|\phi\rangle}{(\langle\epsilon_{\alpha\alpha}|\phi\rangle\langle\epsilon_{\beta\beta}|\phi\rangle)^{1/2}}\right]. \tag{6.109}$$

Note that the molecular diffusivities appearing in the definition of $\epsilon_{\alpha\beta} = (\Gamma_\alpha\Gamma_\beta)^{1/2}(\nabla\phi'_\alpha)\cdot(\nabla\phi'_\beta)$ cancel out in (6.109), and the diagonal components of $\mathbf{G}$ are unity.

Although (6.107) does not uniquely define $\mathbf{B}$, it is possible to write $\mathbf{G}$ as the square of a symmetric matrix:[98]

$$\mathbf{G}(\phi) = \mathbf{C}_{\mathrm{g}}(\phi)\mathbf{C}_{\mathrm{g}}(\phi) = \mathbf{C}_{\mathrm{g}}(\phi)\mathbf{C}_{\mathrm{g}}(\phi)^{\mathrm{T}}, \tag{6.110}$$

so that $\mathbf{B} = \mathbf{S}_{\mathrm{g}}\mathbf{C}_{\mathrm{g}}$. The SDE for the FP model then becomes

$$\mathrm{d}\phi = -\frac{1}{2}\left(\mathbf{S}_\Gamma\boldsymbol{\varepsilon}\mathbf{S}_\Gamma^{-1} + \boldsymbol{\varepsilon}\right)\mathbf{S}_\phi^{-1}\mathbf{U}_\rho\mathbf{S}_\rho\mathbf{U}_\rho^{\mathrm{T}}\mathbf{S}_\phi^{-1}(\phi - \langle\phi\rangle)\,\mathrm{d}t + \mathbf{S}_g(\phi)\mathbf{C}_g(\phi)\,\mathrm{d}\mathbf{W}(t). \tag{6.111}$$

Note that the rank of $\mathbf{C}_{\mathrm{g}}$ determines the number of independent Wiener processes that affect the system.[99] The following two limiting cases are of practical interest.

(1) Rank($\mathbf{C}_{\mathrm{g}}$) = 1. In this case, all components of $\mathbf{C}_{\mathrm{g}}$ are either 1 or −1, and it has only one independent row (column). If the allowable region at $t = 0$ is one-dimensional, then it will remain one-dimensional for all time (assuming that the rank does not change). This limiting case will occur when all scalars can be written as a function of the mixture fraction (e.g., the conditional-moment closure).

97 The strong independence condition proposed by Pope (1983) will not be satisfied by the FP model unless $\mathbf{M}$ is diagonal. This will occur only when all scalar fields are uncorrelated. In this limit, $\boldsymbol{\rho} = \mathbf{I}$, and $\boldsymbol{\varepsilon}$ is diagonal. The FP model thus satisfies only the weak independence condition, and scalar-field correlation is essentially determined by the model for $\boldsymbol{\varepsilon}$.

98 This decomposition is not unique (Gardiner 1990), but yields the same joint PDF as other possible choices. See Fox (1999) for an example with two scalars. Alternative decompositions can be written in terms of the eigenvectors and (non-negative) eigenvalues of $\mathbf{G}$: $\mathbf{G} = \mathbf{U}_{\mathrm{g}}\boldsymbol{\Lambda}_{\mathrm{g}}\mathbf{U}_{\mathrm{g}}^{\mathrm{T}} \Rightarrow \mathbf{C}_{\mathrm{g}} = \mathbf{U}_{\mathrm{g}}\boldsymbol{\Lambda}_{\mathrm{g}}^{1/2}$ or $\mathbf{C}_{\mathrm{g}} = \mathbf{U}_{\mathrm{g}}\boldsymbol{\Lambda}_{\mathrm{g}}^{1/2}\mathbf{U}_{\mathrm{g}}^{\mathrm{T}}$. Note that the rank of $\mathbf{C}_{\mathrm{g}}$ is the same as the rank of $\boldsymbol{\varepsilon}$. Finally, note that the eigenvalue decomposition can be carried out on $\langle\epsilon|\phi\rangle$ to find $\mathbf{B}(\phi)$ directly.

99 The dimension of the allowable region in composition space will thus be determined by the rank of $\mathbf{C}_{\mathrm{g}}$.

(2) $\text{Rank}(\mathbf{C}_g) = N_s$. In this case, the allowable region will be N_s-dimensional for all time. This can occur, for example, when differential diffusion is important, or may be due to how the scalar fields are initialized at time zero.

Before looking at its other properties, we should note that by construction the FP model satisfies constraints (I) and (II) (refer again to Table 6.1), and can be made to satisfy constraint (III) by introducing velocity conditioning in all expected values appearing in (6.111). Of the six desirable properties, we note that (i) will be satisfied if $\mathbf{S}_g(\boldsymbol{\phi})\mathbf{C}_g(\boldsymbol{\phi})$ becomes independent of $\boldsymbol{\phi}$ when the variance matrix $\mathbf{S}_\phi$ approaches zero. Since they are expected to be smooth functions near $\boldsymbol{\phi} = \langle\boldsymbol{\phi}\rangle$, property (i) will be satisfied.

Property (ii) is also controlled by the behavior of $\mathbf{S}_g(\boldsymbol{\phi})\mathbf{C}_g(\boldsymbol{\phi})$. In general, the diffusion matrix should have the property that it does not allow movement in the direction normal to the surface of the allowable region.[100] Defining the surface unit normal vector by $\mathbf{n}(\boldsymbol{\phi}^*)$, property (ii) will be satisfied if $\mathbf{S}_g(\boldsymbol{\phi}^*)\mathbf{C}_g(\boldsymbol{\phi}^*)\mathbf{n}(\boldsymbol{\phi}^*) = \mathbf{0}$, where $\boldsymbol{\phi}^*$ lies on the surface of the allowable region. This condition implies that $\langle\boldsymbol{\epsilon}|\boldsymbol{\phi}^*\rangle\mathbf{n}(\boldsymbol{\phi}^*) = \mathbf{0}$, which Girimaji (1992) has shown to be true for the single-scalar case. Thus, the FP model satisfies property (ii), but *the user must provide the unknown conditional joint scalar dissipation rates that satisfy* $\langle\boldsymbol{\epsilon}|\boldsymbol{\phi}^*\rangle\mathbf{n}(\boldsymbol{\phi}^*) = \mathbf{0}$.

Property (iii) applies in the absence of differential-diffusion effects. In this limit, the FP model becomes

$$\mathrm{d}\boldsymbol{\phi} = -\varepsilon\mathbf{S}_\phi^{-1}\mathbf{U}_\rho\mathbf{S}_\rho\mathbf{U}_\rho^{\mathrm{T}}\mathbf{S}_\phi^{-1}(\boldsymbol{\phi} - \langle\boldsymbol{\phi}\rangle)\,\mathrm{d}t + \mathbf{S}_g(\boldsymbol{\phi})\mathbf{C}_g(\boldsymbol{\phi})\,\mathrm{d}\mathbf{W}(t). \tag{6.112}$$

Let $\mathbf{L}$ be an invertible linear transformation matrix such that $\boldsymbol{\varphi}' = \mathbf{L}\boldsymbol{\phi}'$. Premultiplying (6.112) by $\mathbf{L}$, the FP model for $\boldsymbol{\varphi}$ becomes

$$\mathrm{d}\boldsymbol{\varphi} = -\mathbf{L}\varepsilon_\phi\mathbf{S}_\phi^{-1}\mathbf{U}_\rho\mathbf{S}_\rho\mathbf{U}_\rho^{\mathrm{T}}\mathbf{S}_\phi^{-1}\mathbf{L}^{-1}\boldsymbol{\varphi}'\,\mathrm{d}t + \mathbf{L}\mathbf{S}_g(\boldsymbol{\phi})\mathbf{C}_g(\boldsymbol{\phi})\,\mathrm{d}\mathbf{W}(t), \tag{6.113}$$

where ε_ϕ is the joint scalar dissipation rate based on $\boldsymbol{\phi}$. From this expression, we see that linearity of the FP model will require that the following two conditions hold.

(1) Linearity condition for drift coefficient:

$$\mathbf{L}\varepsilon_\phi\mathbf{S}_\phi^{-1}\mathbf{U}_\rho\mathbf{S}_\rho\mathbf{U}_\rho^{\mathrm{T}}\mathbf{S}_\phi^{-1}\mathbf{L}^{-1} = \varepsilon_\varphi\mathbf{S}_\varphi^{-1}\mathbf{U}_\varrho\mathbf{S}_\varrho\mathbf{U}_\varrho^{\mathrm{T}}\mathbf{S}_\varphi^{-1}, \tag{6.114}$$

where all quantities on the right-hand side are based on $\boldsymbol{\varphi}$.

(2) Linearity condition for diffusion coefficient:

$$\langle\boldsymbol{\epsilon}_\varphi|\boldsymbol{\varphi}\rangle = \mathbf{L}\langle\boldsymbol{\epsilon}_\phi|\boldsymbol{\phi}\rangle\mathbf{L}^{\mathrm{T}}. \tag{6.115}$$

Condition (1) follows directly from the definitions of the joint scalar dissipation rates and the covariance matrices, i.e.,

$$\varepsilon_\varphi = \mathbf{L}\varepsilon_\phi\mathbf{L}^{\mathrm{T}}, \tag{6.116}$$

[100] This is the so-called 'natural' boundary condition to the Fokker–Planck equation (Gardiner 1990).

and

$$\mathbf{C}_{\varphi} = \mathbf{L}\mathbf{C}_{\phi}\mathbf{L}^{\mathrm{T}}. \tag{6.117}$$

Condition (2) is more difficult to satisfy, and requires that the functional form of the conditional joint scalar dissipation rates be carefully chosen. For example, one can construct a model of the form

$$\langle \epsilon_{\phi} | \boldsymbol{\phi} \rangle = \mathbf{H}(\boldsymbol{\phi}) \varepsilon_{\phi} \mathbf{H}^{\mathrm{T}}(\boldsymbol{\phi}), \tag{6.118}$$

where $\mathbf{H}(\boldsymbol{\phi})$ is a matrix function with the property[101]

$$\langle \mathbf{H}(\boldsymbol{\phi}) \varepsilon_{\phi} \mathbf{H}^{\mathrm{T}}(\boldsymbol{\phi}) \rangle = \varepsilon_{\phi}. \tag{6.119}$$

This 'shape' matrix[102] must prevent flux through the boundaries of the allowable region in composition space, and must commute with $\mathbf{L}$:

$$\mathbf{H}\mathbf{L} = \mathbf{L}\mathbf{H}. \tag{6.120}$$

The simplest (although not very useful!) model of this type is $\mathbf{H} = \mathbf{I}$. However, this model can only be used for unbounded scalars.

Like the IEM model, the FP model weakly satisfies property (iv). Likewise, property (v) can be built into the model for the joint scalar dissipation rates (Fox 1999), and the Sc dependence in property (vi) is included explicitly in the FP model. Thus, of the three molecular mixing models discussed so far, the FP model exhibits the greatest number of desirable properties *provided suitable functional forms can be found for* $\langle \epsilon | \boldsymbol{\phi} \rangle$.

One method for determining the shape matrix $\mathbf{H}(\boldsymbol{\phi})$ is to appeal to a multi-variate mapping closure (H. Chen *et al.* 1989; Gao 1991; Gao and O'Brien 1991; Pope 1991b) of the form

$$\boldsymbol{\phi} = \mathbf{g}(\mathbf{Z}; \mathbf{x}, t), \tag{6.121}$$

where $\mathbf{Z}$ is a random vector of length N_{s} with a known time-independent, homogeneous PDF $f_{\mathbf{Z}}(\mathbf{z})$ and known conditional joint scalar dissipation rate matrix $\langle \epsilon_{Z} | \mathbf{Z} = \mathbf{z} \rangle$.[103] If the function $\mathbf{g}$ is one-to-one and invertible:

$$\mathbf{Z} = \mathbf{g}^{\dagger}(\boldsymbol{\phi}; \mathbf{x}, t) \quad \Longrightarrow \quad \mathbf{g}(\mathbf{g}^{\dagger}(\boldsymbol{\phi}; \mathbf{x}, t); \mathbf{x}, t) = \boldsymbol{\phi}, \tag{6.122}$$

101 Since ε_{ϕ} is deterministic, the expected value in this definition is with respect to the composition PDF f_{ϕ}. However, due to the matrix form, ε_{ϕ} cannot be factored out to simplify the expression. Another alternative is to define a symmetric 'square-root' matrix $\mathbf{S}_{\phi}$ such that $\varepsilon_{\phi} = \mathbf{S}_{\phi}\mathbf{S}_{\phi}$, and then define $\mathbf{H}$ such that $\langle \epsilon_{\phi} | \boldsymbol{\phi} \rangle = \mathbf{S}_{\varepsilon}\mathbf{H}(\boldsymbol{\phi})\mathbf{H}^{\mathrm{T}}(\boldsymbol{\phi})\mathbf{S}_{\varepsilon}$ and $\langle \mathbf{H}(\boldsymbol{\phi})\mathbf{H}^{\mathrm{T}}(\boldsymbol{\phi}) \rangle = \mathbf{I}$. For this case, condition (2) reduces to $\mathbf{S}_{\varphi} = \mathbf{L}\mathbf{S}_{\phi}$, where $\mathbf{S}_{\varphi}$ is the square-root matrix for ε_{φ}.

102 The functional form of $\mathbf{H}(\boldsymbol{\phi})$ will control the shape of the joint composition PDF. For example, if $\mathbf{H} = \mathbf{I}$ the PDF will evolve to a Gaussian form.

103 The most common choice is for the components of $\mathbf{Z}$ to be uncorrelated standardized Gaussian random variables. For this case, $\langle \epsilon_{Z} | \mathbf{z} \rangle = \varepsilon_{Z} = \mathbf{diag}(\varepsilon_{Z_1}, \dots, \varepsilon_{Z_{N_{\mathrm{s}}}})$, i.e., the conditional joint scalar dissipation rate matrix is constant and diagonal.

we can define a gradient matrix

$$\mathbf{G}(\phi; \mathbf{x}, t) \equiv [G_{\alpha\beta}(\phi; \mathbf{x}, t)], \tag{6.123}$$

where

$$G_{\alpha\beta}(\psi; \mathbf{x}, t) \equiv \frac{\partial g_\alpha}{\partial z_\beta}(\mathbf{z}; \mathbf{x}, t) \quad \text{with} \quad \mathbf{z} = \mathbf{g}^\dagger(\psi; \mathbf{x}, t). \tag{6.124}$$

Note that if $\mathbf{g}$ is invertible, then $\mathbf{G}$ will be full rank. The rank of $\langle \epsilon_Z | \mathbf{Z} \rangle$ will thus determine the rank of $\langle \epsilon_\phi | \phi \rangle$ and the number of linearly independent scalars.[104] The conditional joint scalar dissipation rate matrix is given by[105]

$$\langle \epsilon_\phi | \psi \rangle = \mathbf{S}_\Gamma \mathbf{G}(\psi) \mathbf{S}_\Gamma^{-1} \langle \epsilon_Z | \mathbf{Z} = \mathbf{g}^\dagger(\psi) \rangle \mathbf{S}_\Gamma^{-1} \mathbf{G}(\psi)^{\mathrm{T}} \mathbf{S}_\Gamma. \tag{6.125}$$

The remaining challenge is then to formulate and solve transport equations for the mapping functions $\mathbf{g}(\mathbf{z}; \mathbf{x}, t)$ (Gao and O'Brien 1991; Pope 1991b). Note that if $\mathbf{g}(\mathbf{z}; \mathbf{x}, t)$ is known, then the FP model can be used to describe $\mathbf{Z}(t)$, and $\phi(t)$ will follow from (6.121). Since the PDF of $\mathbf{Z}$ is stationary and homogeneous, the FP model needed to describe it will be particularly simple. With the mapping closure, the difficulties associated with the chemical source terms are thus shifted to the model for $\mathbf{g}(\mathbf{z}; \mathbf{x}, t)$.

There are very few examples of scalar-mixing cases for which an explicit form for $\langle \epsilon_\phi | \phi \rangle$ can be found using the known constraints. One of these is multi-stream mixing of inert scalars with equal molecular diffusivity. Indeed, for bounded scalars that can be transformed to a mixture-fraction vector, a shape matrix can be generated by using the surface normal vector $\mathbf{n}(\boldsymbol{\zeta}^*)$ mentioned above for property (ii). For the mixture-fraction vector, the faces of the allowable region are hyperplanes, and the surface normal vectors are particularly simple. For example, a two-dimensional mixture-fraction vector has three surface normal vectors:

$$\mathbf{n}(0, \zeta_2^*) = \begin{bmatrix} 1 \\ 0 \end{bmatrix}, \quad \mathbf{n}(\zeta_1^*, 0) = \begin{bmatrix} 0 \\ 1 \end{bmatrix}, \quad \mathbf{n}(\zeta_1^*, \zeta_2^*) = \frac{1}{\sqrt{2}} \begin{bmatrix} 1 \\ 1 \end{bmatrix}, \tag{6.126}$$

where the third vector corresponds to the diagonal surface defined by $\zeta_1^* + \zeta_2^* = 1$. For this example, the conditional joint scalar dissipation rate matrix has three unknown components:

$$\langle \epsilon_\xi | \zeta_1, \zeta_2 \rangle = \begin{bmatrix} \langle \epsilon_{11} | \zeta_1, \zeta_2 \rangle & \langle \epsilon_{12} | \zeta_1, \zeta_2 \rangle \\ \langle \epsilon_{12} | \zeta_1, \zeta_2 \rangle & \langle \epsilon_{22} | \zeta_1, \zeta_2 \rangle \end{bmatrix}, \tag{6.127}$$

[104] In other words, if some of the components of ϕ are linearly dependent, then so should an equal number of components of $\mathbf{Z}$. As an example, $\mathbf{G}$ could be diagonal so that ϕ_α depends only on Z_α. In addition, if $\mathbf{Z}$ were chosen to be Gaussian, then $\langle \epsilon_Z | \mathbf{Z} \rangle$ would be independent of $\mathbf{Z}$ and have the same correlation structure as $\langle \epsilon_\phi | \phi \rangle$. The joint dissipation rate matrix ε_Z could be found using the LSR model (Fox 1999).

[105] Note that, by assumption, ϕ_α and Z_α have molecular diffusivity Γ_α.

which are constrained by $\langle\langle \epsilon_\xi | \xi_1, \xi_2 \rangle\rangle = \varepsilon_\xi$. Thus, overall, this constraint and property (ii) provide nine conditions that must be satisfied:

$$
\begin{aligned}
\langle\langle \epsilon_{11} | \xi_1, \xi_2 \rangle\rangle &= \varepsilon_{11}, \\
\langle\langle \epsilon_{12} | \xi_1, \xi_2 \rangle\rangle &= \varepsilon_{12}, \\
\langle\langle \epsilon_{22} | \xi_1, \xi_2 \rangle\rangle &= \varepsilon_{22}, \\
\langle \epsilon_{11} | 0, \zeta_2^* \rangle &= 0 \quad \text{for all } \zeta_2^*, \\
\langle \epsilon_{12} | 0, \zeta_2^* \rangle &= 0 \quad \text{for all } \zeta_2^*, \\
\langle \epsilon_{12} | \zeta_1^*, 0 \rangle &= 0 \quad \text{for all } \zeta_1^*, \\
\langle \epsilon_{22} | \zeta_1^*, 0 \rangle &= 0 \quad \text{for all } \zeta_1^*, \\
\langle \epsilon_{11} | \zeta_1^*, \zeta_2^* \rangle + \langle \epsilon_{12} | \zeta_1^*, \zeta_2^* \rangle &= 0 \quad \text{for all } \zeta_1^* + \zeta_2^* = 1, \\
\langle \epsilon_{12} | \zeta_1^*, \zeta_2^* \rangle + \langle \epsilon_{22} | \zeta_1^*, \zeta_2^* \rangle &= 0 \quad \text{for all } \zeta_1^* + \zeta_2^* = 1.
\end{aligned}
\tag{6.128}
$$

Particular functional forms can be found by using power series:[106]

$$
\begin{aligned}
\langle \epsilon_{11} | \zeta_1, \zeta_2 \rangle &= a_0 + a_1\zeta_1 + a_2\zeta_2 + a_3\zeta_1^2 + a_4\zeta_1\zeta_2 + a_5\zeta_2^2, \\
\langle \epsilon_{12} | \zeta_1, \zeta_2 \rangle &= b_0 + b_1\zeta_1 + b_2\zeta_2 + b_3\zeta_1^2 + b_4\zeta_1\zeta_2 + b_5\zeta_2^2, \\
\langle \epsilon_{22} | \zeta_1, \zeta_2 \rangle &= c_0 + c_1\zeta_1 + c_2\zeta_2 + c_3\zeta_1^2 + c_4\zeta_1\zeta_2 + c_5\zeta_2^2.
\end{aligned}
\tag{6.129}
$$

Applying the constraints in (6.128) eliminates most of the coefficients:

$$
\begin{aligned}
\langle \epsilon_{11} | \zeta_1, \zeta_2 \rangle &= \alpha\zeta_1(1 - \zeta_1 - \zeta_2) - \beta\zeta_1\zeta_2, \\
\langle \epsilon_{12} | \zeta_1, \zeta_2 \rangle &= \beta\zeta_1\zeta_2, \\
\langle \epsilon_{22} | \zeta_1, \zeta_2 \rangle &= \gamma\zeta_2(1 - \zeta_1 - \zeta_2) - \beta\zeta_1\zeta_2,
\end{aligned}
\tag{6.130}
$$

where

$$\alpha = \frac{\varepsilon_{11} + \varepsilon_{12}}{\langle \xi_1(1 - \xi_1 - \xi_2) \rangle}, \tag{6.131}$$

$$\beta = \frac{\varepsilon_{12}}{\langle \xi_1\xi_2 \rangle}, \tag{6.132}$$

and

$$\gamma = \frac{\varepsilon_{22} + \varepsilon_{12}}{\langle \xi_2(1 - \xi_1 - \xi_2) \rangle}. \tag{6.133}$$

Applying the same procedure to higher-dimensional mixture-fraction vectors yields expressions of the same form as (6.130). Note also that for any set of bounded scalars that can be linearly transformed to a mixture-fraction vector, (6.115) can be used to find the corresponding joint conditional scalar dissipation rate matrix starting from $\langle \epsilon_\xi | \boldsymbol{\zeta} \rangle$.

[106] Truncating the power series at second order implies that only the means and covariances will be needed to specify the coefficients. The beta PDF has this property, and thus we can speculate that the stationary PDF predicted by the FP model with these coefficients should be the same as (5.147).

Before leaving the FP model, it is of interest to consider particular limiting cases wherein the form of $\langle \epsilon_\phi | \phi \rangle$ is relatively simple. For example, in many non-premixed flows without differential diffusion, the composition vector is related to the mixture fraction by a linear transformation:[107]

$$\phi(\mathbf{x}, t) = \mathbf{l}(\xi(\mathbf{x}, t) - \langle \xi \rangle) + \langle \phi \rangle, \tag{6.134}$$

where $\mathbf{l}$ is a constant vector. This relation implies that[108]

$$\mathbf{C} = \mathbf{l}\mathbf{l}^{\mathrm{T}} \langle \xi'^2 \rangle, \tag{6.135}$$

$$\rho = \frac{\mathbf{l}\mathbf{l}^{\mathrm{T}}}{\mathbf{l}^{\mathrm{T}}\mathbf{l}}, \tag{6.136}$$

$$\varepsilon = \mathbf{l}\mathbf{l}^{\mathrm{T}} \varepsilon_\xi, \tag{6.137}$$

and

$$\langle \epsilon | \psi \rangle = \mathbf{l}\mathbf{l}^{\mathrm{T}} \langle \epsilon_\xi | \zeta \rangle. \tag{6.138}$$

The eigenvalues and eigenvectors of ρ yield

$$\mathbf{S}_\rho = \mathbf{diag}(1, 0, \ldots, 0), \tag{6.139}$$

and

$$\mathbf{U}_\rho = [\mathbf{l}^* \ \mathbf{U}_2], \tag{6.140}$$

where $\mathbf{l}^* = \mathbf{l}/|\mathbf{l}|$ is orthogonal to $\mathbf{U}_2$. The gradient correlation matrix reduces to

$$\mathbf{G} = \frac{\mathbf{l}\mathbf{l}^{\mathrm{T}}}{\mathbf{l}^{\mathrm{T}}\mathbf{l}} = \rho, \tag{6.141}$$

so that the 'noise' term in (6.112) reduces to

$$\mathbf{S}_{\mathrm{g}}(\phi)\mathbf{C}_{\mathrm{g}}(\phi)\, \mathrm{d}\mathbf{W}(t) = \langle \epsilon_\xi | \xi \rangle^{1/2} \mathbf{l}\, \mathrm{d}W(t). \tag{6.142}$$

The uni-variate Wiener process $\mathrm{d}W(t)$ produces fluctuations only in a one-dimensional sub-space. Moreover, since $\langle \epsilon_\xi | 0 \rangle = \langle \epsilon_\xi | 1 \rangle = 0$, the fluctuations will be zero outside the unit interval, and the allowable region for the joint composition PDF will remain one-dimensional and bounded as required by (6.134).

The limiting case presented above applies only when the transformation between the mixture fraction and the composition vector is linear – a case that is not of much interest in the study of reacting flows. However, the results can be extended to the case where the transformation is non-linear:[109]

$$\phi' = \mathbf{g}_0(\xi) \equiv \langle \phi | \xi \rangle - \langle \phi \rangle, \tag{6.143}$$

[107] This discussion can be easily extended to the mixture-fraction vector $\boldsymbol{\xi}$, or any other conserved linear combination of the composition variables.

[108] Note that this case can be modeled by a shape function $h(\xi)$.

[109] As shown in Chapter 5, the composition vector can be decomposed into a reaction-progress vector φ_{rp} and the mixture-fraction vector $\boldsymbol{\xi}$. Here we will denote the reacting scalars by ϕ, and consider only binary mixing.

and

$$\nabla\phi' = \mathbf{g}_1(\xi)\nabla\xi \equiv \frac{\partial\langle\phi|\xi\rangle}{\partial\xi}\nabla\xi. \tag{6.144}$$

Note that the vector functions $\mathbf{g}_0$ and $\mathbf{g}_1$ will normally be time-dependent, but can be found from the conditional moments $\langle\phi|\xi\rangle$. In the transported PDF context, the latter can be computed directly from the joint composition PDF so that $\mathbf{g}_0$ and $\mathbf{g}_1$ will be well defined functions.[110] The FP model in this limit is thus equivalent to a transported PDF extension of the conditional-moment closure (CMC) discussed in Section 5.8.[111]

The FP model (including the chemical source term $\mathbf{S}(\phi, \xi)$) becomes

$$\mathrm{d}\begin{bmatrix}\phi\\ \xi\end{bmatrix} = \begin{bmatrix}\mathbf{S}(\phi,\xi)\\ 0\end{bmatrix}\mathrm{d}t - \boldsymbol{\varepsilon}\mathbf{C}^{*-1}\begin{bmatrix}\phi - \langle\phi\rangle\\ \xi - \langle\xi\rangle\end{bmatrix}\mathrm{d}t + \langle\epsilon_\xi|\xi\rangle^{1/2}\begin{bmatrix}\mathbf{g}_1(\xi)\\ 1\end{bmatrix}\mathrm{d}W(t), \tag{6.145}$$

where the matrix $\mathbf{C}^*$ is found as in (6.102) starting from the covariance matrix:

$$\mathbf{C} = \begin{bmatrix}\langle\phi'\phi'^{\mathrm{T}}\rangle & \langle\phi'\xi'\rangle\\ \langle\xi'\phi'^{\mathrm{T}}\rangle & \langle\xi'^2\rangle\end{bmatrix} = \begin{bmatrix}\langle\mathbf{g}_0\mathbf{g}_0^{\mathrm{T}}\rangle & \langle\mathbf{g}_0\xi'\rangle\\ \langle\xi'\mathbf{g}_0^{\mathrm{T}}\rangle & \langle\xi'^2\rangle\end{bmatrix}, \tag{6.146}$$

and the joint scalar dissipation rate matrix is given by[112]

$$\boldsymbol{\varepsilon} = \begin{bmatrix}\langle\epsilon_\xi\mathbf{g}_1\mathbf{g}_1^{\mathrm{T}}\rangle & \langle\epsilon_\xi\mathbf{g}_1\rangle\\ \langle\epsilon_\xi\mathbf{g}_1^{\mathrm{T}}\rangle & \varepsilon_\xi\end{bmatrix}. \tag{6.147}$$

Because the conditional scalar Laplacian is approximated in the FP model by a non-linear diffusion process (6.91), (6.145) will not agree exactly with CMC. Nevertheless, since transported PDF methods can be easily extended to inhomogeneous flows,[113] which are problematic for the CMC, the FP model offers distinct advantages.

When applying (6.145) and (6.146), the conditional scalar dissipation rate $\langle\epsilon_\xi|\zeta\rangle$ must be supplied. However, unlike with the CMC, where both $\langle\epsilon_\xi|\zeta\rangle$ and a *consistent* mixture-fraction PDF $f_\xi(\zeta)$ must be provided by the user,[114] the FP model predicts the mixture-fraction PDF. Indeed, the stationary[115] mixture-fraction PDF predicted by (6.145) is (Gardiner 1990)

$$f_\xi^{\mathrm{s}}(\zeta) = \mathcal{N}\frac{\varepsilon_\xi}{\langle\epsilon_\xi|\zeta\rangle}\exp\left[\frac{2\varepsilon_\xi}{\langle\xi'^2\rangle}\int\frac{(\langle\xi\rangle - \zeta)}{\langle\epsilon_\xi|\zeta\rangle}\,\mathrm{d}\zeta\right], \tag{6.148}$$

110 In a 'particle' implementation of transported PDF methods (see Chapter 7), it will be necessary to estimate $\mathbf{g}_0(\zeta)$ using, for example, smoothing splines. $\mathbf{g}_1(\zeta)$ will then be found by differentiating the splines. Note that this implies that estimates for the conditional moments (i.e., $\mathbf{g}_0$) are found only in regions of composition space where the mixture fraction occurs with non-negligible probability.

111 A transported PDF extension of the flamelet model can be derived in a similar manner using the Lagrangian spectral relaxation model (Fox 1999) for the joint scalar dissipation rate.

112 Note that the ranks of $\boldsymbol{\varepsilon}$ and $\mathbf{C}$ are one.

113 For inhomogeneous flows, turbulent transport will bring fluid particles with different histories to a given point in the flow. Thus, it cannot be expected that (6.143) will be exact in such flows. Nonetheless, since the conditional moments will be well defined, the FP model may still provide a useful approximation for molecular mixing.

114 See the discussion in Section 5.8 starting near (5.301) on p. 212.

115 Because the coefficients are time-dependent, the stationary PDF will only approximate f_ξ after an initial transient period.

where $\mathcal{N}$ is a normalization constant found by forcing the total probability to be one. Thus, the user need only supply a functional form for $\langle \epsilon_\xi | \zeta \rangle$ which satisfies $\langle \epsilon_\xi | 0 \rangle = \langle \epsilon_\xi | 1 \rangle = 0$. For example, a beta PDF can be found from (6.148) by using the shape function given by the uni-variate version of (6.130):

$$\langle \epsilon_\xi | \zeta \rangle = \varepsilon_\xi \frac{\zeta(1-\zeta)}{\langle \xi(1-\xi) \rangle}. \tag{6.149}$$

Using this expression and a model for ε_ξ in (6.145) and (6.146) yields a complete description of reactive scalar mixing without the consistency problems associated with CMC. Moreover, since it is not necessary to supply boundary conditions for the conditional moments,[116] the FP model can be applied to 'partially mixed' regions of the flow[117] where the CMC boundary conditions cannot be predicted *a priori*.

6.6.5 Prospects for mixing model improvements

In the application of transported PDF methods, the mixing model remains the weakest link. Despite considerable effort, the elaboration of a simple mixing model that possesses all six desirable properties, and yields satisfactory agreement for all possible initial conditions, has yet to be achieved (nor is it likely to be possible). Despite this pessimistic outlook, improvements in molecular mixing models are still possible. For example, for non-premixed flows length-scale information can be included by using a more detailed model for the scalar dissipation rate. For binary mixing (which encompasses many of the potential applications), existing mixing models based on an extension of the beta presumed PDF method yield excellent predictions for isothermal flows. For non-isothermal flows with binary mixing, progress can be made by including chemistry effects in models for the joint scalar dissipation rate of the reaction-progress variables.

More generally, by using the linear transformation given in (5.107) on p. 167, the mixing model can be decomposed into a non-premixed, inert contribution for $\boldsymbol{\xi}$ and a 'premixed,'[118] reacting contribution for φ_{rp}. It may then be possible to make judicious assumptions concerning the joint scalar dissipation rate. For example, if the spatial gradients of $\boldsymbol{\xi}$ and φ_{rp} are assumed to be uncorrelated, then

$$\langle \epsilon | \boldsymbol{\xi}, \varphi_{\text{rp}} \rangle = \begin{bmatrix} \langle \epsilon_\xi | \boldsymbol{\xi} \rangle & \mathbf{0} \\ \mathbf{0} & \langle \epsilon_{\text{rp}} | \boldsymbol{\xi}, \varphi_{\text{rp}} \rangle \end{bmatrix}. \tag{6.150}$$

The effect of chemistry is thereby isolated in the premixed contribution: $\langle \epsilon_{\text{rp}} | \boldsymbol{\xi}, \varphi_{\text{rp}} \rangle$. However, in order to make progress in model development, it will be necessary to have

116 Indeed, $\langle \phi | 0 \rangle$ and $\langle \phi | 1 \rangle$ are predicted by the FP model.
117 In other words, regions where $f_\xi(0) = 0$ or $f_\xi(1) = 0$ so that no 'pure' fluid exists.
118 The reaction-progress vector is premixed in the sense that variations due to finite-rate chemistry will occur along iso-clines of constant mixture fraction.

access to detailed simulations of the scalar transport equation (6.6) from which conditional statistics (such as $\langle \epsilon | \xi, \varphi_{\text{rp}} \rangle$) can be extracted.[119]

With respect to using detailed simulations, the good news is that – in comparison to turbulence modeling – the simulation requirements for mixing models are not too restrictive. For example, a simple one-dimensional linear-eddy model (Section 4.3, p. 110) is most likely sufficient for generating conditional statistics for model development. Indeed, since the shape of the composition PDF in homogeneous flows is dominated by molecular diffusion, the dependence of $\langle \epsilon | \phi \rangle$ on ϕ can even be studied in purely diffusive systems (Fox 1994), thereby avoiding the need to simulate the velocity field.[120] This approach has been followed by Fox (1997) to derive a Lagrangian mixing model based on the FP model wherein molecular mixing occurs by diffusion only for a given value of the turbulence-conditioned scalar dissipation rate $\langle \epsilon_\phi | \epsilon \rangle$.[121] In any case, since it controls the shape of the composition PDF, detailed information on the ϕ dependence of $\langle \epsilon | \phi \rangle$ will be a key requirement for further improvements of molecular mixing models.

6.7 Lagrangian PDF methods

The PDF transport equations and closures considered up to this point have been Eulerian, e.g., $f_{\mathbf{U},\phi}(\mathbf{V}, \psi; \mathbf{x}, t)$ for fixed $\mathbf{x}$ and t. The large number of independent variables (e.g., $\mathbf{V}$, ψ, $\mathbf{x}$, and t) in the PDF transport equation makes it intractable to solve using standard discretization methods. Instead, Lagrangian PDF methods (Pope 1994a) can be used to express the problem in terms of stochastic differential equations for so-called 'notional' particles. In Chapter 7, we will discuss grid-based Eulerian PDF codes which also use notional particles. However, in the Eulerian context, a notional particle serves only as a discrete representation of the Eulerian PDF and not as a model for a Lagrangian fluid particle. The Lagrangian Monte-Carlo simulation methods discussed in Chapter 7 are based on Lagrangian PDF methods.

6.7.1 Lagrangian notional particles

A Lagrangian notional particle follows a trajectory in velocity–composition–physical space (i.e., $\mathbf{U}^*(t), \phi^*(t), \mathbf{X}^*(t)$)[122] which originates at a random location $\mathbf{Y}$ in the physical

[119] In theory, such information could be extracted from high-resolution experimental data. However, due to the technical challenges, conditional statistics of small-scale quantities are more readily accessible from numerical simulations.

[120] Obviously, the rate of mixing (as measured by the magnitude of joint scalar dissipation rates) will be strongly affected by turbulence, and thus must be modeled separately.

[121] For the FP model, the shape information is contained in the shape matrix $\mathbf{H}(\phi)$, and rate information is contained in the mean joint scalar dissipation rate matrix $\boldsymbol{\varepsilon}$.

[122] We will also discuss Lagrangian PDF models for the composition PDF. In this case, the notional particles follow trajectories in composition–physical space: $\phi^*(t), \mathbf{X}^*(t)$. A gradient-diffusion model is then used to represent conditional velocity fluctuations.

domain:[123]

$$
\begin{aligned}
\mathbf{U}^*(0) &= \mathbf{U}(\mathbf{Y}, t_0),\\
\phi^*(0) &= \phi(\mathbf{Y}, t_0),\\
\mathbf{X}^*(0) &= \mathbf{Y},
\end{aligned}
\tag{6.151}
$$

where t_0 is an arbitrary fixed reference time, and $\mathbf{U}(\mathbf{x}, t)$ and $\phi(\mathbf{x}, t)$ are the Eulerian velocity and composition fields, respectively. A notional-particle trajectory is defined by stochastic differential equations of the form:[124]

$$
\frac{\mathrm{d}\mathbf{X}^*}{\mathrm{d}t}(t) = \mathbf{U}^*(t), \tag{6.152}
$$

$$
\begin{aligned}
\mathrm{d}\mathbf{U}^*(t) = {} & \mathbf{a}_U(\mathbf{U}^*(t), \phi^*(t), \mathbf{X}^*(t), t)\,\mathrm{d}t\\
& + \mathbf{B}_{UU}(\mathbf{U}^*(t), \phi^*(t), \mathbf{X}^*(t), t)\,\mathrm{d}\mathbf{W}_U(t)\\
& + \mathbf{B}_{U\phi}(\mathbf{U}^*(t), \phi^*(t), \mathbf{X}^*(t), t)\,\mathrm{d}\mathbf{W}_\phi(t),
\end{aligned}
\tag{6.153}
$$

and

$$
\begin{aligned}
\mathrm{d}\phi^*(t) = {} & \mathbf{a}_\phi(\mathbf{U}^*(t), \mathbf{X}^*(t), t)\,\mathrm{d}t\\
& + \mathbf{B}_{\phi U}(\mathbf{U}^*(t), \phi^*(t), \mathbf{X}^*(t), t)\,\mathrm{d}\mathbf{W}_U(t)\\
& + \mathbf{B}_{\phi\phi}(\mathbf{U}^*(t), \phi^*(t), \mathbf{X}^*(t), t)\,\mathrm{d}\mathbf{W}_\phi(t).
\end{aligned}
\tag{6.154}
$$

The primary task in Lagrangian PDF modeling is thus to find appropriate functional forms for the drift ($\mathbf{a}_U, \mathbf{a}_\phi$) and diffusion ($\mathbf{B}_{UU}, \mathbf{B}_{U\phi}, \mathbf{B}_{\phi U}, \mathbf{B}_{\phi\phi}$) coefficients. Fortunately, in many applications only a small sub-set of the coefficients will be non-zero. For example, for constant-density flows, $\mathbf{a}_U$ and $\mathbf{B}_{UU}$ are independent of $\phi^*(t)$, and both $\mathbf{B}_{U\phi}$ and $\mathbf{B}_{\phi U}$ are zero.

The stochastic differential equations in (6.152)–(6.154) generate a Lagrangian PDF which is conditioned on the initial location:[125]

$$
f_{\mathrm{L}}^* \equiv f_{\mathbf{U}^*,\phi^*,\mathbf{X}^*|\mathbf{Y}}(\mathbf{V}, \psi, \mathbf{x}|\mathbf{y}; t).
$$

This conditional PDF is governed by the corresponding Fokker–Planck equation (Gardiner 1990):

$$
\frac{\partial f_{\mathrm{L}}^*}{\partial t} + V_i \frac{\partial f_{\mathrm{L}}^*}{\partial x_i} + \frac{\partial}{\partial z_i}[a_i(\mathbf{z}, \mathbf{x}, t) f_{\mathrm{L}}^*] = \frac{1}{2}\frac{\partial^2}{\partial z_i \partial z_j}[b_{ij}(\mathbf{z}, \mathbf{x}, t) f_{\mathrm{L}}^*], \tag{6.155}
$$

[123] Note that, even though $\mathbf{Y}$ and t_0 are fixed, the initial velocity and composition will be random variables, since $\mathbf{U}(\mathbf{x}, t)$ and $\phi(\mathbf{x}, t)$ are random fields.

[124] Alternatively, the white-noise processes $\mathbf{W}(t)$ could be replaced by colored-noise processes. Since the latter have finite auto-correlation times, the resulting Lagrangian correlation functions for $\mathbf{U}^*$ and ϕ^* would be non-exponential. However, it would generally not be possible to describe the Lagrangian PDF by a Fokker–Planck equation. Thus, in order to simplify the comparison with Eulerian PDF methods, we will use white-noise processes throughout this section.

[125] A more precise definition would include conditioning on the random initial velocity and compositions: $f_{\mathbf{U}^*,\phi^*,\mathbf{X}^*|\mathbf{U}_0,\phi_0,\mathbf{Y}}(\mathbf{V}, \psi, \mathbf{x}|\mathbf{V}_0, \psi_0, \mathbf{y}; t)$. However, only the conditioning on initial location is needed in order to relate the Lagrangian and Eulerian PDFs. Nevertheless, the initial conditions ($\mathbf{U}_0$, ϕ_0) for a notional particle must have the same one-point statistics as the random variables $\mathbf{U}(\mathbf{Y}, t_0)$ and $\phi(\mathbf{Y}, t_0)$.

where

$$\mathbf{z} = \begin{bmatrix} \mathbf{V} \\ \boldsymbol{\psi} \end{bmatrix}, \tag{6.156}$$

$$\mathbf{a}(\mathbf{z}, \mathbf{x}, t) = \begin{bmatrix} \mathbf{a}_U \\ \mathbf{a}_\phi \end{bmatrix}, \tag{6.157}$$

and the diffusion matrix $\mathbf{B} = [b_{ij}]$ is given by

$$\mathbf{B}(\mathbf{z}, \mathbf{x}, t) = \begin{bmatrix} \mathbf{B}_{UU}\mathbf{B}_{UU}^{\mathrm{T}} + \mathbf{B}_{U\phi}\mathbf{B}_{U\phi}^{\mathrm{T}} & \mathbf{B}_{UU}\mathbf{B}_{\phi U}^{\mathrm{T}} + \mathbf{B}_{U\phi}\mathbf{B}_{\phi\phi}^{\mathrm{T}} \\ \mathbf{B}_{\phi U}\mathbf{B}_{UU}^{\mathrm{T}} + \mathbf{B}_{\phi\phi}\mathbf{B}_{U\phi}^{\mathrm{T}} & \mathbf{B}_{\phi U}\mathbf{B}_{\phi U}^{\mathrm{T}} + \mathbf{B}_{\phi\phi}\mathbf{B}_{\phi\phi}^{\mathrm{T}} \end{bmatrix}. \tag{6.158}$$

For constant-density flows, (6.155) simplifies to

$$\begin{aligned} \frac{\partial f_{\mathrm{L}}^*}{\partial t} + V_i \frac{\partial f_{\mathrm{L}}^*}{\partial x_i} + \frac{\partial}{\partial V_i}[a_{U,i}(\mathbf{V}, \mathbf{x}, t) f_{\mathrm{L}}^*] + \frac{\partial}{\partial \psi_i}[a_{\phi,i}(\mathbf{V}, \boldsymbol{\psi}, \mathbf{x}, t) f_{\mathrm{L}}^*] \\ = \frac{1}{2}\frac{\partial^2}{\partial V_i \partial V_j}[b_{U,ij}(\mathbf{V}, \mathbf{x}, t) f_{\mathrm{L}}^*] + \frac{1}{2}\frac{\partial^2}{\partial \psi_i \partial \psi_j}[b_{\phi,ij}(\mathbf{V}, \boldsymbol{\psi}, \mathbf{x}, t) f_{\mathrm{L}}^*], \end{aligned} \tag{6.159}$$

where the diffusion matrices are defined by $\mathbf{B}_U \equiv [b_{U,ij}] = \mathbf{B}_{UU}\mathbf{B}_{UU}^{\mathrm{T}}$ and $\mathbf{B}_\phi \equiv [b_{\phi,ij}] = \mathbf{B}_{\phi\phi}\mathbf{B}_{\phi\phi}^{\mathrm{T}}$. By correctly choosing the coefficient matrices ($\mathbf{a}_U$, $\mathbf{a}_\phi$, $\mathbf{B}_U$, and $\mathbf{B}_\phi$), (6.159) can be made to *correspond* with the Eulerian velocity, composition PDF transport equation (6.19). However, it is important to note that $f_{\mathrm{L}}^* \neq f_{\mathbf{U},\phi}$. Thus it remains to determine how the Lagrangian notional-particle PDF f_{L}^* is related to the Eulerian velocity, composition PDF $f_{\mathbf{U},\phi}$. This can be done by considering *Lagrangian fluid particles.*

6.7.2 Lagrangian fluid particles

In addition to notional particles, Lagrangian PDF methods make use of Lagrangian fluid particles (Pope 2000). For example, $\mathbf{X}^+(t, \mathbf{Y})$ and $\mathbf{U}^+(t, \mathbf{Y})$ denote the position and velocity of the fluid particle originating at position $\mathbf{Y}$ at reference time t_0, and corresponding to a *particular realization of the flow* $\mathbf{U}(\mathbf{x}, t)$. In other words, given $\mathbf{U}(\mathbf{x}, t)$ for all $\mathbf{x}$ and t, $\mathbf{X}^+(t, \mathbf{Y})$ and $\mathbf{U}^+(t, \mathbf{Y})$ can be computed exactly (e.g., using DNS). In contrast, the notional particles have no underlying velocity field, and our only requirement will be that they yield the same one-point Lagrangian PDF as the fluid particles:[126]

$$f_{\mathbf{U}^*,\mathbf{X}^*|\mathbf{Y}}(\mathbf{V}, \mathbf{x}|\mathbf{y}; t) = f_{\mathbf{U}^+,\mathbf{X}^+|\mathbf{Y}}(\mathbf{V}, \mathbf{x}|\mathbf{y}; t), \tag{6.160}$$

which is a much weaker requirement than say[127]

$$\mathbf{X}^*(t) = \mathbf{X}^+(t, \mathbf{Y}) \quad \text{and} \quad \mathbf{U}^*(t) = \mathbf{U}^+(t, \mathbf{Y}). \tag{6.161}$$

126 The fluid-particle PDF is found by ensemble averaging over all realizations of the flow.
127 Agreement between the PDF is called 'weak equivalence,' while agreement of the time series is called 'strong equivalence' (Kloeden and Platen 1992).

For constant-density flow, the fundamental relationship (Pope 2000) between the fluid-particle PDF and the Eulerian PDF of the flow is

$$f_{\mathbf{U},\phi}(\mathbf{V},\psi;\mathbf{x},t+t_0) = \int_{\mathcal{V}} f_{\mathbf{U}^+,\phi^+,\mathbf{X}^+|\mathbf{Y}}(\mathbf{V},\psi,\mathbf{x}|\mathbf{y};t)\,\mathrm{d}\mathbf{y}. \tag{6.162}$$

For a Lagrangian PDF model, (6.162) should hold when $f_{\mathbf{U}^*,\phi^*,\mathbf{X}^*|\mathbf{Y}}$ is substituted for $f_{\mathbf{U}^+,\phi^+,\mathbf{X}^+|\mathbf{Y}}$. Thus, we will now look at what conditions are necessary to ensure that (6.162) holds for the notional-particle PDF.

6.7.3 Spatial distribution of notional particles

As shown above in (6.162), the Lagrangian fluid-particle PDF can be related to the Eulerian velocity, composition PDF by integrating over all initial conditions. As shown below in (6.168), for the Lagrangian notional-particle PDF, the same transformation introduces a weighting factor which involves the PDF of the initial positions $f_{\mathbf{Y}}(\mathbf{y})$ and the PDF of the current position $f_{\mathbf{X}^*}(\mathbf{x};t)$. If we let $\mathcal{V}$ denote a closed volume containing a fixed mass of fluid, then, by definition, $\mathbf{x}, \mathbf{y} \in \mathcal{V}$. The first condition needed to reproduce the Eulerian PDF is that the initial locations be uniform:

$$f_{\mathbf{Y}}(\mathbf{y}) = \frac{1}{\mathcal{V}} \quad \text{for } \mathbf{y} \in \mathcal{V}. \tag{6.163}$$

From the continuity equation, it then follows for constant-density flow that the PDF of $\mathbf{X}^*(t)$ is uniform:[128]

$$f_{\mathbf{X}^*}(\mathbf{x};t) = \frac{1}{\mathcal{V}} \quad \text{for } \mathbf{x} \in \mathcal{V}. \tag{6.164}$$

It is important to recognize that (6.164) is a direct result of *choosing* the uniformly distributed initial locations (6.163). In contrast, if one chooses to start all notional particles at the origin: $f_{\mathbf{Y}}(\mathbf{y}) = \delta(\mathbf{y})$, then $f_{\mathbf{X}^*}(\mathbf{x};t)$ will be non-uniform, and the Lagrangian notional-particle PDF will not correspond to the Lagrangian fluid-particle PDF.

6.7.4 Relationship to Eulerian PDF transport equation

The key theoretical concept that makes Lagrangian PDF methods useful is the correspondence between the Eulerian PDF of the flow and the Lagrangian notional-particle PDF. As noted above, in the Lagrangian notional-particle PDF $\mathbf{X}^*$ is a random variable

[128] This is a direct result of forcing the stochastic differential equations to correspond to the Eulerian PDF. From (6.152), it follows that

$$\frac{\partial f_{\mathbf{X}^*}}{\partial t} + \frac{\partial}{\partial x_i}[\langle U_i^*|\mathbf{x}\rangle f_{\mathbf{X}^*}] = 0,$$

where $f_{\mathbf{X}^*}(\mathbf{x};0) = f_{\mathbf{Y}}(\mathbf{x})$ and $\langle \mathbf{U}^*|\mathbf{x}\rangle = \langle \mathbf{U}(\mathbf{x},t+t_0)\rangle$. For constant-density flow, $\nabla \cdot \langle \mathbf{U}\rangle = 0$ so that

$$\frac{\partial f_{\mathbf{X}^*}}{\partial t} + \langle U_i^*|\mathbf{x}\rangle \frac{\partial f_{\mathbf{X}^*}}{\partial x_i} = 0,$$

which implies (6.164) given (6.163).

indicating an arbitrary position in physical space. In contrast, in the Eulerian PDF the spatial location $\mathbf{x}$ is fixed. Thus, in terms of the probability density functions, we must consider the Lagrangian notional-particle PDF of velocity and composition *given that* $\mathbf{X}^*(t) = \mathbf{x}$:

$$f_{\mathbf{U}^*,\phi^*|\mathbf{X}^*}(\mathbf{V}, \psi|\mathbf{x}; t) = \frac{f_{\mathbf{U}^*,\phi^*,\mathbf{X}^*}(\mathbf{V}, \psi, \mathbf{x}; t)}{f_{\mathbf{X}^*}(\mathbf{x}; t)}. \tag{6.165}$$

In words, the Eulerian PDF generated by the notional particles is equal to the Lagrangian PDF for velocity and composition at a *given* spatial location:

$$f^*_{\mathbf{U},\phi}(\mathbf{V}, \psi; \mathbf{x}, t) \equiv f_{\mathbf{U}^*,\phi^*|\mathbf{X}^*}(\mathbf{V}, \psi|\mathbf{x}; t). \tag{6.166}$$

Thus, using (6.163), (6.164), and (6.165), we can relate $f^*_{\mathbf{U},\phi}$ to the Lagrangian notional-particle PDF:

$$\begin{aligned} f^*_{\mathbf{U},\phi}(\mathbf{V}, \psi; \mathbf{x}, t) &= f_{\mathbf{U}^*,\phi^*|\mathbf{X}^*}(\mathbf{V}, \psi|\mathbf{x}; t) \\ &= \frac{f_{\mathbf{U}^*,\phi^*,\mathbf{X}^*}(\mathbf{V}, \psi, \mathbf{x}; t)}{f_{\mathbf{X}^*}(\mathbf{x}; t)} \\ &= \int_{\mathcal{V}} \frac{f_{\mathbf{U}^*,\phi^*,\mathbf{X}^*,\mathbf{Y}}(\mathbf{V}, \psi, \mathbf{x}, \mathbf{y}; t)}{f_{\mathbf{X}^*}(\mathbf{x}; t)} \, \mathrm{d}\mathbf{y} \\ &= \int_{\mathcal{V}} f_{\mathbf{U}^*,\phi^*,\mathbf{X}^*|\mathbf{Y}}(\mathbf{V}, \psi, \mathbf{x}|\mathbf{y}; t) \frac{f_{\mathbf{Y}}(\mathbf{y})}{f_{\mathbf{X}^*}(\mathbf{x}; t)} \, \mathrm{d}\mathbf{y}. \end{aligned} \tag{6.167}$$

If the initial particle locations are uniformly distributed and (6.160) holds, then (6.167) relates the Eulerian notional-particle PDF to the Eulerian PDF (for fixed $\mathbf{x}$ and t with $t_0 = 0$):[129]

$$\begin{aligned} f^*_{\mathbf{U},\phi}(\mathbf{V}, \psi; \mathbf{x}, t) &= \int_{\mathcal{V}} f_{\mathbf{U}^*,\phi^*,\mathbf{X}^*|\mathbf{Y}}(\mathbf{V}, \psi, \mathbf{x}|\mathbf{y}; t) \, \mathrm{d}\mathbf{y} \\ &= \int_{\mathcal{V}} f_{\mathbf{U}^+,\phi^+,\mathbf{X}^+|\mathbf{Y}}(\mathbf{V}, \psi, \mathbf{x}|\mathbf{y}; t) \, \mathrm{d}\mathbf{y} \\ &= f_{\mathbf{U},\phi}(\mathbf{V}, \psi; \mathbf{x}, t). \end{aligned} \tag{6.168}$$

Furthermore, since (6.159) does not depend on $\mathbf{y}$, if the notional particles are uniformly distributed the Fokker–Planck equation for $f^*_{\mathbf{U},\phi}$ is

$$\begin{aligned} &\frac{\partial f^*_{\mathbf{U},\phi}}{\partial t} + V_i \frac{\partial f^*_{\mathbf{U},\phi}}{\partial x_i} + \frac{\partial}{\partial V_i}[a_{U,i}(\mathbf{V}, \mathbf{x}, t) f^*_{\mathbf{U},\phi}] + \frac{\partial}{\partial \psi_i}[a_{\phi,i}(\mathbf{V}, \psi, \mathbf{x}, t) f^*_{\mathbf{U},\phi}] \\ &\quad = \frac{1}{2} \frac{\partial^2}{\partial V_i \partial V_j}[b_{U,ij}(\mathbf{V}, \mathbf{x}, t) f^*_{\mathbf{U},\phi}] + \frac{1}{2} \frac{\partial^2}{\partial \psi_i \partial \psi_j}[b_{\phi,ij}(\mathbf{V}, \psi, \mathbf{x}, t) f^*_{\mathbf{U},\phi}]. \end{aligned} \tag{6.169}$$

Thus, in summary, the two necessary conditions for correspondence between the notional-particle system and the fluid-particle system in constant-density flows are

(1) $f_{\mathbf{X}^*}(\mathbf{x}; t)$ *must be a uniform distribution for all* t,

[129] The first equality follows when $f_{\mathbf{X}^*}(\mathbf{x}; t) = f_{\mathbf{X}^*}(\mathbf{x}; 0) = f_{\mathbf{Y}}(\mathbf{x})$ is uniform. The second equality follows from the correspondence condition in (6.160). The third equality follows from (6.162).

and

(2) *the correspondence condition in (6.160) must hold.*

In order for condition (2) to hold, it will be necessary to choose the coefficients properly in (6.153) and (6.154). We will return to this question below. First, however, note that for constant-density flow, the first line of (6.167) leads to a relationship between the one-point Eulerian moments and the conditional Lagrangian moments. For example, let $Q(\mathbf{V}, \psi)$ be an arbitrary function. It follows that:

$$\begin{aligned}\langle Q(\mathbf{U}(\mathbf{x}, t), \phi(\mathbf{x}, t))\rangle &= \iint_{-\infty}^{+\infty} Q(\mathbf{V}, \psi) f_{\mathbf{U},\phi}(\mathbf{V}, \psi; \mathbf{x}, t)\, \mathrm{d}\mathbf{V}\, \mathrm{d}\psi \\ &= \iint_{-\infty}^{+\infty} Q(\mathbf{V}, \psi) f_{\mathbf{U}^*,\phi^*|\mathbf{X}^*}(\mathbf{V}, \psi|\mathbf{x}; t)\, \mathrm{d}\mathbf{V}\, \mathrm{d}\psi \\ &= \langle Q(\mathbf{U}^*(t), \phi^*(t))|\mathbf{X}^*(t) = \mathbf{x}\rangle.\end{aligned} \tag{6.170}$$

Thus, correspondence between the notional-particle system and the Eulerian PDF of the flow requires agreement at the moment level. In particular, it requires that $\langle \mathbf{U}(\mathbf{x}, t)\rangle = \langle \mathbf{U}^*(t)|\mathbf{X}^*(t) = \mathbf{x}\rangle$ and $\langle \phi(\mathbf{x}, t)\rangle = \langle \phi^*(t)|\mathbf{X}^*(t) = \mathbf{x}\rangle$. It remains then to formulate stochastic differential equations for the notional-particle system which yield the desired correspondence.

6.7.5 Stochastic differential equations for notional particles

Equations (6.169) and (6.170) offer two distinct ways for specifying the stochastic differential equations that determine $\mathbf{U}^*(t)$ and $\phi^*(t)$. The first (and most difficult!) approach is to choose stochastic models that exactly reproduce the Eulerian PDF, $f_{\mathbf{U},\phi}$, from (6.169). However, since the 'exact' form of $f_{\mathbf{U},\phi}$ is only known for a few simple flows (e.g., fully developed homogeneous turbulence), finding a general model for all flows is most likely impossible (or at least as difficult as solving the Navier–Stokes equation directly). The second (and more realistic) approach is to formulate stochastic models that reproduce the lower-order moments found from (6.170). The generalized Langevin model (GLM) discussed in Section 6.5 is an example of the second approach applied to the second-order moments of the velocity field (i.e., $\langle \mathbf{U}(\mathbf{x}, t)\rangle = \langle \mathbf{U}^*(t)|\mathbf{X}^*(t) = \mathbf{x}\rangle$ and $\langle U_i(\mathbf{x}, t)U_j(\mathbf{x}, t)\rangle = \langle U_i^*(t)U_j^*(t)|\mathbf{X}^*(t) = \mathbf{x}\rangle$). With the availability of DNS data for Lagrangian correlation functions (Yeung and Pope 1989; Yeung 1997; Yeung 1998b; Yeung 2001), it is also possible to enforce correspondence at the level of Lagrangian two-time statistics such as $\langle X_i^*(t_1)X_j^*(t_2)\rangle$, $\langle U_i^*(t_1)U_j^*(t_2)\rangle$, and $\langle \phi_\alpha^*(t_1)\phi_\beta^*(t_2)\rangle$. Because the velocity auto-correlation function is nearly exponential, stochastic models of the form of (6.153) perform satisfactorily. However, the scalar auto-correlation function has more structure (Yeung 2001) than can be described by (6.154).

Owing to the sensitivity of the chemical source term to the shape of the composition PDF, the application of the second approach to model $\phi^*(t)$ is problematic. As seen with the molecular mixing models in Section 6.6, a successful model for $\phi^*(t)$ must not only satisfy the moment constraints in Table 6.1, but also possess the desirable properties. In addition, the Lagrangian correlation functions for each pair of scalars ($\langle \phi^*_\alpha(t_1)\phi^*_\beta(t_2)\rangle$) should agree with available DNS data.[130] Some of these requirements (e.g., desirable property (ii)) require models that control the shape of f_ϕ, and for these reasons the development of stochastic differential equations for micromixing is particularly difficult.

Regardless of the approach chosen, there remain two possible choices for determining the coefficients:

(i) *Lagrangian correspondence.* Select coefficients in (6.153) and (6.154) such that $f_{\mathbf{U}^*,\phi^*,\mathbf{X}^*|\mathbf{Y}} = f_{\mathbf{U}^+,\phi^+,\mathbf{X}^+|\mathbf{Y}}$;
(ii) *Eulerian correspondence.* Select coefficients in (6.169) such that $f^*_{\mathbf{U},\phi} = f_{\mathbf{U},\phi}$.

Note that Lagrangian correspondence does not automatically imply Eulerian correspondence, since the latter also requires that $f_{\mathbf{X}^*}(\mathbf{x};t)$ must be a uniform distribution for all t. Nevertheless, Lagrangian correspondence is the preferable choice, and is obtained by working directly with the stochastic differential equations.

Using (6.4) and (6.6), the notional-particle trajectories can be expressed in terms of the conditional fluxes:[131]

$$\begin{aligned}
\frac{\mathrm{d}\mathbf{X}^*}{\mathrm{d}t} &= \mathbf{U}^*,\\
\frac{\mathrm{d}\mathbf{U}^*}{\mathrm{d}t} &= \langle \mathbf{A}|\mathbf{U}^*,\phi^*,\mathbf{X}^*\rangle + \mathcal{N}_U(t),\\
\frac{\mathrm{d}\phi^*}{\mathrm{d}t} &= \langle \mathbf{\Theta}|\mathbf{U}^*,\phi^*,\mathbf{X}^*\rangle + \mathcal{N}_\phi(t),
\end{aligned} \tag{6.171}$$

where $\mathcal{N}_U(t)$ and $\mathcal{N}_\phi(t)$ are 'noise' terms that are added to control the shape of both the Lagrangian PDF and the Lagrangian correlation functions.[132] Comparison of (6.171) with (6.153) and (6.154) for constant-density flow yields

$$[\langle \mathbf{A}|\mathbf{U}^*,\mathbf{X}^*\rangle + \mathcal{N}_U(t)]\,\mathrm{d}t = \mathbf{a}_U(\mathbf{U}^*,\mathbf{X}^*,t)\,\mathrm{d}t + \mathbf{B}_{UU}(\mathbf{U}^*,\mathbf{X}^*,t)\,\mathrm{d}\mathbf{W}_U(t), \tag{6.172}$$

and

$$\begin{aligned}
&[\langle \mathbf{\Theta}|\mathbf{U}^*,\phi^*,\mathbf{X}^*\rangle + \mathcal{N}_\phi(t)]\,\mathrm{d}t\\
&\quad = \mathbf{a}_\phi(\mathbf{U}^*,\phi^*,\mathbf{X}^*,t)\,\mathrm{d}t + \mathbf{B}_{\phi\phi}(\mathbf{U}^*,\phi^*,\mathbf{X}^*,t)\,\mathrm{d}\mathbf{W}_\phi(t).
\end{aligned} \tag{6.173}$$

130 In particular, for scalars with different Schmidt numbers the Lagrangian correlation function must exhibit the correct dependence on Sc and Re.

131 In general, the conditional fluxes are deterministic functions of $\mathbf{V}$, ψ, $\mathbf{x}$, and t. On the right-hand side of these equations, the conditional fluxes are evaluated at the current location of the notional particle in velocity–composition–physical space: $\mathbf{V} = \mathbf{U}^*(t)$, $\psi = \phi^*(t)$, $\mathbf{x} = \mathbf{X}^*(t)$.

132 In a single realization of a turbulent flow, the corresponding fluid-particle fluxes ($\mathbf{A}(\mathbf{X}^+,t)$ and $\mathbf{\Theta}(\mathbf{X}^+,t)$) are deterministic. However, they will be different for separate realizations. The 'noise' components thus represent the variability between different realizations in a large ensemble of turbulent flows.

Lagrangian PDF modeling then consists of finding expressions for the coefficients on the right-hand sides of (6.172) and (6.173) that are consistent with the known behavior of the conditional fluxes.

For the Lagrangian composition PDF, $f_{\phi^*,\mathbf{X}^*}$, the notional-particle trajectories are found from

$$\begin{aligned}\frac{\mathrm{d}\mathbf{X}^*}{\mathrm{d}t} &= \langle \mathbf{U}(\mathbf{X}^*, t)\rangle + \langle \mathbf{u}|\phi^*, \mathbf{X}^*\rangle,\\ \frac{\mathrm{d}\phi^*}{\mathrm{d}t} &= \langle \boldsymbol{\Theta}|\phi^*, \mathbf{X}^*\rangle + \mathcal{N}_\phi(t),\end{aligned} \tag{6.174}$$

wherein the mean velocity $\langle \mathbf{U}(\mathbf{x}, t)\rangle$ must be provided by a turbulence model, and the conditional velocity fluctuations $\langle \mathbf{u}|\phi^*, \mathbf{X}^*\rangle$ must be modeled. The corresponding Fokker–Planck equation is

$$\begin{aligned}&\frac{\partial f_{\phi^*,\mathbf{X}^*}}{\partial t} + \frac{\partial}{\partial x_i}[(\langle U_i\rangle + \langle u_i|\boldsymbol{\psi}\rangle) f_{\phi^*,\mathbf{X}^*}] + \frac{\partial}{\partial \psi_i}[a_{\phi,i}(\boldsymbol{\psi}) f_{\phi^*,\mathbf{X}^*}]\\ &= \frac{1}{2}\frac{\partial^2}{\partial\psi_i\partial\psi_j}[b_{\phi,ij}(\boldsymbol{\psi}) f_{\phi^*,\mathbf{X}^*}].\end{aligned} \tag{6.175}$$

Using the gradient-diffusion model, (6.29), the term involving the conditional velocity fluctuations can be written as[133]

$$\begin{aligned}\langle u_i|\boldsymbol{\psi}\rangle f_{\phi^*,\mathbf{X}^*} &= -\Gamma_\mathrm{T}(\mathbf{x}, t)\frac{\partial f_{\phi^*,\mathbf{X}^*}}{\partial x_i}\\ &= \left(\frac{\partial \Gamma_\mathrm{T}}{\partial x_i}\right) f_{\phi^*,\mathbf{X}^*} - \frac{1}{2}\frac{\partial}{\partial x_i}[2\Gamma_\mathrm{T} f_{\phi^*,\mathbf{X}^*}],\end{aligned} \tag{6.176}$$

and the stochastic differential equation for $\mathbf{X}^*(t)$ becomes

$$\mathrm{d}\mathbf{X}^* = [\langle \mathbf{U}(\mathbf{X}^*, t)\rangle + \boldsymbol{\nabla}\Gamma_\mathrm{T}(\mathbf{X}^*, t)]\,\mathrm{d}t + (2\Gamma_\mathrm{T}(\mathbf{X}^*, t))^{1/2}\,\mathrm{d}\mathbf{W}_x(t). \tag{6.177}$$

Note that the turbulent diffusivity $\Gamma_\mathrm{T}(\mathbf{x}, t)$ must be provided by a turbulence model, and for inhomogeneous flows its spatial gradient appears in the drift term in (6.177). If this term is neglected, the notional-particle location PDF, $f_{\mathbf{X}^*}$, will not remain uniform when $\boldsymbol{\nabla}\Gamma_\mathrm{T} \neq \mathbf{0}$, in which case the Eulerian PDFs will not agree, i.e., $f_\phi^* \neq f_\phi$.

6.7.6 Lagrangian velocity PDF closures

As in Section 6.5, the Lagrangian conditional acceleration can be decomposed into mean and fluctuating components. However, unlike for the Eulerian PDF, the mean fields must be replaced by their conditional counterparts (see (6.170)):

$$\langle A_i|\mathbf{U}^*, \mathbf{X}^*\rangle = \nu\frac{\partial^2\langle U_i^*|\mathbf{X}^*\rangle}{\partial x_j\partial x_j}(\mathbf{X}^*, t) - \frac{1}{\rho}\frac{\partial P}{\partial x_i}(\mathbf{X}^*, t) + g_i + \langle A_i'|\mathbf{U}^*, \mathbf{X}^*\rangle, \tag{6.178}$$

133 Since $\mathbf{X}^*$ is a random variable, this extension of (6.29) to $f_{\phi^*,\mathbf{X}^*}$ is not necessarily obvious. However, by working backwards from (6.177), it is possible to show that this definition is the only choice which permits $f_{\mathbf{X}^*}$ to remain uniform.

where $P(\mathbf{x}, t)$ is the *particle-pressure field*, which obeys a Poisson equation (Pope 2000):

$$\nabla^2 P(\mathbf{x}, t) = -\rho \frac{\partial^2 \langle U_i^* U_j^* | \mathbf{X}^* \rangle}{\partial x_i \partial x_j}(\mathbf{x}, t). \tag{6.179}$$

The notation used above for the derivative terms is meant to emphasize that three steps are required to compute them. For example, in order to compute $\nabla \langle \mathbf{U}^* | \mathbf{X}^* \rangle (\mathbf{X}^*, t)$, one must:

(1) *Compute the conditional field:* given the notional-particle values $\mathbf{U}^*$ and $\mathbf{X}^*$, compute $\langle \mathbf{U}^* | \mathbf{X}^* = \mathbf{x} \rangle = \langle \mathbf{U}^* | \mathbf{X}^* \rangle (\mathbf{x}, t)$ for all $\mathbf{x}$;
(2) *Compute gradients of the conditional field:* given particle field $\langle \mathbf{U}^* | \mathbf{X}^* \rangle (\mathbf{x}, t)$, compute derivatives with respect to $\mathbf{x}$: $\nabla \langle \mathbf{U}^* | \mathbf{X}^* \rangle (\mathbf{x}, t)$;
(3) *Evaluate at notional-particle location:* $\nabla \langle \mathbf{U}^* | \mathbf{X}^* \rangle (\mathbf{X}^*, t)$.

In Section 6.8 we will discuss how particle fields such as $\langle U_i^* U_j^* | \mathbf{X}^* \rangle$ can be estimated from the notional particles. However, it is important to note that since the particle-pressure field is found by solving (6.179), the estimate of $\langle U_i^* U_j^* | \mathbf{X}^* \rangle$ must be accurate enough to allow second-order derivatives. As noted after (6.61), the problem of dealing with noisy estimates of $P(\mathbf{x}, t)$ is one of the key challenges in applying (6.178).[134]

The Lagrangian generalized Langevin model (LGLM) for the fluctuating acceleration follows from (6.50):

$$[\langle A_j' | \mathbf{U}^*, \mathbf{X}^* \rangle + \mathcal{N}_U(t)] \, dt = \mathbf{a}_U'(\mathbf{U}^*, \mathbf{X}^*, t) \, dt + \mathbf{B}_{UU}(\mathbf{U}^*, \mathbf{X}^*, t) \, d\mathbf{W}_U(t), \tag{6.180}$$

where the drift coefficient is given by[135]

$$\mathbf{a}_U'(\mathbf{U}^*, \mathbf{X}^*) = \mathbf{G}^*(\mathbf{X}^*, t)(\mathbf{U}^* - \langle \mathbf{U}^* | \mathbf{X}^* \rangle), \tag{6.181}$$

and the diffusion matrix is given by

$$\mathbf{B}_{UU}(\mathbf{U}^*, \mathbf{X}^*, t) = (C_0 \varepsilon(\mathbf{X}^*, t))^{1/2} \mathbf{I}. \tag{6.182}$$

The second-order particle-field tensor $G_{ij}^*(\mathbf{X}^*, t)$ has the same form as $G_{ij}(\mathbf{x}, t)$ (i.e., (6.57)), but with the particle fields appearing in place of the mean fields:

$$\langle \mathbf{U}^* | \mathbf{X}^* \rangle, \quad \langle u_i^* u_j^* | \mathbf{X}^* \rangle, \quad \text{and} \quad \varepsilon(\mathbf{X}^*, t),$$

where the fluctuating particle velocity is defined by $\mathbf{u}^* \equiv \mathbf{U}^* - \langle \mathbf{U}^* | \mathbf{X}^* \rangle$. Using the LGLM, it can easily be shown (Pope 2000) that $\langle \mathbf{U}^* | \mathbf{X}^* \rangle$ obeys the RANS mean velocity transport equation (6.39) with P in place of the mean pressure field $\langle p \rangle$.

As noted above, in the applications of Lagrangian PDF methods to inhomogeneous flows, evaluation of the particle-pressure field can be problematic. In order to avoid this difficulty, *hybrid PDF methods* have been developed (Muradoglu, *et al.* 1999; Jenny *et al.*

[134] The estimate of $\langle U_i^* | \mathbf{X}^* \rangle$ will also contain statistical error. However, at high Reynolds numbers, the molecular transport term in (6.178) will be small, and thus noise in this term is less problematic.

[135] Choosing the drift coefficient to be linear in $\mathbf{U}^*$ and the diffusion matrix to be independent of $\mathbf{U}^*$ ensures that the Lagrangian velocity PDF will be Gaussian in homogeneous turbulence. Many other choices will yield a Gaussian PDF; however, none have been studied to the same extent as the LGLM.

2001b) that couple an RANS model for the mean velocity field $\langle \mathbf{U}(\mathbf{x}, t)\rangle$ with a Lagrangian PDF model for the fluctuating velocity:

$$\mathbf{u}^*(t) \equiv \mathbf{U}^*(t) - \langle \mathbf{U}(\mathbf{X}^*, t)\rangle. \tag{6.183}$$

Unlike (6.178), the Lagrangian model (LGLM) for $\mathbf{u}^*$ has no explicit dependence on the mean pressure field:

$$\begin{aligned} \mathrm{d}\mathbf{u}^* = {} & [\nabla \cdot \langle \mathbf{u}^*\mathbf{u}^* | \mathbf{X}^*\rangle - \mathbf{u}^* \cdot \nabla \langle \mathbf{U}(\mathbf{X}^*, t)\rangle + \mathbf{G}^*(\mathbf{X}^*, t)\mathbf{u}^*]\,\mathrm{d}t \\ & + (C_0\varepsilon(\mathbf{X}^*, t))^{1/2}\,\mathrm{d}\mathbf{W}_u(t). \end{aligned} \tag{6.184}$$

However, the mean velocity field will depend on conditional moments of the fluctuating velocity through the mean velocity transport equation,

$$\frac{\partial \langle U_i\rangle}{\partial t} + \langle U_j\rangle \frac{\partial \langle U_i\rangle}{\partial x_j} + \frac{\partial a^*_{ji}}{\partial x_j} = \nu \frac{\partial^2 \langle U_i\rangle}{\partial x_j \partial x_j} - \frac{1}{\rho}\frac{\partial p^*}{\partial x_i} + g_i, \tag{6.185}$$

and the Poisson equation for the modified pressure field,

$$\nabla^2 p^*(\mathbf{x}, t) = -\rho \left[\frac{\partial^2 a^*_{ij}}{\partial x_i \partial x_j}(\mathbf{x}, t) + \frac{\partial \langle U_j(\mathbf{x}, t)\rangle}{\partial x_i} \frac{\partial \langle U_i(\mathbf{x}, t)\rangle}{\partial x_j} \right], \tag{6.186}$$

where the particle anisotropy tensor is defined by

$$a^*_{ij}(\mathbf{x}, t) \equiv \langle u^*_i u^*_j | \mathbf{X}^*\rangle(\mathbf{x}, t) - \frac{2}{3} k^*(\mathbf{x}, t)\delta_{ij}, \tag{6.187}$$

and the particle kinetic energy field is defined by

$$k^*(\mathbf{x}, t) \equiv \frac{1}{2}\langle u^*_i u^*_i | \mathbf{X}^*\rangle(\mathbf{x}, t). \tag{6.188}$$

The principal advantage of using (6.184) to determine $\mathbf{u}^*$ is that the feedback of statistical noise through the particle-pressure field in (6.178) will be minimized by solving (6.185) for $\langle \mathbf{U}\rangle$. Indeed, for homogeneous turbulence, (6.184) is independent of $\mathbf{X}^*$:

$$\mathrm{d}\mathbf{u}^* = -\mathbf{u}^* \cdot \nabla \langle \mathbf{U}\rangle\,\mathrm{d}t + \mathbf{G}^*(t)\mathbf{u}^*\,\mathrm{d}t + (C_0\varepsilon(t))^{1/2}\,\mathrm{d}\mathbf{W}_u(t), \tag{6.189}$$

and (since $a^*_{ij}(t)$ is independent of $\mathbf{x}$) (6.185) and (6.186) are uncoupled from the particle fields. The extent of coupling in more complicated flows will depend on the magnitude of the spatial derivatives of $a^*_{ij}(\mathbf{x}, t)$. However, it can be anticipated that many of the numerical stability issues (Xu and Pope 1999) associated with solving (6.179) for the particle-pressure field will be avoided by using methods based on (6.185) and (6.186).

6.7.7 Lagrangian mixing models

Some of the models for the conditional diffusion presented in Section 6.6 can be used directly to close the right-hand side of (6.173). For example, the IEM model in (6.84) yields the Lagrangian IEM (LIEM) model. With the LIEM, the drift and diffusion coefficients become

$$a_{\phi,\alpha}(\mathbf{U}^*, \phi^*, \mathbf{X}^*, t) = \Gamma_\alpha \nabla^2 \langle \phi_\alpha^* | \mathbf{X}^* \rangle (\mathbf{X}^*, t) + S_\alpha(\phi^*) + \frac{C_\phi}{2} \frac{\varepsilon(\mathbf{X}^*, t)}{k^*(\mathbf{X}^*, t)} [\langle \phi_\alpha^* | \mathbf{X}^* \rangle (\mathbf{X}^*, t) - \phi_\alpha^*] \tag{6.190}$$

and

$$\mathbf{B}_{\phi\phi} = \mathbf{0}. \tag{6.191}$$

Note that (6.190) contains a number of conditional expected values that must be evaluated from the particle fields. The Lagrangian VCIEM model follows from (6.86), and has the same form as the LIEM model, but with the velocity, location-conditioned scalar mean $\langle \phi_\alpha^* | \mathbf{U}^*, \mathbf{X}^* \rangle (\mathbf{U}^*, \mathbf{X}^*, t)$ in place of location-conditioned scalar mean $\langle \phi_\alpha^* | \mathbf{X}^* \rangle (\mathbf{X}^*, t)$ in the final term on the right-hand side of (6.190).

The drift and diffusion coefficients for the Lagrangian FP (LFP) model follow from (6.111):

$$\mathbf{a}_\phi(\mathbf{U}^*, \phi^*, \mathbf{X}^*, t) = \mathbf{\Gamma} \nabla^2 \langle \phi^* | \mathbf{X}^* \rangle (\mathbf{X}^*, t) + \mathbf{S}(\phi^*) + \frac{1}{2} \left(\mathbf{S}_\Gamma \varepsilon^* \mathbf{S}_\Gamma^{-1} + \varepsilon^* \right) \mathbf{S}_{\phi^*}^{-1} \mathbf{U}_{\rho^*} \mathbf{S}_{\rho^*}^{-2} \mathbf{U}_{\rho^*}^{\mathrm{T}} \mathbf{S}_{\phi^*}^{-1} [\langle \phi^* | \mathbf{X}^* \rangle (\mathbf{X}^*, t) - \phi^*] \tag{6.192}$$

and

$$\mathbf{B}_{\phi\phi} = \mathbf{S}_g(\phi^*) \mathbf{C}_g(\phi^*). \tag{6.193}$$

The superscript * used in the coefficient matrices in (6.192) is a reminder that the statistics must be evaluated at the notional-particle location. For example, $\varepsilon^* \equiv \varepsilon(\mathbf{X}^*, t)$, and the scalar standard-deviation matrix $\mathbf{S}_{\phi^*}$ and scalar correlation matrix ρ^* are computed from the location-conditioned scalar second moments $\langle \phi_\alpha^* \phi_\beta^* | \mathbf{X}^* \rangle (\mathbf{X}^*, t)$.

As noted earlier, the LIEM model uses the same characteristic mixing time for all scalars. Thus, it is unable to model correctly the Lagrangian correlation function for two scalars with different Schmidt numbers. On the other hand, because the LFP model includes an explicit Schmidt-number dependence, it predicts different time scales for scalars with different Schmidt numbers. This property allows the LFP model to predict the correct Lagrangian auto-correlation time (Fox and Yeung 2003). However, because it uses a white-noise process in the diffusion term, the Lagrangian auto-correlation functions have an incorrect form for small lag times (compare figs. 4 and 5 in Yeung (2002)). This can be corrected by using a colored-noise process to model the diffusion in composition space (Fox and Yeung 2003).

Mixing models based on the CD model have discrete 'jumps' in the composition vector, and thus cannot be represented by a diffusion process (i.e., in terms of $\mathbf{a}_\phi$ and $\mathbf{B}_{\phi\phi}$). Instead, they require a generalization of the theory of Markovian random processes that encompasses *jump processes*[136] (Gardiner 1990). The corresponding governing equation

[136] Examples include birth–death processes, the Poisson process, and the random telegraph process.

for f_{L}^{*} is called the *differential Chapman–Kolmogorov equation*, and has the form of an integro-differential equation. In the context of Lagrangian PDF methods, jump processes are generally avoided because the Lagrangian fluid-particle compositions $\phi^{+}(t, \mathbf{Y})$ are known to vary continuously with time (Yeung 2001). More generally, jump processes are usually avoided in order to ensure that mixing is local in composition space (i.e., desirable property (iv) in Table 6.1). In Section 6.10, we will return to the topic of Lagrangian PDF methods when we look at higher-order models that treat the turbulence and mixing time scales as random processes.

6.8 Particle-field estimation

We have seen that Lagrangian PDF methods allow us to express our closures in terms of SDEs for notional particles. Nevertheless, as discussed in detail in Chapter 7, these SDEs must be simulated numerically and are non-linear and coupled to the mean fields through the model coefficients. The numerical methods used to simulate the SDEs are statistical in nature (i.e., Monte-Carlo simulations). The results will thus be subject to statistical error, the magnitude of which depends on the sample size, and deterministic error or bias (Xu and Pope 1999). The purpose of this section is to present a brief introduction to the problem of particle-field estimation. A more detailed description of the statistical error and bias associated with particular simulation codes is presented in Chapter 7.

6.8.1 Notional particles

In order to simulate the SDEs, we will introduce a large ensemble of notional particles that move through the simulation domain according to the Lagrangian PDF models. As an example, we will consider a single inert-scalar field in a one-dimensional domain. The position and composition of the nth notional particle can be denoted by $X^{(n)}(t)$ and $\phi^{(n)}(t)$, respectively. The SDEs for the Lagrangian composition PDF (with closures) become

$$\mathrm{d}X^{(n)} = \left[\langle U\rangle(t) + \frac{\partial \Gamma_{\text{T}}}{\partial x}\left(X^{(n)}, t\right)\right] \mathrm{d}t + \left(2\Gamma_{\text{T}}\left(X^{(n)}, t\right)\right)^{1/2} \mathrm{d}W^{(n)}(t) \tag{6.194}$$

and

$$\frac{\mathrm{d}\phi^{(n)}}{\mathrm{d}t} = \frac{C_{\phi}}{2} \frac{\varepsilon\left(X^{(n)}, t\right)}{k\left(X^{(n)}, t\right)} \left[\langle \phi^{*} | X^{*}\rangle\left(X^{(n)}, t\right) - \phi^{(n)}\right], \tag{6.195}$$

where the mean velocity $[\langle U\rangle(t)]$ and turbulence fields $[k(x, t)$ and $\varepsilon(x, t)]$ are assumed known. Furthermore, the Wiener processes $W^{(n)}(t)$ are independent for each notional particle, so that each particle represents a statistically independent sample.

As discussed in Section 6.7, initially the particle positions $[X^{(n)}(0)]$ must be distributed uniformly throughout the computational domain (e.g., $x \in [0, 1]$). The initial compositions

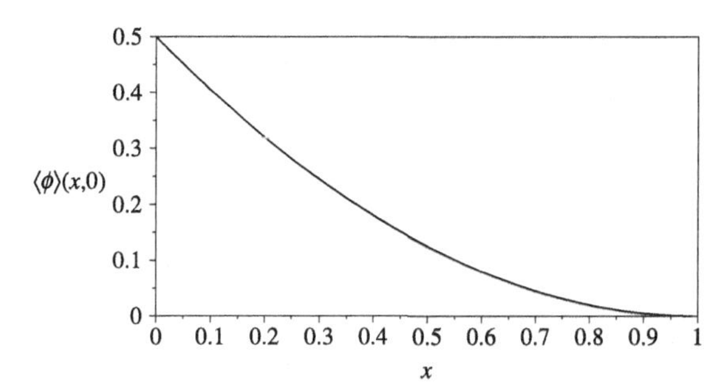

Figure 6.4. Sketch of initial scalar field.

$[\phi^{(n)}(0)]$, on the other hand, can be arbitrarily chosen. For example, we can let

$$\phi^{(n)}(0) = \frac{1}{2} - X^{(n)}(0) + \frac{1}{2}\left(X^{(n)}(0)\right)^2. \tag{6.196}$$

Note that, with this choice, the initial mean composition field is a deterministic function (see Fig. 6.4):[137]

$$\langle\phi\rangle(x, 0) = \frac{1}{2} - x + \frac{1}{2}x^2. \tag{6.197}$$

With this initial condition, a solution to the (turbulent) diffusion equation:[138]

$$\frac{\partial\langle\phi\rangle}{\partial t} = \Gamma_{\mathrm{T}}\frac{\partial^2\langle\phi\rangle}{\partial x^2} \tag{6.198}$$

is

$$\langle\phi\rangle(x, t) = \frac{1}{2} + \Gamma_{\mathrm{T}}t - x + \frac{1}{2}x^2. \tag{6.199}$$

We will use this solution to illustrate estimation errors below.

In order to simulate (6.194) and (6.195) numerically, it will be necessary to *estimate* the location-conditioned mean scalar field $\langle\phi^*|X^*\rangle(x, t)$ from the notional particles $[X^{(n)}(t), \phi^{(n)}(t)]$ for $n \in 1, \ldots, N_{\mathrm{p}}$. In order to distinguish between the estimate and the true value, we will denote the former by $\{\phi^*|X^*\}_{N_{\mathrm{p}},M}$. The subscript N_{p} is a reminder that the estimate will depend on the number of notional particles used in the simulation. Likewise, the subscript M is a reminder that the estimate will depend on the number of grid cells (M) used to resolve the mean fields across the computational domain.

The numerical error in the estimate can be broken into three contributions (Pope 1995; Xu and Pope 1999):

$$\begin{aligned} e_\phi &\equiv \{\phi^*|X^*\}_{N_{\mathrm{p}},M} - \langle\phi^*|X^*\rangle \\ &= \Sigma_\phi + B_\phi + S_\phi. \end{aligned} \tag{6.200}$$

[137] Note that the mean gradient is negative at $x = 0$ and zero at $x = 1$. This implies that $\langle\phi\rangle$ will increase with time.

[138] For simplicity, we have assumed that $\langle U\rangle = 0$ and Γ_{T} is constant. This implies that $\varepsilon(x, t)/k(x, t) \propto k(x, t)/\Gamma_{\mathrm{T}}$ in (6.195).

The statistical error Σ_ϕ is defined in terms of the (unknown) expected value of $\{\phi^*|X^*\}_{N_p,M}$:

$$\Sigma_\phi \equiv \{\phi^*|X^*\}_{N_p,M} - \langle\{\phi^*|X^*\}_{N_p,M}\rangle. \tag{6.201}$$

In general, Σ_ϕ will scale as $N_p^{-1/2}$, and thus can only be eliminated by increasing the number of notional particles. The quantity $\langle\{\phi^*|X^*\}_{N_p,M}\rangle$ can be estimated by running multiple independent simulations with fixed N_p and M (Xu and Pope 1999).

The other two terms in (6.200) are the deterministic errors due to bias (B_ϕ) and discretization (S_ϕ). The former is defined by

$$B_\phi \equiv \langle\{\phi^*|X^*\}_{N_p,M}\rangle - \{\phi^*|X^*\}_{\infty,M}, \tag{6.202}$$

and is a result of using a finite number of particles for estimation. Xu and Pope (1999) have shown that B_ϕ scales as N_p^{-1}. The discretization error is defined by

$$S_\phi \equiv \{\phi^*|X^*\}_{\infty,M} - \langle\phi^*|X^*\rangle, \tag{6.203}$$

and is a result of using a finite number of grid points to represent the continuous function $\langle\phi^*|X^*\rangle(x,t)$.

Ideally, one would like to choose N_p and M large enough that e_ϕ is dominated by statistical error (Σ_ϕ), which can then be reduced through the use of multiple independent simulations. In any case, for fixed N_p and M, the relative magnitudes of the errors will depend on the method used to estimate the mean fields from the notional-particle data. We will explore this in detail below after introducing the so-called 'empirical' PDF.

6.8.2 Empirical PDF

By definition, the mean scalar field $\langle\phi\rangle(x,t)$ can be found from the Eulerian composition PDF:

$$\langle\phi\rangle(x,t) = \int_{-\infty}^{+\infty} \psi f_\phi(\psi; x, t)\,\mathrm{d}\psi. \tag{6.204}$$

As noted in Section 6.7, (6.195) will yield a uniform distribution for $X^{(n)}(t)$,[139] and thus the Eulerian composition PDF will be related to the location-conditioned Lagrangian PDF by $f_\phi(\psi; x, t) = f_{\phi^*|X^*}(\psi|x; t)$. It follows that

$$\begin{aligned}\langle\phi\rangle(x,t) &= \int_{-\infty}^{+\infty} \psi f_{\phi^*|X^*}(\psi|x;t)\,\mathrm{d}\psi \\ &= \langle\phi^*|X^*\rangle(x,t),\end{aligned} \tag{6.205}$$

and thus it suffices to estimate the Lagrangian conditional PDF $f_{\phi^*|X^*}(\psi|x;t)$ from the notional-particle data $[X^{(n)}(t), \phi^{(n)}(t)]$.

Note that, by construction, all notional particles are identically distributed. Thus, in the absence of deterministic errors caused by using $\{\phi^*|X^*\}_{N_p,M}(X^{(n)}, t)$ in place of the true mean field, the Lagrangian PDF ($f_{\phi^{(n)},X^{(n)}}$) found from (6.194) and (6.195) would be equal

[139] For multi-dimensional flows, the mean velocity field may be a function of **x**. For this case, deterministic errors can be introduced through $\langle U\rangle(X^{(n)}, t)$, resulting in a non-uniform distribution for $X^{(n)}(t)$. This is an important source of numerical error in transported PDF codes, and we will look at this problem in more detail in Chapter 7.

to f_{ϕ^*,X^*} at all times. Likewise, we can choose the initial notional-particle concentrations such that $\langle\{\phi^*|X^*\}_{N_p,M}\rangle = \langle\phi^*|X^*\rangle$, so that initially $B_\phi = -S_\phi$ (i.e., the bias and discretization error cancel each other out). All numerical errors in the initial estimated mean field will then be entirely due to statistical error (Σ_ϕ). Nevertheless, as time progresses, both B_ϕ and S_ϕ can change due to estimation errors in the coefficients of the SDEs, leading to significant deterministic errors. We will look at this problem in detail below for the specific example of the estimated scalar mean field.

In order to use the notional particles to estimate $f_{\phi^*|X^*}$, we need a method to identify a finite sample of notional particles 'in the neighborhood' of x on which to base our estimate. In transported PDF codes, this can be done by introducing a kernel function $h_W(s)$ centered at $s = 0$ with bandwidth W. For example, a so-called 'constant' kernel function (Wand and Jones 1995) can be employed:

$$h_W(s) = \begin{cases} 1 & \text{if } |s| \leq W \\ 0 & \text{otherwise.} \end{cases} \tag{6.206}$$

For simplicity, we will let $W = 1/(2M)$, i.e., the bandwidth is inversely proportional to the number of grid cells. Another alternative is to use the M grid cells to define the kernel function:

$$h_l(x) = \begin{cases} 1 & \text{if } x \in \text{cell } l \\ 0 & \text{otherwise.} \end{cases} \tag{6.207}$$

Note that since the cell size is equal to $2W$, the bandwidth for h_l is the same as for h_W. Other widely used kernel functions are described in Chapter 7.

Using the kernel function, we can define the number of notional particles at point x by

$$N_p(x) \equiv \sum_{n=1}^{N_p} h_W\left(x - X^{(n)}\right), \tag{6.208}$$

or in cell l by

$$N_{pl} \equiv \sum_{n=1}^{N_p} h_l\left(X^{(n)}\right). \tag{6.209}$$

Note that, for a uniform grid, both $N_p(x)$ and N_{pl} scale as N_p/M. Using these definitions, the *empirical PDF* is defined as

$$f_{\phi^*|X^*;N_p,M}(\psi|x;t) \equiv \frac{1}{N_p(x)} \sum_{n=1}^{N_p} h_W\left(x - X^{(n)}\right) \delta\left(\psi - \phi^{(n)}\right), \tag{6.210}$$

or

$$f_{\phi^*|X^*;N_p,M}(\psi|x;t) \equiv \frac{1}{N_{pl}} \sum_{n=1}^{N_p} h_l\left(X^{(n)}\right) \delta\left(\psi - \phi^{(n)}\right), \tag{6.211}$$

respectively. As discussed earlier, the subscripts N_p and M are a reminder that the empirical PDF will depend on the value of these parameters. In the limit where both N_p and M

approach infinity, the Glivenko–Cantelli theorem (Billingsley 1979) ensures that

$$f_{\phi^*|X^*;\infty,\infty}(\psi|x;t) = f_{\phi^*|X^*}(\psi|x;t) \tag{6.212}$$

for all x and t.

Given $f_{\phi^*|X^*;N_p,M}$, the estimate for the mean scalar field (or any other statistic) is found in the usual manner:

$$\begin{aligned} \{\phi^*|X^*\}_{N_p,M}(x,t) &\equiv \int_{-\infty}^{+\infty} \psi f_{\phi^*|X^*;N_p,M}(\psi|x;t)\,\mathrm{d}\psi \\ &= \frac{1}{N_p(x)} \sum_{n=1}^{N_p} h_W\left(x - X^{(n)}\right) \int_{-\infty}^{+\infty} \psi\delta\left(\psi - \phi^{(n)}\right) \mathrm{d}\psi \\ &= \frac{1}{N_p(x)} \sum_{n=1}^{N_p} h_W\left(x - X^{(n)}\right) \phi^{(n)}, \end{aligned} \tag{6.213}$$

which is just the so-called 'cloud-in-cell' ensemble average.[140] The mean scalar field appearing in (6.195) (i.e., $\langle\phi^*|X^*\rangle(X^{(n)},t)$) will be approximated by

$$\{\phi^*|X^*\}_{N_p,M}(X^{(n)},t) = \frac{1}{N_p(x)} \sum_{m=1}^{N_p} h_W\left(X^{(n)}(t) - X^{(m)}(t)\right) \phi^{(m)}(t). \tag{6.214}$$

The statistical and deterministic errors resulting from this approximation are discussed next.

6.8.3 Errors in mean-field estimate

As with all statistical methods, the mean-field estimate will have statistical error due to the finite sample size (Σ_ϕ), and deterministic errors due to the finite grid size (S_ϕ) and feedback of error in the coefficients of the SDEs (B_ϕ). Since error control is an important consideration in transported PDF simulations, we will now consider a simple example to illustrate the tradeoffs that must be made to minimize statistical error and bias. The example that we will use corresponds to (6.198), where the exact solution[141] to the SDEs has the form:

$$\phi^{(n)}(t) = \frac{1}{2} + \Gamma_T t - X^{(n)}(t) + \frac{1}{2}\left(X^{(n)}(t)\right)^2 + \xi^{(n)}(X^{(n)}(t),t). \tag{6.215}$$

Time stepping of the SDEs generates the 'noise' terms $\xi^{(n)}(x,t)$, which are independent, identically distributed with mean $\langle\xi^{(n)}|X^{(n)}\rangle = 0$ and variance $\langle\left(\xi^{(n)}\right)^2|X^{(n)} = x\rangle = \sigma_\xi^2(x,t)$. We will now use (6.215) in (6.213) to study errors in the mean-field estimate $\{\phi^*|X^*\}_{N_p,M}(x,t)$.

In order to simplify the discussion, we will consider only the constant kernel h_W and assume that x does not lie too close to the boundaries of the computational domain. We

140 With $h_l(x)$, the right-hand side will just be the sum over all notional particles in cell l divided by N_{pl}.

141 The exact solution occurs when the deterministic errors are null and corresponds to $\phi^*(t)$ written in terms of $X^*(t)$.

will then write the summation over all particles appearing in (6.213) as

$$\begin{aligned}\frac{1}{N_\mathrm{p}(x)}\sum_{n=1}^{N_\mathrm{p}} h_W\left(x - X^{(n)}\right) Q^{(n)} &= \frac{1}{N_\mathrm{p}(x)}\sum_{n=1}^{N_\mathrm{p}(x)} h_W\left(x - X^{(n)}\right) Q^{(n)} \\ &= \frac{1}{N_\mathrm{p}(x)}\sum_{n=1}^{N_\mathrm{p}(x)} Q^{(n)},\end{aligned} \tag{6.216}$$

where $Q^{(n)}$ is an arbitrary function of particle properties. The summations on the right-hand side are over the $N_\mathrm{p}(x)$ particles uniformly distributed across the interval $[x - W, x + W]$. The final equality is a direct result of choosing the constant kernel. For these particles, the mean location is $\langle X^{(n)}\rangle = x$, and the variance is $\sigma_X^2 = 4W^2/12 = 1/(12M^2)$.

Inserting (6.215) into (6.213) yields four terms, as follows.

(i) Location-independent term:

$$\frac{1}{N_\mathrm{p}(x)}\sum_{n=1}^{N_\mathrm{p}(x)}\left(\frac{1}{2} + \Gamma_\mathrm{T} t\right) = \frac{1}{2} + \Gamma_\mathrm{T} t. \tag{6.217}$$

Note that no errors result from this term.

(ii) Linear term:

$$\begin{aligned}-\frac{1}{N_\mathrm{p}(x)}\sum_{n=1}^{N_\mathrm{p}(x)} X^{(n)} &= -x - \frac{1}{N_\mathrm{p}(x)}\sum_{n=1}^{N_\mathrm{p}(x)}\left(X^{(n)} - x\right) \\ &= -x - \xi_X(x,t),\end{aligned} \tag{6.218}$$

where $\xi_X(x,t)$ is a statistical 'noise' term with mean zero and variance $\sigma_{\xi_X}^2 = 1/(12M^2 N_\mathrm{p}(x)) \approx 1/(12MN_\mathrm{p})$.[142] This term is generated by the random movement of the particles across the computational domain, and contributes only to Σ_ϕ.

(iii) Quadratic term:

$$\begin{aligned}&\frac{1}{2}\frac{1}{N_\mathrm{p}(x)}\sum_{n=1}^{N_\mathrm{p}(x)}\left(X^{(n)}\right)^2 \\ &= \frac{1}{2}x^2 + \frac{x}{N_\mathrm{p}(x)}\sum_{n=1}^{N_\mathrm{p}(x)}\left(X^{(n)} - x\right) + \frac{1}{2N_\mathrm{p}(x)}\sum_{n=1}^{N_\mathrm{p}(x)}\left(X^{(n)} - x\right)^2 \\ &= \frac{1}{2}x^2 + x\xi_X(x,t) + \frac{1}{24M^2}U(x,t),\end{aligned} \tag{6.219}$$

where $U(x,t)$ is a random variable with $\langle U\rangle = 1$ and variance $\sigma_U^2 \approx M/N_\mathrm{p}$. The final term in (6.219) contributes both to Σ_ϕ and S_ϕ, and generates bias when used in the SDEs.

[142] $N_\mathrm{p}(x)$ is a random variable with mean N_p/M. For simplicity, we will assume that the variance is small so that $N_\mathrm{p}(x) \approx N_\mathrm{p}/M$.

(iv) Noise term:

$$\frac{1}{N_{\mathrm{p}}(x)}\sum_{n=1}^{N_{\mathrm{p}}(x)}\xi^{(n)}(X^{(n)}(t),t)=\xi_{\mathrm{S}}(x,t), \tag{6.220}$$

where $\langle\xi_{\mathrm{S}}\rangle=0$, and the variance scales as $\sigma_{\mathrm{S}}^2\sim M/N_{\mathrm{p}}$. This term is generated by the random movement of the particles, and contributes only to Σ_ϕ.

Collecting the terms, the estimated mean scalar field can be written as

$$\begin{aligned}\{\phi^*|X^*\}_{N_{\mathrm{p}},M}(x,t)&\\=\frac{1}{2}+\Gamma_{\mathrm{T}}t-x+\frac{1}{2}x^2-(1-x)\xi_X(x,t)+\frac{1}{24M^2}U(x,t)+\xi_{\mathrm{S}}(x,t)&\\=\langle\phi^*|X^*\rangle(x,t)-(1-x)\xi_X(x,t)+\frac{1}{24M^2}U(x,t)+\xi_{\mathrm{S}}(x,t).&\end{aligned} \tag{6.221}$$

The estimation error is thus

$$e_\phi(x,t)=\Sigma_\phi(x,t)+S_\phi(x,t), \tag{6.222}$$

where

$$\Sigma_\phi(x,t)=-(1-x)\xi_X(x,t)+\frac{1}{24M^2}[U(x,t)-1]+\xi_{\mathrm{S}}(x,t) \tag{6.223}$$

and

$$S_\phi(x,t)=\frac{1}{24M^2}. \tag{6.224}$$

Note that the bias $B_\phi(x,t)$ is null. However, because bias can be generated when the estimated mean field, (6.221), is used in the SDE to find $\phi^{(n)}(t+\Delta t)$, we shall see that $B_\phi(x,t+\Delta t)$ is non-zero.

As is shown below, in order to generate bias, the mixing time in (6.195) must depend on x. Assuming that the turbulent kinetic energy is quadratic in x,

$$k(x,t)=kx^2, \tag{6.225}$$

and that $C_\phi=2$, the SDEs for notional-particle properties can be approximated by[143]

$$X^{(n)}(t+\Delta t)=X^{(n)}(t)+(2\Gamma_{\mathrm{T}}\Delta t)^{1/2}\xi^{(n)} \tag{6.226}$$

and

$$\phi^{(n)}(t+\Delta t)=\phi^{(n)}(t)+\frac{C_1}{\tau_u}\left(X^{(n)}(t)\right)^2\left[\{\phi^*|X^*\}_{N_{\mathrm{p}},M}\left(X^{(n)}(t),t\right)-\phi^{(n)}(t)\right]\Delta t, \tag{6.227}$$

[143] The error will depend on the numerical approximation used for the SDEs. Here we use a simple Euler scheme to allow us to find analytical expressions for the errors. In addition to these equations, the behavior of the notional particles at the boundaries of the domain must also be specified. Here we avoid this difficulty by considering only notional particles 'far' from the domain boundaries.

where $\xi^{(n)}$ is a Gaussian random number with zero mean and unit variance, C_1 is a constant of order one, and τ_u is a (constant) integral time scale. Note that (6.226) does not involve any estimated quantities, and hence can be thought of as 'exact.'[144] On the other hand, the true composition $\phi^{(n)*}$ obeys

$$\phi^{(n)*}(t+\Delta t) = \phi^{(n)*}(t) + \frac{C_1}{\tau_u}\left(X^{(n)}(t)\right)^2\left[\langle\phi^{(n)*}|X^*\rangle\left(X^{(n)}(t),t\right) - \phi^{(n)*}(t)\right]\Delta t, \tag{6.228}$$

wherein $\langle\phi^*|X^*\rangle(x,t) = \frac{1}{2} + \Gamma_{\mathrm{T}}t - x + (x^2/2)$. Note that, by assumption, the initial conditions are identical, $\phi^{(n)*}(t) = \phi^{(n)}(t)$.

We can now define the time-stepping error by

$$e^{(n)}(t+\Delta t) \equiv \phi^{(n)}(t+\Delta t) - \phi^{(n)*}(t+\Delta t), \tag{6.229}$$

which can be found by subtracting (6.228) from (6.227):

$$e^{(n)}(t+\Delta t) = \frac{C_1}{\tau_u}\left(X^{(n)}(t)\right)^2 e_\phi\left(X^{(n)}(t),t\right)\Delta t, \tag{6.230}$$

where e_ϕ is given by (6.222). Note that $e^{(n)}(t) = 0$ in order to be consistent with the assumptions made in deriving (6.222).

Likewise, we can initialize the notional-particle properties so that both $\xi_X(x,t)$ and $\xi_S(x,t)$ are null.[145] The estimation error in the initial conditions, (6.222), is then due only to discretization error:

$$e_\phi(x,t) = \frac{1}{2N_{\mathrm{p}}(x)}\sum_{m=1}^{N_{\mathrm{p}}(x)}\left[\left(X^{(m)}(t)\right)^2 - x^2\right], \tag{6.231}$$

i.e., $\langle e_\phi(x,t)\rangle = S_\phi(x,t)$. Using (6.230), the change in the deterministic error at time $t+\Delta t$ can be defined by[146]

$$\begin{aligned} D_\phi(x,t+\Delta t) &= \frac{M}{N_{\mathrm{p}}}\sum_{n=1}^{N_{\mathrm{p}}/M}\left\langle e^{(n)}(t+\Delta t)\right\rangle \\ &= \frac{C_1\Delta t}{\tau_u}\frac{M}{N_{\mathrm{p}}}\sum_{n=1}^{N_{\mathrm{p}}/M}\left\langle\left(X^{(n)}(t)\right)^2 e_\phi\left(X^{(n)}(t),t\right)\right\rangle. \end{aligned} \tag{6.232}$$

Note that the spatial dependence of the mixing time enters the deterministic error through the expected value. Using (6.231), we can rewrite (6.232) as

$$D_\phi(x,t+\Delta t) = \frac{C_1\Delta t}{2\tau_u}\frac{M^2}{N_{\mathrm{p}}^2}\sum_{m=1}^{N_{\mathrm{p}}/M}\sum_{n=1}^{N_{\mathrm{p}}/M}\left[\left\langle\left(X^{(m)}\right)^2\left(X^{(n)}\right)^2\right\rangle - \left\langle\left(X^{(n)}\right)^4\right\rangle\right]. \tag{6.233}$$

[144] In other words, $X^{(n)}(t)$ will be the same as $X^*(t)$.

[145] In contrast, it is impossible to initialize the notional-particle locations such that $U(x,t) = 1$.

[146] For simplicity, we again assume that the number of notional particles used in the estimate is exactly N_{p}/M.

From this expression, it is now evident that, without the spatial dependence of the mixing time, $D_\phi(x, t + \Delta t)$ would be null, and the deterministic errors would be independent of time.[147]

Since $X^{(m)}$ and $X^{(n)}$ are uncorrelated when $m \neq n$, and identically distributed, (6.233) can be manipulated to find

$$D_\phi(x, t + \Delta t) = \frac{C_1 \Delta t}{2\tau_u} \left(1 - \frac{M}{N_\mathrm{p}}\right) \left[\langle X^2 \rangle_M^2 - \langle X^4 \rangle_M\right], \tag{6.234}$$

where the moments of the uniform distribution are

$$\langle X^2 \rangle_M = x^2 + \frac{1}{12M^2} \tag{6.235}$$

and

$$\langle X^4 \rangle_M = x^4 + \frac{x^2}{2M^2} + \frac{1}{80M^4}. \tag{6.236}$$

Thus, the final expression for the change in the deterministic error at time $t + \Delta t$ is

$$D_\phi(x, t + \Delta t) = -\frac{C_1 \Delta t}{2\tau_u} \left(1 - \frac{M}{N_\mathrm{p}}\right) \left[\frac{a_1 x^2}{M^2} + \frac{a_2}{M^4}\right], \tag{6.237}$$

where $a_1 = 1/3$ and $a_2 = (1/80) - (1/144)$.

Separating $D_\phi(x, t + \Delta t)$ into bias and discretization error then yields

$$B_\phi(x, t + \Delta t) = \frac{C_1 \Delta t}{2\tau_u} \frac{M}{N_\mathrm{p}} \left[\frac{a_1 x^2}{M^2} + \frac{a_2}{M^4}\right], \tag{6.238}$$

and, using (6.224),

$$S_\phi(x, t + \Delta t) = \frac{1}{24M^2} - \frac{C_1 \Delta t}{2\tau_u} \left[\frac{a_1 x^2}{M^2} + \frac{a_2}{M^4}\right], \tag{6.239}$$

respectively. As expected, the bias is inversely proportional to N_p and becomes non-zero due to time stepping. On the other hand, the discretization error is initially non-zero and actually decreases with time stepping.[148]

By again neglecting statistical errors due to ξ_X and ξ_S, the deterministic errors at time $t + 2\Delta t$ could be estimated by using

$$\begin{aligned} e_\phi(x, t + \Delta t) &= \frac{1}{24M^2} U(x, t + \Delta t) + D_\phi(x, t + \Delta t) \\ &= \frac{M}{2N_\mathrm{p}} \sum_{m=1}^{N_\mathrm{p}/M} \left[\left(X^{(m)}(t + \Delta t)\right)^2 - x^2\right] \\ &\quad - \frac{C_1 \Delta t}{2\tau_u} \left(1 - \frac{M}{N_\mathrm{p}}\right) \left[\frac{a_1 x^2}{M^2} + \frac{a_2}{M^4}\right] \end{aligned} \tag{6.240}$$

147 The form of the spatial dependence is not particularly important. However, by choosing a simple form we are able to find analytical expressions for the deterministic errors.

148 Although these results strictly hold only for the constant kernel estimator, similar conclusions can be drawn for other kernels.

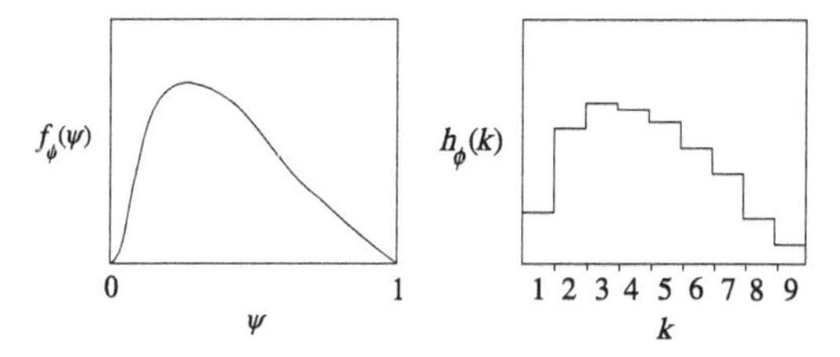

Figure 6.5. Estimation of $f_\phi(\psi)$ using a histogram.

in

$$D_\phi(x, t + 2\Delta t) = \frac{C_1 \Delta t}{\tau_u} \frac{M}{N_\mathrm{p}} \sum_{n=1}^{N_\mathrm{p}/M} \left\langle \left(X^{(n)}(t + \Delta t)\right)^2 e_\phi\left(X^{(n)}(t + \Delta t), t + \Delta t\right)\right\rangle. \quad (6.241)$$

However, it should be obvious to the reader that the deterministic errors will continue to grow as time stepping proceeds. Eventually, these errors may reach statistically stationary values[149] that can most easily be determined by numerical 'experiments.' In order to control these errors, it will be necessary to choose N_p and M sufficiently large. Similar conclusions can be drawn for more complicated examples (Pope 1995; Welton and Pope 1997; Xu and Pope 1999), and should be carefully considered when developing a transported PDF code.

6.8.4 PDF estimation

For the moment estimates, we have seen that the composition PDF, $f_\phi(\psi)$, can be approximated by a sum of delta functions (i.e., the empirical PDF in (6.210)). However, it should be intuitively apparent that this representation is unsatisfactory for understanding the behavior of $f_\phi(\psi)$ as a function of ψ. In practice, the delta-function representation is replaced by a histogram using finite-sized bins in composition space (see Fig. 6.5). The histogram $h_\phi(k)$ for the kth cell in composition space is defined by

$$h_\phi(k; \mathbf{x}, t) \equiv \int_{-\infty}^{+\infty} I_k(\psi) f_\phi(\psi; \mathbf{x}, t)\, \mathrm{d}\psi, \quad (6.242)$$

where the indicator function is given by

$$I_k(\psi) = \begin{cases} \frac{1}{\Delta_k} & \text{if } \psi \text{ is in composition cell } k \\ 0 & \text{otherwise,} \end{cases} \quad (6.243)$$

and the kth cell in composition space has width Δ_k.

Using the spatial grid kernel function, (6.207), the *estimated* histogram is given by

$$\hat{h}_\phi(k; l, t) = \frac{1}{N_{\mathrm{p}l}} \sum_{n=1}^{N_{\mathrm{p}l}} \int_{-\infty}^{+\infty} I_k(\psi) \delta\left(\psi - \phi^{(n)}(t)\right) \mathrm{d}\psi = \frac{1}{\Delta_k} \frac{N_{\mathrm{p}l,k}(t)}{N_{\mathrm{p}l}}, \quad (6.244)$$

where $N_{\mathrm{p}l,k}(t)$ is just the number of notional particles in spatial cell l and composition cell k at time t. Notice that, for a given k, the sample size is relatively small (i.e., $N_{\mathrm{p}l,k} \ll N_{\mathrm{p}l}$),

[149] By adding a zero-order scalar source term, the scalar statistics can be made time-independent.

and thus the statistical error in $\hat{h}_\phi(k;l,t)$ will be relatively large: $\varepsilon_h \sim 1/(N_{\mathrm{p}l,k})^{1/2}$. Histogram estimates of PDFs will therefore require much larger sample sizes than estimates of mean fields. Moreover, since the dimension of the sample space increases, the problem is compounded when estimating the joint PDF. Fortunately, estimation of the model coefficients in the SDEs governing the notional particles, and validation with experimental data, usually can be done at the level of moment estimates.

6.9 Chemical source term

For reacting-flow applications, the principal attraction of transported PDF methods is the fact that the chemical source term can be treated exactly. In principle, this means that, given the chemical kinetic scheme, kinetic parameters, and species thermodynamic data, we can simulate any single-phase turbulent reacting flow. However, in practice, the computational requirements of transported PDF codes are strongly dependent on the number of chemical species appearing in the kinetic scheme, and on how difficult the chemical kinetic scheme is to treat numerically. For example, if the chemical time scales are widely separated, the chemical source term will generate a system of stiff differential equations, and a specially adapted numerical integrator will be required. In this section, we give an overview of the main difficulties encountered when implementing detailed kinetic schemes in transported PDF simulations, and a brief overview of the methods available to deal with them.

6.9.1 Stiff kinetics

If one goes to the trouble of using transported PDF methods to model a turbulent reacting flow, it can be assumed that some of the reaction rates are fast compared with the micromixing time, while others are of the same order of magnitude. Indeed, if all of the reaction rates are much slower or much faster than micromixing, equivalent results can be found using much more economical methods, as described in Chapter 5. For simplicity, we will consider in the discussion below a homogeneous reacting flow,[150] and we will use the LIEM model for the conditional-diffusion term in (6.190). In order to advance the composition variables, it will thus be necessary to solve

$$\frac{\mathrm{d}\phi^*}{\mathrm{d}t} = \frac{C_\phi}{2}\frac{1}{\tau_u}(\langle\phi\rangle(t) - \phi^*) + \mathbf{S}(\phi^*), \tag{6.245}$$

where the turbulence integral time scale τ_u is assumed to be constant. This system of N_s non-linear ordinary differential equations (ODEs) must be solved subject to initial conditions $\phi^*(t)$ in order to determine $\phi^*(t+\Delta t)$, where $0 < \Delta t \ll \tau_u$ is a finite time step.[151] For large N_s and stiff chemical kinetics, the solution of (6.245) requires a stiff

[150] Technically, there is no reason to limit consideration to homogeneous flows. However, since the micromixing time and location-conditioned expected values will be independent of particle position, we do so here in order to simplify the notation and to isolate the problems associated with the chemical kinetics.

[151] As discussed in Chapter 7, when modeling inhomogeneous flows the value of Δt must be chosen also to be smaller than the minimum convective and diffusive time scales of the flow.

ODE solver. Relative to non-stiff ODE solvers, stiff ODE solvers typically use implicit methods, which require the numerical inversion of an $N_s \times N_s$ Jacobian matrix, and thus are considerably more expensive. In a transported PDF simulation lasting T time units, the composition variables must be updated $N_{\text{sim}} = T/\Delta t \sim 10^6$ times for each notional particle. Since the number of notional particles will be of the order of $N_p \sim 10^6$, the total number of times that (6.245) must be solved during a transported PDF simulation can be as high as $N_p \times N_{\text{sim}} \sim 10^{12}$. Thus, the computational cost associated with treating the chemical source term becomes *the* critical issue when dealing with detailed chemistry.

6.9.2 Decoupling from transport terms

In a transported PDF simulation a large ensemble of notional particles is employed in order to estimate the mean fields accurately:[152]

$$\langle \phi \rangle(t) = \frac{1}{N_p} \sum_{n=1}^{N_p} \phi^{(n)}(t), \tag{6.246}$$

where $\phi^{(n)}$ is the composition vector for the nth notional particle. Thus, the system, (6.245), of stiff ODEs will be large ($N_p \times N_s$) due to the coupling through the mean compositions $\langle \phi \rangle$. In transported PDF simulations, this problem can be circumvented by employing *fractional time stepping* as follows.

(1) In the first fractional time step, the micromixing term is solved separately for each notional particle:[153]

$$\frac{d\phi}{ds} = \frac{C_\phi}{2\tau_u} (\langle \phi \rangle(t) - \phi) \quad \text{with} \quad \phi(0) = \phi^{(n)}(t), \tag{6.247}$$

yielding $\phi^{(n)\dagger}(t + \Delta t) = \phi(\Delta t)$. Since in the absence of chemical reactions $\langle \phi \rangle(t)$ is constant, the notional-particle compositions at the end of the first fractional time step are given by[154]

$$\phi^{(n)\dagger}(t + \Delta t) = \langle \phi \rangle(t) + \left[\phi^{(n)}(t) - \langle \phi \rangle(t)\right] e^{-C_\phi \Delta t/(2\tau_u)}. \tag{6.248}$$

(2) In the next fractional time step, the change due to the chemical source term is found separately for each notional particle using $\phi^{(n)\dagger}(t + \Delta t)$ as the initial condition:

$$\frac{d\phi}{ds} = \mathbf{S}(\phi) \quad \text{with} \quad \phi(0) = \phi^{(n)\dagger}(t + \Delta t), \tag{6.249}$$

yielding $\phi^{(n)}(t + \Delta t) = \phi(\Delta t)$. This step will require a stiff ODE solver, and must be repeated for $n = 1, \dots, N_p$. At the end of the second fractional time step, the

[152] In the statistics literature, one usually distinguishes between the estimated mean $\langle \hat{\phi} \rangle$ and the true (unknown) mean $\langle \phi \rangle$. Here, in order to keep the notation as simple as possible, we will not make such distinctions. However, the reader should be aware of the fact that the estimate will be subject to statistical error (bias, variance, etc.) that can be reduced by increasing the number of notional particles N_p.

[153] Because this ODE is autonomous (i.e., no explicit dependence on time), the initial condition can be arbitrarily set to $s = 0$ (instead of $s = t$).

[154] By applying (6.246), the reader should convince themselves that this expression leaves the mean unchanged.

notional-particle compositions have all been updated, and the procedure can be repeated for the next time step.

The overall fractional-time-stepping process can be represented by

$$\phi^{(n)}(t) \xrightarrow{\text{mixing}} \phi^{(n)\dagger}(t + \Delta t) \xrightarrow{\text{reaction}} \phi^{(n)}(t + \Delta t),$$

and can be generalized to inhomogeneous flows by added fractional convection and diffusion steps. As a general rule, the reaction step should be completed last, reflecting the fact that some of the chemical time scales will usually be much smaller than the flow time scales (see Fig. 5.3, p. 153).[155]

Fractional time stepping is widely used in reacting-flow simulations (Boris and Oran 2000) in order to isolate terms in the transport equations so that they can be treated with the most efficient numerical methods. For non-premixed reactions, the fractional-time-stepping approach will yield acceptable accuracy if $\Delta t \ll \tau_u$. Note that since the exact solution to the mixing step is known (see (6.248)), the stiff ODE solver is only needed for (6.249), which, because it can be solved independently for each notional particle, is *uncoupled*. This fact can be exploited to treat the chemical source term efficiently using chemical lookup tables.

6.9.3 Pre-computed lookup tables

In a transported PDF simulation, the chemical source term, (6.249), is integrated over and over again with each new set of initial conditions. For fixed inlet flow conditions, it is often the case that, for most of the time, the initial conditions that occur in a particular simulation occupy only a small sub-volume of composition space. This is especially true with fast chemical kinetics, where many of the reactions attain a quasi-steady state within the small time step Δt. Since solving the stiff ODE system is computationally expensive, this observation suggests that it would be more efficient first to solve the chemical source term for a set of representative initial conditions in composition space,[156] and then to store the results in a pre-computed chemical lookup table. This operation can be described mathematically by a non-linear *reaction map*:

$$\phi_{\Delta t} = \mathbf{R}(\phi_0; \Delta t), \tag{6.250}$$

which is found by integrating (numerically) (6.249) over a time step Δt starting from initial composition ϕ_0. As we shall see below, the properties of the reaction map determine the accuracy of the tabulation scheme and the number of points that must be tabulated for a particular kinetic scheme.

In order to construct the chemical lookup table, the reaction map $\mathbf{R}(\phi_0; \Delta t)$ must be found for a set of representative points in composition space $\phi_0^{[i]}$ $(i \in 1, \dots, N_{\text{tab}})$. The

[155] If, on the other hand, the mixing time step were done last, the fast chemical reactions would be left in an unphysical 'non-equilibrium' state. Taking the reaction step last avoids this problem.

[156] For example, on an evenly spaced grid that covers the allowable region in composition space. Or, even better, on a specially adapted grid covering only the *accessed* region of composition space.

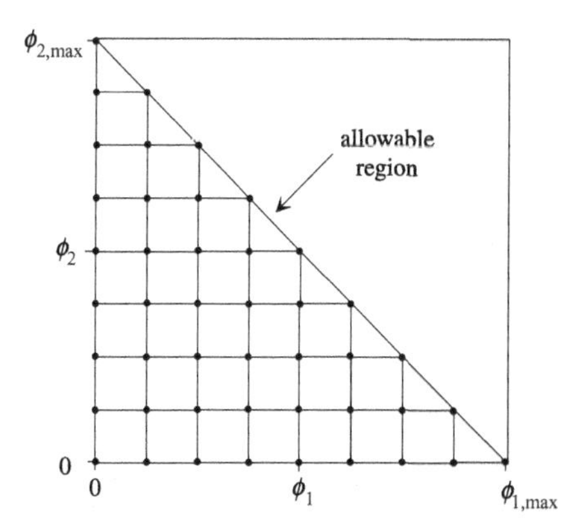

Figure 6.6. Sketch of representative points ($N_s = 2$) used in pre-computed lookup tables.

chemical lookup table stores the updated values $\phi_{\Delta t}^{[i]}$. Standard *pre-computed* lookup tables choose the representative points by computing the allowable region in composition space based on the inlet flow conditions and the kinetic scheme. The simplest (although not necessarily most efficient) method is to place these points on an (N_s)-dimensional grid in composition space (see Fig. 6.6). Because the allowable region can be determined beforehand, a pre-computed chemical lookup table can be constructed before carrying out the transported PDF simulation.[157] However, an important disadvantage of using a pre-computed lookup table is that it will contain many representative points that are never used in a particular transported PDF simulation.[158] In contrast, *in situ* lookup tables (Pope 1997) store only the representative points that actually occur during the transported PDF simulations (i.e., a sub-set of the allowable region called the *accessed region*). An *in situ* lookup table must thus be constructed dynamically, and typically contains a much smaller set of representative points.[159] We will briefly describe both methods below.

After the chemical lookup table has been constructed, it is no longer necessary to solve (6.249) to find $\phi^{(n)}(t + \Delta t)$. Instead, numerical interpolation[160] is used to find $\phi^{(n)}(t + \Delta t)$ based on the tabulated points in the 'neighborhood' of $\phi^{(n)}(t)$ in composition space. In pre-computed lookup tables the representative points are usually located at the nodes of 'hypercubes' in composition spaces. It is thus a relatively straightforward operation to identify the hypercube wherein $\phi^{(n)}(t)$ is located. The neighboring tabulated points are typically taken to be the node points of the hypercube (see Fig. 6.6). The interpolation error can be controlled by using an *adaptive* grid (see, for example, Norris and Pope (1995)),

[157] A similar idea is described for equilibrium chemistry in Chapter 5.

[158] In other words, a pre-computed lookup table must cover the entire allowable region, while only a small sub-set (the accessed region) is used in a particular simulation.

[159] The *in situ* tabulation method proposed by Pope (1997) is also *adaptive*. The latter is used to control tabulation errors and leads to a modest increase in the amount of information that must be stored for each representative point.

[160] In most applications, linear interpolation has been employed. Higher accuracy can be achieved by using an adaptable grid that places more points in regions where the function is changing quickly.

for which the sizes of the hypercubes are decreased in regions of composition space where $\phi_{\Delta t}^{[i]}$ changes rapidly.[161]

Due to table-storage limitations, the applicability of pre-computed lookup tables will be limited by the dimensions of the allowable region. Standard pre-computed lookup tables (i.e., ones that do not attempt to find a 'low-dimensional' representation of the chemical kinetics) will be limited by computer memory to three to five chemical species. For example, five scalars on a reasonably refined grid yields:

$$\begin{gathered} 100 \text{ grid points in each 'direction'}: \phi_\alpha \text{ with } \alpha = 1, \dots, 5 \\ \Downarrow \\ N_{\text{tab}} = 100^5 = 10^{10} \text{ representative points} \\ \Downarrow \\ 5 \times 10^{10} \text{ real numbers stored in pre-computed lookup table.} \end{gathered}$$

Since the quantity of real numbers that must be stored in the table will increase as 10^{2N_s}, it should be obvious to the reader that detailed kinetics schemes (which often involve tens or even hundreds of species) cannot be treated using pre-computed lookup tables. Considerable effort has thus been directed towards 'smart' tabulation algorithms.

An example of a smart tabulation method is the *intrinsic, low-dimensional manifold* (ILDM) approach (Maas and Pope 1992). This method attempts to reduce the number of dimensions that must be tabulated by projecting the composition vectors onto the non-linear manifold defined by the slowest chemical time scales.[162] In combusting systems far from extinction, the number of 'slow' chemical time scales is typically very small (i.e, one to three). Thus the resulting non-linear 'slow manifold' ILDM will be low-dimensional (see Fig. 6.7), and can be accurately tabulated. However, because the ILDM is non-linear, it is usually difficult to find and to parameterize for a detailed kinetic scheme (especially if the number of slow dimensions is greater than three!). In addition, the shape, location in composition space, and dimension of the ILDM will depend on the inlet flow conditions (i.e., temperature, pressure, species concentrations, etc.). Since the time and computational effort required to construct an ILDM is relatively large, the ILDM approach has yet to find widespread use in transported PDF simulations outside combustion.

6.9.4 *In situ* adaptive tabulation

In situ adaptive tabulation (ISAT) was proposed by Pope (1997), and it overcomes many of the difficulties associated with pre-computed lookup tables. First, the *in situ* nature of the method reduces the tabulation to only those points that occur during a particular simulation (i.e., the accessed region). Secondly, an adaptive algorithm is employed to control interpolation errors while minimizing the number of points that must be tabulated.

161 As discussed below for *in situ* tabulation, the singular values and corresponding orthogonal vectors found from the Jacobian of the reaction mapping can be used to locate regions and directions of rapid change. However, for pre-computed lookup tables this process is usually carried out using local grid refinement.

162 As discussed in Chapter 5, the chemical time scales can be found from the Jacobian of the chemical source term.

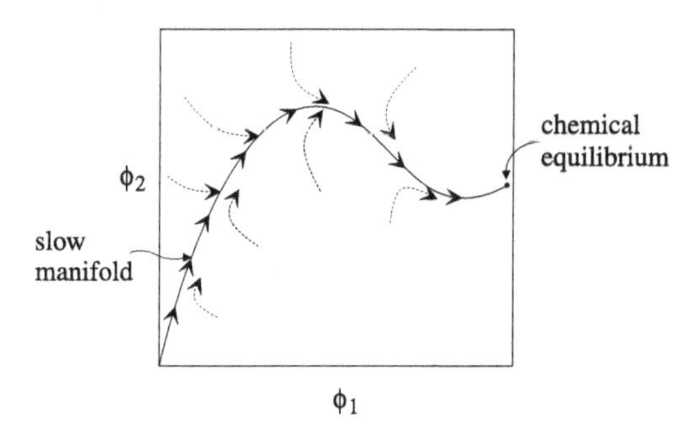

Figure 6.7. Sketch of a one-dimensional, non-linear slow manifold. The dashed curves represent trajectories in composition space that rapidly approach the slow manifold.

Thirdly, the binary-tree tabulation algorithm used in ISAT is very different from the grid-based method described above for pre-computed lookup tables. We will look at each of these aspects in detail below. However, we will begin by briefly reviewing a few points from non-linear systems theory that will be needed to understand ISAT.

Reaction mapping

The reaction mapping defined in (6.250) holds for arbitrary Δt, i.e., $\phi(t) = \mathbf{R}(\phi_0; t)$. Note that $\mathbf{R}$ is a function of both t and ϕ_0,[163] and has the property $\mathbf{R}(\phi_0; 0) = \phi_0$. By definition, the reaction mapping obeys an autonomous system of ODEs:

$$\frac{\partial \mathbf{R}}{\partial t}(\phi_0; t) = \mathbf{S}(\mathbf{R}(\phi_0; t)) \quad \text{with} \quad \mathbf{R}(\phi_0; 0) = \phi_0, \tag{6.251}$$

where $\mathbf{S}$ is the chemical source term. The sensitivity of the reaction mapping to changes in the initial conditions is measured by the *reaction-mapping Jacobian matrix*:[164]

$$\mathbf{A}(\phi_0; t) \equiv \frac{\partial \mathbf{R}}{\partial \phi_0}(\phi_0; t) \equiv \left[\frac{\partial R_i}{\partial \phi_{0j}}(\phi_0; t) \right]. \tag{6.252}$$

By differentiating (6.251), it is easily shown that the $N_s \times N_s$ matrix $\mathbf{A}$ obeys a linear matrix ODE:

$$\frac{\partial \mathbf{A}}{\partial t}(\phi_0; t) = \mathbf{J}(\mathbf{R}(\phi_0; t))\mathbf{A}(\phi_0; t) \quad \text{with} \quad \mathbf{A}(\phi_0; 0) = \mathbf{I}, \tag{6.253}$$

where the chemical-source-term Jacobian matrix $\mathbf{J}$ is defined by

$$\mathbf{J}(\phi) \equiv \frac{\partial \mathbf{S}}{\partial \phi}(\phi) \equiv \left[\frac{\partial S_i}{\partial \phi_j}(\phi) \right]. \tag{6.254}$$

Note that (6.253) is coupled to (6.251) through the dependence on $\mathbf{R}(\phi_0; t)$ in $\mathbf{J}$. Thus a total of $N_s \times (N_s + 1)$ coupled non-linear ODEs must be solved to find

[163] The version of ISAT described here requires that all tabulated points have the same Δt and pressure p. However, by adding extra variables in the definition of ϕ, this restriction can easily be overcome. For example, by defining $\phi_{N_s+1}(t) = t - t_0$ and $S_{N_s+1} = 1$, the last component of the reaction mapping will be $R_{N_s+1}(\phi_0) = \Delta t$. The value of the time step and composition vector will then be tabulated at each representative point in 'time-composition' space.

[164] Recall that $\mathbf{a} = [a_{ij}]$ denotes a matrix with components a_{ij}.

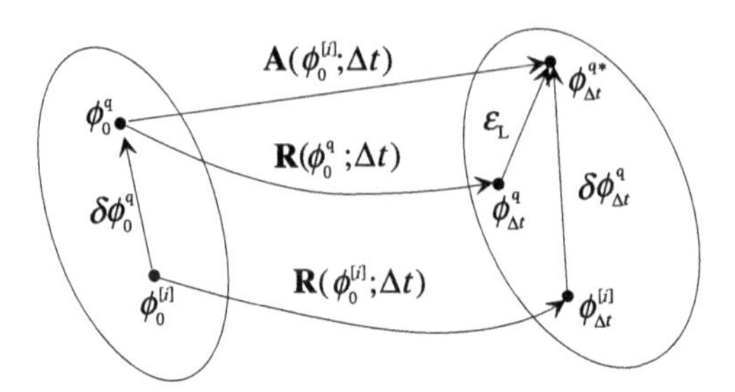

Figure 6.8. Linearized mapping used in ISAT.

the reaction mapping and sensitivity matrix at each tabulated point in composition space.[165]

In summary, given an initial composition ϕ_0 and the time step Δt, (6.251) and (6.253) can be integrated numerically to find $\phi_{\Delta t} = \mathbf{R}(\phi_0; \Delta t)$ and $\mathbf{A}(\phi_0; \Delta t)$. This step will be referred to as *direct integration* (DI), and is implicitly assumed to yield an accurate value for $\phi_{\Delta t}$ given ϕ_0. However, DI is expensive compared with interpolation. Thus, since it must be repeated at every tabulated point, DI represents the dominant computational cost in the ISAT algorithm.

Linear interpolation

The next step in understanding the ISAT algorithm is to define how a *query point* ϕ_0^{q} is interpolated to find $\phi_{\Delta t}^{\mathrm{q}}$ based on a 'neighboring' *tabulated point* $\phi_0^{[i]}$ (see Fig. 6.8). Note that, for each tabulated point, ISAT stores the following information:

$$\phi_0^{[i]},\ \phi_{\Delta t}^{[i]},\ \text{and}\ \mathbf{A}\big(\phi_0^{[i]}; \Delta t\big),$$

which have been previously computed using DI as described above. By definition, $\phi_{\Delta t}^{\mathrm{q}}$ can be found from

$$\phi_{\Delta t}^{\mathrm{q}} \equiv \mathbf{R}\big(\phi_0^{\mathrm{q}}; \Delta t\big) = \mathbf{R}\big(\phi_0^{[i]} + \delta\phi_0^{\mathrm{q}}; \Delta t\big), \tag{6.255}$$

where $\delta\phi_0^{\mathrm{q}} \equiv \phi_0^{\mathrm{q}} - \phi_0^{[i]}$. By expanding the right-hand side of this expression in a Taylor expansion about $\phi_0^{[i]}$, a *linear mapping* approximation results:

$$\delta\phi_{\Delta t}^{\mathrm{q}} \equiv \phi_{\Delta t}^{\mathrm{q}*} - \phi_{\Delta t}^{[i]} = \mathbf{A}\big(\phi_0^{[i]}; \Delta t\big)\delta\phi_0^{\mathrm{q}}. \tag{6.256}$$

The error induced by the linear approximation is

$$\varepsilon_{\mathrm{L}} \equiv \phi_{\Delta t}^{\mathrm{q}*} - \phi_{\Delta t}^{\mathrm{q}}, \tag{6.257}$$

and will scale as $|\delta\phi_0^{\mathrm{q}}|^2$. As will be discussed below, in the ISAT algorithm ε_{L} is controlled by an appropriate definition for what is meant by the 'neighborhood of $\phi_0^{[i]}$.'

The sensitivity matrix $\mathbf{A}(\phi_0^{[i]}; \Delta t)$ appearing in (6.256) contains important information on the *local* chemical time scales (i.e., in the neighborhood of $\phi_0^{[i]}$). By using the

[165] By comparison, for pre-computed tables only N_{s} equations are required. Note, however, that if $\mathbf{J}(\mathbf{R}(\phi_0; t))$ can be approximated by a constant matrix (e.g., $\mathbf{J}_{\Delta t} \equiv \mathbf{J}(\mathbf{R}(\phi_0; \Delta t))$), then $\mathbf{A}(\phi_0; t) = \exp(\mathbf{J}_{\Delta t} t)$. This matrix exponential can be computed more economically than solving (6.253) directly.

singular-value decomposition of the sensitivity matrix:

$$\mathbf{A}(\phi_0^{[i]}; \Delta t) = \mathbf{U}_A \mathbf{\Sigma}_A \mathbf{V}_A^{\mathrm{T}}, \tag{6.258}$$

(6.256) can be rewritten as

$$\delta\phi_{\Delta t}^* = \mathbf{V}_A^{\mathrm{T}} \mathbf{U}_A \mathbf{\Sigma}_A \delta\phi_0^*, \tag{6.259}$$

where $\delta\phi^* \equiv \mathbf{V}_A^{\mathrm{T}} \delta\phi^{\mathrm{q}}$. Since $\mathbf{U}_A$ and $\mathbf{V}_A$ are unitary matrices, the product $\mathbf{V}_A^{\mathrm{T}}\mathbf{U}_A$ has no effect on the magnitude of $\delta\phi_{\Delta t}^*$. Thus the (non-negative) singular values appearing on the diagonal of $\mathbf{\Sigma}_A$ will determine whether the components of $\delta\phi_{\Delta t}^*$ increase, decrease, or remain the same relative to the components of $\delta\phi_0^*$. In view of (6.259), the singular values can be used to define three orthogonal sub-spaces for the difference vector $\delta\phi^*$:[166]

(i) *Conserved manifold.* Vectors on the conserved manifold are unchanged by chemical reactions for all values of Δt. This implies that $\mathbf{\Sigma}_A \delta\phi_{\mathrm{CM}}^* = \delta\phi_{\mathrm{CM}}^*$. Thus the non-zero components of $\delta\phi_{\mathrm{CM}}^*$ must correspond to the singular values that are equal to one as $\Delta t \to \infty$. The dimension of the conserved manifold is equal to the number of such singular values. As discussed in Chapter 5, the conserved scalars can be found directly from the reaction coefficient matrix $\mathbf{\Upsilon}$.

(ii) *Fast manifold.* Vectors on the fast manifold quickly approach chemical equilibrium. This implies that $\mathbf{\Sigma}_A \delta\phi_{\mathrm{FM}}^* \approx \mathbf{0}$. Thus, the non-zero components of $\delta\phi_{\mathrm{FM}}^*$ must correspond to very small singular values, and the dimension of the fast manifold is equal to the number of such singular values.[167]

(iii) *Slow manifold.* Vectors on the slow manifold depend on Δt, but do not attain chemical equilibrium over the time step; thus, they correspond to singular values that are not 'very small,' and arise due to finite-rate chemistry.

Note that the dimensions of the fast and slow manifolds will depend upon the time step. In the limit where Δt is much larger than all chemical time scales, the slow manifold will be zero-dimensional. Note also that the fast and slow manifolds are defined *locally* in composition space. Hence, depending on the location of $\phi_0^{[i]}$, the dimensions of the slow manifold can vary greatly. In contrast to the ILDM method, wherein the dimension of the slow manifold must be globally constant (and less than two or three!), ISAT is applicable to slow manifolds of any dimension. Naturally this flexibility comes with a cost: ISAT does not reduce the number (N_{s}) of scalars that are needed to describe a reacting flow.[168]

166 In other words, the difference vector can be decomposed into the sum of three orthogonal vectors: $\delta\phi^* = \delta\phi_{\mathrm{CM}}^* + \delta\phi_{\mathrm{FM}}^* + \delta\phi_{\mathrm{SM}}^*$, where $\mathbf{\Sigma}_A\delta\phi_{\mathrm{CM}}^* = \delta\phi_{\mathrm{CM}}^*$, $\mathbf{\Sigma}_A\delta\phi_{\mathrm{FM}}^* = \mathbf{0}$, and $\mathbf{\Sigma}_A\delta\phi^* = \delta\phi_{\mathrm{CM}}^* + \mathbf{\Sigma}_A\delta\phi_{\mathrm{SM}}^*$.

167 As with the ILDM method, the definition of 'very small' will depend on the application. However, a cut-off value of around 10^{-6} is usually reasonable.

168 For an extreme example, consider the case where only one singular value is non-zero at every point in composition space. The (non-linear) slow manifold is then one-dimensional, and the ILDM method will parameterize it using a single scalar (e.g., the 'distance' from chemical equilibrium). In the ILDM method, any point not lying on the slow manifold must be 'projected' onto it. Nevertheless, the 'projection operator' has some degree of arbitrariness due to the loss of information concerning the fast manifold. In comparison, the accessed region found by ISAT will correspond (approximately) to the same one-dimensional slow manifold found by ILDM. However, the accessed region will be parameterized by vectors $\phi_0^{[i]}$ of length N_{s}, and the projection operator will be well defined in terms of $\mathbf{A}(\phi_0^{[i]}; \Delta t)$.

In summary, for a query point $\phi_0^{\rm q}$ in the neighborhood of $\phi_0^{[i]}$, ISAT provides a linear approximation of the form

$$\phi_{\Delta t}^{\rm q*} = \phi_{\Delta t}^{[i]} + \mathbf{A}\big(\phi_0^{[i]}; \Delta t\big)\big(\phi_0^{\rm q} - \phi_0^{[i]}\big). \tag{6.260}$$

It thus remains to determine the exact definition of 'neighborhood of $\phi_0^{[i]}$,' and to find an algorithm for determining in which (if any) neighborhood of tabulated points the query point $\phi_0^{\rm q}$ falls. In ISAT, the neighborhood is defined in terms of an *ellipsoid of accuracy* (EOA) found, in part, from the singular-value decomposition of $\mathbf{A}(\phi_0^{[i]}; \Delta t)$. Finding the corresponding EOA for a particular query point is accomplished by storing the tabulated points in a *binary-tree-data structure* consisting of 'nodes' and 'leaves.' Starting from the first node, a binary tree can be rapidly traversed to determine which leaf is 'closest' to the query point. We will look at both of these aspects of ISAT separately, starting with the EOA.

Ellipsoid of accuracy

We will begin by assuming that, for a particular query point $\phi_0^{\rm q}$, an algorithm exists (discussed below) for finding the 'nearest' point in composition space $\phi_0^{[i]}$. It then remains to determine if $\phi_0^{\rm q}$ lies in the EOA of $\phi_0^{[i]}$. The definition of the EOA requires a parameter $\varepsilon_{\rm tol}$ which controls the interpolation error (and consequently the number of tabulated points). Denoting the singular values of $\mathbf{A}(\phi_0^{[i]}; \Delta t)$ by σ_j ($j \in 1, \dots, N_{\rm s}$), the *initial* principal axes matrix for the EOA can be defined by

$$\tilde{\boldsymbol{\Sigma}} \equiv \mathbf{diag}(\sigma_1^*, \dots, \sigma_{N_{\rm s}}^*), \tag{6.261}$$

where

$$\sigma_j^* = \max\left(\sigma_j, \frac{1}{2}\right), \tag{6.262}$$

and the half-length of the principal axis in the jth direction is

$$l_j = \varepsilon_{\rm tol}/\sigma_j^*. \tag{6.263}$$

Note that, by this definition, l_j is inversely proportional to the *rate of increase* of the jth component of $\delta\phi_0^*$ in (6.259). Likewise, in directions of *rapid decrease*, the maximum half-length of a principal axis is limited to two. Using the definition of the difference vector $\delta\phi_0 \equiv \phi_0^{\rm q} - \phi_0^{[i]}$, the *initial* EOA for the tabulated point $\phi_0^{[i]}$ is defined[169] by

$$0 \leq \delta\phi_0^{\rm T}\mathbf{V}_A\tilde{\boldsymbol{\Sigma}}^{\rm T}\tilde{\boldsymbol{\Sigma}}\mathbf{V}_A^{\rm T}\delta\phi_0 \leq \varepsilon_{\rm tol}^2, \tag{6.264}$$

where $\mathbf{V}_A$ is the unitary matrix from the singular-value decomposition of $\mathbf{A}(\phi_0^{[i]}; \Delta t)$. Thus, for a specified $\varepsilon_{\rm tol}$, $\phi_0^{\rm q}$ is in the EOA of $\phi_0^{[i]}$ if (6.264) is satisfied (see Fig. 6.9).

[169] As discussed in Pope (1997), a non-singular scaling matrix $\mathbf{B}$ can be introduced such that $0 \leq \delta\phi_0^{\rm T}\mathbf{V}_A\tilde{\boldsymbol{\Sigma}}^{\rm T}\mathbf{B}^{\rm T}\mathbf{B}\tilde{\boldsymbol{\Sigma}}\mathbf{V}_A^{\rm T}\delta\phi_0 \leq \varepsilon_{\rm tol}^2$ defines the EOA.

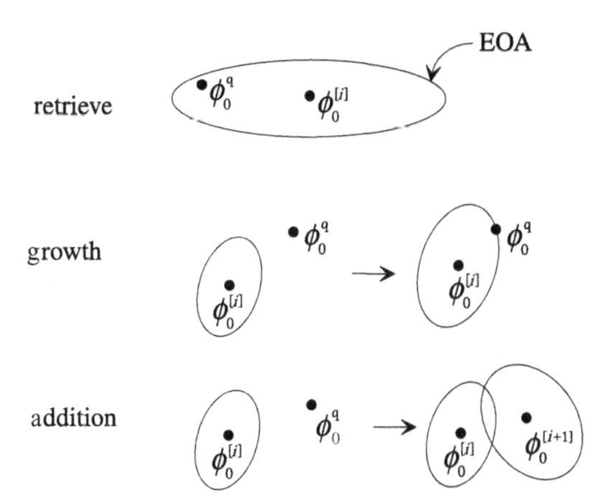

Figure 6.9. Sketch of the ellipsoid of accuracy (EOA) used in ISAT to control the local interpolation error. The *initial* EOA is usually conservative, and thus the EOA is *grown* in subsequent iterations.

The role of the EOA is to control the interpolation error (6.257) such that

$$|\varepsilon_{\mathrm{L}}| = \left|\phi_{\Delta t}^{\mathrm{q}*} - \phi_{\Delta t}^{\mathrm{q}}\right| \leq \varepsilon_{\mathrm{tol}}. \tag{6.265}$$

In practice, the definition of the initial EOA, (6.264), is often overly conservative such that $\delta\phi_0$ does not satisfy (6.264) even though (6.265) is satisfied. In the ISAT algorithm, if ϕ_0^{q} does not satisfy (6.264), then (6.251) must be integrated numerically to find $\phi_{\Delta t}^{\mathrm{q}}$. Since $\phi_{\Delta t}^{\mathrm{q}*}$ can be computed by linear interpolation, the interpolation error $|\varepsilon_{\mathrm{L}}|$ can then be computed and compared to $\varepsilon_{\mathrm{tol}}$. Thus, for a given query point ϕ_0^{q}, one of three possible outcomes will occur as follows.

(i) *Retrieve:* $\delta\phi_0$ satisfies (6.264), and $\phi_{\Delta t}^{\mathrm{q}} \approx \phi_{\Delta t}^{\mathrm{q}*}$ is found by linear interpolation, (6.260).

(ii) *Growth:* $\delta\phi_0$ does not satisfy (6.264), and $\phi_{\Delta t}^{\mathrm{q}}$ is found by DI, but $|\varepsilon_{\mathrm{L}}| \leq \varepsilon_{\mathrm{tol}}$ (i.e., linear interpolation is accurate for query point ϕ_0^{q}). For this case, the EOA is grown to include both ϕ_0^{q} and the original EOA (see Fig. 6.9). This is accomplished by redefining the principal axes matrix $\tilde{\boldsymbol{\Sigma}}$. Note that this procedure retains the center of the EOA at $\phi_0^{[i]}$, and thus adds points to the EOA on the 'side' opposite ϕ_0^{q} that have not been tested for accuracy. However, since the error will be symmetric for small $\delta\phi_0$, by appropriately choosing $\varepsilon_{\mathrm{tol}}$, this method for growing the EOA leads to adequate control of local interpolation errors.

(iii) *Addition:* $\delta\phi_0$ does not satisfy (6.264), $\phi_{\Delta t}^{\mathrm{q}}$ is found by DI, and $|\varepsilon_{\mathrm{L}}| > \varepsilon_{\mathrm{tol}}$ (i.e., linear interpolation is inaccurate for query point ϕ_0^{q}). For this case, a new leaf is added to the binary tree as described below.

The growth step is a critical component of ISAT as it confers the *adaptive* nature to the algorithm. Without it, the number of leaves in the binary tree could potentially be orders of magnitude larger, thereby severely limiting the number of species that could be treated. Indeed, by including the growth step, the computational penalty of choosing an overly conservative initial EOA is small, as each EOA will rapidly adapt itself in order to attain the desired error tolerance. We now turn our attention to the construction of the binary tree.

Binary tree

In the discussion above, we have assumed that $\phi_0^{[i]}$ is known for a given query point ϕ_0^{q}. We now describe the binary-tree-data structure from where $\phi_0^{[i]}$ is found. Recall that $i \in 1, \dots, N_{\mathrm{tab}}$ denotes the record number for a particular leaf in the binary tree. In the ISAT algorithm, the binary tree is initially *empty*, so that *the first call to ISAT leads to an addition*:

$$\phi_0^{[1]} = \phi_0^{\mathrm{q}}$$

and $N_{\mathrm{tab}} = 1$. For the first leaf,

$$\phi_{\Delta t}^{[1]}, \ \mathbf{A}(\phi_0^{[1]}; \Delta t), \ \mathbf{V}_A^{[1]}, \ \text{and } \tilde{\mathbf{\Sigma}}^{[1]}$$

are also computed using DI and stored.

On the *second* call, since the binary tree has only one leaf ($\phi_0^{[1]}$), it is by default the nearest leaf to the *new* query point ϕ_0^{q}.[170] The algorithm described above is applied with $\phi_0^{[1]}$, and leads to retrieve, growth, or addition. If the latter occurs, then the number of tabulated points is increased to $N_{\mathrm{tab}} = 2$, and the new query point is tabulated:

$$\phi_0^{[2]} = \phi_0^{\mathrm{q}},$$

along with

$$\phi_{\Delta t}^{[2]}, \ \mathbf{A}(\phi_0^{[2]}; \Delta t), \ \mathbf{V}_A^{[2]}, \ \text{and } \tilde{\mathbf{\Sigma}}^{[2]}.$$

The two leaves ($\phi_0^{[1]}$ and $\phi_0^{[2]}$) are connected to a common *node* of the binary tree (see Fig. 6.10). This node has an associated *cutting-plane* vector $\mathbf{v}^{[1]}$ and scalar $a^{[1]}$ defined such that (for any composition vector ϕ) $\mathbf{v}^{[1]\mathrm{T}}\phi < a^{[1]}$ when ϕ is 'nearer to' $\phi_0^{[1]}$ and $\mathbf{v}^{[1]\mathrm{T}}\phi > a^{[1]}$ when ϕ is 'nearer to' $\phi_0^{[2]}$. Pope (1997) defines the cutting plane in terms of the perpendicular bisector of the line between $\phi_0^{[1]}$ and $\phi_0^{[2]}$ in the transformed space where the EOA of $\phi_0^{[1]}$ is a hypersphere.[171]

Subsequent calls to ISAT follow the flow diagram shown in Fig. 6.11. With each new addition $\phi_0^{[i]}$ to the binary tree, the procedure described above is repeated. For each leaf $i \in 1, \dots, N_{\mathrm{tab}}$, the ISAT algorithm tabulates

$$\phi_0^{[i]}, \ \phi_{\Delta t}^{[i]}, \ \mathbf{A}(\phi_0^{[i]}; \Delta t), \ \mathbf{V}_A^{[i]}, \ \text{and } \tilde{\mathbf{\Sigma}}^{[i]}.$$

170 The reader must be careful to distinguish between the distinct query points. However, to avoid overly complicated notation, we will continue to use ϕ_0^{q} to represent all query points.

171 In the transformed space, the composition vector is defined by $\psi \equiv \tilde{\mathbf{\Sigma}}^{[1]}\mathbf{V}_A^{[1]\mathrm{T}}\phi$ (e.g., $\psi_0^{[1]} \equiv \tilde{\mathbf{\Sigma}}^{[1]}\mathbf{V}_A^{[1]\mathrm{T}}\phi_0^{[1]}$). The cutting-plane vector between $\psi_0^{[1]}$ and $\psi_0^{[2]}$ can be defined as $\mathbf{v}^{[1]} \equiv (\psi_0^{[2]} - \psi_0^{[1]})/|\psi_0^{[2]} - \psi_0^{[1]}|^2$. The cutting-plane scalar is then $a^{[1]} \equiv 0.5 + \mathbf{v}^{[1]\mathrm{T}}\psi_0^{[1]}$, and the cutting-plane inequalities become $\mathbf{v}^{[1]\mathrm{T}}\psi < a^{[1]}$ and $\mathbf{v}^{[1]\mathrm{T}}\psi > a^{[1]}$, respectively.

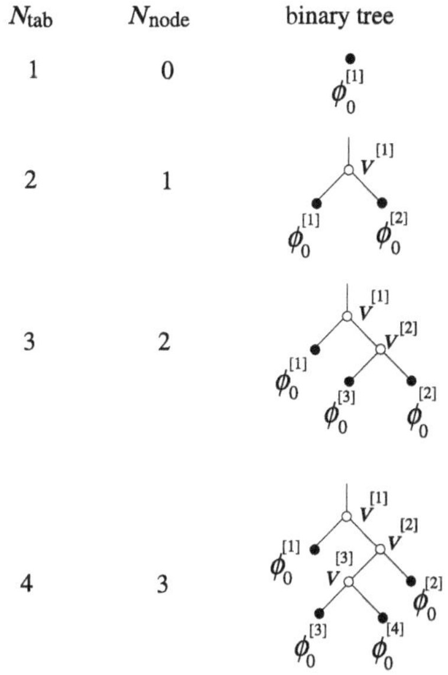

N_{tab}	N_{node}	binary tree
1	0	$\phi_0^{[1]}$
2	1	$V^{[1]}$, $\phi_0^{[1]}$, $\phi_0^{[2]}$
3	2	$V^{[1]}$, $V^{[2]}$, $\phi_0^{[1]}$, $\phi_0^{[3]}$, $\phi_0^{[2]}$
4	3	$V^{[1]}$, $V^{[2]}$, $V^{[3]}$, $\phi_0^{[1]}$, $\phi_0^{[2]}$, $\phi_0^{[3]}$, $\phi_0^{[4]}$

Figure 6.10. Sketch of the binary-tree-data structure used in ISAT. The *initial* tree is empty, and thus the tree is grown by adding leaves and nodes. Traversing the binary tree begins at the first node and proceeds using the cutting-plane vectors until a leaf is reached. The final structure depends on the actual sequence of query points.

Likewise, for each node $j \in 1, \ldots, N_{\text{node}}$, ISAT stores the cutting-plane data:

$$\mathbf{v}^{[j]} \text{ and } a^{[j]},$$

along with information identifying which two leaves/nodes are connected to the jth node. Starting from the first node (see Fig. 6.10), this information is used to traverse the binary tree to find the tabulated leaf $\phi_0^{[i]}$ that is 'closest' to a query point ϕ_0^{q}. The entire process is repeated with each new query point occurring during the simulation. The resulting *in situ* formation of the binary tree ensures that only those points in the accessed region will be stored in the lookup table, thereby drastically reducing the computer memory needed as compared to pre-computed lookup tables.

Further improvements

The description of ISAT presented above follows closely the presentation in Pope (1997), and has been employed successfully in transported PDF studies of combusting systems (Saxena and Pope 1998; Saxena and Pope 1999; Xu and Pope 2000) and vapor-phase chlorination (Shah and Fox 1999; Raman *et al.* 2001; Raman *et al.* 2003). A commercial version of ISAT is described in Masri *et al.* (2003) and has the following additional features:

(1) the restriction to constant Δt is removed;
(2) the restriction to constant pressure is removed;

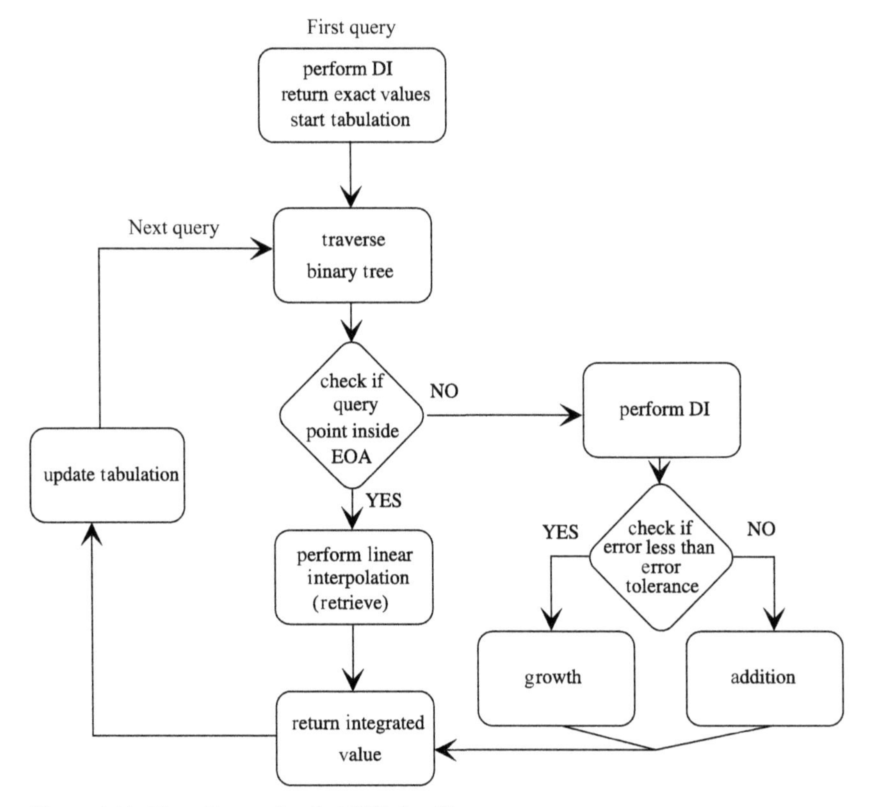

Figure 6.11. Flow diagram for the ISAT algorithm.

(3) radiative heat loss can be accounted for;
(4) multiple binary trees are used to reduce the time needed to locate the nearest leaf.

Nevertheless, the general features of the ISAT algorithm remain unchanged.

It should also be recognized that ISAT is a relatively new concept, and thus further improvements are possible and desirable. Based on current knowledge, future improvements can be expected in the following two areas:

(i) *Dimension reduction.* Unlike with the ILDM method, the current implementation of ISAT does not attempt to reduce the number of dimensions in composition space. In many chemical systems of interest, the number of species is large, and the ability to store the data for a large number of leaves in the binary tree will be prohibitive. Ideally, only information on the conserved and slow manifolds needs to be stored, and often the dimensions of these manifolds are much smaller than the dimensions of composition space. An algorithm which incorporates the current features of ISAT with dimension reduction would allow for the treatment of much larger chemical systems. Work in this direction is reported in Tang and Pope (2002).

(ii) *Parallelization.* Even with ISAT, turbulent-reacting-flow simulations with complex chemistry are computationally expensive as compared with non-reacting-flow simulations. This fact imposes the use of parallel computing hardware in order to reduce computing times. The non-linear data structure (binary tree) employed in ISAT requires special consideration when scaling up to large numbers of processors. One alternative is to produce a separate binary tree on each processor; however, this choice may not be optimal if the tables have a high degree of redundancy. Parallelization schemes that avoid redundancy and are highly scalable will be needed to treat large chemical systems.

6.10 Higher-order PDF models

The transported PDF models discussed so far in this chapter involve the velocity and/or compositions as random variables. In order to include additional physics, other random variables such as acceleration, turbulence dissipation, scalar dissipation, etc., can be added. Examples of higher-order models developed to describe the turbulent velocity field can be found in Pope (2000), Pope (2002a), and Pope (2003). Here, we will limit our discussion to higher-order models that affect the scalar fields.

6.10.1 Turbulence frequency

In the joint velocity, composition PDF description, the user must supply an external model for the turbulence time scale τ_u. Alternatively, one can develop a higher-order PDF model wherein the turbulence frequency ω is treated as a random variable (Pope 2000). In these models, the instantaneous turbulence frequency is defined as

$$\omega(\mathbf{x}, t) \equiv \frac{\epsilon(\mathbf{x}, t)}{k(\mathbf{x}, t)}, \tag{6.266}$$

where $\epsilon(\mathbf{x}, t)$ is the *instantaneous* turbulent dissipation rate, and $k(\mathbf{x}, t)$ is the turbulent kinetic energy. The turbulence frequency is then related to the turbulence time scale by

$$\langle \omega(\mathbf{x}, t) \rangle = \frac{\varepsilon(\mathbf{x}, t)}{k(\mathbf{x}, t)} = \frac{1}{\tau_u(\mathbf{x}, t)}. \tag{6.267}$$

Likewise, the higher-order statistics of $\epsilon(\mathbf{x}, t)$ will determine the higher-order statistics of $\omega(\mathbf{x}, t)$.

Although it is possible to derive a PDF transport equation for $\omega(\mathbf{x}, t)$ as described in Section 6.2, this is not usually done. Instead, a stochastic model for the Lagrangian turbulence frequency $\omega^*(t)$ is developed along the lines of those discussed in Section 6.7. The goal of these models is to reproduce as many of the relevant one-point, two-time statistics of the Lagrangian fluid-particle turbulence frequency, $\omega^+(t)$, as possible. Examples of two such models (log-normal model (Jayesh and Pope 1995) and gamma-distribution model (Pope and Chen 1990; Pope 1991a; Pope 1992)) can be found in Pope (2000). Here we will look at the stretched-exponential model developed by Fox (1997) to describe the DNS data of Overholt and Pope (1996). In comparison to the log-normal and gamma-distribution models, the stretched-exponential model exhibits very long, Reynolds-number-dependent tails for large ω^*.

For statistically stationary isotropic turbulence, the stretched-exponential model has the form

$$\mathrm{d}\omega^* = C_{\omega 1}\langle\omega\rangle\left(1 - \frac{\langle\omega\rangle(\omega^*)^{\gamma_{\omega 1}}}{\langle\omega^{1+\gamma_{\omega 1}}\rangle}\right)\omega^*\mathrm{d}t + \omega^*\left(\frac{2C_{\omega 1}\langle\omega\rangle(\langle\omega\rangle + C_{\omega 2}\omega^*)}{3(\gamma_{\omega 2}\langle\omega\rangle + 0.5C_{\omega 2}\omega^*)}\right)^{1/2}\mathrm{d}W, \tag{6.268}$$

where $\gamma_{\omega 1}$, $C_{\omega 2}$, and $\gamma_{\omega 2}$ are model parameters that control the shape of the PDF of ω^*. In particular, $\gamma_{\omega 1}$ determines the decay rate of the PDF for large ω, and is known to decrease slowly with Reynolds number. In Fox (1997), $\gamma_{\omega 2}$ and $C_{\omega 2}$ were selected as 1.11 and 0.35, respectively. The stationary PDF found from (6.268) is given in Fox (1997). The extension of (6.268) to inhomogeneous flows can be done in the manner described by Pope (2000) for the gamma-distribution model.

Using DNS data for the standardized moments of ϵ from Vedula *et al.* (2001), Fox and Yeung (2003) found a power-law fit for $\gamma_{\omega 1}$:

$$\gamma_{\omega 1} = 1.25\mathrm{Re}_1^{-0.26}. \tag{6.269}$$

The Lagrangian auto-correlation time T_ω found from (6.268) is inversely proportional to $C_{\omega 1}$. The latter can be fit to the DNS data of Yeung (2001) as:

$$C_{\omega 1} = 2.54\mathrm{Re}_1^{0.577}. \tag{6.270}$$

Note that T_ω decreases with increasing Reynolds number. As discussed in Fox and Yeung (2003), the auto-correlation time of ω^* has a direct effect on the model predictions for the scalar dissipation rate (described below). Thus, in applications involving scalar mixing, it is best to use the Reynolds-number-independent value of $C_{\omega 1} = 5$, originally suggested by Dreeben and Pope (1997b).

6.10.2 Lagrangian SR model

Given a stochastic model for the turbulence frequency, it is natural to enquire how fluctuations in ω^* will affect the scalar dissipation rate (Anselmet and Antonia 1985; Antonia and Mi 1993; Anselmet *et al.* 1994). In order to address this question, Fox (1997) extended the SR model discussed in Section 4.6 to account for turbulence frequency fluctuations. The resulting model is called the Lagrangian spectral relaxation (LSR) model. The LSR model has essentially the same form as the SR model, but with all variables conditioned on the current and past values of the turbulence frequency: $\{\omega^*(s),\ s \le t\}$. In order to simplify the notation, this conditioning is denoted by $\langle\cdot\rangle^*$, e.g.,

$$\langle\epsilon_\phi\rangle^*(t) = \langle\epsilon_\phi^*(t)|\{\omega^*(s),\ s \le t\}\rangle,$$

where $\epsilon_{\phi}^{*}(t)$ is the instantaneous Lagrangian scalar dissipation rate.[172] Note that $\varepsilon_{\phi} = \langle\langle\epsilon_{\phi}\rangle^{*}\rangle$, where the outermost brackets denote the expected value with respect to the PDF of $\{\omega^{*}(s),\ s \leq t\}$. However, in practice, ε_{ϕ} is computed by ensemble averages based on notional-particle values for $\langle\epsilon_{\phi}\rangle^{*}$ (see Section 6.8).

The LSR model for $\langle\epsilon_{\phi}\rangle^{*}$ has the same form as (4.117) on p. 131, but with additional terms to account for interactions between the particle value $\langle\epsilon_{\phi}\rangle^{*}$ and the mean ε_{ϕ}:

$$\begin{aligned}\frac{\mathrm{d}\langle\epsilon_{\phi}\rangle^{*}}{\mathrm{d}t} &= \gamma_{\mathrm{D}}\mathcal{P}_{\phi}\frac{\langle\epsilon_{\phi}\rangle^{*}}{\langle\phi'^{2}\rangle_{\mathrm{D}}^{*}} + \frac{\gamma_{\mathrm{D}}\varepsilon_{\phi}}{\langle\phi'^{2}\rangle_{\mathrm{D}}}\left(\varepsilon_{\phi} - \langle\epsilon_{\phi}\rangle^{*}\right) + C_{\mathrm{D}}\left(\frac{\varepsilon}{\nu}\right)^{1/2}\mathcal{T}_{\varepsilon}^{*} \\ &\quad + C_{\mathrm{s}}\left(\frac{\varepsilon}{\nu}\right)^{1/2}\left(\frac{\omega^{*}}{\langle\omega\rangle}\right)^{1/2}\langle\epsilon_{\phi}\rangle^{*} - C_{\mathrm{d}}\frac{\langle\epsilon_{\phi}\rangle^{*}}{\langle\phi'^{2}\rangle_{\mathrm{D}}^{*}}\langle\epsilon_{\phi}\rangle^{*},\end{aligned} \tag{6.271}$$

where (see (4.121), p. 132) the Lagrangian spectral transfer rate is given by

$$\mathcal{T}_{\varepsilon}^{*} = (\alpha_{3\mathrm{D}} + \beta_{3\mathrm{D}})\left(f_{3}\langle\phi'^{2}\rangle_{3}^{*} + f_{3}^{\mathrm{c}}\langle\phi'^{2}\rangle_{3}\right) - \beta_{\varepsilon}\langle\epsilon_{\phi}\rangle^{*}. \tag{6.272}$$

In this expression, f_{3} is the fraction of forward transfer coming from the same particle ($\langle\phi'^{2}\rangle_{3}^{*}$), and $f_{3}^{\mathrm{c}} = 1 - f_{3}$ is the fraction coming from the mean ($\langle\phi'^{2}\rangle_{3}$). Note that the only stochastic component in (6.271) comes from the gradient-amplification term involving ω^{*}. Thus, if the turbulence frequency is non-stochastic, so that $\omega^{*} = \langle\omega\rangle$, then (6.271) reduces to (4.117). Finally note that, in general, large values of ω^{*} will generate large values of $\langle\epsilon_{\phi}\rangle^{*}$, and vice versa.

In the LSR model, $\langle\epsilon_{\phi}\rangle^{*}$ is a random variable (i.e., its value is different in each notional particle). It then follows that the variables denoting the scalar energy in each wavenumber band $\langle\phi'^{2}\rangle_{n}^{*}$ must also be random variables. In the LSR model, these variables are governed by (compare with (4.103) on p. 129)

$$\begin{aligned}\frac{\mathrm{d}\langle\phi'^{2}\rangle_{1}^{*}}{\mathrm{d}t} &= \mathcal{T}_{1}^{*} + \gamma_{1}\mathcal{P}_{\phi} + \frac{\gamma_{\mathrm{D}}\varepsilon_{\phi}}{\langle\phi'^{2}\rangle_{\mathrm{D}}}\left(\langle\phi'^{2}\rangle_{1} - \langle\phi'^{2}\rangle_{1}^{*}\right) \\ &\quad + f_{\mathrm{D}}\left(\langle\epsilon_{\phi}\rangle^{*}\frac{\langle\phi'^{2}\rangle_{1}}{\langle\phi'^{2}\rangle} - \varepsilon_{\phi}\frac{\langle\phi'^{2}\rangle_{1}^{*}}{\langle\phi'^{2}\rangle}\right),\end{aligned} \tag{6.273}$$

$$\begin{aligned}\frac{\mathrm{d}\langle\phi'^{2}\rangle_{2}^{*}}{\mathrm{d}t} &= \mathcal{T}_{2}^{*} + \gamma_{2}\mathcal{P}_{\phi} + \frac{\gamma_{\mathrm{D}}\varepsilon_{\phi}}{\langle\phi'^{2}\rangle_{\mathrm{D}}}\left(\langle\phi'^{2}\rangle_{2} - \langle\phi'^{2}\rangle_{2}^{*}\right) \\ &\quad + f_{\mathrm{D}}\left(\langle\epsilon_{\phi}\rangle^{*}\frac{\langle\phi'^{2}\rangle_{2}}{\langle\phi'^{2}\rangle} - \varepsilon_{\phi}\frac{\langle\phi'^{2}\rangle_{2}^{*}}{\langle\phi'^{2}\rangle}\right),\end{aligned} \tag{6.274}$$

$$\begin{aligned}\frac{\mathrm{d}\langle\phi'^{2}\rangle_{3}^{*}}{\mathrm{d}t} &= \mathcal{T}_{3}^{*} + \gamma_{3}\mathcal{P}_{\phi} + \frac{\gamma_{\mathrm{D}}\varepsilon_{\phi}}{\langle\phi'^{2}\rangle_{\mathrm{D}}}\left(\langle\phi'^{2}\rangle_{3} - \langle\phi'^{2}\rangle_{3}^{*}\right) \\ &\quad + f_{\mathrm{D}}\left(\langle\epsilon_{\phi}\rangle^{*}\frac{\langle\phi'^{2}\rangle_{3}}{\langle\phi'^{2}\rangle} - \varepsilon_{\phi}\frac{\langle\phi'^{2}\rangle_{3}^{*}}{\langle\phi'^{2}\rangle}\right),\end{aligned} \tag{6.275}$$

[172] In words, $\langle\epsilon_{\phi}\rangle^{*}(t)$ is the mean scalar dissipation rate for all particles with the same strain-rate history. The reader should appreciate that two fluid particles can have different values of $\langle\epsilon_{\phi}\rangle^{*}(t)$, even though they share the same value of $\omega^{*}(t)$.

and

$$\frac{\mathrm{d}\langle \phi'^2 \rangle_{\mathrm{D}}^*}{\mathrm{d}t} = \mathcal{T}_{\mathrm{D}}^* + \gamma_{\mathrm{D}} \mathcal{P}_\phi + \frac{\gamma_{\mathrm{D}} \varepsilon_\phi}{\langle \phi'^2 \rangle_{\mathrm{D}}} \left(\langle \phi'^2 \rangle_{\mathrm{D}} - \langle \phi'^2 \rangle_{\mathrm{D}}^* \right) + f_{\mathrm{D}} \left(\langle \epsilon_\phi \rangle^* \frac{\langle \phi'^2 \rangle_{\mathrm{D}}}{\langle \phi'^2 \rangle} - \varepsilon_\phi \frac{\langle \phi'^2 \rangle_{\mathrm{D}}^*}{\langle \phi'^2 \rangle} \right) - \langle \epsilon_\phi \rangle^*, \tag{6.276}$$

where the Lagrangian spectral transfer rates are given by

$$\mathcal{T}_1^* = \beta_{21} \langle \phi'^2 \rangle_2^* - (\alpha_{12} + \beta_{12}) \langle \phi'^2 \rangle_1^*, \tag{6.277}$$

$$\mathcal{T}_2^* = (\alpha_{12} + \beta_{12}) \left(f_1 \langle \phi'^2 \rangle_1^* + f_1^{\mathrm{c}} \langle \phi'^2 \rangle_1 \right) + \beta_{32} \langle \phi'^2 \rangle_3^* - (\alpha_{23} + \beta_{23}) \langle \phi'^2 \rangle_2^* - \beta_{21} \langle \phi'^2 \rangle_2^*, \tag{6.278}$$

$$\mathcal{T}_3^* = (\alpha_{23} + \beta_{23}) \left(f_2 \langle \phi'^2 \rangle_2^* + f_2^{\mathrm{c}} \langle \phi'^2 \rangle_2 \right) + \beta_{\mathrm{D}3} \langle \phi'^2 \rangle_{\mathrm{D}}^* - (\alpha_{3\mathrm{D}} + \beta_{3\mathrm{D}}) \langle \phi'^2 \rangle_3^* - \beta_{32} \langle \phi'^2 \rangle_3^*, \tag{6.279}$$

and

$$\mathcal{T}_{\mathrm{D}}^* = (\alpha_{3\mathrm{D}} + \beta_{3\mathrm{D}}) \left(f_3 \langle \phi'^2 \rangle_3^* + f_3^{\mathrm{c}} \langle \phi'^2 \rangle_3 \right) - \beta_{\mathrm{D}3} \langle \phi'^2 \rangle_{\mathrm{D}}^*. \tag{6.280}$$

The dissipation constant f_{D} appearing in the waveband scalar energy governing equations is given by Vedula *et al.* (2001):

$$f_{\mathrm{D}} = 1 - \exp\left(-\frac{0.466}{\mathrm{Sc}^{1/2}} \right). \tag{6.281}$$

The parameters γ_n and the forward (α) and backscatter (β) parameters appearing in the transfer rates are given in Section 4.6. The transfer fractions f_n and $f_n^{\mathrm{c}} = 1 - f_n$ are defined (for $\mathrm{Sc} \leq 1$) by

$$f_1 = \left(\frac{1}{C_u \mathrm{Re}_1} \right)^{3/2}, \tag{6.282}$$

$$f_2 = \left(\frac{3}{C_u \mathrm{Re}_1 + 2} \right)^{3/2}, \tag{6.283}$$

and

$$f_3 = \mathrm{Sc}^{1/2}. \tag{6.284}$$

Summing together (6.273)–(6.276) yields the LSR model equation for the conditional scalar variance (recall $\gamma_1 + \gamma_2 + \gamma_3 + \gamma_{\mathrm{D}} = 1$):

$$\frac{\mathrm{d}\langle \phi'^2 \rangle^*}{\mathrm{d}t} = \mathcal{T}^* + \mathcal{P}_\phi - \langle \epsilon_\phi \rangle^*, \tag{6.285}$$

where the variance spectral transfer term is defined by

$$
\begin{aligned}
\mathcal{T}^* \equiv\ & (\alpha_{12} + \beta_{12}) f_1^{\mathrm{c}} \left(\langle \phi'^2 \rangle_1 - \langle \phi'^2 \rangle_1^* \right) \\
& + (\alpha_{23} + \beta_{23}) f_2^{\mathrm{c}} \left(\langle \phi'^2 \rangle_2 - \langle \phi'^2 \rangle_2^* \right) + (\alpha_{3\mathrm{D}} + \beta_{3\mathrm{D}}) f_3^{\mathrm{c}} \left(\langle \phi'^2 \rangle_3 - \langle \phi'^2 \rangle_3^* \right) \\
& + \frac{\gamma_{\mathrm{D}} \varepsilon_\phi}{\langle \phi'^2 \rangle_{\mathrm{D}}} \left(\langle \phi'^2 \rangle - \langle \phi'^2 \rangle^* \right) + f_{\mathrm{D}} \left(\langle \epsilon_\phi \rangle^* - \varepsilon_\phi \frac{\langle \phi'^2 \rangle^*}{\langle \phi'^2 \rangle} \right).
\end{aligned}
\tag{6.286}
$$

In general, $\langle \phi'^2 \rangle^*$ will be larger than $\langle \phi'^2 \rangle$ when $\langle \epsilon_\phi \rangle^*$ is smaller than ε_ϕ, and vice versa. The role of $\mathcal{T}^*$ is thus to transfer scalar variance from particles with low scalar dissipation rates to particles with high scalar dissipation rates. On average, the net transfer of scalar energy is null, i.e., $\langle \mathcal{T}^* \rangle = 0$.

The LSR model must be applied in conjunction with a consistent Lagrangian mixing model for $\phi^*(t)$. For example, if the Lagrangian FP model is used, one consistent model has the form

$$
\mathrm{d}\phi^* = \frac{(\mathcal{T}^* - 2\langle \epsilon_\phi \rangle^*)}{2\langle \phi'^2 \rangle^*} (\phi^* - \langle \phi \rangle)\, \mathrm{d}t + (\langle \epsilon_\phi | \phi^* \rangle^*)^{1/2}\, \mathrm{d}W(t),
\tag{6.287}
$$

where $\langle \epsilon_\phi | \phi^* \rangle^*$ is the doubly conditioned scalar dissipation rate. This model is consistent in the sense that it yields the correct expression for the conditional scalar variance:

$$
\frac{\mathrm{d}\langle \phi'^2 \rangle^*}{\mathrm{d}t} = \mathcal{T}^* - \langle \epsilon_\phi \rangle^*.
\tag{6.288}
$$

(Recall that the production term $\mathcal{P}_\phi$ in (6.285) results from the scalar flux, which is not included in (6.287).) As with the FP model discussed in Section 6.6, the doubly conditioned scalar dissipation rate must be supplied by the user. For example, the conditional scalar PDF $f_\phi^*(\psi; t)$ generated by

$$
\langle \epsilon_\phi | \phi^* \rangle^* = \langle \epsilon_\phi \rangle^* \frac{\phi^*(1 - \phi^*)}{\langle \phi(1 - \phi) \rangle^*}
\tag{6.289}
$$

will be a beta PDF with mean $\langle \phi \rangle$ and variance $\langle \phi'^2 \rangle^*$. The unconditional scalar PDF with mean $\langle \phi \rangle$ and variance $\langle \phi'^2 \rangle$ will then be slightly different from a beta PDF due to fluctuations in ω^* (and hence in $\langle \epsilon_\phi \rangle^*$ and $\langle \phi'^2 \rangle^*$).

6.10.3 LSR model with differential diffusion

The extension of the SR model to differential diffusion is outlined in Section 4.7. In an analogous fashion, the LSR model can be used to model scalars with different molecular diffusivities (Fox 1999). The principal changes are the introduction of the conditional scalar covariances in each wavenumber band $\langle \phi'_\alpha \phi'_\beta \rangle_n^*$ and the conditional joint scalar dissipation rate matrix $\langle \epsilon \rangle^*$. For example, for a two-scalar problem, the LSR model involves three covariance components: $\langle \phi'^2_1 \rangle_n^*$, $\langle \phi'_1 \phi'_2 \rangle_n^*$, and $\langle \phi'^2_2 \rangle_n^*$; and three joint dissipation rates: $\langle \epsilon_{11} \rangle^*$, $\langle \epsilon_{12} \rangle^*$, and $\langle \epsilon_{22} \rangle^*$. The form of the LSR model is essentially the same for each component (Fox and Yeung 2003). The Schmidt-number dependence in the equations for the cross terms (e.g., $\langle \phi'_1 \phi'_2 \rangle_n^*$) is found using the arithmetic mean diffusivity: $\Gamma_{12} = (\Gamma_1 + \Gamma_2)/2$. Further details on the model can be found in Fox (1999). A consistent multi-variate Lagrangian mixing model will also be required for ϕ^*. A consistent multi-variate FP model has a form similar to (6.287), but with $\mathcal{T}^*$ replaced by a matrix. More details can be found in Fox and Yeung (2003).

6.10.4 LSR model with reacting scalars

The extension of the LSR model to reacting scalars has yet to be fully explored. However, several technical difficulties can be expected. Two of the most challenging are discussed below.

(i) In the model equations for the covariances $\langle \phi'_\alpha \phi'_\beta \rangle_n^*$, the chemical source terms (see (3.136), p. 90):

$$\langle S_{\alpha\beta} \rangle_n^* = \langle \phi'_\beta S_\alpha(\boldsymbol{\phi}) \rangle_n^* + \langle \phi'_\alpha S_\beta(\boldsymbol{\phi}) \rangle_n^*, \tag{6.290}$$

must be provided. However, in practice, only the sum of the source terms $\langle S_{\alpha\beta} \rangle^*$ can be estimated from the particle compositions ϕ^*. For example, with four wavenumber bands, we have

$$\langle S_{\alpha\beta} \rangle^* = \langle S_{\alpha\beta} \rangle_1^* + \langle S_{\alpha\beta} \rangle_2^* + \langle S_{\alpha\beta} \rangle_3^* + \langle S_{\alpha\beta} \rangle_{\mathrm{D}}^*. \tag{6.291}$$

The terms on the right-hand side of this expression are the unknown covariance source terms, in particular wavenumber bands due to chemical reactions. For example, for a fast non-premixed, one-step reaction one would expect the product to be formed in the scalar dissipation range, so that $\langle S_{\alpha\beta} \rangle_{\mathrm{D}}^*$ is dominant.

In the LSR model, the covariances $\langle \phi'_\alpha \phi'_\beta \rangle_n^*$ are known, but not the higher-order moments. Thus, only for first-order reactions will (6.290) be closed:

$$\langle S_{\alpha\beta} \rangle_n^* = k_\alpha \langle \phi'_\beta \phi'_\alpha \rangle_n^* + k_\beta \langle \phi'_\alpha \phi'_\beta \rangle_n^* = (k_\alpha + k_\beta) \langle \phi'_\alpha \phi'_\beta \rangle_n^*. \tag{6.292}$$

For higher-order reactions, a model must be provided to close the covariance source terms. One possible approach to develop such a model is to extend the FP model to account for scalar fluctuations in each wavenumber band (instead of only accounting for fluctuations in ϕ^*). In any case, correctly accounting for the spectral distribution of the scalar covariance chemical source term is a key requirement for extending the LSR model to reacting scalars.

(ii) The LSR model for the joint scalar dissipation rate (see (4.134), p. 136) will have an unclosed source term:

$$\begin{aligned}\frac{\mathrm{d}\langle\epsilon_{\alpha\beta}\rangle^*}{\mathrm{d}t} &= \mathcal{P}_\varepsilon + \frac{\gamma_\mathrm{D}\gamma_{\alpha\beta}\varepsilon_{\alpha\beta}}{\langle\phi'_\alpha\phi'_\beta\rangle_\mathrm{D}}(\varepsilon_{\alpha\beta} - \langle\epsilon_{\alpha\beta}\rangle^*) + C_\mathrm{D}\left(\frac{\varepsilon}{\nu}\right)^{1/2}\mathcal{T}_c^* \\ &+ C_\mathrm{s}\left(\frac{\varepsilon}{\nu}\right)^{1/2}\left(\frac{\omega^*}{\langle\omega\rangle}\right)^{1/2}\langle\epsilon_{\alpha\beta}\rangle^* - C_\mathrm{d}\frac{\gamma_{\alpha\beta}\langle\epsilon_{\alpha\beta}\rangle^*}{\langle\phi'_\alpha\phi'_\beta\rangle_\mathrm{D}^*}\langle\epsilon_{\alpha\beta}\rangle^* \\ &+ \sum_\gamma\left(\left(\frac{\Gamma_\alpha}{\Gamma_\gamma}\right)^{1/2}\langle J_{\gamma\alpha}(\boldsymbol{\phi})\epsilon_{\beta\gamma}\rangle^* + \left(\frac{\Gamma_\beta}{\Gamma_\gamma}\right)^{1/2}\langle J_{\gamma\beta}(\boldsymbol{\phi})\epsilon_{\alpha\gamma}\rangle^*\right),\end{aligned} \tag{6.293}$$

where the joint scalar dissipation gradient-source term[173] is modeled by

$$\mathcal{P}_\varepsilon \equiv -\frac{\gamma_\mathrm{D}}{\gamma_{\alpha\beta}}\left(\frac{\langle\epsilon_\alpha\rangle^*}{\langle\phi'^2_\alpha\rangle_\mathrm{D}^*}\langle u_i\phi_\alpha\rangle\frac{\partial\langle\phi_\beta\rangle}{\partial x_i} + \frac{\langle\epsilon_\beta\rangle^*}{\langle\phi'^2_\beta\rangle_\mathrm{D}^*}\langle u_i\phi_\beta\rangle\frac{\partial\langle\phi_\alpha\rangle}{\partial x_i}\right), \tag{6.294}$$

and the final term on the right-hand side is the joint scalar dissipation chemical source term derived in Section 3.3. The Jacobian matrix of the chemical source term $\mathbf{J}$ is defined by (3.146) on p. 92. The unclosed terms in (6.293) can be written simply as $\langle\mathbf{J}(\boldsymbol{\phi})\boldsymbol{\epsilon}\rangle^*$. Physically, these terms describe the interaction between the chemistry and the scalar gradients. Since the leading-order terms on the right-hand side of (6.293) have Kolmogorov scaling (i.e., $(\varepsilon/\nu)^{1/2}$), the chemistry term will only be important when the reaction rates have similar scaling (i.e., when the chemistry is fast compared with the Kolmogorov time scale).

In order to close $\langle\mathbf{J}(\boldsymbol{\phi})\boldsymbol{\epsilon}\rangle^*$, we can recognize that because $\mathbf{J}(\boldsymbol{\phi})$ depends only on the $\boldsymbol{\phi}$, it is possible to replace $\boldsymbol{\epsilon}$ by $\langle\boldsymbol{\epsilon}|\boldsymbol{\phi}\rangle^*$. The closure problem then reduces to finding an expression for the doubly conditioned joint scalar dissipation rate matrix. For example, if the FP model is used to describe scalar mixing, then a model of the form

$$\langle\boldsymbol{\epsilon}|\boldsymbol{\phi}\rangle^* = \mathbf{H}(\boldsymbol{\phi})\langle\boldsymbol{\epsilon}\rangle^*\mathbf{H}(\boldsymbol{\phi}) \tag{6.295}$$

can be used. In the FP model, the shape matrix $\mathbf{H}(\boldsymbol{\phi})$ must be provided by the user. Thus, $\langle\mathbf{J}(\boldsymbol{\phi})\boldsymbol{\epsilon}\rangle^* = \langle\mathbf{J}(\boldsymbol{\phi})\mathbf{H}(\boldsymbol{\phi})\langle\boldsymbol{\epsilon}\rangle^*\mathbf{H}(\boldsymbol{\phi})\rangle^*$ will be closed since the expected value with respect to $\boldsymbol{\phi}$ can be estimated from the particle compositions. In summary, we can conclude that the principal modeling challenge for closing (6.293) is the same as for the FP model, namely finding appropriate functional forms for the shape matrix $\mathbf{H}(\boldsymbol{\phi})$.

[173] This term corresponds to $\mathcal{G}_\varepsilon^{\alpha\beta} + \mathcal{C}_\varepsilon^{\alpha\beta}$ in (3.151).

7

Transported PDF simulations

In Chapter 6 we reviewed the theory underlying transported PDF methods. In order to apply this theory to practical flow problems, numerical algorithms are required to 'solve' the PDF transport equation. In general, solving the PDF transport equation using standard finite-difference (FD) or finite-volume (FV) methods is computationally intractable for a number of reasons. For example, the velocity, composition PDF transport equation ((6.19), p. 248) has three space variables ($\mathbf{x}$), three velocity variables ($\mathbf{V}$), N_s composition variables ($\boldsymbol{\psi}$), and time (t). Even for a statistically two-dimensional, steady-state flow with only one scalar, a finite-difference grid in at least *five* dimensions would be required! Add to this the problem of developing numerical techniques that ensure $f_{\mathbf{U},\phi}$ remains non-negative and normalized to unity at every space/time point ($\mathbf{x}$, t), and the technical difficulties quickly become insurmountable.

A tractable alternative to 'solving' the PDF transport equation is to use statistical or Monte-Carlo (MC) simulations. Unlike FV methods, MC simulations can handle a large number of independent variables, and always ensure that the resulting estimate of $f_{\mathbf{U},\phi}$ is well behaved. As noted in Section 6.8, MC simulations employ representative samples or so-called 'notional' particles. The principal challenge in constructing an MC algorithm is thus to define appropriate rules for the rates of change of the notional-particle variables so that they have statistical properties identical to $f_{\mathbf{U},\phi}(\mathbf{V}, \boldsymbol{\psi}; \mathbf{x}, t)$. The reader should, however, keep in mind that the necessarily finite ensemble of notional particles provides only a (poor) *estimate* of $f_{\mathbf{U},\phi}$. When developing MC algorithms, it will thus be important to consider the magnitude of the estimation errors[1] and to develop ways to control them. Indeed, unlike with FV methods, wherein numerical errors can often be easily discerned (Ferziger and Perić 2002), the inherent random fluctuations can mask the magnitude and origin of errors in MC simulations. The primary objective of this chapter is thus to give

[1] As discussed in Section 6.8, the estimation errors can be categorized as statistical, bias, and discretization. In a well designed MC simulation, the statistical error will be controlling. In contrast, in FV methods the dominant error is usually discretization.

the reader an overview of transported PDF simulation methods with particular attention to potential sources of numerical errors.

7.1 Overview of simulation codes

Transported PDF simulations can be categorized as either *Eulerian* or *Lagrangian.* The former is used only for the composition PDF, and the notional particles provide a statistical representation of f_ϕ at each FV cell in the computational domain. In Eulerian PDF simulations, all turbulence fields (e.g., $\langle \mathbf{U} \rangle$) are found using standard FV solutions to the RANS equations on the *same computational grid.* In contrast, Lagrangian PDF simulations can be used with any PDF transport equation and are based on the Lagrangian PDF methods discussed in Section 6.7. In principle, Lagrangian PDF codes are *grid free* and, hence, do not suffer from so-called 'numerical diffusion' due to discretization errors. Thus, unlike Eulerian PDF codes (and most FV codes), Lagrangian PDF codes yield 'exact' solutions for purely convective flows. However, as discussed in Section 6.8, the governing SDEs for the notional particles contain location-dependent coefficients that must be estimated from the notional-particle fields. Because these estimates require a relatively large sample in order to reduce statistical errors, discretization errors will also be present in Lagrangian PDF simulations.

The inherent advantages of Lagrangian PDF methods have led to their widening use for practical applications. As with Eulerian PDF methods, Lagrangian composition PDF codes must be coupled to an FV code, which provides the turbulence fields. On the other hand, Lagrangian velocity, composition PDF codes can be 'stand-alone,' i.e., they can compute all the fields appearing in the governing SDEs from the notional-particle fields. However, as noted in Sections 6.5 and 6.7, estimation of the particle-pressure field is particularly problematic (Xu and Pope 1999). In order to overcome this difficulty, the current trend in Lagrangian velocity, composition PDF codes[2] is towards so-called 'hybrid' finite-volume/particle PDF codes (Jenny *et al.* 2001). In hybrid velocity, composition codes, the FV method is used to solve for the mean velocity field ($\langle \mathbf{U} \rangle$) and the modified pressure field (p^* in (6.186)), and the particle code is used to simulate the *fluctuating* velocity ($\mathbf{u}^*$ in (6.184)). The numerical stability of hybrid codes makes them computationally much more efficient than stand-alone codes (Muradoglu *et al.* 1999). A hybrid composition code has also been developed for large-eddy simulation of turbulent reacting flows (Jaberi *et al.* 1999).

In general, all MC simulation codes enjoy the following favorable attributes.

- Computational cost increases linearly with the number of independent variables. Thus a large number of *inert* scalars can be treated with little additional computational cost. For reacting scalars, the total computational cost will often be dominated by the chemical source term (see Section 6.9).

[2] Most codes also include a stochastic model for the turbulence frequency.

- Realizability and boundedness of all variables are assured. In particular, since the chemical source term is treated exactly, mass and element conservation is guaranteed at the notional-particle level.
- Using particle partitioning, particle codes exhibit excellent scalability on distributed computing platforms (i.e., cluster computers). However, with complex chemistry, care must be taken when implementing chemical lookup tables to avoid scale-up bottlenecks.

On the other hand, as was pointed out above, all MC simulation codes suffer from statistical 'noise' that must be minimized (or at least understood) before valid comparisons can be made with experimental data (or other CFD methods).

As with any CFD code, the basic simulation algorithms are most easily understood for simple geometries with structured grids. Thus, in this work, we will limit our attention to statistically two-dimensional flows with rectangular grids. However, this restriction is not intrinsic to PDF codes, and we will point out to the reader which steps are grid-specific throughout the discussion. Likewise, in keeping with the restriction to constant-density flows[3] used in earlier chapters, we will not discuss in detail algorithms that are specific to handling density variations.[4] Nevertheless, we will point out the (relatively few) steps where density variations will affect the algorithm, in particular in coupling the composition fields to the mean continuity and momentum equations. Finally, even for the treatment of two-dimensional flows with rectangular grids, it is usually desirable to introduce particle weights (Haworth and El Tahry 1991). For example, in two-dimensional axisymmetric flows, grid cells far from the centerline represent much larger volumes than grid cells near the centerline. Thus, if the number of notional particles in each grid cell is nearly equal, then the effective volume (or weight) that each particle represents must increase for particles far from the centerline. Similar remarks hold for flow domains with widely varying cell sizes. We will thus generalize our concept of notional particles to include particle weights in the definition of the empirical PDF ((6.210), p. 301).

The rest of this chapter is arranged as follows. In Section 7.2, we review Eulerian PDF algorithms for simulation of the composition PDF. In Section 7.3, Lagrangian PDF algorithms for the composition PDF are described. Finally, in Section 7.4, hybrid PDF codes for the velocity, composition PDF are described. The primary goal of the discussions in this chapter is to introduce the reader to the various numerical algorithms needed to simulate the transported PDF equation. With this in mind, the discussion is limited to simple flow geometries and the most widely used algorithms. However, since this field is evolving rapidly, readers interested in more in-depth analysis of particular algorithms and knowledge of current state-of-the-art codes should consult the literature. Finally, although they are a critical component of FV–PDF hybrid codes, we will not discuss numerical

[3] By 'constant density' we mean that ρ is the same at every location (i.e., uniform) and independent of time.

[4] For variable-density flows, the transport equation for the density-weighted PDF is used as the starting point. The resulting PDF codes use the particle mass as an intrinsic random variable. The particle density and specific volume can be computed based on the particle properties.

methods for FV codes. Readers interested in the details on FV solvers can consult Ferziger and Perić (2002).

7.2 Eulerian composition PDF codes

In an Eulerian PDF code (Pope 1981b; Pope 1985), the notional particles are associated with cell centers, which we denote by $\mathbf{x}_l$ for the lth grid cell with $l = 1, \dots, M$. The number of notional particles in each grid cell need not be the same.[5] We will denote the (fixed) number in the lth grid cell by N_l. Likewise, the set of composition vectors for the notional particles in the lth grid cell will be denoted by

$$\{\boldsymbol{\phi}\}_l = \left\{\boldsymbol{\phi}^{(1)}, \dots, \boldsymbol{\phi}^{(N_l)}\right\}_l, \tag{7.1}$$

where $\boldsymbol{\phi}^{(n)}$ is the composition vector for the nth notional particle. The set of composition vectors $\{\boldsymbol{\phi}\}_l$ is used to represent the composition PDF at $\mathbf{x}_l$: $f_{\boldsymbol{\phi}}(\boldsymbol{\psi}; \mathbf{x}_l, t)$. Throughout this section, we will assume that the composition PDF transport equation has the form of (6.30), p. 251. The Eulerian PDF code is used to simulate the PDF transport equation. A separate FV code must provide the mean velocity ($\langle \mathbf{U} \rangle$), the turbulent diffusivity (Γ_T), and turbulence frequency ($\omega = 1/\tau_u$). Although it is not strictly necessary, we will assume that the same grid cells are used in both codes.

Because all notional particles are assumed to have equal weight,[6] the empirical PDF for the lth grid cell is given by[7]

$$f_{\boldsymbol{\phi};N_l,M}(\boldsymbol{\psi}; \mathbf{x}_l, t) = \frac{1}{N_l} \sum_{n=1}^{N_l} \delta\left(\boldsymbol{\psi} - \boldsymbol{\phi}^{(n)}\right), \tag{7.2}$$

where the vector delta function is defined by

$$\delta\left(\boldsymbol{\psi} - \boldsymbol{\phi}^{(n)}\right) \equiv \prod_{\alpha=1}^{N_s} \delta\left(\psi_\alpha - \phi_\alpha^{(n)}\right). \tag{7.3}$$

The estimation of statistical quantities for each cell is straightforward. For example, the estimated scalar mean in the lth cell is just

$$\{\phi_\alpha\}_{N_l,M}(\mathbf{x}_l, t) = \frac{1}{N_l} \sum_{n=1}^{N_l} \phi_\alpha^{(n)}(\mathbf{x}_l, t). \tag{7.4}$$

[5] There appears to be some confusion on this point in the literature. In an Eulerian PDF code, the notional particles *do not represent* fluid particles, rather they are a discrete representation of the composition PDF (e.g., a histogram). Thus, the number of notional particles needed in a grid cell is solely determined by the statistical properties of the PDF. For example, if the PDF is a delta function, then only one particle is required to represent it. Note, however, that the problem of determining the number of particles needed in each grid cell for a particular flow is non-trivial (Pfilzner *et al.* 1999).

[6] This assumption can be relaxed and each particle assigned its own weight. It may then be possible to reduce N_l in selected cells, and thereby the overall computational cost (Pfilzner *et al.* 1999).

[7] This formulation can be compared with the multi-environment (ME) presumed PDF method discussed in Section 5.10. The principal difference between the two approaches is the treatment of turbulent convection. In the transported PDF simulation, turbulent convection is simulated by a random process. In the ME–PDF approach, it is handled using standard FV/FD discretization.

Note that the empirical PDF in (7.2) depends explicitly on the number of notional particles N_l. However, it also depends implicitly on the number of grid cells (M) through the particle transport algorithm described below. The statistical error associated with estimation of the mean composition $\langle \phi \rangle$ using (7.4) will scale as $(\langle \phi'^2 \rangle / N_l)^{1/2}$. In zones of the flow where the variance $\langle \phi'^2 \rangle$ is small (e.g., in streams with pure fluid $\langle \phi'^2 \rangle = 0$), N_l can be small. However, in zones where mixing and reactions are important, $N_l = 100$–500 is typically required to obtain particle-number-independent estimates.

7.2.1 Particle transport processes

From the Eulerian composition transport equation, two types of processes can be identified.

(1) *Intra-cell:* chemical reactions and molecular mixing change the components of the particle composition vectors $\phi^{(n)}$, but not the memberships of the sets $\{\phi\}_l$.
(2) *Inter-cell:* spatial transport by the mean velocity and turbulent diffusivity change the memberships of the sets $\{\phi\}_l$, but not the particle composition vectors $\phi^{(n)}$.

The intra-cell processes are common to all PDF codes, and are treated the same in both Eulerian and Lagrangian PDF codes.[8] On the other hand, inter-cell processes are treated differently in Eulerian PDF codes due to the discrete representation of space in terms of $\mathbf{x}_l$. In PDF codes, fractional time stepping is employed to account for each process separately. Methods for treating chemical reactions and mixing are described in Section 6.9. Thus we will focus here on the treatment of inter-cell processes in Eulerian PDF codes.

Inter-cell processes account for flow of notional particles between neighboring cells. In general, numerical diffusion can be significant for flows dominated by convection and should be treated carefully. A first-order FV numerical scheme using all eight adjacent cells has been proposed in Möbus *et al.* (2001) to improve the description of diagonal convection on two-dimensional grids. Here, a simpler scheme (Pope 1981b) based on spatial transport from the four neighboring cells will be discussed.

Keeping only the accumulation and spatial-transport terms, the FV code solves a discretized form of

$$\frac{\partial \phi}{\partial t} + \langle \mathbf{U} \rangle \cdot \nabla \phi = \nabla \cdot (\Gamma_{\mathrm{T}} \nabla \phi), \tag{7.5}$$

where $\nabla \cdot \langle \mathbf{U} \rangle = 0$. Integrating this expression over a grid cell of volume V_l and surface area S_l yields

$$\frac{\mathrm{d}}{\mathrm{d}t} \int_{V_l} \phi \, \mathrm{d}\mathbf{x} = - \int_{S_l} (\langle \mathbf{U} \rangle \phi) \cdot \mathbf{n} \, \mathrm{d}\mathbf{x} + \int_{S_l} (\Gamma_{\mathrm{T}} \nabla \phi) \cdot \mathbf{n} \, \mathrm{d}\mathbf{x}, \tag{7.6}$$

where $\mathbf{n}$ is the outward directed unit-normal vector. A first-order FV discretization of (7.6)

[8] Note that the estimation of particle-field quantities such as $\{\phi\}_{N_l,M}(\mathbf{x}_l, t)$ is more difficult in Lagrangian PDF codes. However, the implementation of mixing models based on the estimates is nearly identical for both types of codes.

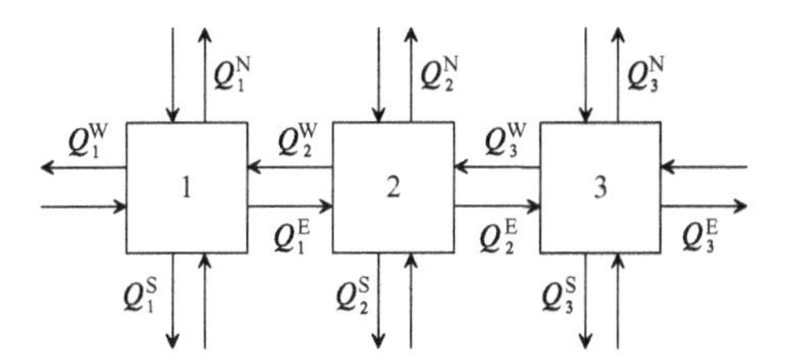

Figure 7.1. Sketch of effective volumetric flow rates between neighboring cells used in the finite-volume code.

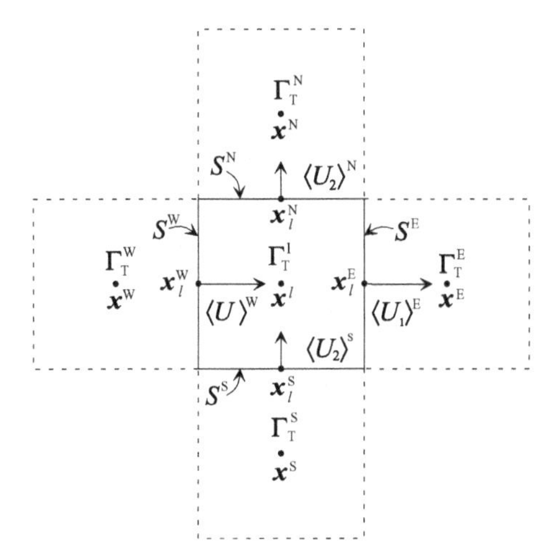

Figure 7.2. Sketch of flow variables for the lth grid cell and its four neighbors.

(see Fig. 7.1) can be expressed as

$$V_l \frac{\mathrm{d}\phi_l}{\mathrm{d}t} = (Q\phi)^{\mathrm{N}} + (Q\phi)^{\mathrm{S}} + (Q\phi)^{\mathrm{E}} + (Q\phi)^{\mathrm{W}} - \left(Q_l^{\mathrm{N}} + Q_l^{\mathrm{S}} + Q_l^{\mathrm{E}} + Q_l^{\mathrm{W}}\right)\phi_l, \tag{7.7}$$

where Q^{N}, Q^{S}, Q^{E}, and Q^{W} are the effective volumetric flow rates from adjacent cells *into* the lth cell,[9] and Q_l^{N}, Q_l^{S}, Q_l^{E}, and Q_l^{W} are the effective volumetric flow rates *out of* the lth cell into adjacent cells. These flow rates will be functions of the mean velocities, turbulent diffusivities, and grid-cell geometry supplied by the FV code.

For example, consider the grid cell shown in Fig. 7.2. Note that we have assumed that the mean velocity is known at the cell faces, and that the mean flow direction is from S–E to N–W (i.e., $\langle U_1\rangle^{\mathrm{E}}$, $\langle U_1\rangle^{\mathrm{W}}$, $\langle U_2\rangle^{\mathrm{N}}$, and $\langle U_2\rangle^{\mathrm{S}}$ are positive). The corresponding cell surface areas are denoted by S^{E}, S^{W}, S^{N}, and S^{S}. For the rectangular grid cell in Fig. 7.2, the surface areas are related to the cell volume by

$$\begin{aligned} S^{\mathrm{N}} = S^{\mathrm{S}} &= \frac{1}{\left|\mathbf{x}_l^{\mathrm{N}} - \mathbf{x}_l^{\mathrm{S}}\right|} V_l, \\ S^{\mathrm{E}} = S^{\mathrm{W}} &= \frac{1}{\left|\mathbf{x}_l^{\mathrm{E}} - \mathbf{x}_l^{\mathrm{W}}\right|} V_l. \end{aligned} \tag{7.8}$$

[9] For example, Q^{N} is the volumetric flow rate from the grid cell at N to the lth grid cell.

Assuming that the turbulent diffusivities are known at the cell centers, the diffusion velocities[10] at the cell faces are found by linear interpolation:[11]

$$D_l^{\mathrm{k}} \equiv \frac{\Gamma_{\mathrm{T}}^{\mathrm{k}} |\mathbf{x}_l^{\mathrm{k}} - \mathbf{x}_l| + \Gamma_{\mathrm{T}}^{l} |\mathbf{x}_l^{\mathrm{k}} - \mathbf{x}^{\mathrm{k}}|}{|\mathbf{x}^{\mathrm{k}} - \mathbf{x}_l|^2} \quad \text{for k} = \text{N, S, E, W.} \tag{7.9}$$

The effective outflow and inflow rates for the example in Fig. 7.2 are then given by[12]

$$\begin{aligned} Q_l^{\mathrm{N}} &= \left(\langle U_2\rangle^{\mathrm{N}} + D_l^{\mathrm{N}}\right) S^{\mathrm{N}}, & Q^{\mathrm{N}} &= D_l^{\mathrm{N}} S^{\mathrm{N}}, \\ Q_l^{\mathrm{S}} &= D_l^{\mathrm{S}} S^{\mathrm{S}}, & Q^{\mathrm{S}} &= \left(\langle U_2\rangle^{\mathrm{S}} + D_l^{\mathrm{S}}\right) S^{\mathrm{S}}, \\ Q_l^{\mathrm{E}} &= D_l^{\mathrm{E}} S^{\mathrm{E}}, & Q^{\mathrm{E}} &= \left(\langle U_1\rangle^{\mathrm{E}} + D_l^{\mathrm{E}}\right) S^{\mathrm{E}}, \\ Q_l^{\mathrm{W}} &= \left(\langle U_1\rangle^{\mathrm{W}} + D_l^{\mathrm{W}}\right) S^{\mathrm{W}}, & Q^{\mathrm{W}} &= D_l^{\mathrm{W}} S^{\mathrm{W}}, \end{aligned} \tag{7.10}$$

where the convective terms have been approximated by a first-order up-wind scheme.

Note that continuity requires a balance between the outflow and inflow rates:

$$Q_l^{\mathrm{N}} + Q_l^{\mathrm{S}} + Q_l^{\mathrm{E}} + Q_l^{\mathrm{W}} = Q^{\mathrm{N}} + Q^{\mathrm{S}} + Q^{\mathrm{E}} + Q^{\mathrm{W}}. \tag{7.11}$$

This constraint will be satisfied if

$$\langle U_2\rangle^{\mathrm{N}} S^{\mathrm{N}} + \langle U_1\rangle^{\mathrm{W}} S^{\mathrm{W}} = \langle U_2\rangle^{\mathrm{S}} S^{\mathrm{S}} + \langle U_1\rangle^{\mathrm{E}} S^{\mathrm{E}}, \tag{7.12}$$

which is just the FV discretization of $\nabla \cdot (\langle\rho\rangle\langle\mathbf{U}\rangle) = 0$ with $\langle\rho\rangle$ constant. Thus, the effective flow rates in the Eulerian PDF will satisfy continuity if the FV code provides a mass-conserving discretized mean velocity field.

The effective inflow rates Q^{k} for k = N, S, E, W can be computed for every grid cell as described above. Using the cell volume V_l, a characteristic time step can be computed for each cell:

$$\Delta t_l^q \equiv \frac{V_l}{Q^{\mathrm{N}} + Q^{\mathrm{S}} + Q^{\mathrm{E}} + Q^{\mathrm{W}}}. \tag{7.13}$$

The time step Δt used in the Eulerian PDF simulation[13] can be at most equal to the smallest cell time step. Likewise, a characteristic turbulence time step can be computed for each cell:

$$\Delta t_l^u \equiv \frac{1}{\omega(\mathbf{x}_l)}, \tag{7.14}$$

and is used to characterize the rate of micromixing in the lth cell. Thus, the time step used in the Eulerian PDF code must satisfy:

$$\Delta t \le \min\left[\min_l \left(\Delta t_l^q\right), \min_l \left(\Delta t_l^u\right)\right]. \tag{7.15}$$

10 Recall that the units of Γ_{T} are (length^2/time).

11 For variable-density flows, $\Gamma_{\mathrm{T}}^{\mathrm{k}}$ is replaced by $\langle\rho\rangle(\mathbf{x}^{\mathrm{k}})\Gamma_{\mathrm{T}}^{\mathrm{k}}/\langle\rho\rangle(\mathbf{x}_l)$.

12 For variable-density flows, $\langle U_2\rangle^{\mathrm{k}}$ is replaced by $\langle\rho\rangle(\mathbf{x}^{\mathrm{k}})\langle U_2\rangle^{\mathrm{k}}/\langle\rho\rangle(\mathbf{x}_l)$, and $\langle U_2\rangle^{\mathrm{k}}$ by $\langle\rho\rangle(\mathbf{x}^{\mathrm{k}})\langle U_2\rangle^{\mathrm{k}}/\langle\rho\rangle(\mathbf{x}_l)$.

13 This condition assumes that the time evolution of the composition PDF is of interest. If only the stationary PDF is required, local time stepping can be employed as discussed below.

Furthermore, due to integer round-up, it may be necessary to use an even smaller Δt as noted below.

The simplest numerical implementation of inter-cell transport in an Eulerian PDF code consists of the following steps.

(i) The effective inflow rates for every cell in the flow domain are computed. The simulation time step is found from (7.15).

(ii) The numbers of notional particles that flow into the lth cell from neighboring cells are computed as follows:

$$N_l^{\mathrm{k}} = \mathrm{int}\left(\frac{Q^{\mathrm{k}}\Delta t}{V_l} N_l\right) \quad \text{for k} = \mathrm{N, S, E, W}, \tag{7.16}$$

where the function int(·) rounds its argument to the nearest integer. If $Q^{\mathrm{k}} > 0$ but $N_l^{\mathrm{k}} = 0$ for any k, then N_l must be increased until all $N_l^{\mathrm{k}} \geq 1$. The number of particles that remain in the lth cell at $t + \Delta t$ is then

$$N_l^{\mathrm{R}} = N_l - N_l^{\mathrm{N}} - N_l^{\mathrm{S}} - N_l^{\mathrm{E}} - N_l^{\mathrm{W}}. \tag{7.17}$$

If $N_l^{\mathrm{R}} \leq 0$, then Δt must be reduced, and the entire step must be restarted. This step is repeated for every cell in the flow domain.

(iii) The new set of notional particles at the lth grid cell $\{\phi\}_l^{\mathrm{new}}$ is randomly selected *with replacement*[14] from the old sets of notional particles:

$$\{\phi\}_l^{\mathrm{new}} = \{\phi\}_l^{\mathrm{R}} + \{\phi\}_l^{\mathrm{N}} + \{\phi\}_l^{\mathrm{S}} + \{\phi\}_l^{\mathrm{E}} + \{\phi\}_l^{\mathrm{W}}, \tag{7.18}$$

where $\{\phi\}_l^{\mathrm{k}}$ is the set of N_l^{k} particles randomly selected from the old set $\{\phi\}_l^{\mathrm{k,old}}$.

The random selection in step (iii) is carried out by generating uniform random numbers $U \in [0, 1]$. For example, the index of a random particle selected from a set of N particles will be $n = \mathrm{int}_{\mathrm{up}}(UN)$ where $\mathrm{int}_{\mathrm{up}}(\cdot)$ rounds the argument up to the nearest integer. Note that for constant-density, statistically stationary flow, the effective flow rates will be constant. In this case, steps (i) and (ii) must be completed only once, and the MC simulation is advanced in time by repeating step (iii) and intra-cell processes. For variable-density flow, the mean density field ($\langle\rho\rangle$) must be estimated from the notional particles and passed back to the FV code. In the FV code, the non-uniform density field is held constant when solving for the mean velocity field.[15]

[14] Random selection with replacement means that elements from the set are returned to the selection pool after they are selected. This implies that members of the set can be selected more than once. There appears to be some confusion in the literature on the use of random sampling *with* or *without* replacement. However, because the notional particles represent a composition PDF with fixed statistical properties (and *not* fluid particles), selection with replacement is the correct choice. Moreover, it is the only choice that will allow N_l to be different in every cell. For example, in the extreme case where $N_1 = 1$ and $N_2 = 100$, (7.16) might yield $N_2^{\mathrm{W}} = 10$ so that ten notional particles must be selected from the first grid cell. Since $N_1 = 1$, this can only be accomplished with replacement.

[15] Thus, for statistically stationary flow, the Favre-averaged velocity field satisfies $\nabla \cdot (\langle\rho\rangle\langle\mathbf{U}\rangle) = 0$.

7.2.2 Numerical diffusion

As described above, spatial transport in an Eulerian PDF code is simulated by random jumps of notional particles between grid cells. Even in the simplest case of one-dimensional purely convective flow with equal-sized grids, so-called 'numerical diffusion' will be present. In order to show that this is the case, we can use the analysis presented in Möbus *et al.* (2001), simplified to one-dimensional flow in the domain $[0, L]$ (Möbus *et al.* 1999). Let $X(m\Delta t)$ denote the random location of a notional particle at time step m. Since the location of the particle is discrete, we can denote it by a random integer i: $X(m\Delta t) = i\Delta x$, where the grid spacing is related to the number of grid cells (M) by $\Delta x = L/M$. For purely convective flow, the time step is related to the mean velocity $\langle U \rangle$ by[16]

$$\alpha \equiv \frac{\langle U \rangle \Delta t}{\Delta x} \leq 1. \tag{7.19}$$

Note that $\alpha = 1$ corresponds to the case where all notional particles are moved to the down-wind grid cell at each time step. In more general flows, it will not be possible (or desirable) to move all notional particles in every grid cell. Thus, $0 < \alpha \ll 1$ will commonly be required to control the time step.

If a notional particle is initially located at $X(0) = 0$, the probability that it will be located at $X(t) = i\Delta x$ for time $t = m\Delta t$ is given by

$$P(X(m\Delta t) = i\Delta x | X(0) = 0) = \begin{cases} \binom{m}{i}(1-\alpha)^{m-i}\alpha^i & \text{if } i \leq m \\ 0 & m < i. \end{cases} \tag{7.20}$$

The expected location at time $t = m\Delta t$ follows by summation:

$$\begin{aligned} \langle X(m\Delta t) \rangle &= \sum_{i=0}^{m} (i\Delta x) \binom{m}{i} (1-\alpha)^{m-i}\alpha^i \\ &= m\alpha\Delta x \\ &= m\Delta t \langle U \rangle. \end{aligned} \tag{7.21}$$

Thus, as expected for plug flow, $\langle X(t) \rangle = \langle U \rangle t$.

Numerical diffusion for the purely convective case is measured by the variance of the particle location. In the absence of numerical diffusion, $X(t) = \langle X(t) \rangle$, and the variance is zero. From (7.20), the location variance can be computed as

$$\begin{aligned} \sigma^2(m\Delta t) &= \sum_{i=0}^{m} (i\Delta x - m\alpha\Delta x)^2 \binom{m}{i} (1-\alpha)^{m-i}\alpha^i \\ &= m\alpha(1-\alpha)(\Delta x)^2 \end{aligned} \tag{7.22}$$

or

$$\frac{\sigma^2(t)}{L^2} = \frac{\langle X(t) \rangle}{L} \frac{(1-\alpha)}{M}. \tag{7.23}$$

[16] The time step used here is based on the application of (7.13) with Δx constant.

At $t = \tau_{\rm pfr} = L/\langle U \rangle$ (which corresponds to the convective front reaching the end of the domain), the relative magnitude of the numerical diffusion will depend only on α and M:

$$\frac{\sigma^2(\tau_{\rm pfr})}{L^2} = \frac{(1-\alpha)}{M}. \tag{7.24}$$

For this case (Δx constant), it would be possible to eliminate numerical diffusion by setting $\alpha = 1$ (Roekaerts 1991). However, in more general cases, the value of $\alpha < 1$ will be controlled by the smallest characteristic flow time in (7.13), and thus numerical diffusion cannot be eliminated in an Eulerian PDF code.

In contrast, we shall see in Section 7.3 that Lagrangian PDF codes do not suffer from numerical diffusion. Likewise, by using higher-order discretization schemes, FV codes can greatly reduce the effects of numerical diffusion. Since Eulerian PDF codes are restricted to first-order up-wind schemes, the only method available to reduce numerical diffusion is to increase the grid-cell number (M). As a consequence (see example in Möbus *et al.* (2001)), a much finer grid is required to obtain grid-independent solutions with an Eulerian PDF code as compared with a Lagrangian PDF code. Thus, even though the Lagrangian PDF code may take longer to update the notional particles in a single grid cell, its overall computing time for obtaining grid-independent solutions can be less than Eulerian PDF codes. Moreover, the effect of numerical diffusion can be enhanced by the chemical source term in reacting flow (Möbus *et al.* 1999). Primarily for these reasons, the use of Lagrangian PDF codes has increased in recent years at the expense of Eulerian PDF codes.

7.2.3 Other considerations

For simple flows where the mean velocity and/or turbulent diffusivity depend only weakly on the spatial location, the Eulerian PDF algorithm described above will perform adequately. However, in many flows of practical interest, there will be strong spatial gradients in turbulence statistics. In order to resolve such gradients, it will be necessary to use local grid refinement. This will result in widely varying values for the cell time scales found from (7.13). The simulation time step found from (7.15) will then be much smaller than the characteristic cell time scales for many of the cells. When the simulation time step is applied in (7.16), one will find that N_l must be made unrealistically large in order to satisfy the constraint that $N_l^{\rm k} \geq 1$ for all k.

In principle, using a large value of N_l only adds linearly to the computing cost (and decreases the statistical error!) and thus could be tolerated. However, it usually happens that cells with a large characteristic cell time scale make very little contribution to the flow statistics (e.g., the composition PDF is close to a delta function). In this case, adding more notional particles to improve the empirical PDF is numerically very wasteful. In such cases, it would be preferable to represent the empirical PDF in the grid cell by a single notional particle, and then to concentrate the numerical effort on cells where mixing and reaction are important.

In order to overcome the shortcoming of fixed time steps, Roekaerts (1991) has proposed a modified Eulerian PDF algorithm that keeps track of the cell time scales. A similar algorithm is described in Pipino and Fox (1994). The basic idea is to fix N_l to control the statistical error, and then to keep track of cells where $N_l^{\rm k} < 1$ for some k. For example, if $N_1^{\rm E} = 0.1$ for a given Δt and N_1, then transfer of a notional particle into cell 1 from cell E occurs once every ten time steps. The algorithm thus keeps track of the 'local' transfer rates and, for sufficiently large simulation times, all cells will receive particles from neighboring cells at the statistically correct rate. In this manner, the simulation time step can be chosen to resolve the transport processes in important zones of the flow without wasting particles in zones of little interest.

Although the Eulerian PDF code described above is applicable to time-dependent flows,[17] most applications will require only the statistically stationary statistics. In these cases, the statistical errors in the estimated fields can be reduced by time averaging. This is accomplished by running the simulation forward in time until the flow statistics become time-independent, and then collecting running averages. For example, if we denote the running time average of $\langle \phi \rangle$ at step m by $\langle \phi \rangle_m$, then the running time average at step $m + 1$ is given by

$$\begin{aligned} \langle \phi \rangle_{m+1}(\mathbf{x}_l) &= \frac{1}{m+1} \sum_{i=1}^{m+1} \{\phi\}(\mathbf{x}_l, i\Delta t) \\ &= \frac{m}{m+1} \langle \phi \rangle_m + \frac{1}{m+1} \{\phi\}\,(\mathbf{x}_l, (m+1)\Delta t)\,, \end{aligned} \tag{7.25}$$

where $\{\phi\}(\mathbf{x}_l, t)$ is the estimated mean field (using (7.4)) in the lth cell at time t. This can be generalized to

$$\langle \phi \rangle_{m+1}(\mathbf{x}_l) = \frac{K-1}{K} \langle \phi \rangle_m + \frac{1}{K} \{\phi\}\,(\mathbf{x}_l, (m+1)\Delta t)\,, \tag{7.26}$$

where $K\Delta t$ is the time-average time scale (Jenny *et al.* 2001). The same method can be used to compute running time averages for other statistics.

For statistically stationary flows, the simulation time and numerical diffusion can be further reduced by using *local time stepping*. For a statistically stationary flow, the composition PDF will be independent of time. This implies that the statistical properties of the notional particles will be independent of the time step Δt. Moreover, it implies that we need not use the same time step for every grid cell when applying (7.16). Instead, the local time step Δt_l^q found from (7.13) can be used.[18] Note that this implies that every notional particle in a grid cell will be replaced on each iteration (i.e., $N_l^{\rm R} = 0$). Example results computed using local time stepping can be found in Biagioli (1997), Anand *et al.* (1998), and Möbus *et al.* (1999).

[17] For example, mixing of two streams in a reactor that is initially empty.

[18] If the characteristic micromixing time scale is much smaller than Δt_l^q, then care must be taken in implementing the intra-cell processes. For example, several smaller time steps may be required to represent mixing and chemical reactions at each iteration.

In summary, the advantages and disadvantages of the Eulerian PDF algorithm described above are as follows:

Advantages

- Notional-particle locations correspond to the grid cells used in the FV code.
- All notional particles have equal weight, and estimated statistical quantities are found using cell averages.
- Implementation of mixing and chemical reactions is straightforward, and involves interactions between notional particles in the same cell.
- The spatial-transport algorithm is relatively easy to implement on orthogonal grids using information provided by the FV code.
- The total computational cost is proportional to the number of notional particles ($N_{\rm p}$), and the algorithm is trivial to parallelize.

Disadvantages

- The effective flow rates are grid-dependent and relatively difficult to compute for arbitrary non-orthogonal grids.
- Spatial transport is limited to first-order, up-wind schemes, and is thus strongly affected by numerical diffusion.
- In order to control statistical error, the number of notional particles per grid cell must be relatively large ($N_l = 100$–500).
- In order to obtain grid-independent solutions, relatively fine grids are required (i.e., large M). Since $N_{\rm p} \propto M \times N_l$, this greatly increases the computational cost.
- Flow rates can vary significantly over the computational domain, making it necessary to implement special algorithms to ensure that $N_l^{\rm k} \geq 1$.

While some of the disadvantages listed above can be overcome by modifying the algorithm, the problem of numerical diffusion remains as the principal shortcoming of all Eulerian PDF codes.

Eulerian PDF codes have been used in a large number of turbulent-reacting-flow studies (e.g., Chen and Kollmann 1988; Lakatos and Varga 1988; J. Y. Chen *et al.* 1989; J. Y. Chen *et al.* 1990; Hsu *et al.* 1990; Roekaerts 1991; Vervisch 1991; Roekaerts 1992; Fox and Grier 1993; Pipino and Fox 1994; Raju 1996; Jones and Kakhi 1997; Anand *et al.* 1998; Jones and Kakhi 1998; Möbus *et al.* 1999; Pfilzner *et al.* 1999; Domingo and Benazzouz 2000; Möbus *et al.* 2001). Many of these studies have considered variable-density flow, and some have used non-orthogonal grids. Barlow *et al.* (1999) have compared CMC and Eulerian PDF predictions for jet flames. The reader interested in these topics is thus encouraged to consult the literature. Despite these advances in the development of Eulerian PDF codes, the current trend is towards Lagrangian PDF codes. (Due mainly to the lack of efficient methods for reducing numerical diffusion.) Thus, in the next section, we present an overview of Lagrangian PDF codes for the composition PDF.

7.3 Lagrangian composition PDF codes

Lagrangian PDF codes are based on Lagrangian PDF methods described in Section 6.7. In a Lagrangian composition PDF code, each notional particle is described by its position ($\mathbf{X}^{(n)}(t)$) and composition ($\boldsymbol{\phi}^{(n)}(t)$). In addition, it is convenient to assign a non-unity statistical weight to each particle ($w^{(n)}(t)$). As discussed below, the particle weights will appear in the empirical PDF used for particle-field estimation. A Lagrangian composition PDF code must be coupled to an FV code, which supplies the mean velocity and turbulence fields.[19] This information is usually available at grid nodes and must be interpolated to the particle locations (e.g., $\langle \mathbf{U} \rangle (\mathbf{X}^{(n)}(t), t))$ for use in the MC simulation of the governing stochastic differential equations (SDEs). Likewise, particle-field estimates (e.g., $\{\boldsymbol{\phi}^* | \mathbf{X}^*\}(\mathbf{x}, t))$ will be required for the micromixing model, and for outputting the simulation results. Unless stated otherwise, we will assume that the same grid is used in both the FV and PDF codes, and that the grid is orthogonal (although all grid cells need not be of equal volume). Throughout this section, we will denote individual grid nodes by the subscript α (e.g., $\mathbf{x}_\alpha$) and individual grid cells by the subscript l. The summation over all grid nodes will be denoted by $\sum_\alpha$.

As noted in Chapter 1, the composition PDF description utilizes the concept of turbulent diffusivity (Γ_{T}) to model the scalar flux. Thus, it corresponds to closure at the level of the k–ε and gradient-diffusion models, and should be used with caution for flows that require closure at the level of the RSM and scalar-flux equation. In general, the velocity, composition PDF codes described in Section 7.4 should be used for flows that require second-order closures. On the other hand, Lagrangian composition codes are well suited for use with an LES description of turbulence.

7.3.1 Notional-particle representation

In a (constant-density) Lagrangian composition PDF code, the position of a notional particle is governed by

$$\mathrm{d}\mathbf{X}^{(n)} = \left[\langle \mathbf{U} \rangle \left(\mathbf{X}^{(n)}, t\right) + \nabla \Gamma_{\mathrm{T}} \left(\mathbf{X}^{(n)}, t\right)\right] \mathrm{d}t + \left[2\Gamma_{\mathrm{T}} \left(\mathbf{X}^{(n)}, t\right)\right]^{1/2} \mathrm{d}\mathbf{W}(t), \tag{7.27}$$

and its composition (using the LIEM model; see (6.190))[20] is governed by

$$\frac{\mathrm{d}\boldsymbol{\phi}^{(n)}}{\mathrm{d}t} = \frac{C_\phi}{2} \omega \left(\mathbf{X}^{(n)}, t\right) \left(\{\boldsymbol{\phi}^* | \mathbf{X}^*\} \left(\mathbf{X}^{(n)}, t\right) - \boldsymbol{\phi}^{(n)}\right) + \mathbf{S}\left(\boldsymbol{\phi}^{(n)}\right). \tag{7.28}$$

In the MC simulation, these equations are treated numerically using fractional time stepping. For unsteady flow, the simulation time step Δt is determined from the FV code by

[19] Typically, the turbulence fields supplied by the FV code are $k(\mathbf{x}, t)$ and $\varepsilon(\mathbf{x}, t)$. However, the actual fields required by the PDF code are $\Gamma_{\mathrm{T}}(\mathbf{x}, t)$ and $\omega(\mathbf{x}, t)$. Throughout this section we will assume that the required turbulence fields are available from the FV code.

[20] This choice is made to simplify the discussion which follows. Any other model could be substituted without changing the basic algorithm.

applying a local Courant, Freidrich, Lewy (CFL) condition (Haworth and El Tahry 1991; Jenny *et al.* 2001), which depends on the grid cells and turbulence fields. For stationary flow, local time stepping can be used (Möbus *et al.* 2001; Muradoglu and Pope 2002).

At each time step, the coefficients appearing in (7.27) and (7.28) must be recomputed based on the new position (i.e., $\mathbf{X}^{(n)}(t + \Delta t)$). In order to simplify this task, it is usually desirable to sort the notional particles according to the FV grid cell in which they are located. Efficient particle-sorting algorithms for rectangular grids are straightforward, and can be based on multiple one-dimensional sorting. However, for arbitrarily complex grids, efficient sorting algorithms are a critical component of a Lagrangian PDF code. A modified search-and-locate algorithm for arbitrary quadrilateral grid cells is discussed in Möbus *et al.* (2001). A similar approach – the convex polyhedron method – for highly skewed unstructured three-dimensional deforming grids is presented in Subramaniam and Haworth (2000). An efficient particle-locating algorithm in arbitrary two- and three-dimensional grids is described in Chordá *et al.* (2002). In general, the sorting algorithm will represent a non-trivial fraction of the total computational cost, and thus should be implemented with care. For the remainder of this section, we will denote the number of particles in the lth grid cell by N_{pl},[21] and (after calling the sorting algorithm) assume that the particles in each cell are numbered consecutively: $n = 1, \dots, N_{pl}$.

An FV code provides the turbulence fields ($\langle \mathbf{U} \rangle(\mathbf{x}, t)$, $\Gamma_T(\mathbf{x}, t)$, and $\omega(\mathbf{x}, t)$) by solving an RANS turbulence model.[22] Usually these fields will be available at cell centers or on cell faces (see Fig. 7.2), and must be interpolated to the particle positions. For constant-density flow, the correspondence between the notional-particle and fluid-particle systems requires that the weight of the particles be uniformly distributed throughout the flow domain. This condition will be satisfied by (7.27) if the interpolated mean velocity field is divergence-free at every $\mathbf{X}^{(n)}$, i.e., $\nabla \cdot \langle \mathbf{U} \rangle(\mathbf{X}^{(n)}, t) = 0$. In general, simple interpolation schemes for $\langle \mathbf{U} \rangle(\mathbf{x}, t)$ will not ensure that the mean velocity is divergence-free at the particle locations. However, a two-dimensional divergence-free interpolation scheme for rectangular cells can be formulated and is described in Jenny *et al.* (2001). For more complicated grids, particle number density must be controlled by implementing an algorithm to shift particles in physical space (Haworth and El Tahry 1991).

Although it is not strictly necessary, in most Lagrangian PDF codes the turbulence frequency $\omega = 1/\tau_u$ appearing in the mixing term of (7.28) is not interpolated to the particle positions. Instead, the cell-center value $\omega(\mathbf{x}_l, t)$ is used for all particles in the grid cell. The principal justification for using the grid-cell value is to reduce statistical estimation errors. Note also that using $\omega(\mathbf{x}_l, t)$ will tend to reduce the deterministic error (see (6.233), p. 305), within each grid cell.[23] We will look at this subject in more detail when discussing particle-field estimation below.

21 Since particles move by a random walk, $N_{pl}(t)$ is in fact a random process. Thus, at the end of each time step, the sorting algorithm must determine N_{pl} and renumber the particles.

22 Simply by substituting the corresponding fields, the Lagrangian PDF code described in this section could be coupled to an LES code (Jaberi *et al.* 1999).

23 However, use of the grid-cell kernel induces a deterministic error similar to numerical diffusion due to the piece-wise constant approximation.

In a Lagrangian PDF simulation, each notional particle represents a fluid element with mass $w^{(n)}\Delta m$. In a constant-density system, the 'unit mass' is defined by

$$\Delta m \equiv \frac{\rho V}{N_\mathrm{p}}, \tag{7.29}$$

where ρ is the fluid density, V is the total volume of the flow domain, and N_p is the number of unit-weight particles needed to describe the flow.[24] If all notional particles have equal weight (i.e., $w^{(n)} = 1$), then the particle positions $\mathbf{X}^{(n)}$ will remain uniformly distributed throughout the flow domain. This implies that the expected number of particles in a grid cell will be proportional to the grid-cell volume V_l. Thus, unless all grid-cell volumes are nearly identical, the number of particles in a grid cell ($N_{\mathrm{p}l}$) will vary throughout the flow domain. In particular, in zones where small grid cells are needed to resolve rapid changes, $N_{\mathrm{p}l}$ will be small, leading to large statistical errors. Likewise, in axisymmetric flow domains, the grid-cell volumes far from the centerline will be large ($V_l(r) \propto r\Delta r\Delta z$) and thus would contain a large number of unit-weight particles. This again would lead to large statistical errors near the centerline, where accurate estimates are usually desired.

In order to maintain uniform statistical error, the particle weights $w^{(n)}(t)$ can be initialized according to the grid-cell volumes, and modified during the course of the MC simulation (Haworth and El Tahry 1991). For example, for axisymmetric flow, the initial weights $w^{(n)}(0)$ are set proportional to the initial radial location $X_2^{(n)}(0) = R^{(n)}(0)$. As the simulation advances, notional particles that were initially far from the centerline may move close to the centerline, in which case their weights would be much larger than the other notional particles in the grid cell. In order to avoid this problem, an algorithm must be developed to 'clone' new particles from heavy particles and to 'cluster' light particles in such a way that the total particle weight in the grid cell remains unchanged. For example, the following algorithm could be applied for constant-density flow.

(i) Compute the estimated mean particle weight in the lth grid cell:[25]

$$\{w\}_l \equiv \frac{1}{N_{\mathrm{p}l}} \sum_{n=1}^{N_{\mathrm{p}l}} w^{(n)}. \tag{7.30}$$

(ii) If $w^{(k)} \geq 2\{w\}_l$ for any $k \in 1, \ldots, N_{\mathrm{p}l}$, then clone a new particle:

$$\begin{pmatrix} w \\ \mathbf{X} \\ \phi \end{pmatrix}^{(k)} \xrightarrow[\text{clone}]{} \begin{pmatrix} w^{(k)}/2 \\ \mathbf{X}^{(k)} \\ \phi^{(k)} \end{pmatrix}^{(k)} \text{ and } \begin{pmatrix} w^{(k)}/2 \\ \mathbf{X}^{(k)} \\ \phi^{(k)} \end{pmatrix}^{(k^*)}, \tag{7.31}$$

[24] Since ρ is constant, each notional particle also represents a volume of $w^{(n)}\Delta V$, where $\Delta V = \Delta m/\rho$. Note, in general, that the estimated mean density in the lth grid cell is related to the weights by

$$\{\rho|\mathbf{X}^*\}(\mathbf{x}_l, t) = \frac{\Delta m}{V_l} \sum_{n=1}^{N_{\mathrm{p}l}} w^{(n)}(t).$$

In variable-density flows, this relation is used to couple the PDF code to the FV code by replacing the mean density predicted by the FV code with $\{\rho|\mathbf{X}^*\}$. Because the convergence behavior of the FV code may be sensitive to errors in the estimated mean density, particle-number control is especially critical in variable-density Lagrangian PDF codes.

[25] From the definition of the particle weights, $\{w\}_l \approx 1$.

where k^* is a new particle. Note that the position and composition of the new particle is the same as the original particle. At the end of a clone step, the sum of all particle weights is unchanged.

(iii) If $w^{(k)} \leq \{w\}_l/2$ for any $k \in 1, \ldots, N_{pl}$, then cluster it with another randomly selected particle:

$$\begin{pmatrix} w \\ \mathbf{X} \\ \phi \end{pmatrix}^{(k)} \text{and} \begin{pmatrix} w \\ \mathbf{X} \\ \phi \end{pmatrix}^{(k^*)} \xrightarrow[\text{cluster}]{} \begin{pmatrix} w^{(k)} + w^{(k^*)} \\ \mathbf{X}^{(m)} \\ \phi^{(m)} \end{pmatrix}^{(k)}, \tag{7.32}$$

where $m = k$ with probability $w^{(k)}/(w^{(k)} + w^{(k^*)})$ and $m = k^*$ with probability $w^{(k^*)}/(w^{(k)} + w^{(k^*)})$. Note that the position and composition of the new particle is the same as the randomly chosen mth particle.[26]

The algorithm described above will control the distribution of weights in the grid cells, but not necessarily the numbers of particles N_{pl}. In order to control both the weights and the numbers, a 'desirable' weight for each cell (w_l^*) can be specified according to the grid-cell mass (ρV_l) and the desired number of particles per cell (Subramaniam and Haworth 2000). w_l^* is then used in place of $\{w\}_l$ in steps (ii) and (iii). Since the PDF of the weights is of no intrinsic interest, the details of the weight-control algorithm are not particularly important, provided that they do not change the statistical properties of $\mathbf{X}^{(n)}$ and $\phi^{(n)}$, and that the sum of all $w^{(n)}$ in a given grid cell remains unchanged.

The flow diagram[27] shown in Fig. 7.3 summarizes the principal steps in the Lagrangian PDF code. The first step is to initialize the particle properties ($w^{(n)}(0)$, $\mathbf{X}^{(n)}(0)$, and $\phi^{(n)}(0)$) and the turbulence fields ($\langle \mathbf{U} \rangle(\mathbf{x}, 0)$, $\Gamma_T(\mathbf{x}, 0)$ and $\omega(\mathbf{x}, 0)$). Based on the turbulence fields, a local CFL condition is used to fix the simulation time step Δt. The FV code then advances the flow field. For stationary flow, the FV code returns $\langle \mathbf{U} \rangle(\mathbf{x})$, $\Gamma_T(\mathbf{x})$, and $\omega(\mathbf{x})$. For unsteady flow, $\langle \mathbf{U} \rangle(\mathbf{x}, \Delta t)$, $\Gamma_T(\mathbf{x}, \Delta t)$, and $\omega(\mathbf{x}, \Delta t)$ are returned. In either case, these fields are then interpolated to find their values at the particle locations $\mathbf{X}^{(n)}(0)$. The next step is to use an MC simulation to compute $\mathbf{X}^{(n)}(\Delta t)$. The sorting algorithm is then applied to determine N_{pl} and to renumber the particles according to their grid cell. For the particles in the same grid cell, the weight-control algorithm is applied to find $w^{(n)}(\Delta t)$. Likewise, the intra-cell processes of micromixing and chemical reactions are used to determine $\phi^{(n)}(\Delta t)$. Using the updated particle properties, particle-field estimates are constructed. The simulation clock is then incremented by Δt. If the simulation time t has not reached the stopping time t_{sim}, the algorithm iterates again to advance the simulation for another Δt. Note that, for unsteady flow, Δt need not be the same for every iteration.

[26] Alternatively, the weighted-average position could be used:

$$\mathbf{X}^{(m)} = \frac{w^{(k)}\mathbf{X}^{(k)} + w^{(k^*)}\mathbf{X}^{(k^*)}}{w^{(k)} + w^{(k^*)}}.$$

However, this would result in an undesirable 'drift' of the particle positions towards the cell center. Likewise, since the weights and compositions will usually be correlated, it would be incorrect to use the weighted-average composition.

[27] Apart from the interpolation step, the flow diagram for an Eulerian PDF is identical to the one shown in Fig. 7.3.

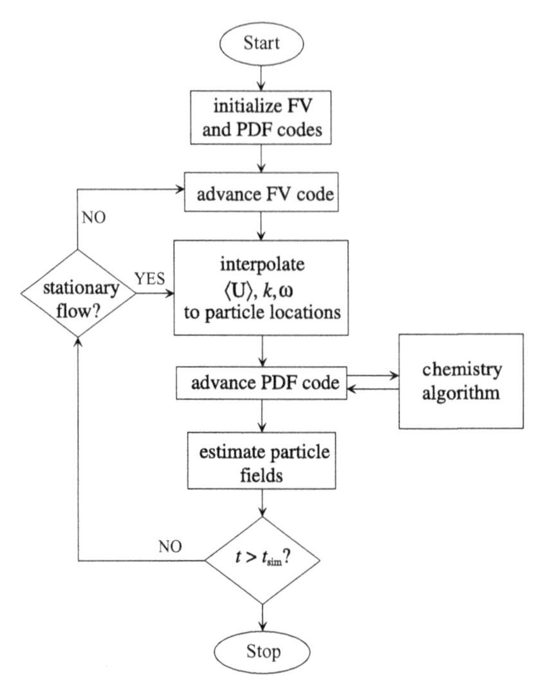

Figure 7.3. Flow diagram for Lagrangian PDF code.

We will look next at the specific algorithms needed to advance the PDF code. In particular, we describe the MC simulation needed to advance the particle position, the application of boundary conditions, and particle-field estimation. We then conclude our discussion of Lagrangian composition PDF codes by considering other factors that can be used to obtain simulation results more efficiently.

7.3.2 Monte-Carlo simulation

The model equation for particle position, (7.27), is a stochastic differential equation (SDE). The numerical solution of SDEs is discussed in detail by Kloeden and Platen (1992).[28] Using a fixed time step Δt, the most widely used numerical scheme for advancing the particle position is the *Euler approximation*:

$$\begin{aligned}\mathbf{X}^{(n)}(t+\Delta t) = \mathbf{X}^{(n)}(t) &+ \left[\langle \mathbf{U}\rangle \left(\mathbf{X}^{(n)}(t), t\right) + \nabla\Gamma_{\mathrm{T}}\left(\mathbf{X}^{(n)}(t), t\right)\right]\Delta t \\ &+ \left[2\Gamma_{\mathrm{T}}\left(\mathbf{X}^{(n)}(t), t\right)\right]^{1/2}\Delta\mathbf{W},\end{aligned} \tag{7.33}$$

where ΔW_i is a Gaussian pseudo-random number with mean $\langle \Delta W_i\rangle = 0$ and covariance $\langle \Delta W_i \Delta W_j\rangle = \Delta t\delta_{ij}$.[29] (See Kloeden and Platen (1992) for a discussion on numerical

[28] Using their nomenclature, the particle position is approximated by a *weakly convergent Ito process*.
[29] Alternatively, one can write $\Delta W_i = \xi_i(\Delta t)^{1/2}$, where ξ_i is an independent, standard-normal, random number.

methods for generating ΔW_i.) Note that the coefficients in (7.33) are evaluated at $\mathbf{X}^{(n)}(t)$, and thus the Euler approximation is explicit.

Higher-order numerical schemes are also available (Kloeden and Platen 1992), but are generally complicated to apply since they involve derivatives of the coefficients. A simpler alternative is to apply a multi-step approach (Pope 1995; Jenny *et al.* 2001). For example, the mid-point position,[30]

$$\mathbf{X}^{(n)}_{\Delta t/2} = \mathbf{X}^{(n)}(t) + \left[\langle \mathbf{U} \rangle \left(\mathbf{X}^{(n)}(t), t\right) + \nabla \Gamma_{\mathrm{T}} \left(\mathbf{X}^{(n)}(t), t\right)\right] \frac{\Delta t}{2} + \left[\Gamma_{\mathrm{T}} \left(\mathbf{X}^{(n)}(t), t\right)\right]^{1/2} \Delta \mathbf{W}, \tag{7.34}$$

can be used to approximate the coefficients:

$$\mathbf{X}^{(n)}(t + \Delta t) = \mathbf{X}^{(n)}(t) + \left[\langle \mathbf{U} \rangle \left(\mathbf{X}^{(n)}_{\Delta t/2}, t\right) + \nabla \Gamma_{\mathrm{T}} \left(\mathbf{X}^{(n)}_{\Delta t/2}, t\right)\right] \Delta t + \left[2\Gamma_{\mathrm{T}} \left(\mathbf{X}^{(n)}_{\Delta t/2}, t\right)\right]^{1/2} \Delta \mathbf{W}, \tag{7.35}$$

where the Gaussian pseudo-random numbers $\Delta \mathbf{W}$ are the same in (7.34) and (7.35). Note that the (interpolated) FV fields $\langle \mathbf{U} \rangle$ and Γ_{T} are assumed to be available for time t, but not for time $t + \Delta t$ (see Fig. 7.3).

After advancing the particle positions, the sorting algorithm must be applied to renumber the particles according to their grid cells. At the same time, the particle weights can be adjusted as described previously. Boundary conditions (described below) can also be applied at this time in order to simulate particles that enter/leave the flow domain. Note that, unlike the Eulerian PDF code, the particle positions vary 'smoothly' in time. The transport of particles between grid cells is thus unaffected by numerical diffusion. For example, in the convective limit ($\Gamma_{\mathrm{T}} = 0$) particles will follow the streamlines of the mean velocity field and never 'inter-mix.'[31]

The final step in the MC simulation is to update the composition fields. As discussed in Section 6.9, this is done by decoupling the chemical source term from the molecular mixing term. Using the fixed time step Δt, the mixing model is advanced first. In general, this will require the estimated mean composition fields $\{\phi^* | \mathbf{X}^*\} (\mathbf{x}, t)$ and, for some mixing models (e.g., the LFP model, see (6.192)), estimates of higher-order statistics. Particle-field estimation methods are discussed in detail below. Here, we shall simply assume that $\{\phi^* | \mathbf{X}^*\} (\mathbf{x}, t)$ is known. Using the LIEM model, the fractional mixing step then yields

$$\phi^{(n)}_{\Delta t} = \{\phi^* | \mathbf{X}^*\} \left(\mathbf{X}^{(n)}(t + \Delta t), t\right) + \left[\phi^{(n)}(t) - \{\phi^* | \mathbf{X}^*\} \left(\mathbf{X}^{(n)}(t + \Delta t), t\right)\right] \times \exp\left(\frac{-C_\phi \Delta t}{2} \omega \left(\mathbf{X}^{(n)}(t + \Delta t), t\right)\right), \tag{7.36}$$

[30] The actual implementation of the mid-point algorithm must account for cases where $\mathbf{X}^{(n)}_{\Delta t/2}$ falls in a different grid cell than $\mathbf{X}^{(n)}(t)$, and where $\mathbf{X}^{(n)}_{\Delta t/2}$ falls outside the boundaries of the computational domain.

[31] Strictly speaking, this can only be true if the interpolated mean velocity field satisfies continuity.

where $\omega(\mathbf{X}^{(n)}(t + \Delta t), t)$ is the turbulence frequency interpolated from the FV code to the current particle position $\mathbf{X}^{(n)}(t + \Delta t)$.[32]

Although (7.36) holds for arbitrary Δt, the upper limit on the time step is limited by the condition that the change in $\phi^{(n)}$ during the fractional mixing step should be 'small.' Typically, this condition is enforced using

$$C_{\text{mix}} = \frac{C_\phi \Delta t}{2} \max_l \omega\left(\mathbf{x}_l, t\right), \tag{7.37}$$

where the maximum is found with respect to all grid cells, and $C_{\text{mix}} = 0.1$–0.2. The value of Δt found from this expression represents the maximum time step with respect to mixing. Thus, it must be compared with the maximum time step found by applying the CFL condition in the FV code. The smaller of these two values is then used as the time step in the MC simulation.

After the fractional mixing step, the particle compositions are advanced by chemical reactions.[33] Formally this step can be written as

$$\phi^{(n)}(t + \Delta t) = \phi^{(n)}_{\Delta t} + \int_t^{t+\Delta t} \mathbf{S}\left(\phi^{(n)}(s)\right) \mathrm{d}s. \tag{7.38}$$

However, in practice, the integral must be evaluated using a stiff ODE solver or chemical lookup tables (see Section 6.9). Because transported PDF simulations are typically used for reacting flows with complex chemistry, the chemical-reaction step will often dominate the overall computational cost. It is thus important to consider carefully the computational efficiency of the chemical-reaction step when implementing a transported PDF simulation.

At the end of the chemical-reaction step, all particle properties ($w^{(n)}$, $\mathbf{X}^{(n)}$, $\phi^{(n)}$) have been advanced in time to $t + \Delta t$. Particle-field estimates of desired outputs can now be constructed, and the MC simulation is ready to perform the next time step. For a constant-density flow, the particle-field estimates are not used in the FV code. Thus, for stationary flow, the particle properties can be advanced without returning to the FV code. For unsteady or variable-density flow, the FV code will be called first to advance the turbulence fields before calling the PDF code (see Fig. 7.3).

7.3.3 Boundary conditions

During the MC simulation, boundary conditions must be applied at the edges of the flow domain. The four most common types are *outflow*, *inflow*, *symmetry*, and a *zero-flux wall*. At an outflow boundary, the mean velocity vector will point out of the flow domain. Thus, there will be a net motion of particles in adjacent grid cells across the outflow boundary. In the MC simulation, these particles are simply eliminated. By keeping track of the weights

[32] In general, the estimated mean composition field will depend on the form of $\omega(\mathbf{x}, t)$ (see, for example, Jenny *et al.* (2001)). Thus, care must be taken when defining $\{\phi^* | \mathbf{X}^*\}$ in order to ensure that (7.36) conserves the mean compositions. We will look at this point in detail when discussing particle-field estimation below.

[33] As noted in Section 6.9, the chemical-reaction step is performed last in order to allow fast chemical reactions to return to their 'local equilibrium' states.

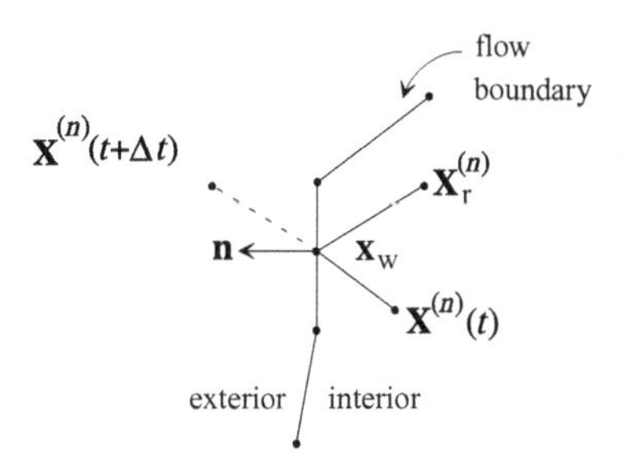

Figure 7.4. Sketch of vectors used in reflective boundary conditions.

and compositions of the outflowing particles, the mass and species outflow rates can be computed at each time step.

At an inflow boundary, the mean velocity vector will point into the flow domain. If we denote the component of the mean velocity normal to the inflow surface (S_{in}) by $\langle U \rangle_{\text{in}}$, then the total mass entering the system in time step Δt is[34]

$$\Delta m_{\text{in}} = \rho_{\text{in}} S_{\text{in}} \langle U \rangle_{\text{in}} \Delta t, \tag{7.39}$$

where ρ_{in} is the density of the incoming fluid. In the MC simulation, inflow can be represented by adding a new particle with weight $w^{(n)} = \Delta m_{\text{in}}/\Delta m$ into the inflow-boundary grid cell.[35] In practice, the initial position of the new particle is not critical (e.g., it can be placed at the grid-cell center: $\mathbf{X}^{(n)} = \mathbf{x}_l$), but $\phi^{(n)}$ must correspond to the inflow compositions.

Although the mean velocity vector points inward at an inflow boundary, as a result of the turbulent diffusion term in (7.33), particles may attempt to leave the flow domain across an inflow boundary. The same phenomenon also occurs with symmetry boundaries and zero-flux walls,[36] and is treated the same way in all three cases by reflecting the particle position back into the flow domain. The location of a particle after reflection (see Fig. 7.4) depends on the point at which the particle path intersects the boundary $\mathbf{x}_{\text{w}}$, the outward-pointing surface-normal vector $\mathbf{n}(\mathbf{x}_{\text{w}})$, and the difference vector $\mathbf{d} \equiv \mathbf{X}^{(n)}(t + \Delta t) - \mathbf{x}_{\text{w}}$, where $\mathbf{X}^{(n)}(t + \Delta t)$ is found from (7.33). Note that $\mathbf{n}^{\text{T}}\mathbf{d} > 0$ when $\mathbf{X}^{(n)}(t + \Delta t)$ lies outside the flow domain. The particle position after reflection is given by $\mathbf{X}_{\text{r}}^{(n)} = \mathbf{x}_{\text{w}} + (\mathbf{I} - 2\mathbf{n}\mathbf{n}^{\text{T}})\mathbf{d}$. In general, it will be necessary to check whether $\mathbf{X}_{\text{r}}^{(n)}$ lies in the flow domain (e.g., in a corner cell with two boundary surfaces). For orthogonal grids, the determination of $\mathbf{X}_{\text{r}}^{(n)}$ is straightforward, and the reflected point almost always lies in the flow domain. However, for non-orthogonal grids, it is more difficult to find $\mathbf{x}_{\text{w}}$, and multiple reflections are more common. Note that the weight and composition of a reflected particle remains unchanged.

34 Note that the 'diffusive' composition fluxes are assumed to be zero on inflow boundaries.

35 Since the weight-control algorithm is applied after the boundary conditions, it is not necessary to add multiple particles in order to limit their weights.

36 The mean velocity and turbulent diffusivity should approach zero at solid walls. In theory, this should be enough to keep particles from crossing wall boundaries. In practice, due to the finite time step, some particles will eventually cross wall boundaries and must be accounted for.

In addition to the four types of boundary conditions mentioned above, situations may arise where the composition flux at a wall is non-zero (e.g., a heat-transfer surface). The implementation of non-zero-flux boundary conditions is complicated by the fact that the scalar boundary layers near the wall are usually not resolved in a PDF simulation. It is thus necessary to use 'wall functions' to estimate the heat/mass-transfer rate based on the grid-cell mean compositions. Examples of using wall functions for the velocity field are described in Anand *et al.* (1989), Dreeben and Pope (1997b), and Minier and Pozorski (1999). Given the total heat/mass fluxes into the grid cell, the particle compositions can be adjusted to match the fluxes. However, unlike the velocity field, the composition PDF will typically be non-Gaussian and will lie on a bounded domain. Therefore, when manipulating the particle compositions, care must be taken to ensure that the appropriate bounds on the composition variables are not violated.

For example, consider the case of heat transfer from the wall to the fluid with wall temperature $T_{\rm w}$. If the particle temperatures before applying the boundary condition are all less than $T_{\rm w}$, then they should be bounded above by $T_{\rm w}$ after applying the boundary condition. One way to ensure that the composition bounds are not violated is to use a linear model of the form

$$\frac{\mathrm{d}T^{(n)}}{\mathrm{d}t} = C_{\rm w}\left(T_{\rm w} - T^{(n)}\right), \tag{7.40}$$

where $C_{\rm w}$ is chosen to yield the desired heat transfer rate. Note that $C_{\rm w}$ will vary from cell to cell and will change with time as the flow evolves. Note also that the maximum heat transfer will occur when $C_{\rm w} = \infty$ (i.e., $T^{(n)} = T_{\rm w}$), and that $C_{\rm w} = 0$ corresponds to a zero-flux wall. Models of the form of (7.40) can be formulated for other composition variables, and, in general, the rate constant ($C_{\rm w}$) will be different for each scalar.

7.3.4 Particle-field estimation

In the Lagrangian composition PDF code, the mixing model requires estimated statistics for the compositions. For example, the LIEM model requires an estimate for the mean composition at the particle location:

$$\{\phi^*|\mathbf{X}^*\}\left(\mathbf{X}^{(n)}, t\right).$$

Although not denoted explicitly, we have seen in Section 6.8 that this estimate will depend on the grid spacing M and the number of particles $N_{\rm p}$. In addition to the mean composition, the output data from the PDF code will usually be various composition statistics estimated at grid-cell centers. We will thus need accurate and efficient statistical estimators for determining particle fields given the ensemble of $N_{\rm p}$ notional particles.

One of the simplest estimation techniques is to use a kernel function $h_W(\mathbf{x})$ (see, for example, (6.206), p. 301). However, care must be taken in choosing the form of the kernel function in order to ensure that 'desirable' physical constraints are not violated. For example, with unit-weight particles, the requirement that the mixing model in (7.28) must leave the estimated mean composition unchanged at every point $\mathbf{x}$ leads to

$$\frac{1}{N_{\mathrm{p}}(\mathbf{x})}\sum_{n=1}^{N_{\mathrm{p}}} h_W\left(\mathbf{x}-\mathbf{X}^{(n)}\right)\left[\frac{C_\phi}{2}\omega\left(\mathbf{X}^{(n)},t\right)\left(\{\phi^*|\mathbf{X}^*\}\left(\mathbf{X}^{(n)},t\right)-\phi^{(n)}\right)\right]$$
$$=\frac{C_\phi\omega}{2}\frac{1}{N_{\mathrm{p}}(\mathbf{x})}\sum_{n=1}^{N_{\mathrm{p}}} h_W\left(\mathbf{x}-\mathbf{X}^{(n)}\right)\left(\{\phi^*|\mathbf{X}^*\}\left(\mathbf{X}^{(n)},t\right)-\phi^{(n)}\right)=0, \tag{7.41}$$

where the first equality holds for constant ω.[37] Using (6.213), p. 302, this expression can be rewritten as[38]

$$\{\phi^*|\mathbf{X}^*\}(\mathbf{x},t)=\frac{1}{N_{\mathrm{p}}(\mathbf{x})}\sum_{n=1}^{N_{\mathrm{p}}} h_W\left(\mathbf{x}-\mathbf{X}^{(n)}\right)\{\phi^*|\mathbf{X}^*\}\left(\mathbf{X}^{(n)},t\right), \tag{7.42}$$

which reduces to a condition on the kernel function:

$$h_W\left(\mathbf{x}-\mathbf{X}^{(n)}\right)=\sum_{m=1}^{N_{\mathrm{p}}}\frac{1}{N_{\mathrm{p}}\left(\mathbf{X}^{(m)}\right)}h_W\left(\mathbf{x}-\mathbf{X}^{(m)}\right)h_W\left(\mathbf{X}^{(m)}-\mathbf{X}^{(n)}\right)$$
$$\text{for all } n\in 1,\ldots,N_{\mathrm{p}}. \tag{7.43}$$

In general, very few kernel functions will satisfy this condition.

For example, it is easily shown that the grid-cell kernel

$$h_l(\mathbf{x})=\begin{cases}1 & \text{if } \mathbf{x} \text{ is in grid cell } l\\ 0 & \text{otherwise}\end{cases} \tag{7.44}$$

satisfies (7.43), but the constant kernel, (6.206), does not. More generally, (7.43) will only be satisfied by kernel functions that partition particles into distinct 'spatial bins' like the grid-cell kernel. For this reason, the grid-cell kernel is often used for estimating composition statistics in Lagrangian PDF codes. For example, the estimated mean composition using (7.44) is the *local constant mean estimate* (LCME):[39]

$$\{\phi^*|\mathbf{X}^*\}(\mathbf{x},t)=\frac{\sum_{n=1}^{N_{\mathrm{p}l}} w^{(n)}\phi^{(n)}}{\sum_{n=1}^{N_{\mathrm{p}l}} w^{(n)}} \tag{7.45}$$

for all $\mathbf{x}$ in the lth grid cell.[40] Note that the LCME is piece-wise constant in each grid cell.

37 Usually, ω will not be constant. We make this assumption here to simplify the notation.
38 In the limit of $N_{\mathrm{p}}=\infty$, this relation can be expressed as an integral equation:

$$\phi(\mathbf{x})=\int_V K(\mathbf{x},\mathbf{x}-\mathbf{y})\phi(\mathbf{y})\,\mathrm{d}\mathbf{y}.$$

The kernel $K(\mathbf{x},\mathbf{x}-\mathbf{y})$ must satisfy this constraint for any integrable function ϕ. This is just the definition of the Dirac delta function: $K(\mathbf{x},\mathbf{x}-\mathbf{y})=\delta(\mathbf{x}-\mathbf{y})$. Note that, in this limit, the kernel function is equivalent to the filter function used in LES. As is well known in LES, filtering a function twice leads to different results unless the integral condition given above is satisfied.
39 This estimate is also known as the *cloud-in-cell* (CIC) mean (Hockney and Eastwood 1988).
40 For variable-density flows, the LCME is appropriate for estimating cell-centered particle fields that are passed back to the FV code.

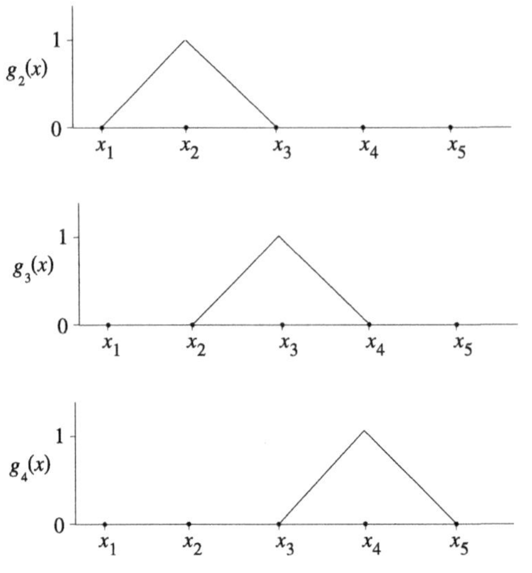

Figure 7.5. Examples of one-dimensional bi-linear basis functions $g_\alpha(x)$.

Alternatively, the mean composition fields can be estimated only at grid-cell centers or grid nodes, and then these 'knot' values can be interpolated to the particle locations (Wouters 1998; Subramaniam and Haworth 2000; Jenny *et al.* 2001). For example, using bi-linear basis functions $g_\alpha(\mathbf{x})$ for each grid node (denoted by α), the estimated mean composition at grid node $\mathbf{x}_\alpha$ is given by (Jenny *et al.* 2001)

$$\{\phi^*|\mathbf{X}^*\}_\alpha \equiv \frac{\sum_{n=1}^{N_\mathrm{p}} g_\alpha\left(\mathbf{X}^{(n)}\right) w^{(n)} \phi^{(n)}}{\sum_{n=1}^{N_\mathrm{p}} g_\alpha\left(\mathbf{X}^{(n)}\right) w^{(n)}}. \tag{7.46}$$

The bi-linear basis functions have the following properties (see Fig. 7.5):

$$\sum_\alpha g_\alpha(\mathbf{x}) = 1 \quad \text{for any } \mathbf{x}, \tag{7.47}$$

where the sum is over all grid nodes, $0 \leq g_\alpha(\mathbf{x}) \leq 1$, and $g_\alpha\left(\mathbf{x}_\beta\right) = \delta_{\alpha\beta}$, where $\delta_{\alpha\beta}$ is the Kronecker delta. The estimated mean composition is then given by the *global linear mean estimate* (GLME):[41]

$$\{\phi^*|\mathbf{X}^*\}(\mathbf{x}, t) = \sum_\alpha g_\alpha(\mathbf{x}) \{\phi^*|\mathbf{X}^*\}_\alpha. \tag{7.48}$$

This two-stage estimation algorithm has been extended to treat the case where $\omega(\mathbf{X}^{(n)}, t)$ (instead of ω constant) is used in the LIEM model (Jenny *et al.* 2001). A more elaborate three-stage estimation algorithm for structured three-dimensional grids is described in Subramaniam and Haworth (2000). As compared to the piece-wise constant estimate found from the LCME, the GLME varies linearly along cell faces. This results in higher spatial accuracy, while at the same time guaranteeing boundedness of the scalar fields.

[41] In one dimension, the cell estimate is piece-wise linear and connects the node-point estimates.

When used in a molecular mixing model, an important property of estimation algorithms is the ability to leave the mean composition unchanged. For example, with the (constant ω) LIEM model, *global* mean conservation requires

$$\sum_{n=1}^{N_p} w^{(n)} \left[\{\phi^* | \mathbf{X}^*\} \left(\mathbf{X}^{(n)}, t \right) - \phi^{(n)} \right] = \mathbf{0}. \tag{7.49}$$

Using the GLME, this expression leads to the identity

$$\sum_{\alpha} \left[g_\alpha (\mathbf{x}) \frac{\sum_{n=1}^{N_p} g_\alpha \left(\mathbf{X}^{(n)} \right)}{\sum_{m=1}^{N_p} g_\alpha \left(\mathbf{X}^{(m)} \right)} \right] = 1 \quad \text{for all } \mathbf{x}, \tag{7.50}$$

and hence the GLME conserves the global means. On the other hand, global estimation algorithms do not satisfy *local* mean conservation (i.e., in each grid cell):

$$\sum_{n=1}^{N_{pl}} w^{(n)} \left[\{\phi^* | \mathbf{X}^*\} \left(\mathbf{X}^{(n)}, t \right) - \phi^{(n)} \right] = \mathbf{0} \quad \text{for the } l\text{th grid cell.} \tag{7.51}$$

With the GLME, application of this condition leads to

$$\sum_{\alpha} \left[g_\alpha (\mathbf{x}) \frac{\sum_{n=1}^{N_{pl}} g_\alpha \left(\mathbf{X}^{(n)} \right)}{\sum_{m=1}^{N_p} g_\alpha \left(\mathbf{X}^{(m)} \right)} \right] < 1, \tag{7.52}$$

instead of (7.50). This problem is again caused by a lack of particle partitioning in global estimation algorithms (i.e., $\{\phi^*\}_\alpha$ depends on particles in more than one grid cell), and will lead to numerical diffusion or 'smoothing out' of the composition fields.[42] The LCME, on the other hand, satisfies both local and global mean conservation.

In order to include spatial variations in the mean composition *inside* each grid cell while enforcing local mean conservation, it is necessary to partition the particles into their respective grid cells (i.e., local mean estimation). For example, Möbus *et al.* (2001) use a local bi-linear least-squares estimate. The principal disadvantage of least squares is that it does not enforce boundedness on the scalar values. Thus, Möbus *et al.* (2001) use an *ad hoc* method to bound the estimated mean compositions based on neighboring cell values, making it difficult to determine whether or not the algorithm satisfies local mean conservation. An alternative local estimation method is to use the bi-linear basis functions, but with the summations restricted to the lth grid cell:

$$\{\phi^* | \mathbf{X}^*\}_{\alpha l} \equiv \frac{\sum_{n=1}^{N_{pl}} g_\alpha \left(\mathbf{X}^{(n)} \right) w^{(n)} \phi^{(n)}}{\sum_{n=1}^{N_{pl}} g_\alpha \left(\mathbf{X}^{(n)} \right) w^{(n)}}. \tag{7.53}$$

The estimated mean composition is then given by the *local linear mean estimate* (LLME):

$$\{\phi^* | \mathbf{X}^*\} (\mathbf{x}, t) = \sum_{\alpha} g_\alpha (\mathbf{x}) \{\phi^* | \mathbf{X}^*\}_{\alpha l} \quad \text{for } \mathbf{x} \text{ in the } l\text{th grid cell.} \tag{7.54}$$

[42] In other words, scalars in some grid cells (i.e., with locally large values) are moved to neighboring cells.

Note that if $l = 1$ and $l = 2$ are two neighboring cells that share grid node α, then $\{\phi^*|\mathbf{X}^*\}_1 \neq \{\phi^*|\mathbf{X}^*\}_2$ at $\mathbf{x}_\alpha$. Thus, like the LCME, the LLME is discontinuous across cell faces.

The lack of continuity at cell faces when using local estimation can be improved (at the cost of higher computing time) by employing higher-order basis functions (Allievi and Bermejo 1997). Because the number of samples ($N_{pl} = 100$–500) is large, second- or third-order basis functions can be employed without introducing significant estimation errors.[43] However, the need for higher-order basis functions is often mitigated by the fact that the particle-field estimates will normally vary only moderately across a grid cell. Indeed, steep composition gradients often coincide with zones of significant mean velocity gradients. In order to resolve the gradients, grid nodes must be spaced sufficiently close to control the discretization error in the FV code. Since the same grid is used in the PDF code, the constant or linear mean estimate will usually offer adequate resolution of the composition gradients.

7.3.5 Other considerations

The Lagrangian PDF code described above is applicable to constant-density, unsteady turbulent reacting flow. As discussed in Section 6.8, the estimated particle fields will be subject to three types of numerical error: statistical, bias, and discretization. The first two can be controlled by increasing the number of particles N_p, while the latter decreases when more grid cells are used. For a particular application, the optimal choice of these parameters will lead to a balance between the various errors. Methods for determining optimal choices are described in Xu and Pope (1999) and Jenny *et al.* (2001). For stationary flow, the statistical error can be effectively reduced by time averaging (see (7.26)). Likewise, in the mixing model, the time-averaged composition means can be used in place of the instantaneous particle-field estimates in order to reduce the bias.[44]

The lack of numerical diffusion in Lagrangian PDF codes for purely convective flow is clearly demonstrated in Möbus *et al.* (2001). These authors also have compared (i) Eulerian PDF, (ii) Lagrangian PDF, (iii) first-order FD, and (iv) second-order FD results for the mean mixture fraction in a convection-diffusion case. The same grid was used for all four codes, and, as expected, the Eulerian PDF and first-order FD results agreed exactly. In contrast, the Lagrangian PDF results were better than the second-order FD predictions on the same grid. Only when the grid for the second-order FD code was increased by a factor of four to obtain a grid-independent solution did it agree with the Lagrangian PDF results. In order to obtain a grid-independent solution with the Eulerian PDF code, the grid had to be increased by a factor of ten. Since the number of particles

[43] In the context of spatial statistics (Cressie 1991), the estimation algorithm in PDF codes is *data dense*. In other words, the number of statistically independent data points is much larger than the number of parameters that must be estimated.

[44] In a variable-density PDF code, passing back the time-averaged particle fields should have an even greater effect on bias. For example, using a Lagrangian velocity, composition PDF code, Jenny *et al.* (2001) have shown that the bias error is inversely proportional to the product $N_p K$.

per grid cell must remain constant in order to obtain the same statistical error, and the time step decreases linearly with the grid-cell size, Möbus *et al.* (2001) concluded that an Eulerian PDF code will be computationally more expensive than a Lagrangian PDF code for the same level of accuracy.

As discussed in Section 7.2, in many flows the local time step Δt_l – determined by applying the CFL condition for the lth grid cell – will vary greatly across the flow domain. In an unsteady simulation, the simulation time step can be no larger than the smallest local time step. Thus, many iterations will be required before the effects of initial conditions disappear. However, for stationary flow, the iterations need not be time-accurate, and local time stepping can be employed (Möbus *et al.* 2001). Muradoglu and Pope (2002) have shown theoretically that if the global time step is Δt and the local time step at point $\mathbf{x}$ is $\Delta t_{\text{loc}}(\mathbf{x})$, it suffices to adjust the particle weights at each time step by

$$w^{(n)}(t + \Delta t) = \frac{\Delta t_{\text{loc}}\left(\mathbf{X}^{(n)}(t + \Delta t)\right)}{\Delta t_{\text{loc}}\left(\mathbf{X}^{n}(t)\right)} w^{(n)}(t), \tag{7.55}$$

and to use $\Delta t_{\text{loc}}\left(\mathbf{X}^{n}(t)\right)$ as the time step to advance the properties of the nth particle. In most cases, the local time step will be the same for all particles in the same grid cell (i.e., Δt_l). Thus, (7.55) is used only when $\mathbf{X}^{n}(t)$ and $\mathbf{X}^{(n)}(t + \Delta t)$ lie in different grid cells. Möbus *et al.* (2001) have reported a dramatic reduction in the overall computational cost by using local time stepping. Likewise, Muradoglu and Pope (2002) report a significant decrease in the number of iterations required to obtain converged stationary solutions with negligible implementation cost. These authors also note that, since the particle mass is equal to $w^{(n)}(t)\Delta m$, the use of (7.55) results in a change in the particle mass. However, they find that this lack of detailed mass conservation does not affect the accuracy of the calculations.

In summary, the advantages and disadvantages of Lagrangian composition PDF codes relative to other methods are as follows:

Advantages

- Relative to Eulerian PDF codes, the spatial-transport algorithm has much higher accuracy. The number of grid cells required for equivalent accuracy is thus considerably smaller.
- As in all transported PDF codes, the numerical algorithm for mixing and chemical reactions is straightforward. If interactions between particles are restricted to the same cell (i.e., local estimation), local mass conservation is guaranteed.
- The total computational cost is proportional to the number of notional particles (N_{p}), and the spatial-transport algorithm is trivial to parallelize.

Disadvantages

- Relative to Eulerian PDF codes, particle tracking and sorting on non-orthogonal grids is computationally intensive, and can represent a large fraction of the total computational cost.

- Since the particles are randomly located in grid cells, interpolation and particle-field-estimation algorithms are required. Special care is needed to ensure local mass conservation (i.e., continuity) and to eliminate bias.
- Relative to velocity, composition PDF codes, the turbulence and scalar transport models have a limited range of applicability. This can be partially overcome by using an LES description of the turbulence. However, consistent closure at the level of second-order RANS models requires the use of a velocity, composition PDF code.
- Local time scales can vary significantly over the computational domain making unsteady simulations expensive. However, local time stepping can be used to reduce drastically convergence for stationary flows.

While some of these disadvantages can be overcome by devising improved algorithms, the problem of level of description of the RANS turbulence model remains as the principal shortcoming of composition PDF code. One thus has the option of resorting to an LES description of the flow combined with a composition PDF code, or a less-expensive second-order RANS model using a velocity, composition PDF code.

Lagrangian composition PDF codes have been used in a growing number of turbulent-reacting-flow studies (e.g., Pope 1981a; Borghi 1986; Borghi 1988; Borghi 1990; Pit 1993; Tsai and Fox 1993; Caillau 1994; Tsai and Fox 1994a; Tsai and Fox 1994b; Tsai and Fox 1995a; Tsai and Fox 1995b; Tsai and Fox 1996a; Tsai and Fox 1996b; Biagioli 1997; Colucci *et al.* 1998; Tsai and Fox 1998; Jaberi *et al.* 1999; Subramaniam and Haworth 2000; Möbus *et al.* 2001; Raman *et al.* 2001; Raman *et al.* 2003). Most of these studies have considered stationary flow on orthogonal grids at the level of a k–ε turbulence model. However, several have considered variable-density flow and block-structured grids, and at least one group has used an LES description of the turbulence. A commercial Lagrangian PDF code is also available, and first results using this code are reported in Masri *et al.* (2003). In order to overcome the limitations of the k–ε and turbulent-diffusivity models (see Sections 4.4 and 4.5), the fluctuating velocity must be included in the transported PDF description (see Section 6.5). Thus, in the next section, we present an overview of Lagrangian velocity, composition PDF codes.

7.4 Velocity, composition PDF codes

As noted in Chapter 1, the inclusion of the fluctuating velocity in the transported PDF description leads to an improved representation of unclosed spatial-transport terms (e.g., $\langle u_i u_j u_k \rangle$ and $\langle u_i u_j \phi \rangle$). The resulting transported PDF models are similar to (although better than) second-order RANS models for the Reynolds stresses and the scalar flux (Pope 1994b). From the point of view of the numerical implementation, one can choose to work either with the velocity $\mathbf{U}$ (Xu and Pope 1999) or the fluctuating velocity $\mathbf{u}$ (Muradoglu *et al.* 1999; Jenny *et al.* 2001). However, the advantages of working with the latter (both in computational efficiency and numerical stability) make it preferable (Muradoglu *et al.*

2001). We will thus restrict our attention to velocity, composition codes written in terms of **u**. The reader should keep in mind that velocity, composition codes based on **u** have only appeared relatively recently in the literature. Advances in the numerical implementation and new applications will most likely be reported after the appearance of this book. Readers interested in applying these methods are thus advised to consult the recent literature.

In general, the treatment of the composition variable in velocity, composition PDF codes is nearly identical to that described in Section 7.3. In particular, (7.28) governs the composition vector and is updated in time as described in (7.36). Thus, in the remainder of this section we will concentrate our attention on the treatment of the Lagrangian fluctuating velocity $\mathbf{u}^*(t)$ and position $\mathbf{X}^*(t)$. In addition, a model for the turbulence frequency (see Section 6.10) is required to close the turbulence model (Pope 2000). In a fully consistent PDF code, an SDE can be solved for $\omega^*(t)$. The numerical algorithm needed to find $\omega^*(t + \Delta t)$ is very similar to (7.33) (Muradoglu *et al.* 2001). Alternatively, a turbulence model can be solved for the dissipation field $\varepsilon(\mathbf{x}, t)$ (e.g., (4.57), p. 118) or the turbulence frequency $\omega(\mathbf{x}, t)$ (e.g., (4.48), p. 116). Although such an approach is 'inconsistent' (i.e., the spatial-transport term in (4.57) is not consistent with the SDE for ω^*), the *ad hoc* nature of the turbulence dissipation model is probably a more serious concern. For simplicity, we shall thus assume that the turbulence frequency field provided by the FV code is adequate. Throughout this section, we will again denote individual grid nodes by the subscript α (e.g., $\mathbf{x}_\alpha$) and individual grid cells by the subscript l. Likewise, the summation over all grid nodes will be denoted by $\sum_\alpha$.

7.4.1 Mean conservation equations

For simplicity, we shall assume that $\langle \mathbf{U}\rangle(\mathbf{x}, t)$ and $\omega(\mathbf{x}, t)$ are available from the FV code for use in the MC simulations. For constant-density flow,[45] the governing equations solved by the FV code are given by

$$\frac{\partial \langle U_i\rangle}{\partial t} + \langle U_j\rangle \frac{\partial \langle U_i\rangle}{\partial x_j} + \frac{1}{\rho}\frac{\partial \langle p\rangle}{\partial x_i} - \nu\nabla^2\langle U_i\rangle = -\frac{\partial \langle u_i u_j\rangle}{\partial x_j} \tag{7.56}$$

and

$$\frac{\partial \omega}{\partial t} + \langle U_i\rangle \frac{\partial \omega}{\partial x_i} - \nu\nabla^2\omega + C_{\omega 2}\omega^2 = \frac{\partial}{\partial x_i}\left(C_\omega \frac{\langle u_i u_j\rangle}{\omega}\frac{\partial \omega}{\partial x_j}\right) - C_{\omega 1}\frac{\omega}{k}\langle u_i u_j\rangle \frac{\partial \langle U_i\rangle}{\partial x_j}, \tag{7.57}$$

where the mean pressure $\langle p\rangle$ is found by invoking continuity: $\nabla \cdot \langle \mathbf{U}\rangle = 0$. All of the terms on the right-hand sides of (7.56) and (7.57) involve unclosed quantities, namely the Reynolds stresses $\langle u_i u_j\rangle$.[46] Thus, the PDF code must provide estimates for the Reynolds-stress tensor at the grid nodes $\{\mathbf{uu}|\mathbf{X}^*\}(\mathbf{x}_\alpha, t)$ for use in the FV code.[47] We will return

[45] For variable-density flow, two additional mean equations are added to solve for the mean density and the energy (Jenny *et al.* 2001; Muradoglu *et al.* 2001).

[46] Recall that k can be found from the diagonal elements of $\langle u_i u_j\rangle$.

[47] In variable-density flow, additional estimated terms are needed for the mean energy equation (Jenny *et al.* 2001).

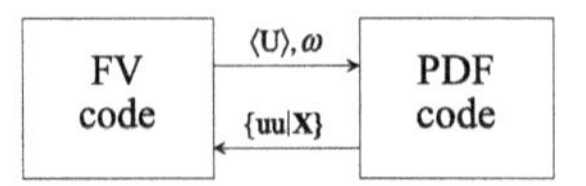

Figure 7.6. Coupling between finite-volume and PDF codes in a velocity, composition PDF simulation.

to this issue below after discussing the notional-particle representation used in the PDF code. In return, the FV code provides interpolated values

$$\langle \mathbf{U} \rangle \left(\mathbf{X}^{(n)}, t\right) \quad \text{and} \quad \omega \left(\mathbf{X}^{(n)}, t\right)$$

at the particle locations. As in Section 7.3, the interpolation algorithm must be carefully implemented to ensure that the particles remain uniformly distributed (Jenny *et al.* 2001; Muradoglu *et al.* 2001). A schematic for the coupling between the two codes is shown in Fig. 7.6. Note that unlike with the Lagrangian composition code in Fig. 7.3, for stationary flow the FV code remains coupled to the PDF code in velocity, composition PDF simulations.

7.4.2 Notional-particle representation

In a velocity, composition PDF code, each notional particle is described by its position $\mathbf{X}^{(n)}(t)$, fluctuating velocity $\mathbf{u}^{(n)}(t)$, composition $\phi^{(n)}(t)$, and weight $w^{(n)}(t)$. The treatment of the particle composition and weight is the same as in Section 7.3, so we shall only concern ourselves here with the position and fluctuating velocity. The governing equations for these variables are

$$\frac{\mathrm{d}\mathbf{X}^{(n)}}{\mathrm{d}t} = \langle \mathbf{U} \rangle \left(\mathbf{X}^{(n)}, t\right) + \mathbf{u}^{(n)} \tag{7.58}$$

and (using the simplified Langevin model – see Section 6.5 and (6.184), p. 296)

$$\begin{aligned} \mathrm{d}\mathbf{u}^{(n)} = {} & \left[\nabla \cdot \{\mathbf{uu}|\mathbf{X}^*\} \left(\mathbf{X}^{(n)}, t\right) - \mathbf{u}^{(n)} \cdot \nabla \langle \mathbf{U} \rangle \left(\mathbf{X}^{(n)}, t\right)\right] \mathrm{d}t \\ & - \left(\frac{1}{2} + \frac{3}{4} C_0\right) \omega \left(\mathbf{X}^{(n)}, t\right) \mathbf{u}^{(n)} \, \mathrm{d}t \\ & + \left[C_0 \{k|\mathbf{X}^*\} \left(\mathbf{X}^{(n)}, t\right) \omega \left(\mathbf{X}^{(n)}, t\right)\right]^{1/2} \mathrm{d}\mathbf{W}(t). \end{aligned} \tag{7.59}$$

In these equations, the mean velocity and turbulence frequency are available from the FV code at (for example) grid-cell centers, and must be interpolated to the particle positions.

As discussed after (6.179), p. 295, the computation of the gradients of estimated particle fields is performed in three steps. As an example, consider the divergence of the estimated Reynolds-stress tensor: $\nabla \cdot \{\mathbf{uu}|\mathbf{X}^*\}(\mathbf{X}^{(n)}, t)$.

(1) First, estimate the particle-field Reynolds-stress tensor at each grid node: $\{\mathbf{uu}|\mathbf{X}^*\}(\mathbf{x}_\alpha, t)$.
(2) Next, use the estimated Reynolds stresses to estimate divergence fields at each grid node: $\nabla \cdot \{\mathbf{uu}|\mathbf{X}^*\}(\mathbf{x}_\alpha, t)$.

(3) Finally, interpolate the estimated divergence fields to each particle location: $\nabla \cdot \{\mathbf{uu}|\mathbf{X}^*\}(\mathbf{X}^{(n)}, t)$.

Each of these steps introduces numerical errors that must be carefully controlled (Jenny *et al.* 2001). In particular, taking spatial gradients in step (2) using noisy estimates from step (1) can lead to numerical instability. We will look at this question more closely when considering particle-field estimation below.

7.4.3 Monte-Carlo simulation

The MC simulation of (7.58) and (7.59) over a time step Δt is done in a two-step process (Jenny *et al.* 2001). In the first half-step the particle positions are advanced to the mid-points:

$$\mathbf{X}^{(n)}_{\Delta t/2} = \mathbf{X}^{(n)}(t) + \left[\langle \mathbf{U} \rangle \left(\mathbf{X}^{(n)}(t), t\right) + \mathbf{u}^{(n)}(t)\right] \frac{\Delta t}{2}, \tag{7.60}$$

and the boundary-condition and sorting algorithms are applied to determine the grid location of $\mathbf{X}^{(n)}_{\Delta t/2}$. The particle fields are then used to estimate the Reynolds-stress tensor, and all fields appearing in the coefficients of (7.59) are interpolated to the particle mid-points.

The first half-step finishes by advancing the fluctuating velocity using the coefficients evaluated at the mid-points:

$$\mathbf{u}^{(n)}(t + \Delta t) = \mathbf{u}^{(n)}(t) + \Delta \mathbf{u}^* + \frac{1}{2} \mathbf{B} \cdot \Delta \mathbf{u}^* \Delta t, \tag{7.61}$$

where

$$\Delta \mathbf{u}^* \equiv \left(\mathbf{a} + \mathbf{B} \cdot \mathbf{u}^{(n)}(t)\right) \Delta t + c \Delta \mathbf{W}, \tag{7.62}$$

and

$$\mathbf{a} \equiv \nabla \cdot \{\mathbf{uu}|\mathbf{X}^*\}\left(\mathbf{X}^{(n)}_{\Delta t/2}, t\right), \tag{7.63}$$

$$\mathbf{B} \equiv -\left[\nabla \langle \mathbf{U} \rangle \left(\mathbf{X}^{(n)}_{\Delta t/2}, t\right)\right]^{\mathrm{T}} - \left(\frac{1}{2} + \frac{3}{4} C_0\right) \omega\left(\mathbf{X}^{(n)}_{\Delta t/2}, t\right) \mathbf{I}, \tag{7.64}$$

$$c \equiv \left[C_0 \{k|\mathbf{X}^*\}\left(\mathbf{X}^{(n)}_{\Delta t/2}, t\right) \omega\left(\mathbf{X}^{(n)}_{\Delta t/2}, t\right)\right]^{1/2}. \tag{7.65}$$

Note that (7.61) is second-order accurate in time. Also, by definition, the estimated mean fluctuating velocity should be null: $\{\mathbf{u}^*|\mathbf{X}^*\} = \mathbf{0}$. This condition will not be automatically satisfied due to numerical errors. Muradoglu *et al.* (2001) propose a simple correction algorithm that consists of subtracting the interpolated value of $\{\mathbf{u}^*|\mathbf{X}^*\}$ (e.g., the LCME found as in (7.45)) from $\mathbf{u}^{(n)}(t + \Delta t)$.

In the second half-step, the particles are moved to their new positions:

$$\mathbf{X}^{(n)}(t + \Delta t) = \mathbf{X}^{(n)}(t) + \left[\langle \mathbf{U} \rangle \left(\mathbf{X}^{(n)}_{\Delta t/2}, t\right) + \frac{1}{2}\left(\mathbf{u}^{(n)}(t) + \mathbf{u}^{(n)}(t + \Delta t)\right)\right] \Delta t. \tag{7.66}$$

The boundary-condition and sorting algorithms are again applied to determine the grid location of each particle. For reflecting boundaries, the component of $\mathbf{u}^{(n)}(t + \Delta t)$ that is normal to the reflection surface must change sign. The particle composition and weights can be updated as described in Section 7.3. The particle fields are then used to estimate the Reynolds-stress tensor at the grid nodes, which is returned to the FV code to start the next time step.

7.4.4 Particle-field estimation and consistency

In constant-density flows,[48] the particle fields must be used to find estimates for the Reynolds stresses $\{\mathbf{uu}|\mathbf{X}^*\}$ and the composition means $\{\phi|\mathbf{X}^*\}$. The latter can be found using any of the methods described earlier (e.g., LCME or LLME). Estimates for the Reynolds-stress tensor are needed at the grid nodes in the FV code. As in (7.46), bi-linear basis functions can be employed to estimate each component at grid node $\mathbf{x}_\alpha$:

$$\{u_i u_j|\mathbf{X}^*\}_\alpha \equiv \frac{\sum_{n=1}^{N_p} g_\alpha\left(\mathbf{X}^{(n)}\right) w^{(n)} u_i^{(n)} u_j^{(n)}}{\sum_{n=1}^{N_p} g_\alpha\left(\mathbf{X}^{(n)}\right) w^{(n)}} \tag{7.67}$$

and

$$\{k|\mathbf{X}^*\}_\alpha = \frac{1}{2}\left(\left\{u_1^2|\mathbf{X}^*\right\}_\alpha + \left\{u_2^2|\mathbf{X}^*\right\}_\alpha + \left\{u_3^2|\mathbf{X}^*\right\}_\alpha\right). \tag{7.68}$$

These values can be returned directly to the FV code, or time-averaged values can be used for stationary flow (Jenny *et al.* 2001).[49] For the PDF code, the interpolated fields are found using the basis functions:

$$\{u_i u_j|\mathbf{X}^*\}(\mathbf{x}, t) = \sum_\alpha g_\alpha(\mathbf{x})\{u_i u_j|\mathbf{X}^*\}_\alpha \tag{7.69}$$

and

$$\{k|\mathbf{X}^*\}(\mathbf{x}, t) = \sum_\alpha g_\alpha(\mathbf{x})\{k|\mathbf{X}^*\}_\alpha, \tag{7.70}$$

evaluated at the particle locations $\mathbf{x} = \mathbf{X}^{(n)}$.

The divergence of the Reynolds stresses, interpolated to the particle locations, is needed in (7.63). The gradients can be approximated at the grid nodes by central differences using (7.67) (Jenny *et al.* 2001). For example, on a rectilinear grid, the gradient in the x_1 direction can be approximated by

$$\left(\frac{\partial\{u_1 u_j|\mathbf{X}^*\}}{\partial x_1}\right)_\alpha = \frac{\{u_1 u_j|\mathbf{X}^*\}_{\alpha+1} - \{u_1 u_j|\mathbf{X}^*\}_{\alpha-1}}{(x_1)_{\alpha+1} - (x_1)_{\alpha-1}}. \tag{7.71}$$

[48] See Jenny *et al.* (2001) for a discussion of particle-field estimation for variable-density flows.

[49] For non-stationary flows, it may be necessary to 'smooth' the estimates before returning them to the FV code in order to improve convergence.

For stationary flows, the time-averaged values should be used in place of $\{u_i u_j|\mathbf{X}^*\}_\alpha$ in the central-difference formula in order to improve the 'smoothness' of the estimated fields. For non-stationary flows, it may be necessary to filter out excess statistical noise in $\{u_1 u_j|\mathbf{X}^*\}_\alpha$ before applying (7.71). In either case, the estimated divergence fields are given by

$$\frac{\partial\{u_i u_j|\mathbf{X}^*\}}{\partial x_i}(\mathbf{x}, t) = \sum_\alpha g_\alpha(\mathbf{x}) \left(\frac{\partial\{u_i u_j|\mathbf{X}^*\}}{\partial x_i}\right)_\alpha, \tag{7.72}$$

where summation over i is implied. The desired values at the particle locations are found by evaluating (7.72) at $\mathbf{x} = \mathbf{X}^{(n)}$.

A velocity, composition PDF code is *consistent* when the fields predicted by the FV code and the PDF code are in agreement (Muradoglu *et al.* 1999; Muradoglu *et al.* 2001). For the constant-density algorithm described above, two independent conditions must be met to ensure consistency.[50]

(1) The location-conditioned velocity fluctuations in each grid cell must be null:

$$\{\mathbf{u}|\mathbf{X}^*\}_l = \mathbf{0}. \tag{7.73}$$

(2) The location-conditioned density in each cell must be equal to the fluid density:

$$\{\rho|\mathbf{X}^*\}(\mathbf{x}_l, t) \equiv \frac{\Delta m}{V_l} \sum_{n=1}^{N_{pl}} w^{(n)}(t) = \rho. \tag{7.74}$$

The first condition may be violated due to statistical error and bias caused by the position-dependent coefficients in (7.61). Likewise, interpolation errors and bias in (7.66) can result in a non-uniform distribution of particle weights. Algorithms for correcting these inconsistencies are discussed in Muradoglu *et al.* (2001). The velocity-correction algorithm simply consists of subtracting $\{\mathbf{u}|\mathbf{X}^*\}_l$ from $\mathbf{u}^{(n)}(t)$ for all particles located in the lth grid cell. On the other hand, the position-correction algorithm is more involved, and is only applicable to stationary flows. For non-stationary flow, a simple 'particle-shifting' algorithm is used by Subramaniam and Haworth (2000).

7.4.5 Other considerations

The algorithms discussed earlier for time averaging and local time stepping apply also to velocity, composition PDF codes. A detailed discussion on the effect of simulation parameters on spatial discretization and bias error can be found in Muradoglu *et al.* (2001). These authors apply a hybrid FV–PDF code for the joint PDF of velocity fluctuations, turbulence frequency, and composition to a piloted-jet flame, and show that the proposed correction algorithms virtually eliminate the bias error in mean quantities. The same code

50 For variable-density flow, Muradoglu *et al.* (2001) identify a third independent consistency condition involving the mean energy equation.

with local time stepping is applied by Muradoglu and Pope (2002) to a non-reacting bluff-body flow (Dally *et al.* 1998). In order to capture the recirculation zone behind the bluff body, a highly stretched grid is required. Even for stretching rates (i.e., the ratio between the largest and smallest grid cells) as high as 160, excellent agreement is obtained using local time stepping. Moreover, the convergence rate to the steady-state solution is shown to be independent of grid stretching so that accurate solutions can be obtained orders of magnitude faster than with uniform time stepping. Similar performance enhancements with local time stepping have been reported by Möbus *et al.* (2001). Muradoglu *et al.* (2001) also find that the mean fields computed with local time stepping are 'smoother' than with uniform time stepping. This has important consequences on the numerical stability of the hybrid algorithm, in particular for the calculation of spatial derivatives (e.g., (7.71)).

In summary, the advantages and disadvantages of Lagrangian velocity, composition PDF codes relative to other methods are as follows.

Advantages

- Relative to Eulerian PDF codes, the spatial-transport algorithm has much higher accuracy. The number of grid cells required for equivalent accuracy is thus considerably smaller.
- Unlike Lagrangian composition codes that use two-equation turbulence models, closure at the level of second-order RANS turbulence models is achieved. In particular, the scalar fluxes are treated in a consistent manner with respect to the turbulence model, and the effect of chemical reactions on the scalar fluxes is treated exactly.
- As in all transported PDF codes, the numerical algorithm for mixing and chemical reactions is straightforward. If interactions between particles are restricted to the same cell (i.e., local estimation), local mass conservation is guaranteed.
- The total computational cost is proportional to the number of notional particles ($N_{\rm p}$), and the spatial-transport algorithm is trivial to parallelize.

Disadvantages

- Relative to Eulerian PDF codes, particle tracking and sorting on non-orthogonal grids is computationally intensive, and can represent a large fraction of the total computational cost.
- Since the particles are randomly located in grid cells, interpolation and particle-field-estimation algorithms are required. Special care is needed to ensure local mass conservation (i.e., continuity) and to eliminate bias.
- Relative to Lagrangian composition PDF codes that use an LES description of the flow, the turbulence models used in velocity, composition PDF codes have a limited range of applicability. However, the computational cost of the latter for reacting flows with detailed chemistry will be considerably lower.
- Local time scales can vary significantly over the computational domain, making unsteady simulations expensive. However, local time stepping can be used to reduce computing time drastically for stationary flows.

Lagrangian velocity, composition PDF codes have been used in a large number of turbulent-reacting-flow studies (e.g., Anand and Pope 1987; Arrojo *et al.* 1988; Anand *et al.* 1989; Masri and Pope 1990; Pope 1990; Pope and Masri 1990; Haworth and El Tahry 1991; Correa and Pope 1992; Norris 1993; Taing *et al.* 1993; Correa *et al.* 1994; Tsai and Fox 1994b; Norris and Pope 1995; Hulek and Lindstedt 1996; Anand *et al.* 1997; Nooren *et al.* 1997; Delarue and Pope 1998; Saxena and Pope 1998; Wouters 1998; Cannon *et al.* 1999; Saxena and Pope 1999; Xu and Pope 2000; Haworth 2001; Jenny *et al.* 2001a; Möbus *et al.* 2001). Most of these studies have considered variable-density flow on orthogonal grids at the level of the SLM turbulence model.

7.5 Concluding remarks

The PDF codes presented in this chapter can be (and have been) extended to include additional random variables. The most obvious extensions are to include the turbulence frequency, the scalar dissipation rate, or velocity acceleration. However, transported PDF methods can also be applied to treat multi-phase flows such as gas–solid turbulent transport. Regardless of the flow under consideration, the numerical issues involved in the accurate treatment of particle convection and coupling with the FV code are essentially identical to those outlined in this chapter. For non-orthogonal grids, the accurate implementation of the particle-convection algorithm is even more critical in determining the success of the PDF simulation.

As with any numerical algorithm, it is important to test transported PDF codes for accuracy using known solutions. Typical test cases for composition PDF codes include

(i) uniform mean velocity both with and without turbulent diffusivity;
(ii) zero mean velocity with a linear turbulent diffusivity (e.g., $\Gamma_T(\mathbf{x}) = \gamma x_1$);
(iii) turbulent scalar mixing layers;
(iv) turbulent round jets.

The same test cases can be used for velocity, composition PDF codes, but with the turbulent kinetic energy and turbulence frequency specified in place of the turbulent diffusivity.

For each test case, a non-reacting scalar (e.g., mixture fraction) should be used to determine the spatial distribution of its mean and variance (i.e., $\langle \xi \rangle$ and $\langle \xi'^2 \rangle$). These results can then be compared with those found by solving the RANS transport equations (i.e., (4.70), p. 120 and (4.90), p. 125) with identical values for $\langle \mathbf{U} \rangle$ and Γ_T. Likewise, the particle-weight distribution should be compared with the theoretical value (i.e., (7.74)). While small fluctuations about the theoretical value are to be expected, a systematic deviation almost always is the result of inconsistencies in the particle-convection algorithm.

Only after the PDF code performs as expected for simple test cases should one attempt to simulate complex flows. Moreover, it is always a good idea to compare the PDF simulation results with grid-independent FV results for a non-reacting scalar before attempting to solve reacting-flow problems. A careful PDF study will also include tests for grid and particle-number effects on the statistics of the velocity and scalar fields. For example, reacting flows near ignition and/or extinction points can be extremely sensitive to the number of particles employed per cell. Thus, in order to obtain reliable predictions, it is imperative to check (by increasing the number of particles) whether the predictions are independent of particle number.

APPENDIX A

Derivation of the SR model

A.1 Scalar spectral transport equation

In the presence of a mean scalar gradient $\nabla\Phi_\alpha$ and a fluctuating (zero-mean) velocity field u_i, the fluctuation field of a passive scalar ϕ_α with molecular diffusion coefficient Γ_α is governed by

$$\frac{\partial \phi_\alpha}{\partial t} + u_i \frac{\partial \phi_\alpha}{\partial x_i} = -u_i \frac{\partial \Phi_\alpha}{\partial x_i} + \Gamma_\alpha \frac{\partial^2 \phi_\alpha}{\partial x_i \partial x_i}. \tag{A.1}$$

The time-evolution equation for the spherically integrated scalar-variance spectrum $E_{\alpha\alpha}(\kappa, t)$ obtained from (A.1) can be written as

$$\frac{\partial}{\partial t} E_{\alpha\alpha}(\kappa, t) = G_{\alpha\alpha}(\kappa, t) + T_{\alpha\alpha}(\kappa, t) - 2\frac{\nu}{\mathrm{Sc}_\alpha}\kappa^2 E_{\alpha\alpha}(\kappa, t), \tag{A.2}$$

where the Schmidt number is defined by

$$\mathrm{Sc}_\alpha \equiv \frac{\nu}{\Gamma_\alpha}. \tag{A.3}$$

$G_{\alpha\alpha}$ is the scalar-variance source term proportional to the uniform mean scalar gradient and the scalar-flux spectrum, and $T_{\alpha\alpha}$ is the scalar-variance transfer spectrum.

Likewise, the time-evolution equation for the scalar-covariance spectrum $E_{\alpha\beta}(\kappa, t)$ can be written as

$$\frac{\partial}{\partial t} E_{\alpha\beta}(\kappa, t) = G_{\alpha\beta}(\kappa, t) + T_{\alpha\beta}(\kappa, t) - 2\frac{\nu}{\mathrm{Sc}_{\alpha\beta}}\kappa^2 E_{\alpha\beta}(\kappa, t), \tag{A.4}$$

where

$$\mathrm{Sc}_{\alpha\beta} \equiv \frac{2\nu}{\Gamma_\alpha + \Gamma_\beta}. \tag{A.5}$$

$G_{\alpha\beta}$ is the corresponding scalar-covariance source term, and $T_{\alpha\beta}$ is the scalar-covariance transfer spectrum. In the following, we will relate the SR model for the scalar variance to (A.2); however, analogous expressions can be derived for the scalar covariance from (A.4) by following the same procedure.

A key assumption in deriving the SR model (as well as earlier spectral models; see Batchelor (1959), Saffman (1963), Kraichnan (1968), and Kraichnan (1974)) is that the transfer spectrum is a linear operator with respect to the scalar spectrum (e.g., a linear convection-diffusion model) which has a characteristic time constant that depends only on the velocity spectrum. The linearity assumption (which is consistent with the linear form of (A.1)) ensures not only that the scalar transfer spectra are conservative, but also that if $\mathrm{Sc}_{\alpha\beta} = \mathrm{Sc}_\gamma$ in (A.4), then $E_{\alpha\beta}(\kappa, t) = E_{\gamma\gamma}(\kappa, t)$ for all t when it is true for $t = 0$. In the SR model, the linearity assumption implies that the forward and backscatter rate constants (defined below) have the same form for both the variance and covariance spectra, and that for the covariance spectrum the rate constants depend on the molecular diffusivities only through $\mathrm{Sc}_{\alpha\beta}$ (i.e., not independently on Sc_α or Sc_β).

Following Yeung (1996) and Yeung (1994), the transfer spectra can be decomposed into contributions from velocity and scalar modes in specified scale ranges. For example, letting $T_{\alpha\alpha}(\kappa|p, q)$ denote the contribution from the velocity mode centered at p and the scalar mode centered at q, the scalar-variance transfer spectrum can be expressed as

$$T_{\alpha\alpha}(\kappa, t) = \int_0^\infty \int_0^\infty T_{\alpha\alpha}(\kappa|p, q)\,\mathrm{d}p\,\mathrm{d}q = \int_0^\infty S_{\alpha\alpha}(\kappa|q)\,\mathrm{d}q. \tag{A.6}$$

The scalar-scalar transfer function $S_{\alpha\alpha}(\kappa|q)$ appearing in the final term on the right-hand side of (A.6) denotes the contribution of scalar mode q to the scalar-variance transfer spectrum at κ. (See Yeung (1994) and Yeung (1996) for examples of these functions extracted from DNS.) Similarly, the scalar-covariance transfer spectrum can be decomposed using $T_{\alpha\beta}(\kappa|p, q)$ and $S_{\alpha\beta}(\kappa|q)$.

The scalar-scalar transfer function can be used to decompose the scalar-variance transfer spectrum into forward-transfer and backscatter contributions:

$$T_{\alpha\alpha}(\kappa, t) = T^{>}_{\alpha\alpha}(\kappa, t) + T^{<}_{\alpha\alpha}(\kappa, t), \tag{A.7}$$

where the forward-transfer contribution is defined by

$$T^{>}_{\alpha\alpha}(\kappa, t) \equiv \int_0^\kappa S_{\alpha\alpha}(\kappa|q)\,\mathrm{d}q, \tag{A.8}$$

and the backscatter contribution is defined by

$$T^{<}_{\alpha\alpha}(\kappa, t) \equiv \int_\kappa^\infty S_{\alpha\alpha}(\kappa|q)\,\mathrm{d}q. \tag{A.9}$$

Consistent with these definitions, DNS data (Yeung 1994; Yeung 1996) show that $T^{>}_{\alpha\alpha}$ is always positive, while $T^{<}_{\alpha\alpha}$ is always negative. Analogous definitions and remarks hold for the scalar-covariance transfer function, i.e., for $T^{>}_{\alpha\beta}$ and $T^{<}_{\alpha\beta}$.

The scalar coherency spectrum is defined in terms of the variance and covariance spectra by

$$\rho_{\alpha\beta}(\kappa, t) \equiv \frac{E_{\alpha\beta}(\kappa, t)}{(E_{\alpha\alpha}(\kappa, t)E_{\beta\beta}(\kappa, t))^{1/2}}. \tag{A.10}$$

The 'banded' coherency spectrum has an analogous definition (Yeung and Pope 1993) where the energy spectra are replaced with the spectra integrated over a finite wavenumber band. From the spectral time-evolution equations, it is easily shown that, in the absence of mean scalar gradients, the time evolution of the coherency spectrum is governed by

$$\frac{1}{\rho_{\alpha\beta}}\frac{\partial \rho_{\alpha\beta}}{\partial t} = \frac{T_{\alpha\beta}}{E_{\alpha\beta}} - \frac{1}{2}\frac{T_{\alpha\alpha}}{E_{\alpha\alpha}} - \frac{1}{2}\frac{T_{\beta\beta}}{E_{\beta\beta}}. \tag{A.11}$$

Thus, as noted by Yeung and Pope (1993), since the molecular diffusivities do not appear on the right-hand side, molecular differential diffusion affects the coherency only indirectly, i.e., through inter-scale transfer processes which propagate incoherency from small scales to large scales. The choice of the model for the scalar transfer spectra thus completely determines the long-time behavior of $\rho_{\alpha\beta}$ in the absence of mean scalar gradients.

A.2 Spectral relaxation model

The SR model can be derived from (A.2) using the cut-off wavenumbers for each wavenumber band. The Kolmogorov wavenumber is defined by

$$\kappa_\eta \equiv \frac{1}{\eta} = \left(\frac{\varepsilon}{\nu^3}\right)^{1/4}, \tag{A.12}$$

and the turbulent Reynolds number is defined by

$$\mathrm{Re}_1 \equiv \frac{k}{(\nu\varepsilon)^{1/2}}. \tag{A.13}$$

For $1 < \mathrm{Sc}_\alpha$, the cut-off wavenumbers[1] are defined by

$$\kappa_0 \equiv 0, \tag{A.14}$$

$$\kappa_j \equiv \left(\frac{C_u 3^{j-1}}{C_u \mathrm{Re}_1 + 3^{j-1} - 1}\right)^{3/2} \kappa_\eta \qquad \text{for } j = 1, \ldots, n_u, \tag{A.15}$$

and

$$\kappa_j \equiv \mathrm{Sc}_\alpha^{(j-n_u-1)/(2n-2n_u-2)} C_u^{3/2} \kappa_\eta \qquad \text{for } j = n_u + 1, \ldots, n, \tag{A.16}$$

where $(n_u - 1)$ is the number of wavenumber bands needed to describe the energy-containing and inertial-convective sub-ranges, and $(n + 1)$ is the total number of wavenumber bands (including the scalar dissipation range). Note that n_u will depend on Re_1, while $(n - n_u)$ will depend only on Sc (Fox 1995). By definition, $\kappa_n = \kappa_D$.

[1] For $\mathrm{Sc}_\alpha \leq 1$, the cut-off wavenumbers are given in Section 4.6. For the covariance spectrum, the same cut-off wavenumbers are used with $\mathrm{Sc}_{\alpha\beta}$ in place of Sc_α.

The scalar energy in each wavenumber band is found by integration:

$$\langle\phi_\alpha^2\rangle_j(t) \equiv \int_{\kappa_{j-1}}^{\kappa_j} E_{\alpha\alpha}(\kappa, t)\,\mathrm{d}\kappa \qquad \text{for } j = 1, \dots, n, \tag{A.17}$$

and

$$\langle\phi_\alpha^2\rangle_{\mathrm{D}}(t) \equiv \int_{\kappa_{\mathrm{D}}}^{\infty} E_{\alpha\alpha}(\kappa, t)\,\mathrm{d}\kappa. \tag{A.18}$$

Integrating (A.2) over each wavenumber band yields the exact time-evolution equation for $\langle\phi_\alpha^2\rangle_j$:

$$\frac{\mathrm{d}\langle\phi_\alpha^2\rangle_j}{\mathrm{d}t} = \gamma_j \mathcal{P}_\alpha + \mathcal{T}_j - \varepsilon_{\alpha j} \qquad \text{for } j = 1, \dots, n, \tag{A.19}$$

and

$$\frac{\mathrm{d}\langle\phi_\alpha^2\rangle_{\mathrm{D}}}{\mathrm{d}t} = \gamma_{\mathrm{D}} \mathcal{P}_\alpha + \mathcal{T}_{\mathrm{D}} - \varepsilon_{\alpha\mathrm{D}}. \tag{A.20}$$

The right-hand sides of these expressions are defined in terms of $G_{\alpha\alpha}$, $S_{\alpha\alpha}(\kappa|q)$, and $\Gamma_\alpha\kappa^2 E_{\alpha\alpha}$ as discussed below.

The scalar-variance source term in (A.19) is defined by

$$\mathcal{P}_\alpha \equiv \int_0^\infty G_{\alpha\alpha}(\kappa, t)\,\mathrm{d}\kappa, \tag{A.21}$$

and the fraction of the scalar-variance source term falling in a particular wavenumber band is defined by

$$\gamma_j \equiv \frac{1}{\mathcal{P}_\alpha} \int_{\kappa_{j-1}}^{\kappa_j} G_{\alpha\alpha}(\kappa, t)\,\mathrm{d}\kappa \qquad \text{for } j = 1, \dots, n, \tag{A.22}$$

and

$$\gamma_{\mathrm{D}} \equiv \frac{1}{\mathcal{P}_\alpha} \int_{\kappa_{\mathrm{D}}}^{\infty} G_{\alpha\alpha}(\kappa, t)\,\mathrm{d}\kappa. \tag{A.23}$$

In the present version of the SR model, the fractions γ_j and γ_{D} are assumed to be time-independent functions of Re_1 and Sc. Likewise, the scalar-variance source term $\mathcal{P}_\alpha$ is closed with a gradient-diffusion model. The SR model could thus be further refined (with increased computational expense) by including an explicit model for the scalar-flux spectrum.

The scalar dissipation term in (A.20) is defined by

$$\varepsilon_{\alpha\mathrm{D}} \equiv \int_{\kappa_{\mathrm{D}}}^{\infty} 2\Gamma_\alpha\kappa^2 E_{\alpha\alpha}(\kappa, t)\,\mathrm{d}\kappa. \tag{A.24}$$

In the SR model, κ_{D} is chosen such that $\varepsilon_{\alpha\mathrm{D}} \approx \varepsilon_\alpha$, i.e., so that the bulk of the scalar dissipation occurs in the final wavenumber band. Thus, the scalar dissipation terms appearing in other wavenumber bands are assumed to be negligible:

$$\varepsilon_{\alpha j} \equiv \int_{\kappa_{j-1}}^{\kappa_j} 2\Gamma_\alpha \kappa^2 E_{\alpha\alpha}(\kappa, t)\,\mathrm{d}\kappa \approx 0 \qquad \text{for } j = 1, \dots, n. \tag{A.25}$$

Note that, unlike with the continuous representation (A.2), the discrete representation used in the SR model requires an explicit model for the scalar dissipation rate ε_α.

The spectral transport term for the first wavenumber band is defined by

$$\begin{aligned} \mathcal{T}_1(t) &\equiv \int_0^{\kappa_1} T^{>}_{\alpha\alpha}(\kappa, t)\,\mathrm{d}\kappa + \int_0^{\kappa_1} T^{<}_{\alpha\alpha}(\kappa, t)\,\mathrm{d}\kappa \\ &= -\int_{\kappa_1}^{\infty} T^{>}_{\alpha\alpha}(\kappa, t)\,\mathrm{d}\kappa - \int_{\kappa_1}^{\infty} T^{<}_{\alpha\alpha}(\kappa, t)\,\mathrm{d}\kappa; \end{aligned} \tag{A.26}$$

for wavenumber bands $j = 2, \dots, n$ it is defined by

$$\begin{aligned} \mathcal{T}_j(t) &\equiv \int_{\kappa_{j-1}}^{\kappa_j} T^{>}_{\alpha\alpha}(\kappa, t)\,\mathrm{d}\kappa + \int_{\kappa_{j-1}}^{\kappa_j} T^{<}_{\alpha\alpha}(\kappa, t)\,\mathrm{d}\kappa, \\ &= \int_{\kappa_{j-1}}^{\infty} T^{>}_{\alpha\alpha}(\kappa, t)\,\mathrm{d}\kappa - \int_{\kappa_j}^{\infty} T^{>}_{\alpha\alpha}(\kappa, t)\,\mathrm{d}\kappa \\ &\quad + \int_{\kappa_{j-1}}^{\infty} T^{<}_{\alpha\alpha}(\kappa, t)\,\mathrm{d}\kappa - \int_{\kappa_j}^{\infty} T^{<}_{\alpha\alpha}(\kappa, t)\,\mathrm{d}\kappa; \end{aligned} \tag{A.27}$$

and for the dissipation range it is defined by

$$\mathcal{T}_{\mathrm{D}}(t) \equiv \int_{\kappa_{\mathrm{D}}}^{\infty} T^{>}_{\alpha\alpha}(\kappa, t)\,\mathrm{d}\kappa + \int_{\kappa_{\mathrm{D}}}^{\infty} T^{<}_{\alpha\alpha}(\kappa, t)\,\mathrm{d}\kappa. \tag{A.28}$$

In the SR model, forward and backscatter rate constants are employed to model the spectral transport terms:

$$\mathcal{T}_1(t) = -\alpha^*_{12}\langle\phi_\alpha^2\rangle_1 + \beta_{21}\langle\phi_\alpha^2\rangle_2, \tag{A.29}$$

$$\begin{aligned} \mathcal{T}_j(t) &= \alpha^*_{(j-1)j}\langle\phi_\alpha^2\rangle_{j-1} - \left(\alpha^*_{j(j+1)} + \beta_{j(j-1)}\right)\langle\phi_\alpha^2\rangle_j \\ &\quad + \beta_{(j+1)j}\langle\phi_\alpha^2\rangle_{j+1} \quad \text{for } j = 2, \dots, n-1, \end{aligned} \tag{A.30}$$

$$\mathcal{T}_n(t) = \alpha^*_{(n-1)n}\langle\phi_\alpha^2\rangle_{n-1} - \left(\alpha^*_{n\mathrm{D}} + \beta_{n(n-1)}\right)\langle\phi_\alpha^2\rangle_n + \beta_{\mathrm{D}n}\langle\phi_\alpha^2\rangle_{\mathrm{D}}, \tag{A.31}$$

$$\mathcal{T}_{\mathrm{D}}(t) = \alpha^*_{n\mathrm{D}}\langle\phi_\alpha^2\rangle_n - \beta_{\mathrm{D}n}\langle\phi_\alpha^2\rangle_{\mathrm{D}}, \tag{A.32}$$

where $\alpha^*_{j(j+1)} \equiv \alpha_{j(j+1)} + \beta_{j(j+1)}$.

The forward and backscatter rate constants can be expressed explicitly in terms of the scalar-variance transfer spectrum decomposed into forward and backscatter contributions:

$$\alpha^*_{(j-1)j} \equiv \frac{1}{\langle\phi_\alpha^2\rangle_{j-1}} \int_{\kappa_{j-1}}^{\infty} T^{>}_{\alpha\alpha}(\kappa, t)\,\mathrm{d}\kappa \quad \text{for } j = 2, \dots, n, \tag{A.33}$$

$$\alpha^*_{n\mathrm{D}} \equiv \frac{1}{\langle\phi_\alpha^2\rangle_n} \int_{\kappa_{\mathrm{D}}}^{\infty} T^{>}_{\alpha\alpha}(\kappa, t)\,\mathrm{d}\kappa \tag{A.34}$$

and

$$\beta_{j(j-1)} \equiv -\frac{1}{\langle \phi_\alpha^2 \rangle_j} \int_{\kappa_{j-1}}^{\infty} T_{\alpha\alpha}^{<}(\kappa, t)\, \mathrm{d}\kappa \quad \text{for } j = 2, \dots, n, \tag{A.35}$$

$$\beta_{\mathrm{D}n} \equiv -\frac{1}{\langle \phi_\alpha^2 \rangle_{\mathrm{D}}} \int_{\kappa_{\mathrm{D}}}^{\infty} T_{\alpha\alpha}^{<}(\kappa, t)\, \mathrm{d}\kappa. \tag{A.36}$$

Note that the right-hand sides of these expressions can be extracted from DNS data for homogeneous turbulence in order to explore the dependence of the rate constants on Re_1 and Sc. Results from a preliminary investigation (Fox and Yeung 1999) for $R_\lambda = 90$ have revealed that the backscatter rate constant from the dissipative range has a Schmidt-number dependence like $\beta_{\mathrm{D}n} \sim \mathrm{Sc}^{1/2}$ for Schmidt numbers in the range $[1/8, 1]$. On the other hand, for cut-off wavenumbers in the inertial-convective sub-range, one would expect $\alpha_{j(j+1)}$ and $\beta_{(j+1)j}$ for $j = 1, \dots, n-1$ to be independent of Sc. This is the assumption employed in the SR model, but it can be validated (and modified) using DNS data for the scalar spectrum and the scalar-scalar transfer function. The linearity assumption discussed earlier implies that the rate constants will be unchanged (for the same Reynolds and Schmidt numbers) when they are computed using the scalar-covariance transfer spectrum.

Note that at spectral equilibrium the integral in (A.33) will be constant and proportional to ε_α (i.e., the scalar spectral energy transfer rate in the inertial-convective sub-range will be constant). The forward rate constants α_j^* will thus depend on the chosen cut-off wavenumbers through their effect on $\langle \phi_\alpha^2 \rangle_j$. In the SR model, in order to obtain the most computationally efficient spectral model possible, the total number of wavenumber bands is minimized subject to the condition that

$$R_0 \le \frac{k}{\varepsilon}\alpha_{12}^* \le \frac{k}{\varepsilon}\alpha_{23}^* \le \cdots \le \frac{k}{\varepsilon}\alpha_{n\mathrm{D}}^*, \tag{A.37}$$

where R_0 is the steady-state mechanical-to-scalar time-scale ratio in the absence of a mean scalar gradient ($\mathcal{P}_\alpha = 0$). Assuming a fully developed velocity spectrum, this process results (Fox 1995; Fox 1997) in the cut-off wavenumbers given in (A.14)–(A.16). An obvious extension of the model would thus be to include a dynamical model for the velocity spectrum (Besnard *et al.* 1990; Besnard *et al.* 1992; Clark and Zemach 1995; Canuto and Dubovikov 1996a; Canuto and Dubovikov 1996b; Canuto *et al.* 1996). However, including a more detailed model for the turbulence/scalar-flux spectra would most likely make the extension to a Lagrangian PDF formulation computationally intractable for practical reacting-flow calculations.

A.3 Scalar dissipation rate

The final expression needed to complete the SR model is a closure for $\varepsilon_{\alpha\mathrm{D}}$ appearing in (A.20). In order to develop a model starting from (A.2), we will first introduce the scalar dissipation spectrum defined by

$$D_{\alpha\alpha} \equiv 2\Gamma_\alpha \kappa^2 E_{\alpha\alpha}, \tag{A.38}$$

the scalar spectral energy transfer rate $\mathcal{T}_{\alpha\alpha}$ defined by

$$\mathcal{T}_{\alpha\alpha}(\kappa, t) \equiv \int_\kappa^\infty T_{\alpha\alpha}(s, t)\, \mathrm{d}s, \tag{A.39}$$

and a characteristic spectral time scale $\tau_{\rm st}$ defined by

$$\tau_{\rm st}(\kappa, t) \equiv \frac{\kappa E_{\alpha\alpha}(\kappa, t)}{\mathcal{T}_{\alpha\alpha}(\kappa, t)}. \tag{A.40}$$

Note that in the viscous sub-range (Batchelor 1959; Chasnov 1998), $\tau_{\rm st}$ is proportional to $(\nu/\varepsilon)^{1/2}$.

Multiplication of (A.2) by $2\Gamma_\alpha \kappa^2$, and integration of the resultant expression over the wavenumber range $[\kappa_{\rm D}, \infty)$ using (A.39) and (A.40), yields

$$\frac{\mathrm{d}\varepsilon_{\alpha\rm D}}{\mathrm{d}t} = \int_{\kappa_{\rm D}}^\infty 2\Gamma_\alpha \kappa^2 G_{\alpha\alpha}(\kappa, t)\, \mathrm{d}\kappa + 2\Gamma_\alpha \kappa_{\rm D}^2 \mathcal{T}_{\alpha\alpha}(\kappa_{\rm D}, t) + 2 \int_{\kappa_{\rm D}}^\infty \frac{D_{\alpha\alpha}(\kappa, t)}{\tau_{\rm st}(\kappa, t)}\, \mathrm{d}\kappa - \mathcal{D}_{\alpha\alpha}, \tag{A.41}$$

where the dissipation term is defined by

$$\mathcal{D}_{\alpha\alpha} \equiv \int_{\kappa_{\rm D}}^\infty 2\Gamma_\alpha \kappa^2 \mathcal{D}_{\alpha\alpha}(\kappa, t)\, \mathrm{d}\kappa. \tag{A.42}$$

The right-hand side of (A.41) contains unclosed terms that must be modeled in order to arrive at the transport equation for the scalar dissipation rate. This process is explained next.

The first term on the right-hand side of (A.41) is a source term due to the mean scalar gradient. The following model is employed:

$$\frac{\varepsilon_{\alpha\rm D}}{\langle \phi_\alpha^2 \rangle_{\rm D}} = \frac{\int_{\kappa_{\rm D}}^\infty 2\Gamma_\alpha \kappa^2 G_{\alpha\alpha}(\kappa, t)\, \mathrm{d}\kappa}{\int_{\kappa_{\rm D}}^\infty G_{\alpha\alpha}(\kappa, t)\, \mathrm{d}\kappa}. \tag{A.43}$$

Note that this model is exact if $G_{\alpha\alpha} \propto E_{\alpha\alpha}$ in the scalar dissipation range.

The second term on the right-hand side of (A.41) represents the flux of scalar dissipation into the scalar dissipation range, and can be rewritten in terms of known quantities. From (A.39), it can be seen that $\mathcal{T}_{\alpha\alpha}(\kappa_{\rm D}, t) = \mathcal{T}_{\rm D}(t)$. Likewise, using the definition of $\kappa_{\rm D}$, it follows that $2\Gamma_\alpha \kappa_{\rm D}^2 = C_{\rm D}(\varepsilon/\nu)^{1/2}$. The scalar-dissipation flux term can thus be expressed as

$$C_{\rm D} \left(\frac{\varepsilon}{\nu}\right)^{1/2} \mathcal{T}_{\rm D}(t) = C_{\rm D} \left(\frac{\varepsilon}{\nu}\right)^{1/2} \alpha_{n\rm D}^* \langle \phi_\alpha^2 \rangle_n - C_{\rm D} \left(\frac{\varepsilon}{\nu}\right)^{1/2} \beta_{{\rm D}n} \langle \phi_\alpha^2 \rangle_{\rm D}. \tag{A.44}$$

However, to ensure that the model is numerically stable,[2] the last term on the right-hand side of this expression is replaced with its spectral equilibrium value:

$$C_{\rm D}\left(\frac{\varepsilon}{\nu}\right)^{1/2}\mathcal{T}_{\rm D}(t) = C_{\rm D}\left(\frac{\varepsilon}{\nu}\right)^{1/2}\alpha^{*}_{n{\rm D}}\langle\phi^2_\alpha\rangle_n - C_{\rm D}{\rm Re}_1\beta_{{\rm D}n}\gamma_{\rm D}\varepsilon_{\alpha{\rm D}}/R_0, \tag{A.45}$$

where the steady-state fraction of scalar energy in the dissipation range is

$$\gamma_{\rm D} = \lim_{t\to\infty}\frac{\langle\phi^2_\alpha\rangle_{\rm D}(t)}{\langle\phi^2_\alpha\rangle(t)} \tag{A.46}$$

for $\text{Sc} = 1$ in the absence of a mean scalar gradient.

The next term in (A.41) is modeled by assuming that $\tau_{\rm st}(\kappa, t) \sim (\nu/\varepsilon)^{1/2}$ for all wavenumbers in the range $[\kappa_{\rm D}, \infty)$. The validity of this assumption depends on the Schmidt number, but it is strictly valid for $1 \leq \text{Sc}$. Using this assumption, we can extract $\tau_{\rm st}$ from the integral in the second term on the right-hand side of (A.41) so that the entire term becomes $C_{\rm s}(\varepsilon/\nu)^{1/2}\varepsilon_{\alpha{\rm D}}$. The proportionality constant can be split into two contributions, $C_{\rm s} = C_{\rm B} - C_{\rm D}$, where

$$C_{\rm B} = \frac{1}{\varepsilon_{\alpha{\rm D}}}\left(\frac{\nu}{\varepsilon}\right)^{1/2}\int_{\kappa_{\rm D}}^{\infty}\Gamma_\alpha\kappa^2 T_{\alpha\alpha}(\kappa, t)\,{\rm d}\kappa. \tag{A.47}$$

In the SR model, $C_{\rm s} = 1$, which can be validated using DNS data. Note that at spectral equilibrium $\mathcal{T}_{\rm D} \sim \varepsilon_\alpha$ so that the first two terms on the right-hand side of (A.41) function as a linear source term of the form $C_{\rm B}(\varepsilon/\nu)^{1/2}\varepsilon_{\alpha{\rm D}}$.

The final term in (A.41) ($\mathcal{D}_{\alpha\alpha}$) is modeled by the product of the inverse of a characteristic time scale for the scalar dissipation range and $\varepsilon_{\alpha{\rm D}}$. The characteristic time scale is taken to be proportional to $\langle\phi^2_\alpha\rangle_{\rm D}/\varepsilon_{\alpha{\rm D}}$ so that the final term has the form $C_{\rm d}\varepsilon^2_{\alpha{\rm D}}/\langle\phi^2_\alpha\rangle_{\rm D}$. The proportionality constant $C_{\rm d}$ is thus defined by

$$C_{\rm d} = \frac{\langle\phi^2_\alpha\rangle_{\rm D}}{\varepsilon^2_{\alpha{\rm D}}}\int_{\kappa_{\rm D}}^{\infty}4\Gamma^2_\alpha\kappa^4 E_{\alpha\alpha}(\kappa, t)\,{\rm d}\kappa. \tag{A.48}$$

In the SR model, $C_{\rm d} = 3$ is chosen to agree with passive scalar decay from isotropic initial conditions in the absence of turbulent mixing (i.e., pure diffusion).

The final form of transport equation for the scalar dissipation rate is

$$\begin{aligned}\frac{{\rm d}\varepsilon_{\alpha{\rm D}}}{{\rm d}t} &= \frac{\varepsilon_{\alpha{\rm D}}}{\langle\phi^2_\alpha\rangle_{\rm D}}\gamma_{\rm D}\mathcal{P}_\alpha + C_{\rm D}\left(\frac{\varepsilon}{\nu}\right)^{1/2}\alpha_{n{\rm D}}\langle\phi^2_\alpha\rangle_n - C_{\rm D}{\rm Re}_1\beta^{*}_{{\rm D}n}\varepsilon_{\alpha{\rm D}}\\ &\quad + C_{\rm s}\left(\frac{\varepsilon}{\nu}\right)^{1/2}\varepsilon_{\alpha{\rm D}} - C_{\rm d}\frac{\varepsilon_{\alpha{\rm D}}}{\langle\phi^2_\alpha\rangle_{\rm D}}\varepsilon_{\alpha{\rm D}}.\end{aligned} \tag{A.49}$$

Using the fact that $\kappa_{\rm D}$ was chosen so that $\varepsilon_{\alpha{\rm D}} \approx \varepsilon_\alpha$, we see that this expression is the same as (4.117) on p. 131.

[2] For example, when $\varepsilon_{\alpha{\rm D}} = 0$ the backscatter term should be null. This may not be the case when $\langle\phi^2_\alpha\rangle_{\rm D}$ is used in place of $\varepsilon_{\alpha{\rm D}}$.

Starting from (A.4), analogous arguments/assumptions as those leading to the model for ε_{α} can be used to derive the SR model equation for $\varepsilon_{\alpha\beta}$. Note that DNS data for passive scalar mixing in homogeneous turbulence can be employed to validate all SR model constants, and to explore possible dependencies on Re_1 and Sc (e.g., due to low-Reynolds-number effects). For example, Vedula (2001) and Vedula *et al.* (2001) have shown using DNS that (A.49) holds for all values of Re and Sc investigated, and have proposed functional forms for $C_{\text{D}}(\varepsilon)$ and $C_{\text{s}}(\varepsilon)$.

APPENDIX B

Direct quadrature method of moments

B.1 Quadrature method of moments

The quadrature method of moments (QMOM) is a presumed PDF approach that determines the unknown parameters by forcing the lower-order moments of the presumed PDF to agree with the moment transport equations (McGraw 1997; Barrett and Webb 1998; Marchisio *et al.* 2003a; Marchisio *et al.* 2003b). As with the multi-environment presumed PDF method discussed in Section 5.10, the form of the presumed PDF is

$$f_{\phi}(\psi; \mathbf{x}, t) = \sum_{n=1}^{N_e} p_n(\mathbf{x}, t) \prod_{\alpha=1}^{N_s} \delta\left[\psi_\alpha - \langle\phi_\alpha\rangle_n(\mathbf{x}, t)\right], \tag{B.1}$$

where $N_s = K + 1$ is the number of scalars in the composition vector ϕ. However, unlike with the multi-environment model, the probabilities p_n and conditional means $\langle\phi_\alpha\rangle_n$ are found by forcing (B.1) to yield known values for the moments:

$$\left\langle \phi_1^{m_1} \phi_2^{m_2} \cdots \phi_{N_s}^{m_{N_s}} \right\rangle = \sum_{n=1}^{N_e} p_n \prod_{\alpha=1}^{N_s} \langle\phi_\alpha\rangle_n^{m_\alpha}, \tag{B.2}$$

where the m_i are typically chosen to be non-negative integers.

As an example, consider a uni-variate case ($N_s = 1$) where

$$f_{\phi}(\psi; \mathbf{x}, t) = \sum_{n=1}^{N_e} p_n(\mathbf{x}, t) \delta\left[\psi - \langle\phi\rangle_n(\mathbf{x}, t)\right]. \tag{B.3}$$

The right-hand side of this equation involves $2N_e$ unknowns: $(p_1, \ldots, p_{N_e})$ and $(\langle\phi\rangle_1, \ldots, \langle\phi\rangle_{N_e})$. In the QMOM approach, these unknowns can be found from a unique set of $2N_e$ moments. For example, integer moments with $m = 0, 1, \ldots, 2N_e - 1$ can be employed:

$$
\begin{aligned}
1 &= \sum_{n-1}^{N_e} p_n, \\
\langle \phi \rangle &= \sum_{n=1}^{N_e} p_n \langle \phi \rangle_n, \\
\langle \phi^2 \rangle &= \sum_{n=1}^{N_e} p_n \langle \phi \rangle_n^2, \\
&\vdots \\
\langle \phi^{2N_e-1} \rangle &= \sum_{n=1}^{N_e} p_n \langle \phi \rangle_n^{2N_e-1}.
\end{aligned} \tag{B.4}
$$

The modeling problem then reduces to finding $(p_1, \ldots, p_{N_e})$ and $(\langle \phi \rangle_1, \ldots, \langle \phi \rangle_{N_e})$ assuming the left-hand sides of (B.4) are known (or can be found by solving transport equations as discussed below). For the uni-variate case, an efficient product-difference algorithm exists to determine uniquely the unknown probabilities and conditional means (Gordon 1968; McGraw 1997). Unfortunately, this algorithm does not extend to multi-variate cases (Wright *et al.* 2001). However, a direct method can be derived to generate transport equations for $p_n(\mathbf{x}, t)$ and $\langle \phi_\alpha \rangle_n(\mathbf{x}, t)$. This method is discussed in detail below.

B.2 Direct QMOM

The direct quadrature method of moments (DQMOM) begins with a *closed*[1] joint composition PDF transport equation (see Section 6.3). For simplicity, we will consider the high-Reynolds-number form of (6.30) on p. 251 with the IEM mixing model:

$$
\frac{\partial f_\phi}{\partial t} + \langle U_i \rangle \frac{\partial f_\phi}{\partial x_i} - \frac{\partial}{\partial x_i}\left(\Gamma_{\mathrm{T}} \frac{\partial f_\phi}{\partial x_i}\right) = -\frac{\partial}{\partial \psi_i}\left[\left(C_\phi \frac{\varepsilon}{k}\left(\langle \phi_i \rangle - \psi_i\right) + S_i(\psi)\right) f_\phi\right]. \tag{B.5}
$$

However, the application of DQMOM does not depend on the form of the mixing model. Indeed, the terms on the right-hand side of (B.5) (which describe the flux in composition space) can have almost any form. The only essential requirement is that they result in well defined moments for the composition PDF. The terms on the left-hand side of (B.5) will determine the form of the transport equations for p_n and $\langle \mathbf{s} \rangle_n = p_n \langle \phi \rangle_n$. For simplicity, we will consider the uni-variate case first.

[1] DQMOM does not provide a model for the unclosed terms. It is thus a *computational method* for approximating an Eulerian joint PDF given its closed transport equation. In this sense, DQMOM is an alternative to the Eulerian Monte-Carlo simulations introduced in Chapter 7.

B.2.1 Uni-variate case

The composition PDF transport equation for the uni-variate case has the following form:

$$\frac{\partial f_\phi}{\partial t} + \langle U_i \rangle \frac{\partial f_\phi}{\partial x_i} - \frac{\partial}{\partial x_i}\left(\Gamma_{\mathrm{T}} \frac{\partial f_\phi}{\partial x_i} \right) = R(\psi; \mathbf{x}, t), \tag{B.6}$$

where the right-hand side is given by

$$R(\psi; \mathbf{x}, t) = -\frac{\partial}{\partial \psi}\left[\left(C_\phi \frac{\varepsilon}{k} (\langle \phi \rangle - \psi) + S(\psi) \right) f_\phi \right]. \tag{B.7}$$

We begin the derivation by formally inserting (B.3) for f_ϕ into (B.6). The left-hand side of (B.6) can then be written as

$$\begin{aligned}
&\sum_{n=1}^{N_{\mathrm{e}}} \delta(\psi - \langle \phi \rangle_n) \left[\frac{\partial p_n}{\partial t} + \langle U_i \rangle \frac{\partial p_n}{\partial x_i} - \frac{\partial}{\partial x_i}\left(\Gamma_{\mathrm{T}} \frac{\partial p_n}{\partial x_i} \right) \right] \\
&\quad - \sum_{n=1}^{N_{\mathrm{e}}} \delta^{(1)}(\psi - \langle \phi \rangle_n) \left\{ \frac{\partial \langle s \rangle_n}{\partial t} + \langle U_i \rangle \frac{\partial \langle s \rangle_n}{\partial x_i} - \frac{\partial}{\partial x_i}\left(\Gamma_{\mathrm{T}} \frac{\partial \langle s \rangle_n}{\partial x_i} \right) \right. \\
&\quad \left. - \langle \phi \rangle_n \left[\frac{\partial p_n}{\partial t} + \langle U_i \rangle \frac{\partial p_n}{\partial x_i} - \frac{\partial}{\partial x_i}\left(\Gamma_{\mathrm{T}} \frac{\partial p_n}{\partial x_i} \right) \right] \right\} \\
&\quad - \sum_{n=1}^{N_{\mathrm{e}}} \delta^{(2)}(\psi - \langle \phi \rangle_n) p_n \Gamma_{\mathrm{T}} \frac{\partial \langle \phi \rangle_n}{\partial x_i} \frac{\partial \langle \phi \rangle_n}{\partial x_i} = R(\psi; \mathbf{x}, t),
\end{aligned} \tag{B.8}$$

where $\langle s \rangle_n = p_n \langle \phi \rangle_n$, and $\delta^{(m)}(x)$ denotes the mth derivative of $\delta(x)$. The derivatives of the delta function are defined such that (Pope 2000)

$$\int_{-\infty}^{+\infty} \delta^{(m)}(x - s) g(x) \, \mathrm{d}x = (-1)^m g^{(m)}(s), \tag{B.9}$$

where $g^{(m)}(x)$ is the mth derivative of $g(x)$.

Defining the transport equations for p_n and $\langle s \rangle_n$ by

$$\frac{\partial p_n}{\partial t} + \langle U_i \rangle \frac{\partial p_n}{\partial x_i} - \frac{\partial}{\partial x_i}\left(\Gamma_{\mathrm{T}} \frac{\partial p_n}{\partial x_i} \right) = a_n \tag{B.10}$$

and

$$\frac{\partial \langle s \rangle_n}{\partial t} + \langle U_i \rangle \frac{\partial \langle s \rangle_n}{\partial x_i} - \frac{\partial}{\partial x_i}\left(\Gamma_{\mathrm{T}} \frac{\partial \langle s \rangle_n}{\partial x_i} \right) = b_n, \tag{B.11}$$

we can observe that (B.8) represents a non-constant-coefficient linear equation for a_n and b_n:

$$\begin{aligned}
&\sum_{n=1}^{N_{\mathrm{e}}} \left[\delta(\psi - \langle \phi \rangle_n) + \langle \phi \rangle_n \delta^{(1)}(\psi - \langle \phi \rangle_n) \right] a_n - \sum_{n=1}^{N_{\mathrm{e}}} \delta^{(1)}(\psi - \langle \phi \rangle_n) b_n \\
&\quad = \sum_{n=1}^{N_{\mathrm{e}}} \delta^{(2)}(\psi - \langle \phi \rangle_n) p_n c_n + R(\psi; \mathbf{x}, t),
\end{aligned} \tag{B.12}$$

where

$$c_n \equiv \Gamma_{\mathrm{T}} \frac{\partial \langle \phi \rangle_n}{\partial x_i} \frac{\partial \langle \phi \rangle_n}{\partial x_i}. \tag{B.13}$$

Note that c_n has the same form as the spurious dissipation term discussed in Section 5.10. These terms arise due to diffusion in real space in the presence of a mean scalar gradient, and will thus be non-zero for inhomogeneous scalar mixing.

All steps in the derivation up to this point have been exact in the sense that no arbitrary choices (besides the form of the presumed PDF) have been made. At this point, however, we must choose the set of $2N_{\mathrm{e}}$ moments that will be used to define a_n and b_n. The values of these moments will be computed 'exactly,' i.e., they will be consistent with using (B.3) to represent the PDF in (B.7). For example, if the integer moments are chosen, then we can multiply (B.12) by ψ^m and integrate:

$$\begin{aligned} &\sum_{n=1}^{N_{\mathrm{e}}} \left\{ \int_{-\infty}^{+\infty} \psi^m [\delta(\psi - \langle \phi \rangle_n) + \langle \phi \rangle_n \delta^{(1)}(\psi - \langle \phi \rangle_n)] \, \mathrm{d}\psi \right\} a_n \\ &\quad - \sum_{n=1}^{N_{\mathrm{e}}} \left\{ \int_{-\infty}^{+\infty} \psi^m \delta^{(1)}(\psi - \langle \phi \rangle_n) \, \mathrm{d}\psi \right\} b_n \\ &\quad = \sum_{n=1}^{N_{\mathrm{e}}} \left\{ \int_{-\infty}^{+\infty} \psi^m \delta^{(2)}(\psi - \langle \phi \rangle_n) \, \mathrm{d}\psi \right\} p_n c_n + \int_{-\infty}^{+\infty} \psi^m R(\psi; \mathbf{x}, t) \, \mathrm{d}\psi. \end{aligned} \tag{B.14}$$

Applying (B.9), this yields

$$(1 - m) \sum_{n=1}^{N_{\mathrm{e}}} \langle \phi \rangle_n^m a_n + m \sum_{n=1}^{N_{\mathrm{e}}} \langle \phi \rangle_n^{m-1} b_n = m(m-1) \sum_{n=1}^{N_{\mathrm{e}}} \langle \phi \rangle_n^{m-2} p_n c_n + R_m, \tag{B.15}$$

where the source terms for the moments are given by

$$R_m(\mathbf{x}, t) \equiv \int_{-\infty}^{+\infty} \psi^m R(\psi; \mathbf{x}, t) \, \mathrm{d}\psi. \tag{B.16}$$

Note that if the moments of the PDF are well defined, then the right-hand side of (B.16) will also be well defined.

In order to understand the connection between (B.15) and the moments ($m = 0, 1, \ldots, 2N_{\mathrm{e}} - 1$),

$$\langle \phi^m \rangle = \sum_{n=1}^{N_{\mathrm{e}}} p_n \langle \phi \rangle_n^m, \tag{B.17}$$

we can derive the transport equation for $\langle \phi^m \rangle$ directly from (B.6):

$$\frac{\partial \langle \phi^m \rangle}{\partial t} + \langle U_i \rangle \frac{\partial \langle \phi^m \rangle}{\partial x_i} - \frac{\partial}{\partial x_i} \left(\Gamma_{\mathrm{T}} \frac{\partial \langle \phi^m \rangle}{\partial x_i} \right) = R_m. \tag{B.18}$$

Comparing this result with (B.15), we can observe that the terms in (B.15) involving a_n, b_n, and c_n represent the transport terms for the moments, (B.17), predicted by the presumed PDF, (B.3). In other words, if a_n, b_n, and c_n obey (B.15) for a particular set of moments ($m = 0, 1, \ldots, 2N_e - 1$), then these same moments will obey (B.18). Note, however, that the same cannot be said for other moments not contained in this set. For example, non-integer moments, or integer moments of order $2N_e$ and higher, cannot be constrained to follow (B.18) exactly.

At this point, we can use our model for the flux in composition space, (B.7), to evaluate R_m:

$$\begin{aligned}
R_m &= -\int_{-\infty}^{+\infty} \psi^m \frac{\partial}{\partial \psi} \left[\left(C_\phi \frac{\varepsilon}{k} (\langle\phi\rangle - \psi) + S(\psi) \right) f_\phi \right] \mathrm{d}\psi \\
&= m \int_{-\infty}^{+\infty} \psi^{m-1} \left(C_\phi \frac{\varepsilon}{k} (\langle\phi\rangle - \psi) + S(\psi) \right) f_\phi \, \mathrm{d}\psi \\
&= m \int_{-\infty}^{+\infty} \psi^{m-1} \left(C_\phi \frac{\varepsilon}{k} (\langle\phi\rangle - \psi) + S(\psi) \right) \sum_{n=1}^{N_e} p_n \delta\left(\psi - \langle\phi\rangle_n\right) \mathrm{d}\psi \\
&= m \sum_{n=1}^{N_e} p_n \langle\phi\rangle_n^{m-1} \left(C_\phi \frac{\varepsilon}{k} (\langle\phi\rangle - \langle\phi\rangle_n) + S(\langle\phi\rangle_n) \right).
\end{aligned} \tag{B.19}$$

Note that, because we have assumed that the PDF can be represented by a sum of delta functions, (B.19) is closed. Intuitively, we can expect that the accuracy of the closure for R_m will increase with increasing N_e. It thus remains to solve (B.15) for $m = 0, 1, \ldots, 2N_e - 1$ to find a_n and b_n for $n = 1, 2, \ldots, N_e$.

In order to find a_n and b_n, we can rewrite (B.15) as a linear system of the form

$$\mathbf{A}\boldsymbol{\alpha} = \boldsymbol{\beta}, \tag{B.20}$$

where

$$\boldsymbol{\alpha}^{\mathrm{T}} = [a_1 \quad \cdots \quad a_{N_e} \quad b_1 \quad \cdots \quad b_{N_e}] = \begin{bmatrix} \mathbf{a} \\ \mathbf{b} \end{bmatrix}^{\mathrm{T}}, \tag{B.21}$$

and $\mathbf{A}$ is a square matrix of the form $\mathbf{A} = [\mathbf{A}_1 \;\; \mathbf{A}_2]$. The matrices $\mathbf{A}_1$ and $\mathbf{A}_2$ have the same size and are defined by

$$\mathbf{A}_1 = \begin{bmatrix}
1 & \cdots & 1 \\
0 & \cdots & 0 \\
-\langle\phi\rangle_1^2 & \cdots & -\langle\phi\rangle_{N_e}^2 \\
\vdots & \vdots & \vdots \\
(1-m)\langle\phi\rangle_1^m & \cdots & (1-m)\langle\phi\rangle_{N_e}^m \\
\vdots & \vdots & \vdots \\
2(1-N_e)\langle\phi\rangle_1^{2N_e-1} & \cdots & 2(1-N_e)\langle\phi\rangle_{N_e}^{2N_e-1}
\end{bmatrix} \tag{B.22}$$

and

$$\mathbf{A}_2 = \begin{bmatrix} 0 & \cdots & 0 \\ 1 & \cdots & 1 \\ 2\langle\phi\rangle_1 & \cdots & 2\langle\phi\rangle_{N_e} \\ \vdots & \vdots & \vdots \\ m\langle\phi\rangle_1^{m-1} & \cdots & m\langle\phi\rangle_{N_e}^{m-1} \\ \vdots & \vdots & \vdots \\ (2N_e-1)\langle\phi\rangle_1^{2N_e-2} & \cdots & (2N_e-1)\langle\phi\rangle_{N_e}^{2N_e-2} \end{bmatrix}. \tag{B.23}$$

Note that $\mathbf{A}$ depends only on the conditional scalar means $\langle\phi\rangle_n$, and not on the probabilities p_n.

The right-hand side of the linear system (B.20) can be written as

$$\beta = \mathbf{A}_3\mathbf{P}\mathbf{c} + \mathbf{A}_2\mathbf{P}\mathbf{r}, \tag{B.24}$$

where[2]

$$\mathbf{P} = \mathbf{diag}(p_1, \ldots, p_{N_e}), \tag{B.25}$$

$$\mathbf{c}^{\mathrm{T}} = [c_1 \quad \cdots \quad c_{N_e}], \tag{B.26}$$

$$\mathbf{r} = \begin{bmatrix} C_\phi \frac{\varepsilon}{k}(\langle\phi\rangle - \langle\phi\rangle_1) + S(\langle\phi\rangle_1) \\ \vdots \\ C_\phi \frac{\varepsilon}{k}(\langle\phi\rangle - \langle\phi\rangle_{N_e}) + S(\langle\phi\rangle_{N_e}) \end{bmatrix}, \tag{B.27}$$

and

$$\mathbf{A}_3 = \begin{bmatrix} 0 & \cdots & 0 \\ 0 & \cdots & 0 \\ 2 & \cdots & 2 \\ 6\langle\phi\rangle_1 & \cdots & 6\langle\phi\rangle_{N_e} \\ \vdots & \vdots & \vdots \\ m(m-1)\langle\phi\rangle_1^{m-2} & \cdots & m(m-1)\langle\phi\rangle_{N_e}^{m-2} \\ \vdots & \vdots & \vdots \\ 2(2N_e-1)(N_e-1)\langle\phi\rangle_1^{2N_e-3} & \cdots & 2(2N_e-1)(N_e-1)\langle\phi\rangle_{N_e}^{2N_e-3} \end{bmatrix}. \tag{B.28}$$

Thus, the ranks of the coefficient matrix $\mathbf{A} = [\mathbf{A}_1 \quad \mathbf{A}_2]$ and the augmented matrix $\mathbf{W} = [\mathbf{A}_1 \quad \mathbf{A}_2 \quad \mathbf{A}_3\mathbf{P}\mathbf{c}]$ will determine whether (B.20) has a solution.

[2] The term $\mathbf{A}_2\mathbf{P}\mathbf{r}$ is a direct result of employing the IEM model. If a different mixing model were used, then additional terms would result. For example, with the FP model the right-hand side would have the form $\beta = \mathbf{A}_3\mathbf{P}\mathbf{c} + \mathbf{A}_2\mathbf{P}\mathbf{r} + \mathbf{A}_3\mathbf{P}\mathbf{r}_{\mathrm{d}}$, where $\mathbf{r}_{\mathrm{d}}$ results from the diffusion term in the Fokker–Planck equation.

For the IEM model, it is well known that for a homogeneous system (i.e., when p_n and $\langle\phi\rangle_n$ do not depend on $\mathbf{x}$) the heights of the peaks (p_n) remain constant and the locations ($\langle\phi\rangle_n$) move according to the rates r_n. Using the matrices defined above, we can rewrite the linear system as

$$\begin{aligned}
\mathbf{A}\boldsymbol{\alpha} &= \boldsymbol{\beta} \\
&\Downarrow \\
[\mathbf{A}_1 \quad \mathbf{A}_2]\begin{bmatrix}\mathbf{a}\\ \mathbf{b}\end{bmatrix} &= \boldsymbol{\beta} \\
&\Downarrow \\
\mathbf{A}_1\mathbf{a} + \mathbf{A}_2\mathbf{b} &= \mathbf{A}_3\mathbf{P}\mathbf{c} + \mathbf{A}_2\mathbf{P}\mathbf{r} \\
&\Downarrow \\
\mathbf{A}_1\mathbf{a} + \mathbf{A}_2(\mathbf{b} - \mathbf{P}\mathbf{r}) &= \mathbf{A}_3\mathbf{P}\mathbf{c} \\
&\Downarrow \\
[\mathbf{A}_1 \quad \mathbf{A}_2]\begin{bmatrix}\mathbf{a}\\ \mathbf{b}^*\end{bmatrix} &= \mathbf{A}_3\mathbf{P}\mathbf{c} \\
&\Downarrow \\
\mathbf{A}\boldsymbol{\alpha}^* &= \boldsymbol{\beta}^*,
\end{aligned} \tag{B.29}$$

where

$$\mathbf{b} = \mathbf{b}^* + \mathbf{P}\mathbf{r}. \tag{B.30}$$

Thus, when $\mathbf{c} = \mathbf{0}$, we find $\boldsymbol{\alpha}^* = \mathbf{0}$, so that $\mathbf{a} = \mathbf{0}$ and $\mathbf{b} = \mathbf{P}\mathbf{r}$ as required for the IEM model.[3] The vector $\boldsymbol{\beta}^*$ will generally be non-zero for inhomogeneous systems. In this case, $\boldsymbol{\alpha}^*$ will be non-zero, even when $\mathbf{r} = \mathbf{0}$.

As an example, consider the case of a uniform mean scalar gradient with $p_1 = p_2 = 0.5$, $\mathbf{r} = \mathbf{0}$ and $\mathbf{c}^{\mathrm{T}} = [1 \quad 1]^{\mathrm{T}}$ (i.e., $N_{\mathrm{e}} = 2$). For this case, the scalar variance will grow with time due to the lack of scalar variance dissipation (i.e., $\mathbf{r} = \mathbf{0}$ implies that $\varepsilon_\phi = 0$). Moreover, the higher-order moments (up to $2N_{\mathrm{e}} - 1 = 3$) should approach the Gaussian values.[4] The DQMOM representation for this case yields[5]

$$\begin{bmatrix} 1 & 1 & 0 & 0 \\ 0 & 0 & 1 & 1 \\ -\langle\phi\rangle_1^2 & -\langle\phi\rangle_2^2 & 2\langle\phi\rangle_1 & 2\langle\phi\rangle_2 \\ -2\langle\phi\rangle_1^3 & -2\langle\phi\rangle_2^3 & 3\langle\phi\rangle_1^2 & 3\langle\phi\rangle_2^2 \end{bmatrix} \begin{bmatrix} a_1 \\ a_2 \\ b_1 \\ b_2 \end{bmatrix} = \begin{bmatrix} 0 \\ 0 \\ 2 \\ 3(\langle\phi\rangle_1 + \langle\phi\rangle_2) \end{bmatrix}, \tag{B.31}$$

3 If $\mathbf{A}$ were singular, then non-zero solutions for $\boldsymbol{\alpha}^*$ would also exist. However, only the zero solution will be of interest when $\boldsymbol{\beta}^* = \mathbf{0}$. The treatment of cases where $\mathbf{A}$ is singular and $\boldsymbol{\beta}^*$ is non-zero is discussed below.

4 If $N_{\mathrm{e}} = 3$ is used, then $2N_{\mathrm{e}} - 1 = 5$. In this case, the Gaussian value (3) for the kurtosis (i.e., the fourth-order moment) will be recovered.

5 The right-hand side of (B.31) will be different when $p_1 = 1 - p_2 \neq 0.5$. This will then lead to $\mathbf{a} \neq \mathbf{0}$ and a more complicated expression for $\mathbf{b}$. Nevertheless, the singularity associated with $\langle\phi\rangle_1 = \langle\phi\rangle_2$ will still exist, and can still be handled using perturbations as described below. In fact, a singularity at $t = 0$ will force $p_1(t) = p_2(t) = 0.5$ for $t > 0$.

and thus

$$\begin{bmatrix} a_1 \\ a_2 \\ b_1 \\ b_2 \end{bmatrix} = \frac{1}{(\langle\phi\rangle_1 - \langle\phi\rangle_2)} \begin{bmatrix} 0 \\ 0 \\ 1 \\ -1 \end{bmatrix}. \tag{B.32}$$

This leads to the following ODEs for p_n and $\langle s\rangle_n$ (see (B.10) and (B.11)):

$$\frac{\mathrm{d}p_n}{\mathrm{d}t} = 0, \quad \frac{\mathrm{d}\langle s\rangle_1}{\mathrm{d}t} = \frac{1}{(\langle\phi\rangle_1 - \langle\phi\rangle_2)}, \quad \frac{\mathrm{d}\langle s\rangle_2}{\mathrm{d}t} = -\frac{1}{(\langle\phi\rangle_1 - \langle\phi\rangle_2)}. \tag{B.33}$$

The solution to (B.33) can be written as

$$\begin{aligned} \langle\phi\rangle_1(t) &= \langle\phi\rangle + (2t + \langle\phi'^2\rangle_0)^{1/2}, \\ \langle\phi\rangle_2(t) &= \langle\phi\rangle - (2t + \langle\phi'^2\rangle_0)^{1/2}, \end{aligned} \tag{B.34}$$

where $\langle\phi'^2\rangle_0$ is the scalar variance at $t = 0$. Note that if the initial scalar variance is null, $\langle\phi\rangle_1$ and $\langle\phi\rangle_2$ will spread apart at the rate $(2t)^{1/2}$, which corresponds to diffusion in composition space.

In general, if all $\langle\phi\rangle_n$ $(n = 1, \ldots, N_e)$ are distinct, then $\mathbf{A}$ will be full rank, and thus $\boldsymbol{\alpha}^* = \mathbf{A}^{-1}\boldsymbol{\beta}^*$ as shown in (B.32). However, if any two (or more) $\langle\phi\rangle_n$ are the same, then two (or more) columns of $\mathbf{A}_1$, $\mathbf{A}_2$, and $\mathbf{A}_3$ will be linearly dependent. In this case, the rank of $\mathbf{A}$ and the rank of $\mathbf{W}$ will usually not be the same and the linear system has no consistent solutions. This case occurs most often due to initial conditions (e.g., binary mixing with initially only two non-zero probability peaks in composition space). The example given above, (B.31), illustrates what can happen for $N_e = 2$. When $\langle\phi\rangle_1 = \langle\phi\rangle_2$, the right-hand sides of the ODEs in (B.33) will be singular; nevertheless, the ODEs yield well defined solutions, (B.34). This example also points to a simple method to overcome the problem of the singularity of $\mathbf{A}$ due to repeated $\langle\phi\rangle_n$: it suffices simply to add small perturbations to the non-distinct $\langle\phi\rangle_n$ so that $\mathbf{A}$ will be full rank. Note that the perturbed values need only be used in the definition of $\mathbf{A}$, and that the perturbations should leave the scalar mean $\langle\phi\rangle$ unchanged.

B.2.2 Bi-variate case

The joint composition PDF transport equation for the bi-variate case has the following form:

$$\frac{\partial f_\phi}{\partial t} + \langle U_i\rangle \frac{\partial f_\phi}{\partial x_i} - \frac{\partial}{\partial x_i}\left(\Gamma_{\mathrm{T}} \frac{\partial f_\phi}{\partial x_i}\right) = R(\boldsymbol{\psi}; \mathbf{x}, t), \tag{B.35}$$

where the right-hand side is given by

$$\begin{aligned} R(\boldsymbol{\psi}; \mathbf{x}, t) = &- \frac{\partial}{\partial \psi_1}\left[\left(C_\phi \frac{\varepsilon}{k}(\langle\phi_1\rangle - \psi_1) + S_1(\boldsymbol{\psi})\right) f_\phi\right] \\ &- \frac{\partial}{\partial \psi_2}\left[\left(C_\phi \frac{\varepsilon}{k}(\langle\phi_2\rangle - \psi_2) + S_2(\boldsymbol{\psi})\right) f_\phi\right]. \end{aligned} \tag{B.36}$$

The derivation begins as before by formally inserting (B.1) for f_ϕ into (B.35). After some algebra, we find[6]

$$
\begin{aligned}
&\sum_{n=1}^{N_e} \delta(\psi_1 - \langle\phi_1\rangle_n)\delta(\psi_2 - \langle\phi_2\rangle_n) a_n \\
&\quad - \sum_{n=1}^{N_e} \delta^{(1)}(\psi_1 - \langle\phi_1\rangle_n)\delta(\psi_2 - \langle\phi_2\rangle_n)(b_{1n} - \langle\phi_1\rangle_n a_n) \\
&\quad - \sum_{n=1}^{N_e} \delta(\psi_1 - \langle\phi_1\rangle_n)\delta^{(1)}(\psi_2 - \langle\phi_2\rangle_n)(b_{2n} - \langle\phi_2\rangle_n a_n) \\
&= R(\psi; \mathbf{x}, t) + \sum_{n=1}^{N_e} \delta^{(2)}(\psi_1 - \langle\phi_1\rangle_n)\delta(\psi_2 - \langle\phi_2\rangle_n) p_n c_{11n} \\
&\quad + 2\sum_{n=1}^{N_e} \delta^{(1)}(\psi_1 - \langle\phi_1\rangle_n)\delta^{(1)}(\psi_2 - \langle\phi_2\rangle_n) p_n c_{12n} \\
&\quad + \sum_{n=1}^{N_e} \delta(\psi_1 - \langle\phi_1\rangle_n)\delta^{(2)}(\psi_2 - \langle\phi_2\rangle_n) p_n c_{22n},
\end{aligned} \tag{B.37}
$$

where

$$
c_{\alpha\beta n} \equiv \Gamma_T \frac{\partial \langle\phi_\alpha\rangle_n}{\partial x_i} \frac{\partial \langle\phi_\beta\rangle_n}{\partial x_i}, \tag{B.38}
$$

and the transport equation for $\langle s_\alpha\rangle_n = p_n \langle\phi_\alpha\rangle_n$ is given by

$$
\frac{\partial \langle s_\alpha\rangle_n}{\partial t} + \langle U_i\rangle \frac{\partial \langle s_\alpha\rangle_n}{\partial x_i} - \frac{\partial}{\partial x_i}\left(\Gamma_T \frac{\partial \langle s_\alpha\rangle_n}{\partial x_i}\right) = b_{\alpha n}. \tag{B.39}
$$

We can again observe that (B.37) represents a non-constant-coefficient linear equation for the $3N_e$ components a_n, b_{1n}, and b_{2n}.

The next step is to select the moments used to define a_n and $b_{\alpha n}$. Choosing the integer moments $\langle\phi_1^{m_1}\phi_2^{m_2}\rangle$, where m_1 and m_2 are non-negative integers, (B.37) yields

$$
\begin{aligned}
&\sum_{n=1}^{N_e} (1 - m_1 - m_2)\langle\phi_1\rangle_n^{m_1}\langle\phi_2\rangle_n^{m_2} a_n \\
&\quad + \sum_{n=1}^{N_e} m_1 \langle\phi_1\rangle_n^{m_1-1}\langle\phi_2\rangle_n^{m_2} b_{1n}^* + \sum_{n=1}^{N_e} m_2 \langle\phi_1\rangle_n^{m_1}\langle\phi_2\rangle_n^{m_2-1} b_{2n}^* \\
&= \sum_{n=1}^{N_e} m_1(m_1 - 1)\langle\phi_1\rangle_n^{m_1-2}\langle\phi_2\rangle_n^{m_2} p_n c_{11n} \\
&\quad + 2\sum_{n=1}^{N_e} m_1 m_2 \langle\phi_1\rangle_n^{m_1-1}\langle\phi_2\rangle_n^{m_2-1} p_n c_{12n} \\
&\quad + \sum_{n=1}^{N_e} m_2(m_2 - 1)\langle\phi_1\rangle_n^{m_1}\langle\phi_2\rangle_n^{m_2-2} p_n c_{22n},
\end{aligned} \tag{B.40}
$$

[6] The extension of this expression to the multi-variate case follows by analogy to the bi-variate case. Note that each scalar pair (α, β) generates a 'diffusion' term $c_{\alpha\beta n}$ on the right-hand side, which will usually be non-zero for flows with mean scalar gradients.

where $b^*_{\alpha n} \equiv b_{\alpha n} - p_n r_{\alpha n}$, and

$$r_{\alpha n} \equiv C_\phi \frac{\varepsilon}{k} \left(\langle \phi_\alpha \rangle - \langle \phi_\alpha \rangle_n \right) + S_\alpha \left(\langle \phi \rangle_n \right). \tag{B.41}$$

Note that (B.40) has the form $\mathbf{A}\boldsymbol{\alpha} = \boldsymbol{\beta}$, where the components of $\mathbf{A}$ depend on the choice of m_1 and m_2. A solution $\boldsymbol{\alpha} = \mathbf{A}^{-1}\boldsymbol{\beta}$ will exist if $\mathbf{A}$ is full rank. We must thus check to see what conditions are necessary for this to occur.

First consider the case with $N_e = 2$. Theoretically, $3N_e = 6$ moments will suffice to determine

$$\boldsymbol{\alpha}^T = [a_1 \quad a_2 \quad b^*_{11} \quad b^*_{12} \quad b^*_{21} \quad b^*_{22}].$$

A logical choice of moments would be $(m_1, m_2) = (0, 0), (1, 0), (0, 1), (1, 1), (2, 0)$ and $(0, 2)$ (i.e., the mean and covariance matrices). This choice leads to

$$\mathbf{A} = \begin{bmatrix} 1 & 1 & 0 & 0 & 0 & 0 \\ 0 & 0 & 1 & 1 & 0 & 0 \\ 0 & 0 & 0 & 0 & 1 & 1 \\ -\langle \phi_1 \rangle_1 \langle \phi_2 \rangle_1 & -\langle \phi_1 \rangle_2 \langle \phi_2 \rangle_2 & \langle \phi_2 \rangle_1 & \langle \phi_2 \rangle_2 & \langle \phi_1 \rangle_1 & \langle \phi_1 \rangle_2 \\ -\langle \phi_1 \rangle_1^2 & -\langle \phi_1 \rangle_2^2 & 2\langle \phi_1 \rangle_1 & 2\langle \phi_1 \rangle_2 & 0 & 0 \\ -\langle \phi_2 \rangle_1^2 & -\langle \phi_2 \rangle_2^2 & 0 & 0 & 2\langle \phi_2 \rangle_1 & 2\langle \phi_2 \rangle_2 \end{bmatrix}, \tag{B.42}$$

which has rank$(\mathbf{A}) = 5$. In hindsight, this result is not surprising when one considers that the $N_e = 2$ points $(\langle \phi_1 \rangle_1, \langle \phi_2 \rangle_1)$ and $(\langle \phi_1 \rangle_2, \langle \phi_2 \rangle_2)$ lie on a straight line in composition space. It is thus impossible to force the covariance between ϕ_1 and ϕ_2 to take on any arbitrary realizable value. Indeed, we can deduce from this example that for the bi-variate case N_e must be greater than two in order to predict the covariance matrix correctly.[7] Note, however, that by choosing other higher-order moments (e.g., $(m_1, m_2) = (2, 2)$) in place of $(m_1, m_2) = (1, 1)$, the rank of $\mathbf{A}$ can be increased to six. Thus, although the covariance cannot be predicted exactly, it is possible to use $N_e \leq 2$ for the bi-variate case.

With $N_e = 3$, a total of $3N_e = 9$ moments are required to determine

$$\boldsymbol{\alpha}^T = [a_1 \quad a_2 \quad a_3 \quad b^*_{11} \quad b^*_{12} \quad b^*_{13} \quad b^*_{21} \quad b^*_{22} \quad b^*_{23}].$$

In addition to the six moments used to find (B.42), we can choose from the third-order moments: $(m_1, m_2) = (3, 0), (2, 1), (1, 2)$, and $(0, 3)$. Note that only three additional moments are needed. Thus, one can either arbitrarily choose any three of the four third-order moments, or one can form three linearly independent combinations of the four moment equations. In either case, $\mathbf{A}$ will be full rank (provided that the points in composition space are distinct), or can be made so by perturbing the non-distinct compositions as discussed for the uni-variate case.

In summary, the bi-variate case is handled in essentially the same manner as the uni-variate case. The numerical implementation will require an algorithm for choosing linearly

[7] It also follows that the extension to the N_s-variate case will require $N_e > N_s$ in order to predict all of the second-order moments correctly.

independent moments needed for the definition of **A**, and a method to test for non-distinct points in composition space. The extension to $N_s > 2$ will use the same numerical implementation.

B.2.3 Multi-variate case

The bi-variate case given in (B.40) can be easily extended by inspection to N_s scalars:[8]

$$\sum_{n=1}^{N_e}\left[\left(1-\sum_{\alpha=1}^{N_s} m_\alpha\right)\prod_{\alpha=1}^{N_s}\langle\phi_\alpha\rangle_n^{m_\alpha}\right]a_n + \sum_{n=1}^{N_e}\sum_{\alpha=1}^{N_s}\frac{\partial}{\partial\langle\phi_\alpha\rangle_n}\left(\prod_{\beta=1}^{N_s}\langle\phi_\beta\rangle_n^{m_\beta}\right)b^*_{\alpha n} = \sum_{n=1}^{N_e}\sum_{\alpha=1}^{N_s}\sum_{\beta=1}^{N_s}\frac{\partial^2}{\partial\langle\phi_\alpha\rangle_n\partial\langle\phi_\beta\rangle_n}\left(\prod_{\gamma=1}^{N_s}\langle\phi_\gamma\rangle_n^{m_\gamma}\right)p_n c_{\alpha\beta n}, \tag{B.43}$$

where $N_e \geq N_s + 1$ will be required to predict the covariance matrix correctly.[9] The $N_e(N_s + 1)$ unknowns $(a_n, b^*_{1n}, \ldots, b^*_{N_s n})$ are found by solving (B.43) with the same number of unique combinations of $(m_1, \ldots, m_{N_s})$. If the minimum number of nodes needed to predict the covariance matrix is employed ($N_e = N_s + 1$), then the zero-order ($m_\alpha = 0$), first-order (only one $m_\alpha = 1$), second-order (either only one $m_\alpha = 2$, or $m_\alpha = 1$ and $m_\beta = 1$ with $\alpha \neq \beta$), and some of the third-order moments will be required to obtain a full-rank coefficient matrix. Otherwise, if $N_e \leq N_s$, the user must find a linearly independent set of moments for which the rank of **A** is equal to $N_e(N_s + 1)$. In order to ensure that the components of **p** sum to unity and the scalar means are conserved, this set must include the zero-order ($m_\alpha = 0$) and first-order (only one $m_\alpha = 1$) moments, and $(N_e - 1)(N_s + 1)$ independent higher-order moments. The optimal choice of the latter is most likely problem-dependent.

Given the solution to (B.43), the transport equations for the vector $\mathbf{p} \equiv [p_n]$ and the matrix $\langle S\rangle \equiv [\langle s_\alpha\rangle_n]$ become

$$\frac{\partial \mathbf{p}}{\partial t} + \langle U_i\rangle\frac{\partial \mathbf{p}}{\partial x_i} = \frac{\partial}{\partial x_i}\left(\Gamma_T\frac{\partial \mathbf{p}}{\partial x_i}\right) + \mathbf{a} \tag{B.44}$$

8 Using the properties of the delta function, the partial derivative terms in (B.43) can also be written as

$$\prod_{\alpha=1}^{N_s}\langle\phi_\alpha\rangle_n^{m_\alpha} = \left\langle\prod_{\alpha=1}^{N_s}\phi_\alpha^{m_\alpha}\right\rangle_n,$$

$$\frac{\partial}{\partial\langle\phi_\alpha\rangle_n}\left(\prod_{\beta=1}^{N_s}\langle\phi_\beta\rangle_n^{m_\beta}\right) = \left\langle\frac{\partial}{\partial\phi_\alpha}\prod_{\beta=1}^{N_s}\phi_\beta^{m_\beta}\right\rangle_n,$$

and

$$\frac{\partial^2}{\partial\langle\phi_\alpha\rangle_n\partial\langle\phi_\beta\rangle_n}\left(\prod_{\gamma=1}^{N_s}\langle\phi_\gamma\rangle_n^{m_\gamma}\right) = \left\langle\frac{\partial^2}{\partial\phi_\alpha\partial\phi_\beta}\prod_{\gamma=1}^{N_s}\phi_\gamma^{m_\gamma}\right\rangle_n,$$

where $\langle\cdot\rangle_n$ is the conditional expected value for quadrature node n.

9 As noted earlier, it is possible to use $N_e \leq N_s$ by choosing a linearly independent set of higher-order moments. For example, the multi-environment models discussed in Section 5.10 use $N_e = 2$–4.

and

$$\frac{\partial \langle \mathcal{S} \rangle}{\partial t} + \langle U_i \rangle \frac{\partial \langle \mathcal{S} \rangle}{\partial x_i} = \frac{\partial}{\partial x_i} \left(\Gamma_{\mathrm{T}} \frac{\partial \langle \mathcal{S} \rangle}{\partial x_i} \right) + \mathcal{B} + \mathcal{R}, \tag{B.45}$$

where the matrix $\mathcal{B} \equiv [b^*_{\alpha n}]$ will be non-zero only for inhomogeneous flows, and the matrix $\mathcal{R} \equiv [r_{\alpha n} p_n]$ contains the micromixing and reaction terms. This is a truly remarkable result as it provides an Eulerian CFD formulation of the joint composition PDF transport equation (B.5): an equation that otherwise can only be simulated using the Monte-Carlo methods described in Chapter 7.

The DQMOM results for the IEM model can be compared with the multi-environment presumed PDF models in Section 5.10. In particular, (5.374) on p. 226 can be compared with (B.44), and (5.375) can be compared with (B.45). First, we can note that for the IEM model $\mathbf{G} = \mathbf{0}$ and $\mathbf{G}_{\mathrm{s}} = \mathbf{a}$. Likewise, $\gamma M^{(n)}_{\alpha} + p_n S_{\alpha}(\langle \phi \rangle_n) = \mathcal{R}_{\alpha n}$ and $M^{(n)}_{\mathrm{s}\alpha} = \mathcal{B}_{\alpha n}$. Of the four models introduced in Tables 5.1–5.5, only the symmetric two-environment model in Table 5.2 has $\mathbf{G} = \mathbf{0}$ and $\gamma M^{(n)}_{\alpha} + p_n S_{\alpha} = \mathcal{R}_{\alpha n}$. However, because the spurious dissipation terms only ensure that the mixture-fraction variance is correctly predicted, the symmetric two-environment model does not have $\mathbf{G}_{\mathrm{s}} = \mathbf{a}$ and $M^{(n)}_{\mathrm{s}\alpha} = \mathcal{B}_{\alpha n}$. Thus, the covariance matrix is not predicted correctly, as it would be if (B.43) were used. We can thus conclude that the multi-environment presumed PDF models are incomplete in the sense that they do not control as many of the moments as possible for a given choice of N_{e}.

The results from (B.43) can be combined with the multi-environment presumed PDF models to define a new class of micromixing models that agree with the DQMOM result for inhomogeneous flows. Formally, (B.43) can be written as

$$\mathbf{A} \begin{bmatrix} \mathbf{a} \\ \mathbf{b}^*_1 \\ \vdots \\ \mathbf{b}^*_{N_{\mathrm{e}}} \end{bmatrix} = \mathbf{B}\mathbf{c} \quad \Longrightarrow \quad \mathbf{A} \begin{bmatrix} \mathbf{a} \\ \mathbf{b}_1 \\ \vdots \\ \mathbf{b}_{N_{\mathrm{e}}} \end{bmatrix} = \mathbf{B}\mathbf{c} + \mathbf{A} \begin{bmatrix} \mathbf{0} \\ p_1 \mathbf{r}_1 \\ \vdots \\ p_{N_{\mathrm{e}}} \mathbf{r}_{N_{\mathrm{e}}} \end{bmatrix}, \tag{B.46}$$

where $\mathbf{c}$ denotes the vectorized form of the components $c_{\alpha\beta n}$. The final term on the right-hand side of (B.46) contains the IEM model and the chemical source term. It thus suffices to replace the IEM model with the multi-environment model:

$$\mathbf{A} \begin{bmatrix} \mathbf{a} \\ \mathbf{b}_1 \\ \vdots \\ \mathbf{b}_{N_{\mathrm{e}}} \end{bmatrix} = \mathbf{B}\mathbf{c} + \mathbf{A} \begin{bmatrix} \gamma \mathbf{G} \\ \gamma \mathbf{M}^{(1)} + p_1 \mathbf{S}(\langle \phi \rangle_1) \\ \vdots \\ \gamma \mathbf{M}^{(N_{\mathrm{e}})} + p_{N_{\mathrm{e}}} \mathbf{S}(\langle \phi \rangle_{N_{\mathrm{e}}}) \end{bmatrix}, \tag{B.47}$$

or (assuming that a complete set of linearly independent moments can be found):

$$\begin{bmatrix} \mathbf{a} \\ \mathbf{b}_1 \\ \vdots \\ \mathbf{b}_{N_{\mathrm{e}}} \end{bmatrix} = \mathbf{A}^{-1}\mathbf{B}\mathbf{c} + \begin{bmatrix} \gamma \mathbf{G} \\ \gamma \mathbf{M}^{(1)} + p_1 \mathbf{S}(\langle \phi \rangle_1) \\ \vdots \\ \gamma \mathbf{M}^{(N_{\mathrm{e}})} + p_{N_{\mathrm{e}}} \mathbf{S}(\langle \phi \rangle_{N_{\mathrm{e}}}) \end{bmatrix}. \tag{B.48}$$

For homogeneous flows, $\mathbf{c} = \mathbf{0}$ and (B.48) reduces to the homogeneous multi-environment model ((5.342) and (5.343) on p. 222). The corresponding spurious dissipation terms are thus defined by[10]

$$\begin{bmatrix} \mathbf{G}_s \\ \mathbf{M}_s^{(1)} \\ \vdots \\ \mathbf{M}_s^{(N_e)} \end{bmatrix} = \begin{bmatrix} \mathbf{a}^* \\ \mathbf{b}_1^* \\ \vdots \\ \mathbf{b}_{N_e}^* \end{bmatrix} = \mathbf{A}^{-1}\mathbf{B}\mathbf{c}, \tag{B.49}$$

and can be found by solving (B.43). These definitions for $\mathbf{G}_s$ and $\mathbf{M}_s^{(n)}$ are then used in (5.374) and (5.375) on p. 226 in place of the definitions given in Tables 5.1–5.5. It is interesting to note that because $\mathbf{A}$, $\mathbf{B}$, and $\mathbf{c}$ are defined independently of the micromixing model (i.e., of $\mathbf{G}$ and $\mathbf{M}^{(n)}$), the spurious dissipation terms will not depend directly on the choice of the micromixing model. Instead, they are properties of the joint composition PDF transport equation and of the choice of the linearly independent set of moments used to define $\mathbf{A}$. Nevertheless, $\mathbf{G}_s$ and $\mathbf{M}_s^{(n)}$ will be functions of $\mathbf{x}$ and t, and thus must be updated at every time step for every point in the spatial domain.

Note that the RANS formulation used in (B.44) and (B.45) can easily be extended to the LES, as outlined in Section 5.10. Moreover, by following the same steps as outlined above, DQMOM can be used with the joint velocity, composition PDF transport equation. Finally, the reader can observe that the same methodology is applicable to more general distribution functions than probability density functions. Indeed, DQMOM can be applied to general population balance equations such as those used to describe multi-phase flows.

B.3 DQMOM–IEM model

The multi-variate DQMOM method, (B.43), ensures that the mixed moments used to determine the unknowns $(a_n, b_{1n}^*, \ldots, b_{N_s n}^*)$ are exactly reproduced for the IEM model in the absence of chemical reactions.[11] As discussed earlier, for the homogeneous case $(c_{\alpha\beta n} = 0)$ the solution to (B.43) is trivial $(a_n = 0, b_{\alpha n}^* = 0)$ and exactly reproduces the IEM model for moments of *arbitrary* order. On the other hand, for inhomogeneous cases the IEM model will not be exactly reproduced. Thus, since many multi-variate PDFs exist for a given set of lower-order mixed moments, we cannot be assured that every choice of mixed moments used to solve (B.43) will lead to satisfactory results.

In the first application of (B.40) to an inhomogeneous bi-variate inert-scalar-mixing case (i.e., the so-called 'three-stream mixing problem' (Juneja and Pope 1996)), it was found that, although the lower-order mixed moments are exactly reproduced, the conditional means $\langle \xi_\alpha \rangle_n$ become unrealizable (Marchisio and Fox 2003). Indeed, for every possible choice of the lower-order moments, the sum of the conditional mixture-fraction

[10] The vectors $\mathbf{a}$ and $\mathbf{b}_n^*$ denote the unknowns in (B.43), not the source terms for $\mathbf{p}$ and $\langle \mathbf{s} \rangle_n$.
[11] For this case, the mixed-moment transport equations found from (B.5) will be closed.

components $(\langle \xi_1 \rangle_n + \langle \xi_2 \rangle_n)$ is greater than unity at intermediate times. This clear violation of boundedness (see Section 6.6) seriously compromises the direct application of DQMOM to inhomogeneous scalar mixing problems.

In order to overcome this difficulty and still satisfy (B.43), additional assumptions are required when applying DQMOM to inhomogeneous cases. In particular, if we let $a_n = 0$ the number of unknowns reduces to $N_e \times N_s$, and (B.43) is still satisfied for $m_1 = m_2 = \cdots = m_{N_s} = 0$. If we then specify that only *unmixed* moments can be used to determine $b^*_{\alpha n}$, (B.43) reduces to

$$\sum_{n=1}^{N_e} \langle \phi_\alpha \rangle_n^{m_\alpha - 1} b^*_{\alpha n} = \sum_{n=1}^{N_e} (m_\alpha - 1) \langle \phi_\alpha \rangle_n^{m_\alpha - 2} p_n c_{\alpha\alpha n}, \tag{B.50}$$

where $m_\alpha = 1, 2, \ldots, N_e$ for each $\alpha = 1, 2, \ldots, N_s$. Note that with these additional constraints, the scalars are uncoupled and (B.50) is solved separately for each scalar. In order to distinguish (B.43) from (B.50), we will refer to the latter as the DQMOM–IEM model.

In matrix notation, the DQMOM–IEM model can be written as

$$\mathbf{V}_{1\alpha} \begin{bmatrix} b^*_{\alpha 1} \\ \vdots \\ b^*_{\alpha N_e} \end{bmatrix} = \mathbf{V}_{2\alpha} \mathbf{P} \begin{bmatrix} c_{\alpha\alpha 1} \\ \vdots \\ c_{\alpha\alpha N_e}, \end{bmatrix}, \tag{B.51}$$

where

$$\mathbf{V}_{1\alpha} = \begin{bmatrix} 1 & \cdots & 1 \\ \langle \phi_\alpha \rangle_1 & \cdots & \langle \phi_\alpha \rangle_{N_e} \\ \vdots & & \vdots \\ \langle \phi_\alpha \rangle_1^{N_e - 1} & \cdots & \langle \phi_\alpha \rangle_{N_e}^{N_e - 1} \end{bmatrix} \tag{B.52}$$

is the $N_e \times N_e$ Vandermonde matrix, and

$$\mathbf{V}_{2\alpha} = \begin{bmatrix} 0 & \cdots & 0 \\ 1 & \cdots & 1 \\ 2\langle \phi_\alpha \rangle_1 & \cdots & 2\langle \phi_\alpha \rangle_{N_e} \\ \vdots & & \vdots \\ (N_e - 1)\langle \phi_\alpha \rangle_1^{N_e - 2} & \cdots & (N_e - 1)\langle \phi_\alpha \rangle_{N_e}^{N_e - 2} \end{bmatrix}. \tag{B.53}$$

Vandermonde matrices often arise when reconstructing PDFs from their moments, and efficient numerical methods exist for solving (B.51) (Press *et al.* 1992). Although Vandermonde matrices are notoriously ill-conditioned for large N_e, this will usually not be a problem in applications of the DQMOM–IEM model since it is most attractive (relative to Monte-Carlo methods) when N_e is small.

In summary, DQMOM is a numerical method for solving the Eulerian joint PDF transport equation using standard numerical algorithms (e.g., finite-difference or finite-volume codes). The method works by forcing the lower-order moments to agree with the corresponding transport equations. For unbounded joint PDFs, DQMOM can be applied directly to determine $(a_n, b^*_{1n}, \ldots, b^*_{N_s n})$. However, for joint PDFs defined on bounded domains, care must be taken to ensure that the boundedness conditions are not violated for inhomogeneous cases. With the IEM model, this can be accomplished by using additional constraints, which results in the DQMOM–IEM model. For other micromixing models (e.g., multi-environment presumed PDF models) similar constraints will most likely be required to ensure realizability. In general, these constraints can be developed using assumptions similar to those used to generate (B.50). In particular, constraints that lead to uncoupled scalar equations appear to be a necessary condition for realizability.

References

Adumitroaie, V., D. B. Taulbee, and P. Givi (1997). Explicit algebraic scalar-flux models for turbulent reacting flows. *AIChE Journal* **43**, 1935–1946.

Akselvoll, K. and P. Moin (1996). Large eddy simulation of turbulent confined coannular jets. *Journal of Fluid Mechanics* **315**, 387–411.

Allievi, A. and R. Bermejo (1997). A generalized particle search-locate algorithm for arbitrary grids. *Journal of Computational Physics* **132**, 157–166.

Amerja, P. V., M. Singh, and H. L. Toor (1976). Reactive mixing in turbulent gases. *Chemical Engineering Communications* **2**, 115–120.

Anand, M. S. and S. B. Pope (1987). Calculations of premixed turbulent flames by PDF methods. *Combustion and Flame* **67**, 127–142.

Anand, M. S., S. B. Pope, and H. C. Mongia (1989). A PDF method for turbulent recirculating flows. In *Turbulent Reactive Flows, Lecture Notes in Engineering*, pp. 672–693. Berlin: Springer-Verlag.

Anand, M. S., A. T. Hsu, and S. B. Pope (1997). Calculations of swirl combustors using joint velocity-scalar probability density function methods. *AIAA Journal* **35**, 1143–1150.

Anand, M. S., S. James, and M. K. Razdan (1998). A scalar PDF combustion model for the national combustion code. Paper 98-3856, AIAA.

Anselmet, F. and R. A. Antonia (1985). Joint statistics between temperature and its dissipation in a turbulent jet. *The Physics of Fluids* **28**, 1048–1054.

Anselmet, F., H. Djeridi, and L. Fulachier (1994). Joint statistics of a passive scalar and its dissipation in turbulent flows. *Journal of Fluid Mechanics* **280**, 173–197.

Antonia, R. A. and J. Mi (1993). Temperature dissipation in a turbulent round jet. *Journal of Fluid Mechanics* **250**, 531–551.

Arnold, L. (1974). *Stochastic Differential Equations: Theory and Applications*. New York: John Wiley.

Arrojo, P., C. Dopazo, L. Valiño, and W. P. Jones (1988). Numerical simulation of velocity and concentration fields in a continuous flow reactor. *Chemical Engineering Science* **43**, 1935–1940.

Ashurst, W. T., A. R. Kerstein, R. M. Kerr, and C. H. Gibson (1987). Alignment of vorticity and scalar gradient with strain rate in simulated Navier–Stokes turbulence. *The Physics of Fluids* **30**, 2343–2353.

Bakker, R. A. and H. E. A. van den Akker (1996). A Lagrangian description of micromixing in a stirred tank reactor using 1D-micromixing models in a CFD flow field. *Chemical Engineering Science* **51**, 2643–2648.

Bałdyga, J. (1989). Turbulent mixer model with application to homogeneous, instantaneous chemical reactions. *Chemical Engineering Science* **44**, 1175–1182.

(1994). A closure model for homogeneous chemical reactions. *Chemical Engineering Science* **49**, 1985–2003.

Bałdyga, J. and J. R. Bourne (1984a). A fluid-mechanical approach to turbulent mixing and chemical reaction. Part I. Inadequacies of available methods. *Chemical Engineering Communications* **28**, 231–241.

(1984b). A fluid-mechanical approach to turbulent mixing and chemical reaction. Part II. Micromixing in the light of turbulence theory. *Chemical Engineering Communications* **28**, 243–258.

(1984c). A fluid-mechanical approach to turbulent mixing and chemical reaction. Part III. Computational and experimental results for the new micromixing model. *Chemical Engineering Communications* **28**, 259–281.

(1989). Simplification of micromixing calculations. I. Derivation and application of new model. *The Chemical Engineering Journal* **42**, 83–92.

(1999). *Turbulent Mixing and Chemical Reactions*. New York: John Wiley & Sons, Inc.

Bałdyga, J. and M. Henczka (1995). Closure problem for parallel chemical reactions. *The Chemical Engineering Journal* **58**, 161–173.

(1997). The use of a new model of micromixing for determination of crystal size in precipitation. *Récent Progrès en Génie des Procédés* **11**, 341–348.

Bałdyga, J., J. R. Bourne, B. Dubuis, A. W. Etchells, R. V. Gholap, and B. Zimmerman (1995). Jet reactor scale-up for mixing-controlled reactions. *Transactions of the Institution of Chemical Engineers* **73**, 497–502.

Barlow, R. S., R. W. Dibble, J.-Y. Chen, and R. P. Lucht (1990). Effect of Damköhler number on superequilibrium OH concentration in turbulent nonpremixed jet flames. *Combustion and Flame* **82**, 235–251.

Barlow, R. S., N. S. A. Smith, J.-Y. Chen, and R. W. Bilger (1999). Nitric oxide formation in dilute hydrogen jet flames: Isolation of the effects of radiation and turbulence-chemistry submodels. *Combustion and Flame* **117**, 4–31.

Barrett, J. C. and N. A. Webb (1998). A comparison of some approximate methods for solving the aerosol general dynamic equation. *Journal of Aerosol Science* **29**, 31–39.

Batchelor, G. K. (1953). *The Theory of Homogeneous Turbulence*. Cambridge: Cambridge University Press.

(1959). Small-scale variation of convected quantities like temperature in turbulent fluid. Part 1. General discussion and the case of small conductivity. *Journal of Fluid Mechanics* **5**, 113–133.

Batchelor, G. K., I. D. Howells, and A. A. Townsend (1959). Small-scale variation of convected quantities like temperature in turbulent fluid. Part 2. The case of large conductivity. *Journal of Fluid Mechanics* **5**, 134–139.

Béguier, C., I. Dekeyser, and B. E. Launder (1978). Ratio of scalar and velocity dissipation time scales in shear flow turbulence. *The Physics of Fluids* **21**, 307–310.

Bennani, A., J. N. Gence, and J. Mathieu (1985). The influence of a grid-generated turbulence on the development of chemical reactions. *AIChE Journal* **31**, 1157–1166.

Besnard, D. C., F. H. Harlow, R. M. Rauenzahn, and C. Zemach (1990). Spectral transport model of turbulence. Report LA-11821-MS, Los Alamos National Laboratory.

(1992). Spectral transport model for turbulence. Report LA-UR92-1666, Los Alamos National Laboratory.

Biagioli, F. (1997). Modeling of turbulent nonpremixed combustion with the PDF transport method: comparison with experiments and analysis of statistical error. In G. D. Roy, S. M. Frolow, and P. Givi (eds.), *Advanced Computation and Analysis of Combustion*. ENAS Publishers.

Biferale, L., A. Crisanti, M. Vergassola, and A. Vulpiani (1995). Eddy diffusivities in scalar transport. *Physics of Fluids* **7**, 2725–2734.

Bilger, R. W. (1982). Molecular transport effects in turbulent diffusion flames at moderate Reynolds number. *AIAA Journal* **20**, 962–970.

(1989). Turbulent diffusion flames. *Annual Reviews of Fluid Mechanics* **21**, 101–135.

(1993). Conditional moment closure for turbulent reacting flow. *Physics of Fluids A: Fluid Dynamics* **5**, 436–444.

Bilger, R. W. and R. W. Dibble (1982). Differential molecular diffusion effects in turbulent mixing. *Combustion Science and Technology* **28**, 161–169.

Bilger, R. W., L. R. Saetran, and L. V. Krishnamoorthy (1991). Reaction in a scalar mixing layer. *Journal of Fluid Mechanics* **233**, 211–242.

Billingsley, P. (1979). *Probability and Measure*. New York: Wiley.

Bird, R. B., W. E. Stewart, and E. N. Lightfoot (2002). *Transport Phenomena* (2nd edn). New York: John Wiley & Sons.

Bischoff, K. B. (1966). Mixing and contacting in chemical reactors. *Industrial and Engineering Chemistry* **11**, 18–32.

Bogucki, D., J. A. Domaradzki, and P. K. Yeung (1997). Direct numerical simulations of passive scalars with $Pr > 1$ advected by turbulent flow. *Journal of Fluid Mechanics* **343**, 111–130.

Borghi, R. (1986). Application of Lagrangian models to turbulent combustion. *Combustion and Flame* **63**, 239–250.

(1988). Turbulent combustion modelling. *Progress in Energy and Combustion Science* **14**, 245–292.

(1990). Turbulent premixed combustion: Further discussions on the scales of fluctuations. *Combustion and Flame* **80**, 304–312.

Boris, J. and E. Oran (2000). *Numerical Simulation of Reacting Flows*. New York: Cambridge University Press.

Bourne, J. R. (1983). Mixing on the molecular scale (micromixing). *Chemical Engineering Science* **38**, 5–8.

Bourne, J. R. and H. L. Toor (1977). Simple criteria for mixing effects in complex reactions. *AIChE Journal* **23**, 602–604.

Bourne, J. R., E. Crivelli, and P. Rys (1977). Chemical selectivities disguised by mass diffusion. V. Mixing-disguised azo coupling reactions. *Helvetica Chimica Acta* **60**, 2944–2957.

Branley, N. and W. P. Jones (2001). Large eddy simulation of a turbulent non-premixed flame. *Combustion and Flame* **127**, 1914–1934.

Bray, K. N. C. and N. Peters (1994). Laminar flamelets in turbulent flames. In P. A. Libby and F. A. Williams (eds.), *Turbulent Reacting Flows*, pp. 63–113. New York: Academic Press.

Breidenthal, R. E. (1981). Structure in turbulent mixing layers and wakes using a chemical reaction. *Journal of Fluid Mechanics* **109**, 1–24.

Brodkey, R. S. (1966). Turbulent motion and mixing in a pipe. *AIChE Journal* **12**, 403–404.

(1984). *The Phenomena of Fluid Motions* (4th edn). New York: Addison-Wesley.

Brown, R. J. and R. W. Bilger (1996). An experimental study of a reactive plume in grid turbulence. *Journal of Fluid Mechanics* **312**, 373–407.

Buch, K. A. and W. J. A. Dahm (1996). Experimental study of the fine-scale structure of conserved scalar mixing in turbulent shear flows. Part 1. $Sc \gg 1$. *Journal of Fluid Mechanics* **317**, 21–71.

(1998). Experimental study of the fine-scale structure of conserved scalar mixing in turbulent shear flows. Part 2. $Sc \approx 1$. *Journal of Fluid Mechanics* **364**, 1–29.

Burke, S. P. and T. E. W. Schumann (1928). Diffusion flames. *Industrial and Engineering Chemistry* **20**, 998–1004.

Bushe, W. K. and H. Steiner (1999). Conditional moment closure for large eddy simulation of nonpremixed turbulent reacting flows. *Physics of Fluids* **11**, 1896–1906.

Cai, X. D., E. E. O'Brien, and F. Ladeinde (1996). Uniform mean scalar gradient in grid turbulence: Asymptotic probability distribution of a passive scalar. *Physics of Fluids* **8**, 2555–2557.

Caillau, P. (1994). *Modelisation et Simulation de la Combustion Turbulente par une Approche Probabiliste Eulerienne Lagrangienne.* Ph. D. thesis, Université de Rouen, France.

Calmet, I. and J. Magnaudet (1997). Large-eddy simulation of high-Schmidt number mass transfer in a turbulent channel flow. *Physics of Fluids* **9**, 438–455.

Cannon, S. M., B. S. Brewster, and L. D. Smoot (1999). PDF modeling of lean premixed combustion using in situ tabulated chemistry. *Combustion and Flame* **119**, 233–252.

Canuto, V. M. and M. S. Dubovikov (1996a). A dynamical model for turbulence. I. General formalism. *Physics of Fluids* **8**, 571–586.

(1996b). A dynamical model for turbulence. II. Shear-driven flows. *Physics of Fluids* **8**, 587–598.

Canuto, V. M., M. S. Dubovikov, Y. Cheng, and A. Dienstfrey (1996). A dynamical model for turbulence. III. Numerical results. *Physics of Fluids* **8**, 599–613.

Cha, C. M., G. Kosály, and H. Pitsch (2001). Modeling extinction and reignition in turbulent nonpremixed combustion using a doubly-conditional moment closure approach. *Physics of Fluids* **13**, 3824–3834.

Chasnov, J. R. (1991). Simulation of inertial-conductive subrange. *Physics of Fluids A: Fluid Dynamics* **3**, 188–200.

(1994). Similarity states of passive scalar transport in isotropic turbulence. *Physics of Fluids* **6**, 1036–1051.

(1998). The viscous-convective subrange in nonstationary turbulence. *Physics of Fluids* **10**, 1191–1205.

Chen, H., S. Chen, and R. H. Kraichnan (1989). Probability distribution of a stochastically advected scalar field. *Physical Review Letters* **63**, 2657–2660.

Chen, J. Y. and W. Kollmann (1988). PDF modeling of chemical nonequilibrium effects in turbulent nonpremixed hydrocarbon flames. In *Twenty-second Symposium (International) on Combustion*, pp. 645–653. Pittsburgh, PA: The Combustion Institute.

Chen, J. Y., W. Kollmann, and R. W. Dibble (1989). PDF modeling of turbulent nonpremixed methane jet flames. *Combustion Science and Technology* **64**, 315–346.

Chen, J. Y., R. W. Dibble, and R. W. Bilger (1990). PDF modeling of turbulent nonpremixed $CO/H_2/N_2$ jet flames with reduced mechanisms. In *23rd International Symposium on Combustion*, pp. 775–780. Pittsburgh, PA: The Combustion Institute.

Ching, E. S. C. (1996). General formula for stationary or statistically homogeneous probability density functions. *Physical Review E* **53**, 5899–5903.

Chordá, R., J. A. Blasco, and N. Fueyo (2002). An efficient particle-locating algorithm for application in arbitrary 2D and 3D grids. *International Journal of Multiphase Flow* **28**, 1565–1580.

Christiansen, D. E. (1969). Turbulent liquid mixing: associated scalar spectra and light-scattering measurements. *Industrial and Engineering Chemistry Fundamentals* **8**, 263–271.

Chung, P. M. (1969). A simplified statistical model for turbulent, chemically reacting shear flows. *AIAA Journal* **7**, 1982–1991.

(1970). Chemical reaction in a turbulent flow field with uniform velocity gradient. *The Physics of Fluids* **13**, 1153–1165.

(1976). A kinetic-theory approach to turbulent chemically reacting flows. *Combustion Science and Technology* **13**, 123–153.

Clark, T. T. and C. Zemach (1995). A spectral model applied to homogeneous turbulence. *Physics of Fluids* **7**, 1674–1694.

Colucci, P. J., F. A. Jaberi, P. Givi, and S. B. Pope (1998). Filtered density function for large eddy simulation of turbulent reacting flows. *Physics of Fluids* **10**, 499–515.

Cook, A. W. and J. J. Riley (1994). A subgrid model for equilibrium chemistry in turbulent flows. *Physics of Fluids* **6**, 2868–2870.

(1996). Direct numerical simulation of a turbulent reactive plume on a parallel computer. *Journal of Computational Physics* **129**, 263–283.

(1998). Subgrid-scale modeling for turbulent reacting flows. *Combustion and Flame* **112**, 593–606.

Cook, A. W., J. J. Riley, and G. Kosály (1997). A laminar flamelet approach to subgrid-scale chemistry in turbulent flows. *Combustion and Flame* **109**, 332–341.

Cooper, D., D. C. Jackson, and B. E. Launder (1993a). Impinging jet studies for turbulence model assessment – I. Flow-field experiments. *International Journal of Heat and Mass Transfer* **36**, 2675–2684.

(1993b). Impinging jet studies for turbulence model assessment – II. An examination of the performance of four turbulence models. *International Journal of Heat and Mass Transfer* **36**, 2685–2696.

Correa, S. M. and S. B. Pope (1992). Comparison of a Monte Carlo PDF/finite-volume mean flow model with bluff-body Raman data. In *Twenty-fourth Symposium (International) on Combustion*, pp. 279–285. Pittsburgh, PA: The Combustion Institute.

Correa, S. M., A. Gulati, and S. B. Pope (1994). Raman measurements and joint pdf modeling of a nonpremixed bluff-body-stabilized methane flame. In *Twenty-fifth International Symposium on Combustion*, pp. 1167–1173. Pittsburgh, PA: The Combustion Institute.

Corrsin, S. (1951a). The decay of isotropic temperature fluctuations in an isotropic turbulence. *Journal of Aeronautical Science* **18**, 417–423.

(1951b). On the spectrum of isotropic temperature fluctuations in isotropic turbulence. *Journal of Applied Physics* **22**, 469–473.

(1957). Simple theory of an idealized turbulent mixer. *AIChE Journal* **3**, 329–330.

(1958). Statistical behavior of a reacting mixture in isotropic turbulence. *The Physics of Fluids* **1**, 42–47.

(1961). The reactant concentration spectrum in turbulent mixing with a first-order reaction. *Journal of Fluid Mechanics* **11**, 407–416.

(1964). The isotropic turbulent mixer: Part II. Arbitrary Schmidt number. *AIChE Journal* **10**, 870–877.

(1968). Effect of passive chemical reaction on turbulent dispersion. *AIAA Journal* **6**, 1797–1798.
Cremer, M. A., P. A. McMurtry, and A. R. Kerstein (1994). Effects of turbulence length-scale distribution on scalar mixing in homogeneous turbulent flow. *Physics of Fluids* **6**, 2143–2153.
Cressie, N. A. C. (1991). *Statistics for Spatial Data*. New York: John Wiley & Sons, Inc.
Curl, R. L. (1963). Dispersed phase mixing: 1. Theory and effects in simple reactors. *AIChE Journal* **9**, 175–181.
Curtis, W. K., R. O. Fox, and K. Halasi (1992). Numerical stability analysis of a class of functional differential equations. *SIAM Journal of Applied Mathematics* **52**, 810–834.
Dahm, W. J. A. and P. E. Dimotakis (1990). Mixing at large Schmidt number in the self-similar far field of turbulent jets. *Journal of Fluid Mechanics* **217**, 299–330.
Dahm, W. J. A., K. B. Southerland, and K. A. Buch (1991). Direct, high resolution, four-dimensional measurements of the fine scale structure of $Sc \gg 1$ molecular mixing in turbulent flows. *Physics of Fluids A: Fluid Dynamics* **3**, 1115–1127.
Dally, B. B., D. F. Fletcher, and A. R. Masri (1998). Measurements and modeling of a bluff-body stabilized flame. *Combustion Theory and Modelling* **2**, 193–219.
Daly, B. J. and F. H. Harlow (1970). Transport equations in turbulence. *The Physics of Fluids* **13**, 2634–2649.
Danckwerts, P. V. (1953). The definition and measurement of some characteristics of mixtures. *Applied Scientific Research* **A3**, 279–296.
(1957). Measurement of molecular homogeneity in a mixture. *Chemical Engineering Science* **7**, 116–117.
(1958). The effect of incomplete mixing on homogeneous reactions. *Chemical Engineering Science* **8**, 93–101.
Delarue, B. J. and S. B. Pope (1997). Application of PDF methods to compressible turbulent flows. *Physics of Fluids* **9**, 2704–2715.
(1998). Calculations of subsonic and supersonic turbulent reacting mixing layers using probability density function methods. *Physics of Fluids* **10**, 497–498.
Desjardin, P. E. and S. H. Frankel (1996). Assessment of turbulent combustion submodels using the linear eddy model. *Combustion and Flame* **104**, 343–357.
(1998). Large eddy simulation of a nonpremixed reacting jet: Application and assessment of subgrid-scale combustion models. *Physics of Fluids* **10**, 2298–2314.
Dimotakis, P. E. and P. L. Miller (1990). Some consequences of the boundedness of scalar fluctuations. *Physics of Fluids A: Fluid Dynamics* **2**, 1919–1920.
Domingo, P. and T. Benazzouz (2000). Direct numerical simulation and modeling of a nonequilibrium turbulent plasma. *AIAA Journal* **38**, 73–78.
Dopazo, C. (1975). Probability density function approach for a turbulent axisymmetric heated jet. Centerline evolution. *The Physics of Fluids* **18**, 397–404.
(1979). Relaxation of initial probability density functions in the turbulent convection of scalar fields. *The Physics of Fluids* **22**, 20–30.
(1994). Recent developments in PDF methods. In P. A. Libby and F. A. Williams (eds.), *Turbulent Reacting Flows*, pp. 375–474. New York: Academic Press.
Dopazo, C. and E. E. O'Brien (1973). Isochoric turbulent mixing of two rapidly reacting chemical species with chemical heat release. *The Physics of Fluids* **16**, 2075–2081.
(1974). An approach to the autoignition of a turbulent mixture. *Acta Astronautica* **1**, 1239–1266.

(1976). Statistical treatment of non-isothermal chemical reactions in turbulence. *Combustion Science and Technology* **13**, 99–119.
Dowling, D. R. (1991). The estimated scalar dissipation rate in gas-phase turbulent jets. *Physics of Fluids A: Fluid Dynamics* **3**, 2229–2246.
(1992). Erratum: The estimated scalar dissipation rate in gas-phase turbulent jets [*Phys. Fluids A* **3**, 2229 (1991)]. *Physics of Fluids A: Fluid Dynamics* **4**, 453.
Dowling, D. R. and P. E. Dimotakis (1990). Similarity of the concentration field of gas-phase turbulent jets. *Journal of Fluid Mechanics* **218**, 109–141.
Drake, M. C., R. W. Pitz, and W. Shyy (1986). Conserved scalar probability density functions in a turbulent jet diffusion flame. *Journal of Fluid Mechanics* **171**, 27–51.
Dreeben, T. D. and S. B. Pope (1997a). Probability density function and Reynolds-stress modeling of near-wall turbulent flows. *Physics of Fluids* **9**, 154–163.
(1997b). Wall-function treatment in PDF methods for turbulent flows. *Physics of Fluids* **9**, 2692–2703.
(1998). PDF/Monte Carlo simulation of near-wall turbulent flows. *Journal of Fluid Mechanics* **357**, 141–166.
Durbin, P. A. (1982). Analysis of the decay of temperature fluctuations in isotropic turbulence. *The Physics of Fluids* **25**, 1328–1332.
Durbin, P. A., N. N. Mansour, and Z. Yang (1994). Eddy viscosity transport model for turbulent flow. *Physics of Fluids* **6**, 1007–1015.
Dutta, A. and J. M. Tarbell (1989). Closure models for turbulent reacting flows. *AIChE Journal* **35**, 2013–2027.
Echekki, T., A. R. Kerstein, T. D. Dreeben, and J.-Y. Chen (2001). 'One-dimensional turbulence' simulation of turbulent jet diffusion flames: Model formulation and illustrative applications. *Combustion and Flame* **125**, 1083–1105.
Eidson, T. M. (1985). Numerical simulation of the turbulent Rayleigh-Benard problem using subgrid modelling. *Journal of Fluid Mechanics* **158**, 245–268.
Elgobashi, S. E. and B. E. Launder (1983). Turbulent time scales and the dissipation rate of temperature variance in the thermal mixing layer. *The Physics of Fluids* **26**, 2415–2419.
Eswaran, V. and S. B. Pope (1988). Direct numerical simulations of the turbulent mixing of a passive scalar. *The Physics of Fluids* **31**, 506–520.
Feller, W. (1971). *An Introduction to Probability Theory and Its Applications* (2nd edn). New York: Wiley.
Ferziger, J. H. and M. Perić (2002). *Computational Methods for Fluid Dynamics* (3rd edn). New York: Springer.
Flagan, R. C. and J. P. Appleton (1974). A stochastic model of turbulent mixing with chemical reaction: Nitric oxide formation in a plug-flow burner. *Combustion and Flame* **23**, 249–267.
Fogler, H. S. (1999). *Elements of Chemical Reaction Engineering* (3rd edn). New York: Prentice Hall.
Fox, R. O. (1989). Steady-state IEM model: Singular perturbation analysis near perfect-micromixing limit. *Chemical Engineering Science* **44**, 2831–2842.
(1991). Micromixing effects in the Nicolis-Puhl reaction: Numerical bifurcation and stability analysis of the IEM model. *Chemical Engineering Science* **46**, 1829–1847.
(1992). The Fokker-Planck closure for turbulent molecular mixing: Passive scalars. *Physics of Fluids A: Fluid Dynamics* **4**, 1230–1244.
(1994). Improved Fokker-Planck model for the joint scalar, scalar gradient PDF. *Physics of Fluids* **6**, 334–348.

(1995). The spectral relaxation model of the scalar dissipation rate in homogeneous turbulence. *Physics of Fluids* **7**, 1082–1094.
(1996a). Computational methods for turbulent reacting flows in the chemical process industry. *Revue de l'Institut Français du Pétrole* **51**, 215–243.
(1996b). On velocity-conditioned scalar mixing in homogeneous turbulence. *Physics of Fluids* **8**, 2678–2691.
(1997). The Lagrangian spectral relaxation model of the scalar dissipation in homogeneous turbulence. *Physics of Fluids* **9**, 2364–2386.
(1998). On the relationship between Lagrangian micromixing models and computational fluid dynamics. *Chemical Engineering and Processing* **37**, 521–535.
(1999). The Lagrangian spectral relaxation model for differential diffusion in homogeneous turbulence. *Physics of Fluids* **11**, 1550–1571.
Fox, R. O. and M. R. Grier (1993). Numerical simulation of turbulent reacting flows using PDF methods. In G. B. Tatterson (ed.), *Process Mixing: Chemical and Biochemical Applications – Part II*, pp. 49–54. New York: AIChE.
Fox, R. O. and J. Villermaux (1990a). Micromixing effects in the $ClO_2^- + I^-$ reaction: Multiple-scale perturbation analysis and numerical simulation of the unsteady-state IEM model. *Chemical Engineering Science* **45**, 2857–2876.
(1990b). Unsteady-state IEM model: Numerical simulation and multiple-scale perturbation analysis near perfect-micromixing limit. *Chemical Engineering Science* **45**, 373–386.
Fox, R. O. and P. K. Yeung (1999). Forward and backward spectral transfer in the modeling of scalar mixing in homogeneous turbulence. In *Proceedings of the 3rd ASME/JSME Joint Fluids Engineering Conference*, San Francisco, CA.
(2003). Improved Lagrangian mixing models for passive scalars in isotropic turbulence. *Physics of Fluids* **15**, 961–985.
Fox, R. O., W. D. Curtis, and K. Halasi (1990). Linear stability analysis of the IEM model of micromixing. *Chemical Engineering Science* **45**, 3571–3583.
Fox, R. O., G. Erjaee, and Q. Zou (1994). Bifurcation and stability analysis of micromixing effects in the chlorite-iodide reaction. *Chemical Engineering Science* **49**, 3465–3484.
Fox, R. O., C. M. Cha, and P. Trouillet (2002). Lagrangian PDF mixing models of reacting flows. In *Proceedings of the 2002 CTR Summer Program*, Stanford, CA, pp. 369–380. Center for Turbulence Research.
Gao, F. (1991). Mapping closure and non-Gaussianity of the scalar probability density function in isotropic turbulence. *Physics of Fluids A: Fluid Dynamics* **3**, 2438–2444.
Gao, F. and E. E. O'Brien (1991). A mapping closure for multispecies Fickian diffusion. *Physics of Fluids A: Fluid Dynamics* **3**, 956–959.
(1993). A large-eddy simulation scheme for turbulent reacting flows. *Physics of Fluids A: Fluid Dynamics* **5**, 1282–1284.
(1994). Erratum: A large-eddy simulation scheme for turbulent reacting flows [*Phys. Fluids A* **5**, 1282 (1993)]. *Physics of Fluids* **6**, 1621.
Gardiner, C. W. (1990). *Handbook of Stochastic Methods for Physics, Chemistry and the Natural Sciences* (2nd edn). New York: Springer.
Gegner, J. P. and R. S. Brodkey (1966). Dye injection at the centerline of a pipe. *AIChE Journal* **12**, 817–819.
Germano, M., U. Piomelli, P. Moin, and W. H. Cabot (1991). A dynamic subgrid-scale eddy viscosity model. *Physics of Fluids* **7**, 1760–1765.
Gibson, C. H. (1968a). Fine structure of scalar fields mixed by turbulence. I. Zero-gradient points and minimal-gradient surfaces. *The Physics of Fluids* **11**, 2305–2315.

(1968b). Fine structure of scalar fields mixed by turbulence. II. Spectral theory. *The Physics of Fluids* **11**, 2316–2327.

Gibson, C. H. and P. A. Libby (1972). On turbulent flows with fast chemical reactions. Part II – The distribution of reactants and products near a reacting surface. *Combustion Science and Technology* **8**, 29–35.

Gibson, C. H. and W. H. Schwarz (1963a). Detection of conductivity fluctuations in a turbulent flow field. *Journal of Fluid Mechanics* **16**, 357–364.

(1963b). The universal equilibrium spectra of turbulent velocity and scalar fields. *Journal of Fluid Mechanics* **16**, 365–384.

Gibson, C. H., G. R. Stegen, and R. B. Williams (1970). Statistics of the fine structure of turbulent velocity and temperature fields measured at high Reynolds numbers. *Journal of Fluid Mechanics* **41**, 153–167.

Gicquel, L. Y. M., P. Givi, F. A. Jaberi, and S. B. Pope (2002). Velocity filtered density function for large eddy simulation of turbulent flows. *Physics of Fluids* **14**, 1196–1213.

Girimaji, S. S. (1992). On the modeling of scalar diffusion in isotropic turbulence. *Physics of Fluids A: Fluid Dynamics* **4**, 2529–2537.

(1993). A study of multiscalar mixing. *Physics of Fluids A: Fluid Dynamics* **5**, 1802–1809.

Givi, P. (1989). Model-free simulation of turbulent reactive flows. *Progress in Energy and Combustion Science* **15**, 1–107.

Givi, P. and P. A. McMurtry (1988). Nonpremixed reaction in homogeneous turbulence: Direct numerical simulations. *AIChE Journal* **34**, 1039–1042.

Gonzalez, M. (2000). Study of the anisotropy of a passive scalar field at the level of dissipation. *Physics of Fluids* **12**, 2302–2310.

Gordon, R. G. (1968). Error bounds in equilibrium statistical mechanics. *Journal of Mathematical Physics* **9**, 655–663.

Grant, H. L., B. A. Hughes, W. M. Vogel, and A. Moilliet (1968). The spectrum of temperature fluctuations in turbulent flow. *Journal of Fluid Mechanics* **34**, 423–442.

Guiraud, P., J. Bertrand, and J. Costes (1991). Laser measurements of local velocity and concentration in a turbulent jet-stirred tubular reactor. *Chemical Engineering Science* **46**, 1289–1297.

Hamba, F. (1987). Statistical analysis of chemically reacting passive scalars in turbulent shear flows. *Journal of the Physical Society of Japan* **56**, 79–96.

Hanjalić, K. (1994). Advanced turbulence closure models: A view of current status and future prospects. *International Journal of Heat and Fluid Flow* **15**, 178–203.

Hanjalić, K., S. Jakirlić, and I. Hadzić (1997). Expanding the limits of "equilibrium" second-moment turbulence closures. *Fluid Dynamics Research* **20**, 25–41.

Harris, C. K., D. Roekaerts, F. J. J. Rosendal, F. G. J. Buitendijk, P. Daskopoulos, A. J. N. Vreenegoor, and H. Wang (1996). Computational fluid dynamics for chemical reactor engineering. *Chemical Engineering Science* **51**, 1569–1594.

Haworth, D. C. (2001). Application of turbulent combustion modeling. In J. P. A. J. van Beeck, L. Vervisch, and D. Veynante (eds.), *Turbulence and Combustion*, Lecture Series 2001–03. Rhode-Saint-Genèse, Belgium: Von Karman Institute for Fluid Dynamics.

Haworth, D. C. and S. H. El Tahry (1991). Probability density function approach for multidimensional turbulent flow calculations with application to in-cylinder flows in reciprocating engines. *AIAA Journal* **29**, 208–218.

Haworth, D. C. and S. B. Pope (1986). A generalized Langevin model for turbulent flows. *The Physics of Fluids* **29**, 387–405.

Heeb, T. G. and R. S. Brodkey (1990). Turbulent mixing with multiple second-order chemical reactions. *AIChE Journal* **36**, 1457–1470.

Hewson, J. and A. R. Kerstein (2001). Stochastic simulation of transport and chemical kinetics in turbulent $CO/H_2/N_2$ flames. *Combustion Theory and Modelling* **5**, 669–697.

Hill, C. G. (1977). *An Introduction to Chemical Engineering Kinetics and Reactor Design*. New York: John Wiley & Sons.

Hill, J. C. (1976). Homogeneous turbulent mixing with chemical reaction. *Annual Reviews of Fluid Mechanics* **8**, 135–161.

Hinze, J. O. (1975). *Turbulence* (2nd edn). New York: McGraw-Hill.

Hockney, R. W. and J. W. Eastwood (1988). *Computer Simulations Using Particles*. New York: Adam Hilger.

Hsu, A. T., M. S. Anand, and M. K. Razdan (1990). An assessment of PDF versus finite-volume methods for turbulent reacting flow calculations. In Paper 96-0523, AIAA.

Hughes, T. J. R., A. A. Oberai, and L. Mazzei (2001a). Large eddy simulation of turbulent channel flows by the variational multiscale method. *Physics of Fluids* **13**, 1784–1799.

Hughes, T. J. R., L. Mazzei, A. A. Oberai, and A. A. Wray (2001b). The multiscale formulation of large eddy simulation: Decay of homogeneous isotropic turbulence. *Physics of Fluids* **13**, 505–512.

Hulek, T. and R. P. Lindstedt (1996). Computations of steady-state and transient premixed turbulent flames using pdf methods. *Combustion and Flame* **104**, 481–504.

Jaberi, F. A., R. S. Miller, C. K. Madnia, and P. Givi (1996). Non-Gaussian scalar statistics in homogeneous turbulence. *Journal of Fluid Mechanics* **313**, 241–282.

Jaberi, F. A., P. J. Colucci, S. James, P. Givi, and S. B. Pope (1999). Filtered mass density function for large-eddy simulation of turbulent reacting flows. *Journal of Fluid Mechanics* **401**, 85–121.

Janicka, J., W. Kolbe, and W. Kollmann (1979). Closure of the transport equation for the probability density function of turbulent scalar fields. *Journal of Non-Equilibrium Thermodynamics* **4**, 47–66.

Jayesh and S. B. Pope (1995). Stochastic model for turbulent frequency. Technical Report FDA 95-05, Cornell University.

Jayesh and Z. Warhaft (1992). Probability distribution, conditional dissipation, and transport of passive temperature fluctuations in grid-generated turbulence. *Physics of Fluids A: Fluid Dynamics* **4**, 2292–2307.

Jenny, P., M. Muradoglu, S. B. Pope, and D. A. Caughey (2001a). PDF simulations of a bluff-body stabilized flow. *Journal of Computational Physics* **169**, 1–23.

Jenny, P., S. B. Pope, M. Muradoglu, and D. A. Caughey (2001b). A hybrid algorithm for the joint PDF equation of turbulent reactive flows. *Journal of Computational Physics* **166**, 218–252.

Jiang, T.-L. and E. E. O'Brien (1991). Simulation of scalar mixing by stationary isotropic turbulence. *Physics of Fluids A: Fluid Dynamics* **3**, 1612–1624.

Jiang, T.-L., P. Givi, and F. Gao (1992). Binary and trinary scalar mixing by Fickian diffusion – Some mapping closure results. *Physics of Fluids A: Fluid Dynamics* **4**, 1028–1035.

Jiménez, J., A. Liñan, M. M. Rogers, and F. J. Higuera (1997). *A priori* testing of subgrid models for chemically reacting non-premixed turbulent flows. *Journal of Fluid Mechanics* **349**, 149–171.

Jones, W. P. and M. Kakhi (1997). Application of the transported PDF approach to hydrocarbon turbulent jet diffusion flames. *Combustion Science and Technology* **129**, 393–430.

(1998). Pdf modeling of finite-rate chemistry effects in turbulent nonpremixed jet flames. *Combustion and Flame* **115**, 210–229.

Jou, W. H. and J. J. Riley (1989). Progress in direct numerical simulation of turbulent reacting flows. *AAIA Journal* **27**, 1543–1557.

Juneja, A. and S. B. Pope (1996). A DNS study of turbulent mixing of two passive scalars. *Physics of Fluids* **8**, 2161–2184.

Keeler, R. N., E. E. Petersen, and J. M. Prausnitz (1965). Mixing and chemical reaction in turbulent flow reactors. *AIChE Journal* **11**, 221–227.

Kerstein, A. R. (1988). A linear-eddy model of turbulent scalar transport and mixing. *Combustion Science and Technology* **60**, 391–421.

(1989). Linear-eddy modeling of turbulent transport. II: Application to shear layer mixing. *Combustion and Flame* **75**, 397–413.

(1990). Linear-eddy modelling of turbulent transport. Part 3. Mixing and differential diffusion in round jets. *Journal of Fluid Mechanics* **216**, 411–435.

(1991a). Linear-eddy modelling of turbulent transport. Part 6. Microstructure of diffusive scalar mixing fields. *Journal of Fluid Mechanics* **231**, 361–394.

(1991b). Linear-eddy modelling of turbulent transport. Part V. Geometry of scalar interfaces. *Physics of Fluids A: Fluid Dynamics* **3**, 1110–1114.

(1992). Linear-eddy modelling of turbulent transport. Part 7. Finite rate chemistry and multi-stream mixing. *Journal of Fluid Mechanics* **240**, 289–313.

(1999a). One-dimensional turbulence: Model formulation and application to homogeneous turbulence, shear flows, and buoyant stratified flows. *Journal of Fluid Mechanics* **392**, 277–334.

(1999b). One-dimensional turbulence: Part 2. Staircases in double-diffusive convection. *Dynamics of Atmospheres and Oceans* **30**, 25–46.

(2002). One-dimensional turbulence: A new approach to high-fidelity subgrid closure of turbulent flow simulations. *Computer Physics Communications* **148**, 1–16.

Kerstein, A. R. and T. D. Dreeben (2000). Prediction of turbulent free shear flow using a simple stochastic model. *Physics of Fluids* **12**, 418–424.

Kerstein, A. R., M. A. Cremer, and P. A. McMurtry (1995). Scaling properties of differential molecular effects in turbulence. *Physics of Fluids* **7**, 1999–2007.

Klimenko, A. Y. (1990). Multicomponent diffusion of various admixtures in turbulent flow. *Fluid Dynamics* **25**, 327–334.

(1995). Note on the conditional moment closure in turbulent shear flows. *Physics of Fluids* **7**, 446–448.

Klimenko, A. Y. and R. W. Bilger (1999). Conditional moment closure for turbulent combustion. *Progress in Energy and Combustion Science* **25**, 595–687.

Kloeden, P. E. and E. Platen (1992). *Numerical Solution of Stochastic Differential Equations*. Berlin: Springer-Verlag.

Kollmann, W. (1990). The PDF approach to turbulent flow. *Theoretical and Computational Fluid Dynamics* **1**, 249–285.

Kollmann, W. and J. Janicka (1982). The probability density function of a passive scalar in turbulent shear flows. *The Physics of Fluids* **25**, 1755–1769.

Kolmogorov, A. N. (1941a). Dissipation of energy in locally isotropic turbulence. *Dokl. Akad. Nauk SSSR* **32**, 19–21.

(1941b). The local structure of turbulence in incompressible viscous fluid for very large Reynolds numbers. *Dokl. Akad. Nauk SSSR* **30**, 299–303.
(1962). A refinement of previous hypotheses concerning the local structure of turbulence in viscous incompressible fluid for high Reynolds numbers. *Journal of Fluid Mechanics* **13**, 82–85.
Komori, S., T. Kanzaki, Y. Murakami, and H. Ueda (1989). Simultaneous measurements of instantaneous concentrations of two species being mixed in a turbulent flow by using a combined laser-induced fluorescence and laser-scattering technique. *Physics of Fluids A: Fluid Dynamics* **1**, 349–352.
Komori, S., T. Kanzaki, and Y. Murakami (1991a). Simultaneous measurements of instantaneous concentrations of two reacting species in a turbulent flow with a rapid reaction. *Physics of Fluids A: Fluid Dynamics* **3**, 507–510.
Komori, S., J. C. R. Hunt, K. Kanzaki, and Y. Murakami (1991b). The effects of turbulent mixing on the correlation between two species and on concentration fluctuations in non-premixed reacting flows. *Journal of Fluid Mechanics* **228**, 629–659.
Koochesfahani, M. M. and P. E. Dimotakis (1986). Mixing and chemical reactions in a turbulent liquid mixing layer. *Journal of Fluid Mechanics* **170**, 83–112.
Kosály, G. (1989). Scalar mixing in isotropic turbulence. *Physics of Fluids A: Fluid Dynamics* **1**, 758–760.
Kraichnan, R. H. (1968). Small-scale structure of a scalar field convected by turbulence. *The Physics of Fluids* **11**, 945–953.
(1974). Convection of a passive scalar by a quasi-uniform random straining field. *Journal of Fluid Mechanics* **64**, 737–762.
Kronenburg, A. and R. W. Bilger (1997). Modeling of differential diffusion effects in nonpremixed nonreacting turbulent flow. *Physics of Fluids* **9**, 1435–1447.
Kuznetsov, V. R. and V. A. Sabel'nikov (1990). *Turbulence and Combustion*. New York: Hemisphere.
Lakatos, B. and E. Varga (1988). Probability density functions based micromixing model of chemical reactors. *Computers in Chemical Engineering* **12**, 165–169.
Langford, J. A. and R. D. Moser (1999). Optimal LES formulations for isotropic turbulence. *Journal of Fluid Mechanics* **398**, 321–346.
Launder, B. E. (1991). Current capabilities for modelling turbulence in industrial flows. *Applied Scientific Research* **48**, 247–269.
(1996). An introduction to single-point closure methodology. In T. B. Gatski, M. Y. Hussaini, and J. L. Lumley (eds.), *Simulation and Modeling of Turbulent Flows*, chap. 6, pp. 243–310. New York: Oxford University Press.
Launder, B. E. and D. B. Spalding (1972). *Mathematical Models of Turbulence*. London: Academic Press.
Lee, J. (1966). Isotropic turbulent mixing under a second-order chemical reaction. *The Physics of Fluids* **9**, 1753–1763.
Lee, J. and R. S. Brodkey (1964). Turbulent motion and mixing in a pipe. *AIChE Journal* **10**, 187–193.
Leonard, A. D. and J. C. Hill (1988). Direct numerical simulation of turbulent flows with chemical reaction. *Journal of Scientific Computing* **3**, 25–43.
(1991). Scalar dissipation and mixing in turbulent reacting flows. *Physics of Fluids A: Fluid Dynamics* **3**, 1286–1299.
(1992). Mixing and chemical reaction in sheared and nonsheared homogeneous turbulence. *Fluid Mechanics Research* **10**, 273–297.

Lesieur, M. (1997). *Turbulence in Fluids* (3rd edn). Dordrecht: Kluwer Academic Publishers.
Levenspiel, O. (1998). *Chemical Reaction Engineering* (3rd edn). New York: John Wiley & Sons.
Lin, C.-H. and E. E. O'Brien (1972). Two species isothermal reactions in homogeneous turbulence. *Astronautica Acta* **17**, 771–781.
Lin, C.-H. and E. E. O'Brien (1974). Turbulent shear flow mixing and rapid chemical reactions: an analogy. *Journal of Fluid Mechanics* **64**, 195–206.
Lundgren, T. S. (1985). The concentration spectrum of the product of a fast bimolecular reaction. *Chemical Engineering Science* **40**, 1641–1652.
Maas, U. and S. B. Pope (1992). Simplifying chemical kinetics: Intrinsic low-dimensional manifolds in composition space. *Combustion and Flame* **88**, 239–264.
McComb, W. D. (1990). *The Physics of Fluid Turbulence*. Oxford: Oxford University Press.
McGraw, R. (1997). Description of aerosol dynamics by the quadrature method of moments. *Aerosol Science and Technology* **27**, 255–265.
McKelvey, K. N., H.-N. Yieh, S. Zakanycz, and R. S. Brodkey (1975). Turbulent motion, mixing, and kinetics in a chemical reactor configuration. *AIChE Journal* **21**, 1165–1176.
McMurtry, P. A., S. Menon, and A. R. Kerstein (1993). Linear eddy modeling of turbulent combustion. *Energy and Fuels* **7**, 817–826.
Mann, R., P. Mavros, and J. C. Middleton (1981). A structured stochastic flow model for interpreting flow follower data from a stirred vessel. *Transactions of the Institution of Chemical Engineers* **59**, 127.
Mann, R., D. Vlaev, V. Lossev, S. D. Vlaev, J. Zahradnik, and P. Seichter (1997). A network of zones analysis of the fundamentals of gas liquid mixing in an industrial stirred bioreactor. *Récent Progrès dans le Génie des Procédes* **11**, 223–230.
Mantel, T. and R. Borghi (1994). A new model of premixed wrinkled flame propagation based on a scalar dissipation equation. *Combustion and Flame* **96**, 443–457.
Mao, K. W. and H. L. Toor (1971). Second-order chemical reactions with turbulent mixing. *Industrial and Engineering Chemistry Fundamentals* **10**, 192–197.
Marchisio, D. L. and R. O. Fox (2003). Direct quadrature method of moments: Derivation, analysis and applications. *Journal of Computational Physics* (in press).
Marchisio, D. L., J. T. Pikturna, R. O. Fox, R. D. Vigil, and A. A. Barresi (2003). Quadrature method of moments for population-balance equations. *AIChE Journal* **49**, 1266–1276.
Marchisio, D. L., R. D. Vigil, and R. O. Fox (2003). Quadrature method of moments for aggregation-breakage processes. *Journal of Colloid and Interface Science* **258**, 322–334.
Masri, A. R., R. Cao, S. B. Pope, and G. M. Goldin (2003). PDF calculations of turbulent lifted flames of H_2/N_2 issuing into a vitiated co-flow. *Combustion Theory and Modelling* (in press).
Masri, A. R. and S. B. Pope (1990). PDF calculations of piloted turbulent nonpremixed flames of methane. *Combustion and Flame* **81**, 13–29.
Mehta, R. V. and J. M. Tarbell (1983a). A four-environment model of mixing and chemical reaction. Part I – Model development. *AIChE Journal* **29**, 320–328.
(1983b). A four-environment model of mixing and chemical reaction. Part II – Comparison with experiments. *AIChE Journal* **29**, 329–337.
Mell, W. E., G. Kosály, and J. J. Riley (1991). The length-scale dependence of scalar mixing. *Physics of Fluids A: Fluid Dynamics* **3**, 2474–2476.
Mell, W. E., V. Nilsen, G. Kosály, and J. J. Riley (1994). Investigation of closure models for nonpremixed turbulent reacting flows. *Physics of Fluids* **6**, 1331–1356.

Meneveau, C. and J. Katz (2000). On the Lagrangian nature of the turbulence energy cascade. *Annual Reviews of Fluid Mechanics* **32**, 1–32.

Meneveau, C., T. S. Lund, and W. Cabot (1996). A Lagrangian dynamic subgrid-scale model of turbulence. *Journal of Fluid Mechanics* **319**, 353–385.

Meyers, R. E. and E. E. O'Brien (1981). The joint PDF of a scalar and its gradient at a point in a turbulent flow. *Combustion Science and Technology* **26**, 123–134.

Miller, P. L. (1991). *Mixing in High Schmidt Number Turbulent Jets*. Ph. D. thesis, California Institute of Technology.

Miller, P. L. and P. E. Dimotakis (1991). Reynolds number dependence of scalar fluctuations in a high Schmidt number turbulent jet. *Physics of Fluids A: Fluid Dynamics* **3**, 1156–1163.

(1996). Measurements of scalar power spectra in high Schmidt number turbulent jets. *Journal of Fluid Mechanics* **308**, 129–146.

Miller, R. S., S. H. Frankel, C. K. Madnia, and P. Givi (1993). Johnson-Edgeworth translation for probability modeling of binary scalar mixing in turbulent flows. *Combustion Science and Technology* **91**, 21–52.

Minier, J.-P. and J. Pozorski (1997). Derivation of a PDF model for turbulent flows based on principles from statistical physics. *Physics of Fluids* **9**, 1748–1753.

(1999). Wall-boundary conditions in probability density function methods and application to a turbulent channel flow. *Physics of Fluids* **11**, 2632–2644.

Möbus, H., P. Gerlinger, and D. Brüggemann (1999). Monte Carlo PDF simulation of compressible turbulent diffusion flames using detailed chemical kinetics. In Paper 99-0198, AIAA.

(2001). Comparison of Eulerian and Lagrangian Monte Carlo PDF methods for turbulent diffusion flames. *Combustion and Flame* **124**, 519–534.

Moin, P. and K. Mahesh (1998). Direct numerical simulation: A tool for turbulence research. *Annual Reviews of Fluid Mechanics* **30**, 539–578.

Muradoglu, M. and S. B. Pope (2002). Local time stepping algorithm for solving PDF turbulence model equations. *AIAA Journal* **40**, 1755–1763.

Muradoglu, M., P. Jenny, S. B. Pope, and D. A. Caughey (1999). A consistent hybrid finite-volume/particle method for the PDF equations of turbulent reactive flows. *Journal of Computational Physics* **154**, 342–371.

Muradoglu, M., S. B. Pope, and D. A. Caughey (2001). The hybrid method for the PDF equations of turbulent reactive flows: Consistency conditions and correction algorithms. *Journal of Computational Physics* **172**, 841–878.

Nauman, E. B. and B. A. Buffham (1983). *Mixing in Continuous Flow Systems*. New York: John Wiley and Sons.

Nilsen, V. and G. Kosály (1997). Differentially diffusing scalars in turbulence. *Physics of Fluids* **9**, 3386–3397.

(1998). Differential diffusion in turbulent reacting flows. *Combustion and Flame* **117**, 493–513.

Nishimura, Y. and M. Matsubara (1970). Micromixing theory via the two-environment model. *Chemical Engineering Science* **25**, 1785–1797.

Nomura, K. K. and S. E. Elgobashi (1992). Mixing characteristics of an inhomogeneous scalar in isotropic homogeneous sheared turbulence. *Physics of Fluids A: Fluid Dynamics* **4**, 606–625.

Nooren, P. A., H. A. Wouters, T. W. J. Peeters, D. Roekaerts, U. Maas, and D. Schmidt (1997). Monte Carlo PDF modelling of a turbulent natural-gas diffusion flame. *Combustion Theory and Modelling* **1**, 79–96.

Norris, A. T. (1993). *The Application of PDF Methods to Piloted Diffusion Flames*. Ph. D. thesis, Cornell University.
Norris, A. T. and S. B. Pope (1991). Turbulent mixing model based on ordered pairing. *Combustion and Flame* **83**, 27–42.
(1995). Modeling of extinction in turbulent diffusion flames by the velocity-dissipation-composition PDF method. *Combustion and Flame* **100**, 211–220.
Nye, J. O. and R. S. Brodkey (1967a). Light probe for measurement of turbulent concentration fluctuations. *Review of Scientific Instrumentation* **38**, 26–28.
(1967b). The scalar spectrum in the viscous-convective subrange. *Journal of Fluid Mechanics* **29**, 151–163.
O'Brien, E. E. (1966). Closure approximations applied to stochastically distributed second-order reactions. *The Physics of Fluids* **9**, 1561–1565.
(1968a). Closure for stochastically distributed second-order reactants. *The Physics of Fluids* **11**, 1883–1888.
(1968b). Lagrangian history direct interaction equations for isotropic turbulent mixing under a second-order chemical reaction. *The Physics of Fluids* **11**, 2328–2335.
(1971). Turbulent mixing of two rapidly reacting chemical species. *The Physics of Fluids* **14**, 1326–1331.
(1980). The probability density function approach to reacting turbulent flows. In P. A. Libby and F. A. Williams (eds.), *Turbulent Reacting Flows*, pp. 185–203. Berlin: Springer.
O'Brien, E. E. and T.-L. Jiang (1991). The conditional dissipation rate of an initially binary scalar in homogeneous turbulence. *Physics of Fluids A: Fluid Dynamics* **3**, 3121–3123.
Obukhov, A. M. (1949). Structure of the temperature field in a turbulent flow. *Izv. Akad. Nauk SSSR, Ser. Geogr. Geofiz.* **13**, 58–69.
Ott, R. J. and P. Rys (1975). Chemical selectivities disguised by mass diffusion. I. A simple model of mixing-disguised reactions in solution. *Helvetica Chimica Acta* **58**, 2074–2093.
Ottino, J. M. (1980). Lamellar mixing models for structured chemical reactions and their relationship to statistical models; macro- and micromixing and the problem of averages. *Chemical Engineering Science* **35**, 1377–1391.
(1981). Efficiency of mixing from data on fast reactions in multi-jet reactors and stirred tanks. *AIChE Journal* **27**, 184–192.
(1982). Description of mixing with diffusion and reaction in terms of the concept of material interfaces. *Journal of Fluid Mechanics* **114**, 83–103.
Overholt, M. R. and S. B. Pope (1996). DNS of a passive scalar with imposed mean scalar gradient in isotropic turbulence. *Physics of Fluids* **8**, 3128–3148.
Paul, E. L. and R. E. Treybal (1971). Mixing and product distribution for a liquid-phase, second-order, competitive-consecutive reaction. *AIChE Journal* **17**, 718–724.
Peters, N. (1984). Laminar diffusion flamelet models in non-premixed turbulent combustion. *Progress in Energy and Combustion Science* **10**, 319–339.
(2000). *Turbulent Combustion*. Cambridge Monographs on Mechanics. Cambridge: Cambridge University Press.
Pfilzner, M., A. Mack, N. Brehm, A. Leonard, and I. Romaschov (1999). Implementation and validation of a PDF transport algorithm with adaptive number of particles in industrially relevant flows. In *Computational Technologies for Fluid/Thermal/Structural/Chemical Systems with Industrial Applications*, vol. 397-1, pp. 93–104. ASME.
Pipino, M. and R. O. Fox (1994). Reactive mixing in a tubular jet reactor: A comparison of PDF simulations with experimental data. *Chemical Engineering Science* **49**, 5229–5241.

Pit, F. (1993). *Modélisation du mélange pour la simulation d'écoulements réactifs turbulents: Essais de modèles probabilistes eulériens-lagrangiens*. Ph. D. thesis, Université de Rouen, France.

Pitsch, H. and N. Peters (1998). A consistent flamelet formulation for non-premixed combustion considering differential diffusion effects. *Combustion and Flame* **144**, 26–40.

Pitsch, H. and H. Steiner (2000). Large-eddy simulation of a turbulent piloted methane/air diffusion flame (Sandia flame D). *Physics of Fluids* **12**, 2541–2554.

Poinsot, T. and D. Veynante (2001). *Theoretical and Numerical Combustion*. Philadelphia, PA: R. T. Edwards.

Pope, S. B. (1981a). Monte Carlo calculations of premixed turbulent flames. In *Eighteenth Symposium (International) on Combustion*, pp. 1001–1010. Pittsburgh, PA: The Combustion Institute.

(1981b). A Monte Carlo method for the PDF equations of turbulent reactive flow. *Combustion Science and Technology* **25**, 159–174.

(1982). An improved turbulent mixing model. *Combustion Science and Technology* **28**, 131–135.

(1983). Consistent modeling of scalars in turbulent flows. *The Physics of Fluids* **26**, 404–408.

(1985). PDF methods for turbulent reactive flows. *Progress in Energy and Combustion Science* **11**, 119–192.

(1990). Computations of turbulent combustion: Progress and challenges. Invited Plenary Lecture. In *Twenty-third Symposium (International) on Combustion*, pp. 591–612. Pittsburgh, PA: The Combustion Institute.

(1991a). Application of the velocity-dissipation probability density function model to inhomogeneous turbulent flows. *Physics of Fluids A: Fluid Dynamics* **3**, 1947–1957.

(1991b). Mapping closures for turbulent mixing and reaction. *Theoretical and Computational Fluid Dynamics* **2**, 255–270.

(1992). Erratum: Application of the velocity-dissipation probability density function model to inhomogeneous turbulent flows [*Phys. Fluids A* **3**, 1947 (1991)]. *Physics of Fluids A: Fluid Dynamics* **4**, 1088.

(1994a). Lagrangian PDF methods for turbulent flows. *Annual Reviews of Fluid Mechanics* **26**, 23–63.

(1994b). On the relationship between stochastic Lagrangian models of turbulence and second-moment closures. *Physics of Fluids* **6**, 973–985.

(1995). Particle method for turbulent flows: Integration of stochastic model equations. *Journal of Computational Physics* **117**, 332–349.

(1997). Computationally efficient implementation of combustion chemistry using *in situ* adaptive tabulation. *Combustion Theory and Modelling* **1**, 41–63.

(1998). The vanishing effect of molecular diffusivity on turbulent dispersion: Implications for turbulent mixing and the scalar flux. *Journal of Fluid Mechanics* **359**, 299–312.

(2000). *Turbulent Flows*. Cambridge: Cambridge University Press.

(2002a). A stochastic Lagrangian model for acceleration in turbulent flow. *Physics of Fluids* **14**, 2360–2375.

(2002b). Stochastic Lagrangian models of velocity in homogeneous turbulent shear flow. *Physics of Fluids* **14**, 1696–1702.

(2003). Erratum: A stochastic Lagrangian model for acceleration in turbulent flow [*Phys. Fluids* **14**, 2360 (2002)]. *Physics of Fluids* **15**, 269.

Pope, S. B. and Y. L. Chen (1990). The velocity-dissipation probability density function model for turbulent flows. *Physics of Fluids A: Fluid Dynamics* **2**, 1437–1449.
Pope, S. B. and E. S. C. Ching (1993). The stationary probability density function: An exact result. *Physics of Fluids A: Fluid Dynamics* **5**, 1529–1531.
Pope, S. B. and A. R. Masri (1990). PDF calculations of a piloted turbulent nonpremixed flame of methane. *Combustion and Flame* **81**, 13–29.
Press, W. H., S. A. Teukolsky, W. T. Vetterling, and B. P. Flannery (1992). *Numerical Recipes in Fortran 77: The Art of Scientific Computing* (2nd edn). Cambridge University Press.
Rajagopalan, A. G. and C. Tong (2003). Experimental investigation of scalar-scalar-dissipation filtered joint density function and its transport equation. *Physics of Fluids* **15**, 227–244.
Raju, M. S. (1996). Application of scalar Monte Carlo probability density function method for turbulent spray flames. *Numerical Heat Transfer, Part A* **30**, 753–777.
Raman, V., R. O. Fox, A. D. Harvey, and D. H. West (2001). CFD analysis of premixed chlorination reactors with detailed chemistry. *Industrial & Engineering Chemistry Research* **40**, 5170–5176.
Raman, V., R. O. Fox, A. D. Harvey, and D. H. West (2003). Effect of feed-stream configuration on gas-phase chlorination reactor performance. *Industrial & Engineering Chemistry Research* **42**, 2544–2557.
Ranada, V. V. (2002). *Process Systems Engineering*, vol. 5: *Computational Flow Modeling for Chemical Reactor Engineering*. New York: Academic Press.
Risken, H. (1984). *The Fokker-Planck Equation*. Berlin: Springer-Verlag.
Ritchie, B. W. and A. H. Tobgy (1979). A three-environment micromixing model for chemical reactors with arbitrary separate feedstreams. *Chemical Engineering Journal* **17**, 173–182.
Roekaerts, D. (1991). Use of a Monte Carlo PDF method in a study of the influence of turbulent fluctuations on selectivity in a jet-stirred reactor. *Applied Scientific Research* **48**, 271–300.
(1992). Monte Carlo PDF method for turbulent reacting flow in a jet-stirred reactor. *Computers and Fluids* **21**, 97–108.
Rogallo, R. S. and P. Moin (1984). Numerical simulation of turbulent flows. *Annual Reviews of Fluid Mechanics* **16**, 99–137.
Rogers, M. M., P. Moin, and W. C. Reynolds (1986). The structure and modeling of the hydrodynamic and passive scalar fields in homogeneous turbulent shear flow. Report TF-25, Department of Mechanical Engineering, Stanford University.
Rosner, D. E. (1986). *Transport Processes in Chemically Reacting Flow Systems*. New York: Butterworths.
Rotta, J. C. (1951). Statistiche Theorie nichthomogener Turbulenz. *Zeitscrift für Physik* **129**, 547–572.
Saffman, P. G. (1963). On the fine-scale structure of vector fields convected by a turbulent fluid. *Journal of Fluid Mechanics* **16**, 545–572.
Sanders, J. P. H. and I. Gökalp (1998). Scalar dissipation rate modelling in variable density turbulent axisymmetric jets and diffusion flames. *Physics of Fluids* **10**, 938–948.
Saxena, V. and S. B. Pope (1998). PDF calculations of major and minor species in a turbulent piloted jet flame. In *Twenty-seventh Symposium (International) on Combustion*, pp. 1081–1086. Pittsburgh, PA: The Combustion Institute.
(1999). PDF simulation of turbulent combustion incorporating detailed chemistry. *Combustion and Flame* **117**, 340–350.

Saylor, J. R. and K. R. Sreenivasan (1998). Differential diffusion in low Reynolds number water jets. *Physics of Fluids* **10**, 1135–1146.

Schiestel, R. (1987). Multiple-time-scale modeling of turbulent flows in one-point closures. *The Physics of Fluids* **30**, 722–731.

Shah, J. J. and R. O. Fox (1999). CFD simulation of chemical reactors: Application of in situ adaptive tabulation to methane thermochlorination chemistry. *Industrial & Engineering Chemistry Research* **38**, 4200–4212.

Shenoy, U. V. and H. L. Toor (1990). Unifying indicator and instantaneous reaction methods of measuring micromixing. *AIChE Journal* **36**, 227–232.

Smagorinsky, J. (1963). General circulation experiments with the primitive equations: I. The basic equations. *Monthly Weather Review* **91**, 99–164.

Smith, L. L. (1994). *Differential Molecular Diffusion in Reacting and Non-Reacting Turbulent Jets of H_2 CO_2 Mixing with Air.* Ph. D. thesis, University of California, Berkeley.

Smith, L. L., R. W. Dibble, L. Talbot, R. S. Barlow, and C. D. Carter (1995). Laser Raman scattering measurements of differential molecular diffusion in nonreacting turbulent jets of H_2/CO_2 mixing with air. *Physics of Fluids* **7**, 1455–1466.

Spalding, D. B. (1971). Concentration fluctuations in a round turbulent free jet. *Chemical Engineering Science* **26**, 95–107.

Subramaniam, S. and D. C. Haworth (2000). A pdf method for turbulent mixing and combustion on three-dimensional unstructured deforming meshes. *International Journal of Engine Research* **1**, 171–190.

Subramaniam, S. and S. B. Pope (1998). A mixing model for turbulent reactive flows based on Euclidean minimum spanning trees. *Combustion and Flame* **115**, 487–514.

(1999). Comparison of mixing model performance for nonpremixed turbulent reactive flow. *Combustion and Flame* **117**, 732–754.

Swaminathan, N. and R. W. Bilger (1997). Direct numerical simulation of turbulent nonpremixed hydrocarbon reaction zones using a two-step reduced mechanism. *Combustion Science and Technology* **127**, 167–196.

Taing, S., A. R. Masri, and S. B. Pope (1993). PDF calculations of turbulent nonpremixed flames of H_2/CO_2 using reduced chemical mechanisms. *Combustion and Flame* **95**, 133–150.

Tang, Q. and S. B. Pope (2002). Implementation of combustion chemistry by *in situ* adaptive tabulation of rate-controlled constrained equilibrium manifolds. In *Proceedings of the Combustion Institute*, vol. 29, pp. 1411–1417. Pittsburgh, PA: The Combustion Institute.

Taylor, G. I. (1921). Diffusion by continuous movements. *Proceedings of the London Mathematical Society* **20**, 196–212.

Tennekes, H. and J. L. Lumley (1972). *A First Course in Turbulence* (2nd edn). Cambridge, MA: MIT Press.

Tong, C. (2001). Measurements of conserved scalar filtered density function in a turbulent jet. *Physics of Fluids* **13**, 2923–2937.

Toor, H. L. (1962). Mass transfer in dilute turbulent and non-turbulent systems with rapid irreversible reactions and equal diffusivities. *AIChE Journal* **8**, 70–78.

(1969). Turbulent mixing of two species with and without chemical reactions. *Industrial and Engineering Chemistry Fundamentals* **8**, 655–659.

Toor, H. L. and M. Singh (1973). The effect of scale on turbulent mixing and chemical reaction rates during turbulent mixing in a tubular reactor. *Industrial and Engineering Chemistry Fundamentals* **12**, 448–451.

Torrest, R. S. and W. E. Ranz (1970). Concentration fluctuations and chemical conversion associated with mixing in some turbulent flows. *AIChE Journal* **16**, 930–942.
Tsai, K. and R. O. Fox (1993). PDF modeling of free-radical polymerization in an axisymmetric reactor. EES Report 254, Kansas State University, Manhattan, Kansas.
(1994a). Modeling the effect of turbulent mixing on a series-parallel reaction in a tubular reactor. ICRES Report 9403, Kansas State University, Manhattan, Kansas.
(1994b). PDF simulation of a turbulent series-parallel reaction in an axisymmetic reactor. *Chemical Engineering Science* **49**, 5141–5158.
(1995a). Modeling multiple reactive scalar mixing with the generalized IEM model. *Physics of Fluids* **7**, 2820–2830.
(1995b). PDF modeling of turbulent mixing and chemical reactions in a tubular jet reactor. In G. B. Tatterson (ed.), *Process Mixing: Industrial Mixing Fundamentals*, pp. 31–38 New York: AIChE.
(1996a). Modeling the scalar dissipation rate for a series-parallel reaction. *Chemical Engineering Science* **51**, 1929–1938.
(1996b). PDF modeling of turbulent mixing effects on initiator efficiency in a tubular LDPE reactor. *AIChE Journal* **42**, 2926–2940.
(1998). The BMC/GIEM model for micromixing in non-premixed turbulent reacting flows. *Industrial and Engineering Chemistry Research* **37**, 2131–2141.
Tsai, K. and E. E. O'Brien (1993). A hybrid one- and two-point approach for isothermal reacting flows in homogeneous turbulence. *Physics of Fluids A: Fluid Dynamics* **5**, 2901–2910.
Valiño, L. and C. Dopazo (1990). A binomial sampling model for scalar turbulent mixing. *Physics of Fluids A: Fluid Dynamics* **2**, 1204–1212.
(1991). A binomial Langevin model for turbulent mixing. *Physics of Fluids A: Fluid Dynamics* **3**, 3034–3037.
Van Slooten, P. R. and S. B. Pope (1997). PDF modeling of inhomogeneous turbulence with exact representation of rapid distortions. *Physics of Fluids* **9**, 1085–1105.
(1999). Application of PDF modeling to swirling and non-swirling turbulent jets. *Flow, Turbulence and Combustion* **62**, 295–333.
Van Slooten, P. R., Jayesh, and S. B. Pope (1998). Advances in PDF modeling for inhomogeneous flows. *Physics of Fluids* **10**, 246–265.
Vassilatos, G. and H. L. Toor (1965). Second-order chemical reaction in a nonhomogeneous turbulent fluid. *AIChE Journal* **11**, 666–673.
Vedula, P. (2001). *Study of Scalar Transport in Turbulent Flows Using Direct Numerical Simulations*. Ph. D. thesis, Georgia Institute of Technology, Atlanta.
Vedula, P., P. K. Yeung, and R. O. Fox (2001). Dynamics of scalar dissipation in isotropic turbulence: A numerical and modeling study. *Journal of Fluid Mechanics* **433**, 29–60.
Verman, B., B. Geurts, and H. Kuertan (1994). Realizability conditions for the turbulent stress tensor in large-eddy simulations. *Journal of Fluid Mechanics* **278**, 351–362.
Vervisch, L. (1991). *Prise en compte d'effets de cinétique chimique dans les flammes de diffusion turbulente par l'approche fonction densité de probabilité*. Ph. D. thesis, Université de Rouen, France.
Vervisch, L. and T. Poinsot (1998). Direct numerical simulation of non-premixed turbulent flames. *Annual Reviews of Fluid Mechanics* **30**, 655–691.
Veynante, D. and L. Vervisch (2002). Turbulent combustion modeling. *Progress in Energy and Combustion Science* **28**, 193–266.
Villermaux, J. (1991). Mixing effects on complex chemical reactions in a stirred reactor. *Reviews in Chemical Engineering* **7**, 51–108.

Villermaux, J. and J. C. Devillon (1972). Représentation de la coalescence et de la redispersion des domaines de ségrégation dans un fluide par un modèle d'interaction phénoménologique. In *Proceedings of the 2nd International Symposium on Chemical Reaction Engineering*, pp. 1–13. New York: Elsevier.

Villermaux, J. and L. Falk (1994). A generalized mixing model for initial contacting of reactive fluids. *Chemical Engineering Science* **49**, 5127–5140.

Wall, C., B. J. Boersma, and P. Moin (2000). An evaluation of the assumed beta probability density function subgrid-scale model for large eddy simulation of nonpremixed, turbulent combustion with heat release. *Physics of Fluids* **12**, 2522–2529.

Wand, M. P. and M. C. Jones (1995). *Kernel Smoothing*. London: Chapman & Hall.

Wang, D. and C. Tong (2002). Conditionally filtered scalar dissipation, scalar diffusion, and velocity in a turbulent jet. *Physics of Fluids* **14**, 2170–2185.

Warhaft, Z. (2000). Passive scalars in turbulent flows. *Annual Reviews of Fluid Mechanics* **32**, 203–240.

Warnatz, J., U. Maas, and R. W. Dibble (1996). *Combustion*. Berlin: Springer-Verlag.

Weinstein, H. and R. J. Adler (1967). Micromixing effects in continuous chemical reactors. *Chemical Engineering Science* **22**, 65–75.

Welton, W. C. and S. B. Pope (1997). PDF model calculations of compressible turbulent flows using smoothed particle hydrodynamics. *Journal of Computational Physics* **134**, 150–168.

Wen, C. Y. and L. T. Fan (1975). *Models for Flow Systems and Chemical Reactors*. New York: Marcel Dekker.

Wilcox, D. C. (1993). *Turbulence Modeling for CFD*. La Cañada, California: DCW Industries Inc.

Wouters, H. A. (1998). *Lagrangian Models for Turbulent Reacting Flows*. Ph. D. thesis, Technische Universiteit Delft, The Netherlands.

Wouters, H. A., T. W. J. Peters, and D. Roekaerts (1996). On the existence of a generalized Langevin model representation for second-moment closures. *Physics of Fluids* **8**, 1702–1704.

Wright, D. L., R. McGraw, and D. E. Rosner (2001). Bivariate extension of the quadrature method of moments for modeling simultaneous coagulation and sintering particle populations. *Journal of Colloid and Interface Science* **236**, 242–251.

Xu, J. and S. B. Pope (1999). Assessment of numerical accuracy of PDF/Monte Carlo methods for turbulent reactive flows. *Journal of Computational Physics* **152**, 192–230.

(2000). PDF calculations of turbulent nonpremixed flames with local extinction. *Combustion and Flame* **123**, 281–307.

Yeung, P. K. (1994). Spectral transport of self-similar passive scalar fields in isotropic turbulence. *Physics of Fluids* **6**, 2245–2247.

(1996). Multi-scalar triadic interactions in differential diffusion with and without mean scalar gradients. *Journal of Fluid Mechanics* **321**, 235–278.

(1997). One- and two-particle Lagrangian acceleration correlations in numerically simulated homogeneous turbulence. *Physics of Fluids* **9**, 2981–2990.

(1998a). Correlations and conditional statistics in differential diffusion: Scalars with uniform mean gradients. *Physics of Fluids* **10**, 2621–2635.

(1998b). Multi-scalar mixing and Lagrangian approaches. In *Proceedings of the Second Monte Verita Colloquium on Fundamental Problematic Issues in Turbulence*, Ascona, Switzerland.

(2001). Lagrangian characteristics of turbulence and scalar transport in direct numerical simulations. *Journal of Fluid Mechanics* **427**, 241–274.

(2002). Lagrangian investigations of turbulence. *Annual Reviews of Fluid Mechanics* **34**, 115–142.
Yeung, P. K. and C. A. Moseley (1995). Effects of mean scalar gradients on differential diffusion in isotropic turbulence. Paper 95-0866, AIAA.
Yeung, P. K. and S. B. Pope (1989). Lagrangian statistics from direct numerical simulations of isotropic turbulence. *Journal of Fluid Mechanics* **207**, 531–586.
(1993). Differential diffusion of passive scalars in isotropic turbulence. *Physics of Fluids A: Fluid Dynamics* **5**, 2467–2478.
Yeung, P. K., M. C. Sykes, and P. Vedula (2000). Direct numerical simulation of differential diffusion with Schmidt numbers up to 4.0. *Physics of Fluids* **12**, 1601–1604.
Yeung, P. K., S. Xu, and K. R. Sreenivasan (2002). Schmidt number effects on turbulent transport with uniform mean scalar gradient. *Physics of Fluids* **14**, 4178–4191.
Zwietering, T. N. (1959). The degree of mixing in continuous flow systems. *Chemical Engineering Science* **11**, 1–15.
(1984). A backmixing model describing micromixing in single-phase continuous-flow systems. *Chemical Engineering Science* **39**, 1765–1788.

Index

Printed in the United States
By Bookmasters